Absolute value equations and inequalities	If b is a positive number, $\|a\| = b$ if and only if $a = b$ or $a = -b$; $\|a\| < b$ if and only if $-b < a < b$; $\|a\| > b$ if and only if $a < -b$ or $a > b$.
Distance formula	The distance between (x_1, y_1) and (x_2, y_2) is $\sqrt{(x_1 - x_2)^2 + (y_1 - y_2)^2}$.
Function	A function from a set X to a set Y is a correspondence that assigns to each element of X exactly one element of Y.
Domain, range	The domain of a function defined by $y = f(x)$ is the set of all possible values of x; the range is the set of all possible values of y.
Composition of functions	$(g \circ f)(x) = g[f(x)]$
One-to-one function	A function f is one-to-one if $f(a) = f(b)$ implies that $a = b$.
Inverse functions	Let f be a one-to-one function with domain X and range Y. Let g be a function with domain Y and range X. Then g is the inverse function of f if $(f \circ g)(x) = x$ for every x in Y, and $(g \circ f)(x) = x$ for every x in X.
Intercepts	Find x-intercepts by letting $y = 0$; find y-intercepts by letting $x = 0$.
Slope	The slope m of the line through (x_1, y_1) and (x_2, y_2), with $x_2 - x_1 \neq 0$, is $\dfrac{y_2 - y_1}{x_2 - x_1}$. The slope of a vertical line is undefined.
Point-slope form	A line with slope m passing through the point (x_1, y_1) has $y - y_1 = m(x - x_1)$ as an equation.
Slope-intercept form	The line with slope m and y-intercept b has an equation $y = mx + b$.
Logarithm	$y = \log_a x$ if and only if $x = a^y$.
Properties of logarithms	$\log_a xy = \log_a x + \log_a y \qquad \log_a \dfrac{x}{y} = \log_a x - \log_a y$ $\log_a x^r = r \cdot \log_a x \qquad\qquad \log_a a = 1$ $\log_a 1 = 0$
Binomial theorem	For any positive integer n and any complex numbers x and y, $(x + y)^n = x^n + \binom{n}{n-1} x^{n-1}y + \ldots + \binom{n}{n-r} x^{n-r}y^r + \ldots + \binom{n}{1} xy^{n-1} + y^n.$
Arithmetic sequences	nth term: $\ a_n = a_1 + (n - 1)d \qquad S_n = \dfrac{n}{2}(a_1 + a_n) \ $ or $\ S_n = \dfrac{n}{2}[2a_1 + (n-1)d]$
Geometric sequences	nth term: $\ a_n = a_1 r^{n-1} \qquad S_n = \dfrac{a_1(1 - r^n)}{1 - r} \ (r \neq 1) \qquad S_\infty = \dfrac{a_1}{1 - r} \ (-1 < r < 1)$
Permutations	The number of arrangements of n things taken r at a time is $P(n, r) = \dfrac{n!}{(n - r)!}$.
Combinations	The number of ways to choose r things from a group of n things is $\binom{n}{r} = \dfrac{n!}{(n - r)!r!}$.
Properties of probability	For any events E and F: $P(\text{a certain event}) = 1 \qquad\qquad 0 \leq P(E) \leq 1$ $P(\text{an impossible event}) = 0 \qquad P(E') = 1 - P(E)$ $P(E \text{ or } F) = P(E \cup F) = P(E) + P(F) - P(E \cap F).$

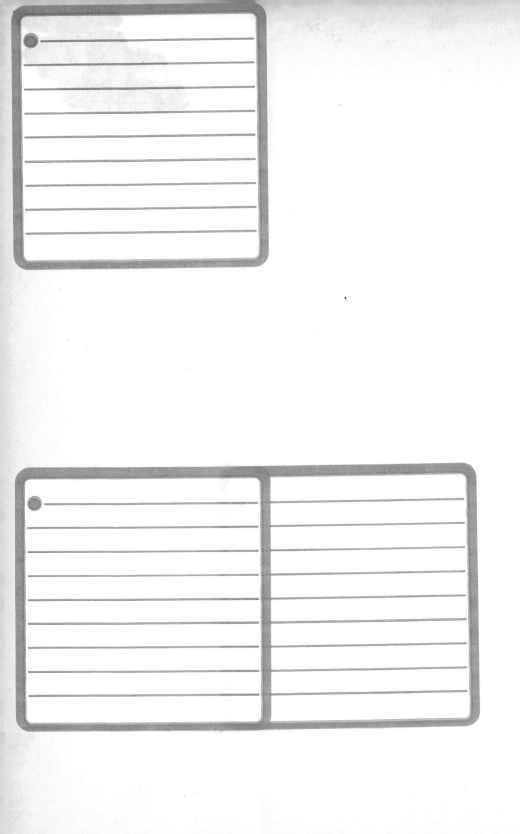

Fundamentals of College Algebra

Fundamentals of College Algebra

Second Edition

Charles D. Miller

Margaret L. Lial
American River College

Scott, Foresman and Company • Glenview, Illinois • London, England

To the Student

A **Solutions and Study Guide** to accompany this textbook is available from your college bookstore. This book contains **additional worked-out examples** beyond what is in the textbook, detailed **step-by-step solutions** to approximately half of the odd-numbered exercises in the textbook, and a **self-test for each chapter**. These features can help you study and understand the course material.

Library of Congress Cataloging-in-Publication Data

Lial, Margaret L.
    Fundamentals of college algebra.

    Includes index.
    1. Algebra. I. Miller, Charles David.
II. Title.
QA154.2.L523   1986   512.9   85-25100
ISBN   0-673-18242-8

Preface

The second edition of *Fundamentals of College Algebra* is designed to give a mathematically sound approach to those topics in algebra needed for success in later courses. Students using this book will be prepared for courses in trigonometry, statistics, finite mathematics, and calculus.

The prerequisite is a high school course in algebra or intermediate algebra at the college level. For students whose algebra background is not as solid as it might be, we have included a review of algebra in Chapter 1. Chapter 2 on equations also may be review for students.

Content Highlights

The **thorough review of algebra** in Chapter 1 discusses operations on the real numbers, absolute value, exponents, polynomials, factoring, rational expressions, radicals, and complex numbers.

Functions, an essential concept in this text, are discussed beginning in Chapter 3 and continuing throughout the book. **Functions and their graphs** are closely related, so the graphs of functions are discussed again and again. The general techniques of graphing also are presented in Chapter 3; later, new graphs are obtained from familiar ones by investigating symmetry and translations.

Exponential and logarithmic functions, presented in Chapter 5, are introduced in such a way that students see the functional aspects, with the numerical aspects downplayed.

Polynomial and rational functions are introduced early in the book through a discussion of graphing, and zeros of polynomials are thoroughly discussed in Chapter 7.

The book features a great **many applications problems,** from the fields of business, engineering, physics, chemistry, biology, demography, and political science. These applications are designed to be optional, so instructors may decide which ones to assign.

Key Features

The **format** of the book has been carefully designed to facilitate learning. **Definitions and rules are set off in boxes with marginal headings,** to help students with review and study, and also to make it easy for instructors to use the text. **Second color has been used pedagogically** to clarify explanations of techniques. **Numerous figures and graphs** illustrate examples and exercises.

The textbook features an **abundance of examples,** which present major points through detailed steps and explanations. The symbol ■ makes it clear when an example ends and the discussion continues. Sections are designed to be covered in one class period.

Exercises   **Extensive exercise sets** have been a strong feature of this textbook. This edition contains nearly 4700 exercises. These problems include a lengthy set of exercises for each section plus review exercises at the end of each chapter (about 900 review problems in all). Answers to odd-numbered exercises are located at the back of the book.

A special set of cumulative review exercises at the end of Chapter 2 can be used to decide which portions of the algebra review in Chapter 1 or the discussion of equations in Chapter 2 need be covered.

Algebra problems similar to those in calculus are included in many exercise sets, and most of the review exercise sets contain exercises from standard calculus textbooks. These exercises help students see the importance of algebra for their future work in calculus.

Supplements   The **Solutions and Study Guide,** by Eldon L. Miller, University of Mississippi, features additional examples for each section in the textbook, as well as solutions to approximately half of the odd-numbered exercises and a self-test for each chapter. This book can help students study and understand the course material by providing additional worked-out examples.

The **Instructor's Guide** contains a diagnostic pretest, a sample minimum assignment for the course, answers to even-numbered exercises, and an extensive test bank that can be used to prepare examinations or to provide additional problems for students to work.

Acknowledgments   A great many people helped with this revision through their suggestions and comments. In particular, we would like to thank those who reviewed the manuscript: Charles Applebaum, Bowling Green State University; Lee H. Armstrong, University of Central Florida; Alan A. Bishop, Western Illinois University; August J. Garver, University of Missouri, Rolla; James Hodge, College of Lake County; John Hornsby, University of New Orleans; Janice McFatter, Gulf Coast Community College; Brenda Marshall, Parkland College; and Eldon L. Miller, University of Mississippi. We also would like to thank the following people who helped us check the answers: James Arnold, University of Wisconsin–Milwaukee; Lewis Blake III, Duke University; Kathleen Pirtle; Marjorie Seachrist; and Priscilla Ware, Columbus Technical College.

We have received a great deal of help from two outstanding people at Scott, Foresman and Company: both Barbara Maring and Linda Youngman helped make the writing of this text a great pleasure.

Charles D. Miller
Margaret L. Lial

Contents

7 Zeros of Polynomials   361

8 Further Topics in Algebra   395

Fundamentals of
College Algebra

1

Fundamentals of Algebra

Today, algebra is required in a great many fields, ranging from accounting to zoology. The reason that algebra is required in all these fields is that algebra is *useful*. Equations, graphs, and functions occur again and again in many different areas. For example, in this text algebra is used to predict population growth, to determine the path of objects orbiting in space, and to investigate the costs versus the benefits of removing pollutants from a substance. To prepare for this study, the book begins with a review of the basics of algebra.

1.1 The Real Numbers

Numbers are the foundation of mathematics. The most common numbers in mathematics are the **real numbers,** the numbers that can be written as a decimal, either repeating, such as

$$\frac{1}{3} = .33333\ldots\ , \qquad \frac{3}{4} = .75000\ldots\ , \qquad \text{or} \qquad 2\frac{4}{7} = 2.571428571428\ldots\ ,$$

or nonrepeating, such as

$$\sqrt{2} = 1.4142135\ldots \qquad \text{or} \qquad \pi = 3.14159\ldots\ .$$

There are four familiar operations on real numbers: addition, subtraction, multiplication, and division. Addition is indicated with the symbol $+$. Subtraction is written with the symbol $-$, as in $8 - 2 = 6$. Multiplication is written in a variety of ways. All the symbols 2×8, $2 \cdot 8$, $2(8)$, and $(2)(8)$ represent the product of 2 and 8, or 16. When writing products involving **variables** (letters used to represent numbers), no operation symbols may be necessary: $2x$ represents the product of 2 and x, while xy indicates the product of x and y. Division of the real numbers a and b is written $a \div b$, or more commonly a/b.

The set of real numbers, together with the relation of equality and the basic operations of addition and multiplication, form the **real number system.** (Informally, a **set** is a collection of objects.) The key properties of the real number system are given below, where a, b, and c are letters used to represent any real number.

Properties of the Real Numbers

Closure properties	$a + b$ is a real number ab is a real number
Commutative properties	$a + b = b + a$ $ab = ba$
Associative properties	$(a + b) + c = a + (b + c)$ $(ab)c = a(bc)$
Identity properties	There exists a unique real number 0 such that $$a + 0 = a \quad \text{and} \quad 0 + a = a.$$ There exists a unique real number 1 such that $$a \cdot 1 = a \quad \text{and} \quad 1 \cdot a = a.$$
Inverse properties	There exists a unique real number $-a$ such that $$a + (-a) = 0 \quad \text{and} \quad (-a) + a = 0.$$ If $a \neq 0$, there exists a unique real number $1/a$ such that $$a \cdot \frac{1}{a} = 1 \quad \text{and} \quad \frac{1}{a} \cdot a = 1.$$
Distributive property	$a(b + c) = ab + ac$

Example I   By the closure properties, the sum or product of two real numbers is a real number. For example,

(a) $4 + 5$ is a real number, 9.   **(b)** $-9(-4)$ is a real number, 36.

(c) $8 \cdot 0$ is a real number, 0.   **(d)** $\sqrt{5} + \sqrt{3}$ is a real number.   ∎

Example 2   The following statements are examples of the commutative property. Notice how the order of the numbers changes from one side of the equals sign to the other.

(a) $6 + x = x + 6$ **(b)** $(6 + x) + 9 = (x + 6) + 9$

(c) $5 \cdot (9 \cdot 8) = (9 \cdot 8) \cdot 5$ **(d)** $5 \cdot (9 \cdot 8) = 5 \cdot (8 \cdot 9)$ ∎

The associative properties are used to add or multiply three or more numbers. For example, the associative property for addition says that the sum $a + b + c$ of the real numbers a, b, and c can be found either by first adding a and b, and then adding c to the result, indicated by the association

$$(a + b) + c,$$

or by first adding b and c, and then adding a to the result, indicated by the association

$$a + (b + c),$$

since, by the associative property, either method gives the same result.

Example 3   As the following statements illustrate, with the associative properties the order of the numbers does not change, but the placement of the parentheses does change.

(a) $4 + (9 + 8) = (4 + 9) + 8$ **(b)** $3(9x) = (3 \cdot 9)x$

(c) $(\sqrt{3} + \sqrt{7}) + 2\sqrt{6} = \sqrt{3} + (\sqrt{7} + 2\sqrt{6})$ ■

The identity properties show that 0 and 1 are special numbers: the sum of 0 and any real number a is the number a, so that 0 preserves the identity of a real number under addition. For this reason, 0 is the **identity element for addition.** In the same way, 1 preserves the identity of a real number under multiplication, making 1 the **identity element for multiplication.**

Example 4   By the identity properties,

(a) $-8 + 0 = -8$ **(b)** $(-9)1 = -9.$ ■

According to the addition inverse property, for any real number a there is a real number, written $-a$, such that the sum of a and $-a$ is 0, or $a + (-a) = 0$. The number $-a$ is called the **addition inverse, opposite,** or **negative** of a. The addition inverse property also says that this number $-a$ is *unique;* that is, a given number has only one addition inverse.

Don't confuse the *negative of a number* with a *negative number.* Since a is a variable, it can represent a positive or a negative number (as well as zero). The negative of a, written $-a$, can also be either a negative or a positive number (or zero). It is a common mistake to think that $-a$ *must* represent a negative number, although, for example, if a is -3, then $-a$ is $-(-3)$ or 3.

For each real number a except 0, there is a real number $1/a$ such that the product of a and $1/a$ is 1, or

$$a \cdot \frac{1}{a} = 1, \quad a \neq 0.$$

The symbol $1/a$ is often written a^{-1}, so that, by definition,

---

For every nonzero real number a,

$$a^{-1} = \frac{1}{a}.$$

The number $1/a$ or a^{-1} is called the **multiplication inverse** or **reciprocal** of the number a. Every real number except 0 has a reciprocal. As with the addition inverse, the multiplication inverse is unique—a given nonzero real number has only one multiplication inverse.

Example 5   By the inverse properties,

(a) $9 + (-9) = 0$

(b) $-15 + 15 = 0$

(c) $6 \cdot \dfrac{1}{6} = 1 \left(\text{so that } 6^{-1} = \dfrac{1}{6} \right)$

(d) $-8 \cdot \left(\dfrac{1}{-8} \right) = 1$

(e) $\dfrac{1}{\sqrt{5}} \cdot \sqrt{5} = 1.$

(f) There is no real number x such that $0 \cdot x = 1$, so there is no multiplication inverse for 0.   ■

One of the most important properties of the real numbers, and the only one that involves both addition and multiplication, is the distributive property. The next example shows how this property is applied.

Example 6   By the distributive property,

(a) $9(6 + 4) = 9 \cdot 6 + 9 \cdot 4$

(b) $3(x + y) = 3x + 3y$

(c) $-\sqrt{5}(m + 2) = -m\sqrt{5} - 2\sqrt{5}$

(The product of $-\sqrt{5}$ and m is often written $-m\sqrt{5}$, since $-\sqrt{5}m$ is too easily confused with $-\sqrt{5m}$.)   ■

The distributive property can be extended to include more than two numbers in the sum, as follows.

$$a(b + c + d + e + \cdots + n) = ab + ac + ad + ae + \cdots + an$$

This form is called the **extended distributive property.**

The **substitution property** is another key property.

Substitution Property	If $a = b$, then a may replace b or b replace a in any expression without affecting the truth or falsity of the statement.

Many further properties of the real numbers can be proven directly from those given above. For example, the following two important properties are used in the next chapter in solving equations.

Addition and Multiplication Properties	For all real numbers a, b, and c, if $a = b$, then $$a + c = b + c, \quad \text{and} \quad ac = bc.$$

In words, these properties say that the same number may be added or multiplied on both sides of a statement of equality.

Two special properties of the number 0 are stated below.

Properties of Zero

> For all real numbers a and b,
>
> $$a \cdot 0 = 0, \quad \text{and} \quad ab = 0 \quad \text{if and only if} \quad a = 0 \text{ or } b = 0.$$

The second property above contains the phrase "if and only if." By definition,

If and Only If

> p if and only if q
>
> means
>
> if p, then q and if q, then p.

Finally, some useful properties of the negatives of numbers follow.

Properties of Negatives

> For all real numbers a and b,
>
> $$-(-a) = a$$
> $$-a(b) = -(ab)$$
> $$a(-b) = -(ab)$$
> $$(-a)(-b) = ab.$$

The proofs of many of these properties are included in the exercises below. Example 7 shows a general style of proof that can be used for many of these properties.

Example 7 If a is a real number, prove that

$$-(-a) = a.$$

We shall prove that both $-(-a)$ and a are addition inverses of $-a$. Since a real number has only one addition inverse, this will show that $-(-a)$ and a are equal. First,

$$-a + [-(-a)] = 0$$

since $-(-a)$ is the addition inverse of $-a$. Also,

$$-a + a = 0,$$

since a is the addition inverse of $-a$. Since both $-(-a)$ and a are addition inverses of $-a$, and since $-a$ has only one addition inverse,

$$-(-a) = a. \quad \blacksquare$$

The properties of real numbers given above apply to addition or multiplication. The two other common operations for the real numbers, subtraction and division, are defined in terms of the operations of addition and multiplication, respectively.

Subtraction is defined by saying that the difference of the numbers a and b, written $a - b$, is found by adding a and the *negative* of b. See the next page.

Subtraction	For all real numbers a and b,
	$$a - b = a + (-b).$$

Example 8   (a) $7 - 2 = 7 + (-2) = 5$

(b) $-8 - (-3) = -8 + (+3) = -5$

(c) $6 - (-15) = 6 + (+15) = 21$ ∎

Division of a real number a by a nonzero real number b is defined in terms of multiplication as follows.

Division	For all real numbers a and b, with $b \neq 0$,
	$$\frac{a}{b} = a \cdot \frac{1}{b} = ab^{-1}.$$

That is, to divide a by b, multiply a by the reciprocal of b.

Example 9   (a) $\dfrac{6}{3} = 6 \cdot \dfrac{1}{3} = 2$ (b) $\dfrac{-8}{4} = -8 \cdot \dfrac{1}{4} = -2$

(c) $\dfrac{-9}{0}$ is meaningless since there is no reciprocal for 0; also, $\dfrac{0}{0}$ is meaningless.   ∎

Several useful properties of quotients are listed below.

Properties of Quotients	For all real numbers a, b, c, and d, with all denominators nonzero,
	$\dfrac{a}{b} = \dfrac{c}{d}$ if and only if $ad = bc$ $\qquad$ $\dfrac{a}{b} + \dfrac{c}{d} = \dfrac{ad + bc}{bd}$
	$\dfrac{ac}{bc} = \dfrac{a}{b}$ $\qquad\qquad\qquad\qquad$ $\dfrac{a}{b} - \dfrac{c}{d} = \dfrac{ad - bc}{bd}$
	$\dfrac{a}{-b} = \dfrac{-a}{b} = -\dfrac{a}{b}$ $\qquad\qquad$ $\dfrac{a}{b} \cdot \dfrac{c}{d} = \dfrac{ac}{bd}$
	$\dfrac{-a}{-b} = \dfrac{a}{b}$ $\qquad\qquad\qquad$ $\dfrac{a}{b} \div \dfrac{c}{d} = \dfrac{a}{b} \cdot \dfrac{d}{c}.$

To avoid possible ambiguity when working problems, use the following **order of operations,** which has been generally agreed upon. (By the way, this order of operations is used by computers and many calculators.)

Order of Operations

If parentheses or square brackets are present:

1. Work separately above and below any fraction bar.
2. Use the rules below within each set of parentheses or square brackets. Start with the innermost and work outward.

If no parentheses are present:

1. Do any multiplications or divisions in the order in which they occur, working from left to right.
2. Do any additions or subtractions in the order in which they occur, working from left to right.

If exponents or roots are involved, they are dealt with before using the order of operations.

Example 10 Use the order of operations given above to simplify the following.

(a) $(6 \div 3) + 2 \cdot 4 = 2 + 2 \cdot 4 = 2 + 8 = 10$

(b) $\dfrac{-9(-3) + (-5)}{2(-8) - 5(3)} = \dfrac{27 + (-5)}{-16 - 15} = \dfrac{22}{-31} = -\dfrac{22}{31}$

(c) $4 \cdot 2^3 = 4 \cdot 8 = 32$ ∎

There are several subsets* of the set of real numbers that come up so often they are given special names, as listed below. Some of the subsets are written with **set-builder notation;** with this notation,

$$\{x \mid x \text{ has property } P\},$$

read "the set of all elements x such that x has property P" represents the set of all elements having some specified property P.

Subsets of the Real Numbers

Natural numbers	$\{1, 2, 3, 4, \ldots\}$
Whole numbers	$\{0, 1, 2, 3, 4, \ldots\}$
Integers	$\{\ldots, -3, -2, -1, 0, 1, 2, 3, \ldots\}$
Rational numbers	$\left\{\dfrac{p}{q} \middle\mid p \text{ and } q \text{ are integers}, q \neq 0\right\}$
Irrational numbers	$\{x \mid x \text{ is a real number that is not rational}\}$

Example 11 Let set $A = \{-8, -6, -3/4, 0, 3/8, 1/2, 1, \sqrt{2}, \sqrt{5}, 6, \sqrt{-1}\}$. List the elements from set A that belong to each of the following sets.

(a) The *natural numbers* in set A are 1 and 6.

(b) The *whole numbers* are 0, 1, and 6.

*Set A is a **subset** of set B if and only if every element of set A is also an element of set B.

(c) The *integers* are -8, -6, 0, 1, and 6.

(d) The *rational numbers* are -8, -6, $-3/4$, 0, 3/8, 1/2, 1, and 6. (The number -8 is rational since -8 can be written as the quotient $-8/1$. Also, 6 is rational since $6 = 6/1$.)

(e) The *irrational numbers* are $\sqrt{2}$ and $\sqrt{5}$. (In further mathematics courses you will see that these numbers do not have repeating or terminating decimal expansions.)

(f) All elements of A are *real numbers* except $\sqrt{-1}$. Square roots of negative numbers are discussed later in this chapter.   ■

The relationships among the various sets of numbers are shown in Figure 1. All the numbers shown are real numbers. (The number π shown in Figure 1 is the ratio of the circumference of a circle to its diameter; π is approximately 3.14159. Also, e is an irrational number discussed in Chapter 5; e is approximately 2.7182818.)

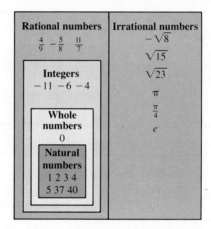

Figure I

I.I Exercises

Identify the properties which are illustrated in each of the following. Some will require more than one property. Assume all variables represent real numbers.

1. $8 \cdot 9 = 9 \cdot 8$

2. $3 + (-3) = 0$

3. $3 + (-3) = (-3) + 3$

4. $0 + (-7) = (-7) + 0$

5. $-7 + 0 = -7$

6. $8 + (12 + 6) = (8 + 12) + 6$

7. $[9(-3)] \cdot 2 = 9[(-3) \cdot 2]$

8. $8(m + 4) = 8m + 8 \cdot 4$

9. If x is a real number, then $x + 2$ is a real number.

10. $(7 - y) + 0 = 7 - y$

11. $8(4 + 2) = (2 + 4)8$

12. $x \cdot \dfrac{1}{x} + x \cdot \dfrac{1}{x} = x \left(\dfrac{1}{x} + \dfrac{1}{x} \right)$ (if $x \neq 0$)

For Exercises 13–22, tell if the statement is true or false. If it is false, tell why.

13. By the identity property, $8 \cdot 0 = 0$.

14. By the identity property, $8 + 1 = 9$.

15. The sum of two whole numbers is always a whole number.

16. The quotient of two whole numbers is always a whole number.

17. The quotient of two rational numbers is always a rational number.

18. The difference of two real numbers is always a real number.

19. The set $\{0, 1\}$ is closed with respect to subtraction.

20. The set $\{0, 1\}$ is closed with respect to multiplication.

21. The set $\{1, -1\}$ is closed with respect to division.

22. The set of irrational numbers contains an identity element for multiplication.

Is each of the following sets closed with respect to the indicated operations?

23. rationals, subtraction

24. rationals, division

25. irrationals, addition

26. reals, subtraction

27. irrationals, multiplication

28. irrationals, division

29. $\{0\}$, addition

30. $\{0\}$, multiplication

31. $\{1\}$, addition

32. $\{1\}$, multiplication

Simplify each of the following expressions using the order of operations given in the text.

33. $9 \div 3 \cdot 4 \cdot 2$

34. $18 \cdot 3 \div 9 \div 2$

35. $8 + 7 \cdot 2 + (-5)$

36. $-9 + 6 \cdot 5 + (-8)$

37. $(-9 + 4 \cdot 3)(-7)$

38. $-15(-8 - 4 \div 2)$

39. $\dfrac{-8 + (-4)(-6) \div 12}{4 - (-3)}$

40. $\dfrac{15 \div 5 \cdot 4 \div 6 - 8}{-6 - (-5) - 8 \div 2}$

41. $\dfrac{17 \div (3 \cdot 5 + 2) \div 8}{-6 \cdot 5 - 3 - 3(-11)}$

42. $\dfrac{-12(-3) + (-8)(-5) + (-6)}{17(-3) + 4 \cdot 8 + (-7)(-2) - (-5)}$

43. $\dfrac{-9.23(5.87) + 6.993}{1.225(-8.601) - 148(.0723)}$

44. $\dfrac{189.4(3.221) - 9.447(-8.772)}{4.889[3.177 - 8.291(3.427)]}$

For Exercises 45–56, choose all words from the following list which apply.

(a) natural number   **(b)** whole number   **(c)** integer   **(d)** rational number
(e) irrational number   **(f)** real number   **(g)** meaningless

45. 12

46. 0

47. -9

48. 3/4

49. $-5/9$

50. $\sqrt{8}$

51. $-\sqrt{2}$

52. $\sqrt{25}$

53. $-\sqrt{36}$

54. 8/0

55. $-9/(0 - 0)$

56. $0/(3 + 3)$

Evaluate each of the following if $p = -2$, $q = 4$, and $r = -5$.

57. $7p - 4q + 10$

58. $8q - r + 3p$

59. $-3(p + 5q)$

60. $2(q - r)$

61. $\dfrac{q + r}{q + p}$

62. $\dfrac{3q}{3p - 2r}$

63. $\dfrac{8q - 3r}{q + r + 1}$

64. $\dfrac{8r + q}{q + 2p}$

65. $\dfrac{\dfrac{q}{4} - \dfrac{r}{5}}{\dfrac{p}{2} + \dfrac{q}{2}}$

66. $\dfrac{\dfrac{3r}{10} - \dfrac{5p}{2}}{q + \dfrac{2r}{5}}$

67. $\dfrac{\dfrac{1}{p} + \dfrac{1}{r}}{\dfrac{1}{p} - \dfrac{1}{q}}$

68. $\dfrac{\dfrac{2}{p+1} + \dfrac{1}{q-2}}{\dfrac{1}{r+2} - \dfrac{2}{q+2}}$

69. The distributive property of multiplication over addition is written $a(b + c) = ab + ac$. Is there a distributive property of addition over multiplication? That is, does

$$a + (b \cdot c) = (a + b)(a + c)$$

for all real numbers a, b, and c? To find out, try various sample values of a, b, and c.

Give a reason for each step in the following:

70. Prove: $a - a = 0$.
$$a - a = a + (-a)$$
$$a + (-a) = 0$$
$$a - a = 0$$

71. Prove: $(a + b)c = ac + bc$.
$$(a + b)c = c(a + b)$$
$$c(a + b) = ca + cb$$
$$(a + b)c = ca + cb$$
$$ca + cb = ac + bc$$
$$(a + b)c = ac + bc$$

72. Prove: $(a + b) + (-a) = b$.
$$(a + b) + (-a) = a + [b + (-a)]$$
$$a + [b + (-a)] = a + [-a + b]$$
$$a + [-a + b] = [a + (-a)] + b$$
$$[a + (-a)] + b = 0 + b$$
$$0 + b = b$$
$$(a + b) + (-a) = b$$

Prove each of the following statements about real numbers a, b, and c.

73. $-a(b) = -(ab)$

74. $(-a)(-b) = ab$

75. $a \cdot 0 = 0$

76. If $a = b$, then $a + c = b + c$.

77. If $a = b$, then $ac = bc$.

78. If $ab = 0$, then $a = 0$ or $b = 0$.

79. $ab(a^{-1} + b^{-1}) = b + a$

80. $(a + b)(a^{-1} + b^{-1}) = 2 + ba^{-1} + ab^{-1}$

81. Show that the sum or product of two irrational numbers can be rational or irrational.

82. Show that the average of two rational numbers is also rational, while the average of two irrational numbers can be either rational or irrational.

1.2   The Number Line and Absolute Value

It is often important to know which of two given real numbers is the smaller. Deciding which is smaller is sometimes easier with a **number line,** a geometric representation of the set of real numbers. Set up a number line as follows: Draw a line and choose any point on the line to represent 0. (See Figure 2.) Then choose any point to the right of 0 and label it 1. The distance from 0 to 1 sets up a unit measure that can be used to locate other points to the right of 1, which are labeled 2, 3, 4, 5, and so on, and points to the left of 0, labeled -1, -2, -3, -4, and so on.

Points representing rational numbers, such as 2/3, 16/9, $-11/7$, and so on, can be located by dividing the intervals between integers. For example, 16/9 (or 1 7/9) can be found by dividing the interval from 1 to 2 into 9 equal parts (using methods given in geometry), and then choosing the correct point. Numbers such as $\sqrt{2}$, $\sqrt{3}$, $\sqrt{5}$, and so on, can be located by other geometric constructions. Points corresponding to other irrational numbers can be found to any desired degree of accuracy by using decimal approximations for the numbers.

Figure 2 shows a number line with the points corresponding to several different numbers marked on it. A number that corresponds to a particular point on a line is called the **coordinate** of the point. For example, the leftmost marked point in Figure 2 has coordinate -4. The correspondence between points on a line and the real numbers is called a **coordinate system** for the line. (From now on, the phrase "the point on a number line with coordinate a" will be abbreviated as "the point with coordinate a," or simply "the point a.")

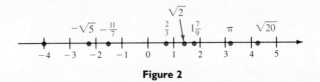

Figure 2

Example 1   Locate the elements of the set $\{-2/3, 0, \sqrt{2}, \sqrt{5}, \pi, 4\}$ on a number line.

The number π is irrational, with $\pi \approx 3.14159$ ($\approx$ means "approximately equal to"). From a calculator, $\sqrt{2} \approx 1.414$ and $\sqrt{5} \approx 2.236$. Using this information, place the given points on a number line as shown in Figure 3.   ■

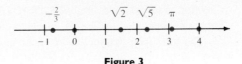

Figure 3

Suppose a and b are two real numbers. If the difference $a - b$ is positive, then a is greater than b, written $a > b$. If the difference $a - b$ is negative, then a is less than b, written $a < b$.

These algebraic statements can be given a geometric interpretation. If $a - b$ is positive, so that $a > b$, then a would be to the *right* of b on a number line. Also, if $a < b$, then a would have to be to the *left* of b. Both the algebraic and geometric statements are summarized below.

Inequality Statements

Statement	Algebraic form	Geometric form
$a > b$	$a - b$ is positive	a is to the right of b
$a < b$	$a - b$ is negative	a is to the left of b

Example 2   **(a)** In Figure 3 above, $-2/3$ is to the left of $\sqrt{2}$, so that $-2/3 < \sqrt{2}$. Also, $\sqrt{2}$ is to the right of $-2/3$, giving $\sqrt{2} > -2/3$.

(b) The difference $-3 - (-8)$ is positive, showing that $-3 > -8$. The difference $-8 - (-3)$ is negative, so that $-8 < -3$.   ■

The following variations on $<$ and $>$ are often used.

Symbol	Meaning
$\leq$	is less than or equal to
$\geq$	is greater than or equal to
$\nless$	is not less than
$\ngtr$	is not greater than

Statements involving these symbols, as well as $<$ and $>$, are called **inequalities.**

Example 3   **(a)** $8 \leq 10$ (since $8 < 10$)   **(b)** $8 \leq 8$ (since $8 = 8$)

(c) $-9 \geq -14$ (since $-9 > -14$)   **(d)** $-8 \ngtr -2$ (since $-8 < -2$)

(e) $4 \nless 2$ (since $4 > 2$)   ■

The expression $a < b < c$ says that b is *between* a and c, since $a < b < c$ means $a < b$ and $b < c$. Also, $a \leq b \leq c$ means $a \leq b$ and $b \leq c$. When writing these "between" statements, make sure that both inequality symbols point in the same direction. For example, both $2 < 7 < 11$ and $5 > -1 > -6$ are true statements, but a statement such as $3 < 5 > -1$ is meaningless.

The following **properties of order** give the basic properties of $<$ and $>$.

Properties of Order

For all real numbers a, b, and c,

Transitive property   If $a < b$ and $b < c$, then $a < c$.

Addition property   If $a < b$, then $a + c < b + c$.

Multiplication property   If $a < b$, and $c > 0$, then $ac < bc$.
        If $a < b$, and $c < 0$, then $ac > bc$.

Trichotomy property   Given the real numbers a and b, either $a < b$, $a > b$, or $a = b$.

The distance on the number line from a number to 0 is called the **absolute value** of that number. The absolute value of the number a is written $|a|$. For example, the distance on the number line from 9 to 0 is 9, as is the distance from -9 to 0 (see Figure 4), so $|9| = 9$ and $|-9| = 9$.

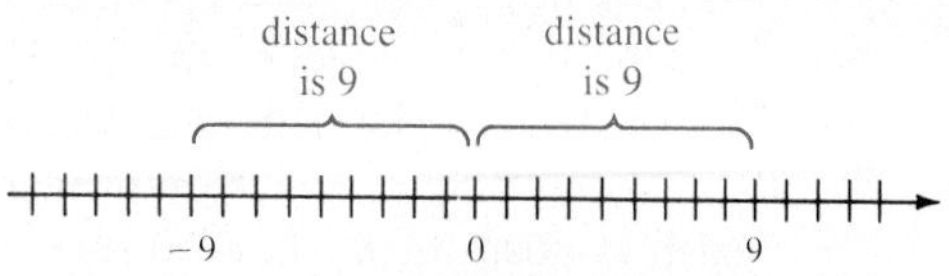

Figure 4

Example 4   **(a)** $|-4| = 4$ **(b)** $|2\pi| = 2\pi$

(c) $-|8| = -(8) = -8$ **(d)** $-|-2| = -(2) = -2$ ∎

The definition of absolute value can be stated as follows:

Absolute Value

> For every real number a,
>
> $$|a| = \begin{cases} a \text{ if } a \geq 0 \\ -a \text{ if } a < 0. \end{cases}$$

The second part of this definition requires some thought. If a is a negative number, that is, if $a < 0$, then $-a$ is positive. Thus, for a *negative a*,

$$|a| = -a.$$

For example, if $a = -5$, then $|a| = |-5| = -(-5) = 5$.

Example 5   Write each of the following without absolute value bars.

(a) $|\sqrt{5} - 2|$
From Figure 3, $\sqrt{5} > 2$, making $\sqrt{5} - 2 > 0$, and $|\sqrt{5} - 2| = \sqrt{5} - 2$.

(b) $|\pi - 4|$
Since $\pi < 4$, then $\pi - 4 < 0$, and $|\pi - 4| = -(\pi - 4) = -\pi + 4 = 4 - \pi$.

(c) $|m - 2|$ if $m < 2$
If $m < 2$, then $m - 2 < 0$, so $|m - 2| = -(m - 2) = 2 - m$.   ∎

The definition of absolute value can be used to prove the following properties of absolute value. (See the exercises below.)

Properties of Absolute Value

> For all real numbers a and b,
>
> $$|a| \geq 0 \qquad\qquad |a| \cdot |b| = |ab|$$
>
> $$|-a| = |a| \qquad\qquad \left|\frac{a}{b}\right| = \frac{|a|}{|b|} \quad b \neq 0$$
>
> $$|a + b| \leq |a| + |b| \qquad \text{(the triangle inequality).}$$

Example 6   Prove that for all real numbers a, $|a| \geq 0$.

If $a \geq 0$, then by the definition of absolute value, $|a| = a$, which is assumed to be greater than or equal to 0. Thus, if $a \geq 0$, then $|a| \geq 0$. If $a < 0$, then $|a| = -a$, which is positive. By these results, if either $a \geq 0$ or $a < 0$, then $|a| \geq 0$, so that $|a| \geq 0$ for every real number a. ∎

The number line of Figure 5 shows the point A, with coordinate -3, and the point B, with coordinate 5. The distance between points A and B is 8 units, which can be found by subtracting the smaller coordinate from the larger. If $d(A, B)$ represents the distance between points A and B, then

$$d(A, B) = 5 - (-3) = 8.$$

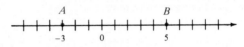

Figure 5

To avoid worrying about which coordinate is smaller, use absolute value as in the following definition.

Distance

> Suppose points A and B have coordinates a and b respectively. The distance between A and B, written $d(A, B)$, is
>
> $$d(A, B) = |a - b|.$$

Example 7   Let points A, B, C, D, and E have coordinates as shown on the number line of Figure 6. Find the indicated distances.

(a) $d(B, E)$

Since B has coordinate -1 and E has coordinate 5,

$$d(B, E) = d(-1, 5) = |-1 - 5| = 6.$$

(b) $d(D, A) = \left| 2\frac{1}{2} - (-3) \right| = 5\frac{1}{2}$

(c) $d(B, C) = |-1 - 0| = 1$

(d) $d(E, E) = |5 - 5| = 0$ ∎

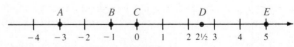

Figure 6

1.2 Exercises

Write the following numbers in numerical order, from smallest to largest. Use a calculator as necessary.

1. $-9, -2, 3, -4, 8$

2. $7, -6, 0, -2, -3$

3. $|-8|, -|9|, -|-6|$

4. $-|-9|, -|7|, -|-2|$

5. $\sqrt{8}, -4, -\sqrt{3}, -2, -5, \sqrt{6}, 3$

6. $\sqrt{2}, -1, 4, 3, \sqrt{8}, -\sqrt{6}, \sqrt{7}$

7. $3/4, \sqrt{2}, 7/5, 8/5, 22/15$

8. $-9/8, -3, -\sqrt{3}, -\sqrt{5}, -9/5, -8/5$

9. $|-8 + 2|, -|3|, -|-2|, -|-2| + (-3), -|-8| - |-6|$

10. $-2 - |-4|, -3 + |2|, -4, -5 + |-3|$

Let $x = -4$ and $y = 2$. Evaluate each of the following.

11. $|2x|$

12. $|-3y|$

13. $|x - y|$

14. $|2x + 5y|$

15. $|3x + 4y|$

16. $|-5y + x|$

17. $|-4x + y| - |y|$

18. $|-8y + x| - |x|$

19. $\dfrac{|x| + 2|y|}{5 + x}$

20. $\dfrac{2|x| - |y + 2|}{|x| \cdot |y|}$

21. $\dfrac{|x| - 2 + y + |x|\,|}{|x + 3|}$

22. $\dfrac{(y - 3)|5 - |x| + 2|y + 4|\,|}{2 - |y - 1|}$

Write an equivalent expression for each of the following without using absolute value bars.

23. $|-6|$

24. $-|-2|$

25. $-|-8| + |-2|$

26. $3 - |-4|$

27. $|8 - \sqrt{50}|$

28. $|2 - \sqrt{3}|$

29. $|\sqrt{7} - 5|$

30. $|\sqrt{2} - 3|$

31. $|\pi - 3|$

32. $|\pi - 5|$

33. $|x - 4|$, if $x > 4$

34. $|y - 3|$, if $y < 3$

35. $|2k - 8|$, if $k < 4$

36. $|3r - 15|$, if $r > 5$

37. $|7m - 56|$, if $m < 4$

38. $|2k - 7|$, if $k > 4$

39. $|-8 - 4m|$, if $m > -2$

40. $|6 - 5r|$, if $r < -2$

41. $|x - y|$, if $x < y$

42. $|x - y|$, if $x > y$

43. $|3 + x^2|$

44. $|x^2 + 4|$

45. $|-1 - p^2|$

46. $|-r^4 - 16|$

47. $|\pi - 5| + 1$

48. $|2 - \pi| + 5$

49. $|\sqrt{7} - 2| + 1$

50. $|3 - \sqrt{11}| + 2$

51. $|m - 3| + |m - 4|$, if $3 < m < 4$

52. $|z - 6| - |z - 5|$, if $5 < z < 6$

In the following exercises, the coordinates of four points are given. Find (a) $d(A, B)$; (b) $d(B, C)$; (c) $d(B, D)$; (d) $d(D, A)$; (e) $d(A, B) + d(B, C)$.

53. $A, -4; B, -3; C, -2; D, 10$

54. $A, -8; B, -7; C, 11; D, 5$

55. $A, -3; B, -5; C, -12; D, -3$

56. $A, 0; B, 6; C, 9; D, -1$

Justify each of the following statements by giving the correct property from this section. Assume that all variables represent real numbers and that no denominators are zero.

57. If $2k < 8$, then $k < 4$.

58. If $x + 8 < 15$, then $x < 7$.

59. If $-4x < 24$, then $x > -6$.

60. If $x < 5$ and $5 < m$, then $x < m$.

61. If $m > 0$, then $9m > 0$.

62. If $k > 0$, then $8 + k > 8$.

63. $|8 + m| \leq |8| + |m|$ **64.** $|k - m| \leq |k| + |-m|$ **65.** $|8| \cdot |-4| = |-32|$

66. $|12 + 11r| \geq 0$ **67.** $\left|\dfrac{-12}{5}\right| = \dfrac{|-12|}{|5|}$ **68.** $\left|\dfrac{6}{5}\right| = \dfrac{|6|}{|5|}$

69. If p is a real number, then $p < 5$, $p > 5$, or $p = 5$.

70. If z is a real number, then $z > -2$, $z < -2$, or $z = -2$.

Under what conditions are the following statements true?

71. $|x| = |y|$ **72.** $|x + y| = |x| + |y|$ **73.** $|x + y| = |x| - |y|$

74. $|x| \leq 0$ **75.** $|x - y| = |x| - |y|$ **76.** $||x + y|| = |x + y|$

Evaluate each of the following for real numbers x and y, if no denominators are equal to 0.

77. $\dfrac{|x|}{x}$ **78.** $\left|\dfrac{x}{|x|}\right|$ **79.** $\left|\dfrac{x - y}{y - x}\right|$ **80.** $\left|\dfrac{x + y}{-x - y}\right|$

Prove each of the following properties. Assume a and b are real numbers, and A and B are points on a number line with A having coordinate a and B having coordinate b.

81. $|-a| = |a|$ **82.** $|a - b| = |b - a|$ **83.** $|a| \cdot |b| = |ab|$

84. $\left|\dfrac{a}{b}\right| = \dfrac{|a|}{|b|}$, if $b \neq 0$ **85.** $-|a| \leq a \leq |a|$ **86.** $|a + b| \leq |a| + |b|$

87. $d(A, B) = d(B, A)$ **88.** $d(A, B) \geq 0$ **89.** $d(A, 0) = |a|$ **90.** $d(A, A) = 0$

91. Suppose $x^2 \leq 81$. Must it then be true that $x \leq 9$?

92. Suppose $x^2 \geq 81$. Must it then be true that $x \geq 9$?

93. Let x be a nonzero real number. Under what conditions is $1/x < x$?

94. Let x be a nonzero real number. When is $x < x^2$?

1.3   Integer Exponents

Exponents are used to write the products of repeated factors. For example, the product $2 \cdot 2 \cdot 2$ can be written as 2^3, where the 3 shows that three factors of 2 appear in the product. Generalizing, the symbol a^n is defined as follows.

Definition of a^n

> If n is any positive integer and a is any real number,
> $$a^n = a \cdot a \cdot a \cdots a,$$
> where a appears as a factor n times.

The integer n is called the **exponent,** and a is the **base.** (Read a^n as "a to the nth power," or just "a to the nth.") For example,

$$(-6)^2 = (-6)(-6) = 36, \qquad 4^3 = 4 \cdot 4 \cdot 4 = 64,$$

$$\text{and} \qquad \left(\dfrac{2}{3}\right)^4 = \dfrac{2}{3} \cdot \dfrac{2}{3} \cdot \dfrac{2}{3} \cdot \dfrac{2}{3} = \dfrac{16}{81}.$$

One common error with exponents occurs with expressions such as $4 \cdot 3^2$. The exponent of 2 applies only to the base 3, so that

$$4 \cdot 3^2 = 4 \cdot 3 \cdot 3 = 36.$$

On the other hand,

$$(4 \cdot 3)^2 = (4 \cdot 3)(4 \cdot 3) = 12 \cdot 12 = 144,$$

so that $\quad\quad\quad 4 \cdot 3^2 \neq (4 \cdot 3)^2.$

Work with exponents can be simplified by using various properties of exponents. By definition, a^m (where m is a positive integer and a is a real number) means a appears as a factor m times. In the same way, a^n (where n is a positive integer) means that a appears as a factor n times. In the product $a^m \cdot a^n$, a will therefore appear as a factor $m + n$ times, so that for any positive integers m and n and any real number a,

$$a^m \cdot a^n = a^{m+n}.$$

Similar arguments can be given for the other properties of exponents listed below.

Properties of Exponents	For any positive integers m and n and for any real numbers a and b,

(a) $a^m \cdot a^n = a^{m+n}$ **(c)** $(a^m)^n = a^{mn}$

(b) $\dfrac{a^m}{a^n} = \begin{cases} a^{m-n} \text{ if } m > n \\ 1 \text{ if } m = n \quad a \neq 0 \\ \dfrac{1}{a^{n-m}} \text{ if } m < n \end{cases}$

(d) $(ab)^m = a^m b^m$

(e) $\left(\dfrac{a}{b}\right)^m = \dfrac{a^m}{b^m} \quad b \neq 0$

Complete proofs of the various properties of exponents require the method of mathematical induction, discussed in the last chapter of this book. Properties (a) and (d) can be generalized: for example,

$$a^m \cdot a^n \cdot a^p = a^{m+n+p} \quad \text{and} \quad (abc)^n = a^n \cdot b^n \cdot c^n.$$

Example 1   Each of the following examples is justified by one or more of the properties of exponents and by the properties of real numbers.

(a) $(4x^3)(-3x^5) = (4)(-3)x^3 \cdot x^5 = -12x^{3+5} = -12x^8$

(b) $\dfrac{m^4 p^2}{m^3 p^5} = \dfrac{m^4}{m^3} \cdot \dfrac{p^2}{p^5} = m \cdot \dfrac{1}{p^3} = \dfrac{m}{p^3} \quad (m \neq 0, p \neq 0)$

(c) $\left(\dfrac{2x^5}{7}\right)^3 = \dfrac{2^3(x^5)^3}{7^3} = \dfrac{8x^{15}}{343}$

(d) $(2z^{2r})(z^{r+1}) = 2z^{2r+(r+1)} = 2z^{3r+1} \quad$ (r is a positive integer)

(e) $\dfrac{k^{r+1}}{k^r} = k^{(r+1)-r} = k \quad$ (r is a positive integer and k is nonzero)   ■

By part (a) of the properties of exponents, $a^m \cdot a^n = a^{m+n}$. If $n = 0$, then

$$a^m \cdot a^n = a^{m+n} \quad \text{becomes} \quad a^m \cdot a^0 = a^{m+0} = a^m.$$

The only way that $a^m \cdot a^0$ can equal a^m is if $a^0 = 1$. Therefore, to be consistent with past work, a^0 is defined as follows.

Zero Exponent

> If a is a nonzero real number, then
> $$a^0 = 1.$$

The symbol 0^0 is not defined.

Example 2   (a) $3^0 = 1$ (b) $(-4)^0 = 1$

(c) $\left(\dfrac{1}{7}\right)^0 = 1$ (d) $(-\sqrt{11})^0 = 1$

(e) $-4^0 = -1$ (f) $-(-4)^0 = -1$ ■

In Section 1.1 the symbol a^{-1} was defined for any nonzero real number a as $a^{-1} = 1/a$. This definition can be extended so as to give meaning to an expression of the form a^{-n}, where n is any positive integer, not just 1. As with zero exponents, any definition for a^{-n} must be consistent with the properties of exponents given earlier. Property (c) from above can be valid only if

$$a^{-n} = a^{(-1)n} = (a^{-1})^n.$$

Since $a^{-1} = \dfrac{1}{a}$,

$$a^{-n} = (a^{-1})^n = \underbrace{\frac{1}{a} \cdot \frac{1}{a} \cdot \frac{1}{a} \cdots \frac{1}{a}}_{n \text{ factors}} = \frac{1}{a^n}.$$

Based on this, for the past properties of exponents to remain valid, a^{-n} must be defined as $1/a^n$. That is,

Negative Exponent

> for all positive integers n and all nonzero real numbers a,
> $$a^{-n} = \frac{1}{a^n}.$$

Example 3   (a) $5^{-3} = \dfrac{1}{5^3} = \dfrac{1}{125}$ (b) $\left(\dfrac{2}{3}\right)^{-3} = \dfrac{1}{\left(\dfrac{2}{3}\right)^3} = \dfrac{1}{\dfrac{8}{27}} = \dfrac{27}{8}$

(c) $-4^{-2} = -\dfrac{1}{4^2} = -\dfrac{1}{16}$ ■

Part (b) of Example 3 involves work with fractions of a type that can lead to errors. For a shortcut useful with such fractions, assume that neither a nor b is zero, so that $(a/b) \cdot (b/a) = 1$, showing that $(a/b)^{-1} = b/a$. Use this, along with properties of exponents, to get

$$\left(\frac{a}{b}\right)^{-n} = \left[\left(\frac{a}{b}\right)^{-1}\right]^n = \left(\frac{b}{a}\right)^n.$$

For all positive integers n and nonzero real numbers a and b,

$$\left(\frac{a}{b}\right)^{-n} = \left(\frac{b}{a}\right)^n.$$

Example 4   **(a)** $\left(\dfrac{1}{3}\right)^{-2} = \left(\dfrac{3}{1}\right)^2 = 9$ **(b)** $\left(\dfrac{2}{3}\right)^{-3} = \left(\dfrac{3}{2}\right)^3 = \dfrac{27}{8}$ ■

It can be shown that all the properties of exponents given above are valid for *all* integer exponents, and not just positive integer exponents. In particular, property (b) from above can now be simplified to just one case:

$$\frac{a^m}{a^n} = a^{m-n}, \quad \text{if } a \neq 0.$$

In the next display all the properties and definitions for exponents are summarized.

Summary for Exponents

For any integers m and n, and any real numbers a and b (assuming no denominators are zero),

(a) $a^m \cdot a^n = a^{m+n}$ **(b)** $\dfrac{a^m}{a^n} = a^{m-n}$

(c) $(a^m)^n = a^{mn}$ **(d)** $(ab)^m = a^m b^m$

(e) $\left(\dfrac{a}{b}\right)^m = \dfrac{a^m}{b^m}$ **(f)** $a^0 = 1$

(g) $a^{-n} = \dfrac{1}{a^n}$ **(h)** $\left(\dfrac{a}{b}\right)^{-n} = \left(\dfrac{b}{a}\right)^n.$

Example 5   **(a)** $2^{-4} \cdot 2^5 = 2^{-4+5} = 2^1 = 2$

(b) $3x^{-2}(4^{-1}x^{-5})^2 = 3x^{-2}(4^{-2}x^{-10}) = 3 \cdot 4^{-2} \cdot x^{-2+(-10)} = 3 \cdot 4^{-2} \cdot x^{-12}$

$$= \frac{3}{1} \cdot \frac{1}{4^2} \cdot \frac{1}{x^{12}} = \frac{3}{16x^{12}}$$

(c) $\dfrac{5m^{-3}}{10m^{-5}} = \dfrac{5}{10}m^{-3-(-5)} = \dfrac{1}{2}m^2$, or $\dfrac{m^2}{2}$ ■

Example 6 Simplify each of the following expressions by writing without negative exponents. Assume all variables represent nonzero real numbers.

(a) $\dfrac{12p^3q^{-1}}{8p^{-2}q}$

Use the properties of exponents, working through the following sequence of steps.

$$\frac{12p^3q^{-1}}{8p^{-2}q} = \frac{12}{8} \cdot \frac{p^3}{p^{-2}} \cdot \frac{q^{-1}}{q}$$

$$= \frac{3}{2} \cdot p^{3-(-2)}\, q^{-1-1} \qquad q = q^1$$

$$= \frac{3}{2}\, p^5\, q^{-2}$$

$$= \frac{3p^5}{2q^2}$$

(b) $\dfrac{(3x^2)^{-1}(3x^5)^{-2}}{(3^{-1}x^{-2})^2} = \dfrac{3^{-1}x^{-2}3^{-2}x^{-10}}{3^{-2}x^{-4}}$

$$= \frac{3^{-1+(-2)}x^{-2+(-10)}}{3^{-2}x^{-4}}$$

$$= \frac{3^{-3}x^{-12}}{3^{-2}x^{-4}}$$

$$= 3^{-3-(-2)}x^{-12-(-4)}$$

$$= 3^{-1}x^{-8}$$

$$= \frac{1}{3x^8} \quad \blacksquare$$

Scientific work often involves numbers that are very large or very small. To avoid working with many zeros, express such numbers in scientific notation.

Scientific Notation | In **scientific notation,** a number is written as the product of a number whose absolute value is at least 1, but less than 10, and an integer power of 10.

For example,

$$125{,}000 = 1.25 \times 10^5,$$
$$7900 = 7.9 \times 10^3,$$
$$.008 = 8 \times 10^{-3},$$
$$.00000563 = 5.63 \times 10^{-6}.$$

One advantage of scientific notation is that some calculations can be simplified by using properties of exponents, as shown in the next example.

Example 7 Find each of the following.

(a) (28,000,000)(.000012)

Writing each number in scientific notation, and using the properties of exponents gives

$$(28{,}000{,}000)(.000012) = (2.8 \times 10^7)(1.2 \times 10^{-5})$$
$$= (2.8 \times 1.2) \times 10^{7+(-5)}$$
$$= 3.36 \times 10^2 = 336.$$

(b) $\dfrac{.041}{.000082} = \dfrac{4.1 \times 10^{-2}}{8.2 \times 10^{-5}}$

$$= \dfrac{4.1}{8.2} \times 10^{-2-(-5)}$$

$$= .5 \times 10^3 = 500 \quad \blacksquare$$

The number $.5 \times 10^3$ is not in scientific notation, since .5 is not between 1 and 10. To write the number in scientific notation, work as follows:

$$.5 \times 10^3 = (5 \times 10^{-1}) \times 10^3 = 5 \times (10^{-1} \times 10^3) = 5 \times 10^2.$$

Alternatively, introduce $10^1 \times 10^{-1} = 10^0 = 1$ as follows:

$$.5 \times 10^3 = (.5 \times 10^1) \times (10^3 \times 10^{-1}) = 5 \times 10^2.$$

Some calculators have the capability of displaying scientific notation. For example, such a calculator might display the number 1.46×10^{12} as

$$1.46 \qquad 12.$$

A common mistake with calculator use is to quote the displayed answer to more accuracy than is warranted by the original data. For example, if we measure a wall to the nearest meter and say that it is 18 meters long, then we are really saying that the wall has a length from 17.5 meters to 18.5 meters. If we measure the wall more accurately, and say that it is 18.3 meters long, then we know that its length is really in the range 18.25 meters to 18.35 meters. A measurement of 18.00 meters would indicate that the wall's length is in the range 17.995 to 18.005 meters. The measurement 18 meters is said to have 2 significant digits of accuracy; 18.3 has 3 significant digits and 18.00 has 4.

Example 8 The following chart shows some numbers, the number of significant digits in each number, and the range represented by each number.

Number	Number of significant digits	Range represented by number
29.6	3	29.55 to 29.65
1.39	3	1.385 to 1.395
.000096	2	.0000955 to .0000965
.03	1	.025 to .035
100.2	4	100.15 to 100.25 $\blacksquare$

There is one possible source of trouble with significant digits. While the measurement 19.00 meters is a measurement to the nearest hundredth meter, what about the measurement 93,000 meters? Does it represent a measurement to the nearest meter? the nearest ten meters? hundred meters? thousand meters? We cannot tell by the way the number is written. To solve this problem, write the number in scientific notation. Depending on what is known about the accuracy of the measurement, write 93,000 using scientific notation as follows.

Measurement to the nearest:	Scientific notation	Number of significant digits
meter	9.3000×10^4	5
ten meters	9.300×10^4	4
hundred meters	9.30×10^4	3
thousand meters	9.3×10^4	2

1.3 Exercises

Simplify each of the following. Write all answers without exponents.

1. 15^3
2. 21^4
3. 2^{-3}
4. 3^{-2}

5. $4^{-1} + 3^{-3}$
6. $5^{-2} - 2^{-1}$
7. $(-4)^{-3}$
8. $(-5)^{-2}$

9. $\left(\dfrac{1}{2}\right)^{-3}$
10. $\left(\dfrac{2}{7}\right)^{-2}$
11. $(.02)^{-3}$
12. $(.25)^{-6}$

13. $\dfrac{12}{10^{-2}}$
14. $\dfrac{5}{2^{-4}}$
15. $(9.864)^{-3}$
16. $(14.259)^{-2}$

Simplify each of the following by writing each with only positive exponents. Assume that all variables represent nonzero real numbers.

17. $(3m)^2(-2m)^3$
18. $(-2a^4b^7)(3b^2)$
19. $(6^{-2})(6^{-5})(6^3)$
20. $(4^8)(4^{-10})(4^2)$

21. $(5^3)(5^{-5})(5^{-2})$
22. $(m^{10})(m^{-4})(m^{-5})$
23. $\dfrac{(3^5)(3^{-2})}{3^{-4}}$
24. $\dfrac{(2^2)(2^{-5})}{2^7}$

25. $\dfrac{(d^{-1})(d^{-2})}{(d^8)(d^{-3})}$
26. $\dfrac{(t^5)(t^{-3})}{(t^4)(t^{-7})}$
27. $\dfrac{2^{-1}x^3y^{-3}}{xy^{-2}}$
28. $\dfrac{5^{-2}m^2y^{-2}}{5^2m^{-1}y^{-2}}$

29. $\dfrac{(4+s)^3(4+s)^{-2}}{(4+s)^{-5}}$
30. $\dfrac{(m+n)^4(m+n)^{-2}}{(m+n)^{-5}}$
31. $[(m^2n)^{-1}]^4$
32. $(x^5t^{-2})^{-2}$

33. $\left(\dfrac{a^{-1}}{b^2}\right)^{-3}$
34. $\left(\dfrac{2c^2}{d^3}\right)^{-2}$
35. $(-2x^{-3}y^2)^{-2}$
36. $(-3p^2q^{-2}s^{-1})^7$

37. $\dfrac{(3^{-1}m^{-2}n^2)^{-2}}{(mn)^{-1}}$
38. $\dfrac{(-4x^3y^{-2})^{-2}}{(4x^5y^4)^{-1}}$
39. $\dfrac{(r^2s^3t^4)^{-2}(r^3s^2t)^{-1}}{(rst)^{-3}(r^2st^2)^3}$
40. $\dfrac{(ab^2c)^{-3}(a^2bc)^{-2}}{(abc^2)^4}$

41. $\dfrac{[k^2(p+q)^4]^{-1}}{k^{-4}(p+q)^3}$
42. $\dfrac{t^4(x+y)^4t^{-8}}{(x+y)^{-2}t^{-5}}$

Evaluate each expression, letting $a = 2$, $b = -3$, and $c = 0$.

43. $a^3 + b$
44. $b^3 + a^2$
45. $-b^2 + 3(c + 5)$
46. $2a^2 - b^4$

47. $a^7b^9c^5$
48. $12a^{15}b^6c^8$
49. $a^b + b^a$
50. $b^c + a^c$

51. $b^a a^b a^c$

52. $c^a a^b b^a$

53. $(3a^3 + 5b^2)^c$

54. $-(4b^5 - 2a^3)^c$

55. $a^{-1} + b^{-1}$

56. $b^{-2} - a$

57. $\dfrac{2b^{-1} - 3a^{-1}}{a + b^2}$

58. $\dfrac{3a^2 - b^2}{b^{-3} + 2a^{-1}}$

Express each of the following numbers in scientific notation.

59. 69,300

60. 5000

61. 6,000,000,000

62. .0001

63. .00792

64. .054

65. $.4 \times 10^8$

66. $.7 \times 10^4$

67. $.009 \times 10^{-5}$

68. $.014 \times 10^{-7}$

69. 3800×10^{-9}

70. 5280×10^{-11}

Express each of the following numbers without exponents.

71. 8.2×10^5

72. 3.7×10^3

73. 1.7×10^8

74. 3.61×10^{-2}

75. 6.15×10^{-3}

76. 9.3×10^{-6}

77. 11.4×10^3

78. 39.4×10^2

79. 809×10^{-4}

80. 548×10^{-6}

81. $.0096 \times 10^{-1}$

82. $.00083 \times 10^{-2}$

Use scientific notation to evaluate each of the following.

83. $(4600)(.00092)$

84. $(.87)(.0004)$

85. $(.00002)(.00009)$

86. $(.0004)(.0000015)$

87. $\dfrac{.000034}{.017}$

88. $\dfrac{.0000042}{.0006}$

89. $\dfrac{28}{.0004}$

90. $\dfrac{50}{.0000025}$

Use a calculator to evaluate each of the following. Write all answers in scientific notation with the multiple of the power of ten rounded to three significant digits.

91. $(1.66 \times 10^4)(2.93 \times 10^3)$

92. $(6.92 \times 10^7)(8.14 \times 10^5)$

93. $(4.08 \times 10^{-9})(3.172 \times 10^4)$

94. $(9.113 \times 10^{12})(8.72 \times 10^{-11})$

95. $\dfrac{(-4.389 \times 10^4)(2.421 \times 10^{-2})}{1.76 \times 10^{-9}}$

96. $\dfrac{(1.392 \times 10^{10})(5.746 \times 10^{-8})}{4.93 \times 10^{-15}}$

97. $\dfrac{-3.9801 \times 10^{-6}}{(7.4993 \times 10^{-8})(2.117 \times 10^{-4})}$

98. $\dfrac{-5.421 \times 10^{-7}}{(4.2803 \times 10^{-4})(9.982 \times 10^{-8})}$

Use scientific notation to work the following exercises.

99. The distance to the sun is 9.3×10^7 mi. How long would it take a rocket, traveling at 2.9×10^3 mph, to reach the sun?

100. A *light-year* is the distance that light travels in one year. Find the number of miles in a light-year if light travels 1.86×10^5 mi per sec.

101. Use the information given in the previous two exercises to find the number of minutes necessary for light from the sun to reach the earth.

102. A computer can do one addition in 1.4×10^{-7} sec. How long would it take the computer to do a trillion (10^{12}) calculations? Give the answer in seconds and then in hours.

Engineering notation is similar to scientific notation, with the difference that the exponent on 10 must be a multiple of 3, with the multiplier at least 1 but less than 1000. (This notation is used to help with certain units conversions.) Write each of the following numbers in engineering notation.

103. 80,000

104. 122,000

105. .000047

106. .000098

107. 6.9×10^{11} **108.** 4.8×10^{5} **109.** 9.05×10^{-8} **110.** 1.27×10^{-13}

Find each of the following products or quotients. Assume that all variables appearing in exponents represent integers and all variables represent nonzero real numbers.

111. $(2k^m)(k^{1-m})$ **112.** $(y^m)(y^{1+m})$ **113.** $5p^r(6p^{3-2r})$ **114.** $(z^{p-1})(z^{3-p})$

115. $(5x^p)(3x^2)$ **116.** $(-z^{2r})(4z^{r+3})$ **117.** $\dfrac{k^{y-5}}{k^{y+2}}$ **118.** $\dfrac{4a^{m+2}}{2a^{m-1}}$

119. $\dfrac{(b^2)^y}{(2b^y)^3}$ **120.** $\left(\dfrac{3p^a}{2m^b}\right)^c$ **121.** $\dfrac{(2m^n)^2(-4m^{2+n})}{8m^{4n}}$ **122.** $\dfrac{(-5z^p)^3(2z^{1-4p})}{-10z^{-p}}$

123. First, show that $(1 + 2^3)^{-1} + (1 + 2^{-3})^{-1} = 1$. Then show that
$(1 + x^{a-b})^{-1} + (1 + x^{b-a})^{-1} = 1$.

1.4 Polynomials

A variable is a letter used to represent an element from a given set. Unless otherwise specified, in this book variables will represent real numbers. An **algebraic expression** is the result of performing the basic operations of addition, subtraction, multiplication, division (except by 0), or extraction of roots, on any collection of variables and numbers. The simplest algebraic expression, a polynomial, is discussed in this section.

To begin, a **term** is the product of a real number and one or more variables raised to powers. The real number is the **numerical coefficient,** or just the **coefficient.** For example, -3 is the coefficient in $-3m^4$, while -1 is the coefficient in $-p^2$.

A **polynomial** is defined as a finite sum of terms, with only nonnegative integer exponents permitted on the variables. If the terms of a polynomial contain only the variable x, then the polynomial is called a **polynomial in x.** (Polynomials in other variables are defined similarly.) By definition,

Polynomial in x

> a **polynomial in x** is an expression of the form
>
> $$a_n x^n + a_{n-1}x^{n-1} + \cdots + a_1 x + a_0,$$
>
> where a_n (read "a-sub-n"), $a_{n-1}, \cdots, a_1$, and a_0 are real numbers and n is a nonnegative integer. If $a_n \neq 0$, then n is the **degree** of the polynomial, and a_n is called the **leading coefficient.**

By this definition,

$$3x^4 - 5x^2 + 2$$

is a polynomial of degree 4 with leading coefficient 3. A nonzero constant is said to have degree 0, but no degree is assigned to the real number 0. If all the coefficients of a polynomial are 0, the polynomial is called the **zero polynomial.**

A polynomial can have more than one variable. A term containing more than one variable is said to have degree equal to the sum of all the exponents appearing on the variables in the term. For example, $-3x^4y^3z^5$ is of degree 12 because $4 + 3 + 5 = 12$. The degree of a polynomial in more than one variable is the highest degree of any term appearing in the polynomial. With this definition, the polynomial

$$2x^4y^3 - 3x^5y + x^6y^2$$

is of degree 8 because of the x^6y^2 term.

A polynomial containing exactly three terms is called a **trinomial;** one containing exactly two terms is a **binomial;** and a single-term polynomial is called a **monomial.** For example, $7x^9 - \sqrt{2}x^4 + 1$ is a trinomial of degree 9.

Since the variables used in polynomials represent real numbers, a polynomial represents a real number. This means that all the properties of the real numbers mentioned at the beginning of this chapter hold for polynomials. In particular, the distributive property holds, so that

$$3m^5 - 7m^5 = (3 - 7)m^5 = -4m^5.$$

To *add* polynomials of one variable, use the distributive property to add coefficients of the same powers; to *subtract* polynomials, subtract coefficients of the same powers, again with the distributive property.

Example 1   Add or subtract, as indicated.

(a) $(2y^4 - 3y^2 + y) + (4y^4 + 7y^2 + 6y)$
$$= (2 + 4)y^4 + (-3 + 7)y^2 + (1 + 6)y$$
$$= 6y^4 + 4y^2 + 7y$$

(b) $(6r^4 + 2r^2) + (3r^3 + 9r) = 6r^4 + 3r^3 + 2r^2 + 9r$

(c) $(-3m^3 - 8m^2 + 4) - (m^3 + 7m^2 - 3)$
$$= (-3 - 1)m^3 + (-8 - 7)m^2 + [4 - (-3)]$$
$$= -4m^3 - 15m^2 + 7 \quad \blacksquare$$

The distributive property, together with the properties of exponents, can also be used to find the product of two polynomials. For example, to find the product of $3x - 4$ and $2x^2 - 3x + 5$, treat $3x - 4$ as a single expression and use the distributive property as follows.

$$(3x - 4)(2x^2 - 3x + 5) = (3x - 4)(2x^2) - (3x - 4)(3x) + (3x - 4)(5)$$

Now use the distributive property three separate times on the right of the equals sign to get

$$(3x - 4)(2x^2 - 3x + 5) = (3x)(2x^2) - 4(2x^2) - (3x)(3x) - (-4)(3x)$$
$$+ (3x)5 - 4(5)$$
$$= 6x^3 - 8x^2 - 9x^2 + 12x + 15x - 20$$
$$= 6x^3 - 17x^2 + 27x - 20.$$

It is sometimes more convenient to find such a product as follows.

$$2x^2 - 3x + 5$$
$$\underline{ 3x - 4}$$
$$-8x^2 + 12x - 20 \leftarrow -4(2x^2 - 3x + 5)$$
$$\underline{6x^3 - 9x^2 + 15x \leftarrow 3x(2x^2 - 3x + 5)}$$
$$6x^3 - 17x^2 + 27x - 20 \qquad \text{Add in columns.}$$

Example 2   Find each product.

(a) $(6m + 1)(4m - 3) = (6m)(4m) - (6m)(3) + 1(4m) - 1(3)$
$$= 24m^2 - 14m - 3$$

(b) $(2k^n - 5)(k^n + 3) = 2k^{2n} + 6k^n - 5k^n - 15$
$$= 2k^{2n} + k^n - 15$$

(c) $(2x + 7)(2x - 7) = 4x^2 - 14x + 14x - 49$
$$= 4x^2 - 49 \quad \blacksquare$$

In Examples 2(a) and 2(b), the product of two binomials was a trinomial, while in Example 2(c) the product of two binomials was a binomial. The product of two binomials of the form $x + y$ and $x - y$ is a binomial. Multiply $x + y$ and $x - y$ to find that

Difference of Two Squares

$$(x + y)(x - y) = x^2 - y^2.$$

This product is called the **difference of two squares.**

Example 3   Find each product.

(a) $(3p + 11)(3p - 11)$
Using the result above, replace x with $3p$ and y with 11. This gives
$$(3p + 11)(3p - 11) = (3p)^2 - 11^2 = 9p^2 - 121.$$

(b) $(5m^3 - 3)(5m^3 + 3) = (5m^3)^2 - 3^2$
$$= 25m^6 - 9$$

(c) $(9k - 11r^3)(9k + 11r^3) = (9k)^2 - (11r^3)^2 = 81k^2 - 121r^6 \quad \blacksquare$

The square of the binomial $x + y$ is found by multiplying $x + y$ by itself. Finding this product, as well as $(x - y)^2$, gives the next results.

Square of a Binomial

$$(x + y)^2 = x^2 + 2xy + y^2$$
$$(x - y)^2 = x^2 - 2xy + y^2$$

Example 4   (a) $(2m + 5)^2 = (2m)^2 + 2(2m)(5) + (5)^2$
$$= 4m^2 + 20m + 25$$

(b) $(3z - 7y^4)^2 = (3z)^2 - 2(3z)(7y^4) + (7y^4)^2$
$$= 9z^2 - 42zy^4 + 49y^8 \quad \blacksquare$$

Two last special products show how to find the cube of a binomial:

Cube of a Binomial

$$(x + y)^3 = x^3 + 3x^2y + 3xy^2 + y^3$$
$$(x - y)^3 = x^3 - 3x^2y + 3xy^2 - y^3.$$

Example 5   Find each product.

(a) $(m + 4n)^3$

Replace x with m and y with $4n$ in the pattern for $(x + y)^3$ to get

$$(m + 4n)^3 = m^3 + 3m^2(4n) + 3m(4n)^2 + (4n)^3$$
$$= m^3 + 12m^2n + 48mn^2 + 64n^3.$$

(b) $(5k - 2z^5)^3 = (5k)^3 - 3(5k)^2(2z^5) + 3(5k)(2z^5)^2 - (2z^5)^3$
$$= 125k^3 - 150k^2z^5 + 60kz^{10} - 8z^{15}. \quad \blacksquare$$

It is useful to memorize and be able to apply these special products. (Both the square of a binomial and the cube of a binomial are special cases of the binomial theorem, discussed in the last chapter of this book.)

To divide a polynomial by a monomial, divide each term of the polynomial by the monomial.

Example 6   Divide.

(a) $\dfrac{2m^5 - 6m^3}{2m^3} = \dfrac{2m^5}{2m^3} - \dfrac{6m^3}{2m^3} = m^2 - 3$

The polynomial $m^2 - 3$ is the quotient of $2m^5 - 6m^3$ and $2m^3$.

(b) $\dfrac{3y^6x^3 - 6y^3x^6 + 8y^5x}{3y^3x^3} = \dfrac{3y^6x^3}{3y^3x^3} - \dfrac{6y^3x^6}{3y^3x^3} + \dfrac{8y^5x}{3y^3x^3}$

$$= y^3 - 2x^3 + \dfrac{8y^2}{3x^2} \quad \blacksquare$$

In part (b) of Example 6, the result is not a polynomial.

1.4 Exercises

Perform the indicated operations.

1. $(3x^2 - 4x + 5) + (-2x^2 + 3x - 2)$

2. $(4m^3 - 3m^2 + 5) + (-3m^3 - m^2 + 5)$

3. $(x^2 + 4x) + (3x^3 - 4x^2 + 2x + 2)$

4. $(r^5 - r^3 + r) + (3r^5 - 4r^4 + r^3 + 2r)$

5. $(12y^2 - 8y + 6) - (3y^2 - 4y + 2)$

6. $(8p^2 - 5p) - (3p^2 - 2p + 4)$

7. $(p^3 - 4p^2 + p) - (3p^2 + 2p + 7)$

8. $(4y^2 - 2y + 7) - (6y - 9)$

9. $(3a^2 - 2a) + (4a^2 + 3a + 1) - (a^2 + 2)$

10. $(6m^4 - 3m^2 + m) - (2m^3 + 5m^2 + 4m) + (m^2 - m)$

11. $(5b^2 - 4b + 3) - (2b^2 + b) - (3b + 4)$

12. $-(8x^3 + x - 3) + (2x^3 + x^2) - (4x^2 + 3x - 1)$

13. $-3(4q^2 - 3q + 2) + 2(-q^2 + q - 4)$

14. $2(3r^2 + 4r + 2) - 3(-r^2 + 4r - 5)$

15. $(4r - 1)(7r + 2)$

16. $(5m - 6)(3m + 4)$

17. $(6p + 5q)(3p - 7q)$

18. $(2z + y)(3z - 4y)$

19. $\left(3x - \dfrac{2}{3}\right)\left(5x + \dfrac{1}{3}\right)$

20. $\left(2m - \dfrac{1}{4}\right)\left(3m + \dfrac{1}{2}\right)$

21. $\left(\dfrac{2}{5}y + \dfrac{1}{8}z\right)\left(\dfrac{3}{5}y + \dfrac{1}{2}z\right)$

22. $\left(\dfrac{3}{4}r - \dfrac{2}{3}s\right)\left(\dfrac{5}{4}r + \dfrac{1}{3}s\right)$

23. $(5r + 2)(5r - 2)$

24. $(6z + 5)(6z - 5)$

25. $(4x + 3y)(4x - 3y)$

26. $(7m + 2n)(7m - 2n)$

27. $(6k - 3)^2$

28. $(3p + 5)^2$

29. $(4m + 2n)^2$

30. $(a - 6b)^2$

31. $(2z - 1)^3$

32. $(3m + 2)^3$

33. $4x^2(3x^3 + 2x^2 - 5x + 1)$

34. $2b^3(b^2 - 4b + 3)$

35. $5m(3m^3 - 2m^2 + m - 1)$

36. $4y^3(y^3 + 2y^2 - 6y + 3)$

37. $(2z - 1)(-z^2 + 3z - 4)$

38. $(x - 1)(x^2 - 1)$

39. $(3p - 1)(9p^2 + 3p + 1)$

40. $(2p - 1)(3p^2 - 4p + 5)$

41. $(2m + 1)(4m^2 - 2m + 1)$

42. $(k + 2)(12k^3 - 3k^2 + k + 1)$

43. $(m - n + k)(m + 2n - 3k)$

44. $(r - 3s + t)(2r - s + t)$

45. $(a - b + 2c)^2$

46. $(k - y + 3m)^2$

47. $(x^{-1} - 2)^2$

48. $(1 + 2y^{-1})^2$

49. $(3m^{-1} - 2n^{-1})(4m^{-1} + n^{-1})$

50. $(-k^{-1} + 3q^{-1})(4k^{-1} - 3q^{-1})$

51. $(\sqrt{6} + \sqrt{5} + \sqrt{3})(\sqrt{6} + \sqrt{5} - \sqrt{3})$

52. $(\sqrt{7} - \sqrt{10} + \sqrt{11})(\sqrt{7} - \sqrt{10} - \sqrt{11})$

53. $(\sqrt[3]{2} - \sqrt[3]{5})^3$ **54.** $(\sqrt[3]{7} + \sqrt[3]{9})^3$

Perform each of the following divisions. Assume that all variables appearing in denominators represent nonzero real numbers.

55. $\dfrac{4m^3 - 8m^2 + 16m}{2m}$

56. $\dfrac{30k^5 - 12k^3 + 18k^2}{6k^2}$

57. $\dfrac{15x^4 + 30x^3 + 12x^2 - 9}{3x}$

58. $\dfrac{16a^6 + 24a^5 - 48a^4 + 12a}{8a^2}$

59. $\dfrac{25x^2y^4 - 15x^3y^3 + 40x^4y^2}{5x^2y^2}$

60. $\dfrac{-8r^3s - 12r^2s^2 + 20rs^3}{4rs}$

In each of the following, find the coefficient of x^3 without finding the entire product.

61. $(x^2 + 4x)(-3x^2 + 4x - 1)$

62. $(4x^3 - 2x + 5)(x^3 + 2)$

63. $(1 + x^2)(1 + x)$

64. $(3 - x)(2 - x^2)$

65. $x^2(4 - 3x)^2$

66. $-4x^2(2 - x)(2 + x)$

Find each of the following products. Assume all variables used as exponents represent integers.

67. $(k^m + 2)(k^m - 2)$ **68.** $(y^x - 4)(y^x + 4)$ **69.** $(b^r + 3)(b^r - 2)$ **70.** $(q^y + 4)(q^y + 3)$

71. $(3p^x + 1)(p^x - 2)$ **72.** $(2^a + 5)(2^a + 3)$ **73.** $(m^x - 2)^2$ **74.** $(z^r + 5)^2$

75. $(3k^a - 2)^3$ **76.** $(r^x - 4)^3$

77. Suppose one polynomial has degree 3 and another also has degree 3. Find all possible values for the degree of their **(a)** sum **(b)** difference **(c)** product.

78. If one polynomial has degree 3 and another has degree 4, find all possible values for the degree of their **(a)** sum **(b)** difference **(c)** product.

79. Generalize the results of Exercises 77 and 78: Suppose one polynomial has degree m and another has degree n, where m and n are natural numbers with $n < m$. Find all possible values for the degree of their **(a)** sum **(b)** difference **(c)** product.

80. Suppose that for a given polynomial p, there exists a polynomial q such that $pq = 1$, making q the multiplication inverse of p. Show that p must have degree 0.

81. Derive a formula for $(x + y + z)^2$.

82. Derive a formula for $(x + y + z)(x + y - z)$.

1.5 Factoring

The process of finding polynomials whose product equals a given polynomial is called **factoring.** For example, since $4x + 12 = 4(x + 3)$, both 4 and $x + 3$ are **factors** of $4x + 12$. Also, $4(x + 3)$ is the **factored form** of $4x + 12$. A polynomial is **factored completely** when it is written as a product of polynomials, none of which can be written as the product of polynomials of positive degree. A polynomial that cannot be written as a product of two polynomials of positive degree is a **prime** or **irreducible** polynomial.

The first step in factoring a polynomial is to look for any **common factors;** that is, expressions that are factors of each term of the given polynomial.

Example 1   Factor a common factor from each of the following polynomials.

(a) $6x^2y^3 + 9xy^4 + 18y^5$

Each term of this polynomial can be written with a factor of $3y^3$, so that $3y^3$ is a common factor. Use the distributive property to get

$$6x^2y^3 + 9xy^4 + 18y^5 = (3y^3)(2x^2) + (3y^3)(3xy) + (3y^3)(6y^2)$$
$$= 3y^3(2x^2 + 3xy + 6y^2).$$

(b) $6x^2t + 8xt + 12t = 2t(3x^2 + 4x + 6)$ ■

Each of the special patterns of multiplication from the previous section can be used in reverse to get a pattern for factoring. One of the most common of these is the difference of two squares.

Difference of Two Squares

$$x^2 - y^2 = (x + y)(x - y)$$

Example 2 Factor each of the following polynomials.

(a) $4m^2 - 9$

First, recognize that $4m^2 - 9$ is the difference of two squares, since $4m^2 = (2m)^2$ and $9 = 3^2$. Use the pattern for the difference of two squares with $2m$ replacing x and 3 replacing y.

$$\begin{aligned} 4m^2 - 9 &= (2m)^2 - 3^2 \\ &= (2m + 3)(2m - 3) \end{aligned}$$

(b) $144r^2 - 25s^2 = (12r + 5s)(12r - 5s)$

(c) $256k^4 - 625m^4$

Use the difference of two squares pattern twice, as follows:

$$\begin{aligned} 256k^4 - 625m^4 &= (16k^2)^2 - (25m^2)^2 \\ &= (16k^2 + 25m^2)(16k^2 - 25m^2) \\ &= (16k^2 + 25m^2)(4k + 5m)(4k - 5m) \end{aligned}$$

(d) $\begin{aligned}[t] (a + 2b)^2 - 4c^2 &= (a + 2b)^2 - (2c)^2 \\ &= [(a + 2b) + 2c][(a + 2b) - 2c] \\ &= (a + 2b + 2c)(a + 2b - 2c) \quad \blacksquare \end{aligned}$

In this chapter, a polynomial with only integer coefficients will be factored so that all factors have only integer coefficients. This assumption is sometimes summarized by saying that only **factoring over the integers** is permitted. With factoring over the integers, the polynomial $x^2 - 5$, for example, cannot be factored. While it is true that

$$x^2 - 5 = (x + \sqrt{5})(x - \sqrt{5}),$$

the two factors $x + \sqrt{5}$ and $x - \sqrt{5}$ have noninteger coefficients.

Now we need to look at methods of factoring a trinomial of degree 2, such as $kx^2 + mx + n$, where k, m, and n are integers. Any factorization will be of the form $(ax + b)(cx + d)$ with a, b, c, and d integers. Multiplying out the product $(ax + b)(cx + d)$ gives

$$(ax + b)(cx + d) = acx^2 + (ad + bc)x + bd,$$

which equals $kx^2 + mx + n$ if

$$ac = k, \quad ad + bc = m, \quad \text{and} \quad bd = n. \qquad (*)$$

In summary, to factor a trinomial $kx^2 + mx + n$, look for four integers a, b, c, and d satisfying the conditions given in $(*)$. If no such integers exist, the trinomial is prime.

Example 3   Factor each of the following polynomials.

(a) $6p^2 - 7p - 5$

To find integers a, b, c, and d so that

$$6p^2 - 7p - 5 = (ap + b)(cp + d),$$

use the results given in (∗) above and try to find integers satisfying $ac = 6$, $ad + bc = -7$, and $bd = -5$. Look for these integers by trying various possibilities. Since $ac = 6$, we might let $a = 2$ and $c = 3$. Since $bd = -5$, we might let $b = -5$ and $d = 1$, giving

$$(2p - 5)(3p + 1) = 6p^2 - 13p - 5. \qquad \text{incorrect}$$

To make another attempt, try

$$(3p - 5)(2p + 1) = 6p^2 - 7p - 5. \qquad \text{correct}$$

Finally, $6p^2 - 7p - 5$ is factored as $(3p - 5)(2p + 1)$.

(b) $4x^3 + 6x^2r - 10xr^2$

There is a common factor of $2x$.

$$4x^3 + 6x^2r - 10xr^2 = 2x(2x^2 + 3xr - 5r^2).$$

Factoring $2x^2 + 3xr - 5r^2$ requires factors of $2x^2$ and of $-5r^2$ that will yield the correct middle term of $3xr$. By inspection,

$$4x^3 + 6x^2r - 10xr^2 = 2x(2x + 5r)(x - r).$$

(c) $r^2 + 6r + 7$

It is not possible to find integers a, b, c, and d so that

$$r^2 + 6r + 7 = (ar + b)(cr + d);$$

therefore, $r^2 + 6r + 7$ is a prime polynomial.   ■

Two other special types of factoring are listed below.

Difference and Sum of Cubes

$$x^3 - y^3 = (x - y)(x^2 + xy + y^2) \qquad \textbf{Difference of two cubes}$$
$$x^3 + y^3 = (x + y)(x^2 - xy + y^2) \qquad \textbf{Sum of two cubes}$$

Example 4   Factor each polynomial.

(a) $m^3 - 64n^3$

Since $64n^3 = (4n)^3$, the given binomial is a difference of two cubes. To factor, use the first pattern above, replacing x with m and y with $4n$, to get

$$m^3 - 64n^3 = m^3 - (4n)^3$$
$$= (m - 4n)[m^2 + m(4n) + (4n)^2]$$
$$= (m - 4n)(m^2 + 4mn + 16n^2).$$

(b) $8q^6 + 125p^9$

Write $8q^6$ as $(2q^2)^3$ and $125p^9$ as $(5p^3)^3$, showing that the given polynomial is a sum of two cubes. Factor as

$$8q^6 + 125p^9 = (2q^2)^3 + (5p^3)^3$$
$$= (2q^2 + 5p^3)[(2q^2)^2 - (2q^2)(5p^3) + (5p^3)^2]$$
$$= (2q^2 + 5p^3)(4q^4 - 10q^2p^3 + 25p^6).$$

(c) $(2a - 1)^3 + 8$

Use the pattern for the sum of two cubes, with $x = 2a - 1$ and $y = 2$. Doing so gives

$$(2a - 1)^3 + 8 = [(2a - 1) + 2][(2a - 1)^2 - (2a - 1)2 + 2^2]$$
$$= (2a - 1 + 2)(4a^2 - 4a + 1 - 4a + 2 + 4)$$
$$= (2a + 1)(4a^2 - 8a + 7). \quad \blacksquare$$

When a polynomial has more than three terms, it can often be factored by grouping. For example, to factor

$$ax + ay + 6x + 6y,$$

collect the terms into two groups,

$$ax + ay + 6x + 6y = (ax + ay) + (6x + 6y),$$

and then factor each group, getting

$$ax + ay + 6x + 6y = a(x + y) + 6(x + y).$$

The quantity $(x + y)$ is now a common factor, which can be factored out, producing

$$ax + ay + 6x + 6y = (x + y)(a + 6).$$

It is not always obvious which terms should be grouped. Experience is the best teacher for techniques of factoring, along with repeated trials.

Example 5 Factor by grouping.

(a) $mp^2 + 7m + 3p^2 + 21$

Group the terms as follows.

$$mp^2 + 7m + 3p^2 + 21 = (mp^2 + 7m) + (3p^2 + 21)$$

Find the common factor for each part.

$$(mp^2 + 7m) + (3p^2 + 21) = m(p^2 + 7) + 3(p^2 + 7)$$
$$= (p^2 + 7)(m + 3)$$

(b) $2y^2 - 2z - ay^2 + az$

Grouping terms as above gives

$$2y^2 - 2z - ay^2 + az = (2y^2 - 2z) + (-ay^2 + az)$$
$$= 2(y^2 - z) + a(-y^2 + z).$$

The expression $-y^2 + z$ is the negative of $y^2 - z$, so the terms should be grouped as follows.

$$2y^2 - 2z - ay^2 + az = (2y^2 - 2z) - (ay^2 - az)$$
$$= 2(y^2 - z) - a(y^2 - z)$$
$$= (y^2 - z)(2 - a)$$

(c) $6p^2 - 15p + 16p - 40 = (6p^2 - 15p) + (16p - 40)$
$$= 3p(2p - 5) + 8(2p - 5)$$
$$= (2p - 5)(3p + 8) \quad \blacksquare$$

1.5 Exercises

Factor as completely as possible. Assume all variables appearing as exponents represent integers.

1. $12mn - 8m$
2. $3pq - 18pqr$
3. $12r^3 + 6r^2 - 3r$
4. $-5k^2g - 25kg - 30kg^2$
5. $6px^2 - 8px^3 - 12px$
6. $9m^2n^3 - 18m^3n^2 + 27m^2n^4$
7. $2(a + b) + 4m(a + b)$
8. $4(y - 2)^2 + 3(y - 2)$
9. $x^2 - 11x + 24$
10. $y^2 - 2y - 35$
11. $4p^2 + 3p - 1$
12. $6x^2 + 7x - 3$
13. $2z^2 + 7z - 30$
14. $4m^2 + m - 3$
15. $12r^2 + 24r - 15$
16. $12p^2 + p - 20$
17. $18r^2 - 3rs - 10s^2$
18. $12m^2 + 16mn - 35n^2$
19. $15x^2 - 14xy - 8y^2$
20. $12t^2 - 25tv + 12v^2$
21. $9x^2 - 6x^3 + x^4$
22. $30a + am - am^2$
23. $2r^2z - rz - 3z$
24. $3m^4 + 7m^3 + 2m^2$
25. $4m^2 - 25$
26. $25a^2 - 16$
27. $144r^2 - 81s^2$
28. $81m^2 - 16n^2$
29. $121p^4 - 9q^4$
30. $4z^4 - w^8$
31. $16x^4 - y^4$
32. $81q^4 - 256m^4$
33. $p^8 - 1$
34. $y^{16} - 1$
35. $8m^3 - 27n^3$
36. $125x^3 - 1$
37. $64 - x^6$
38. $m^6 - 1$
39. $4x + 4y + mx + my$
40. $x^2 + xy + 5x + 5y$
41. $q^2 + 6q + 9 - p^2$
42. $4b^2 + 4bc + c^2 - 16$
43. $a^2 + 2ab + b^2 - x^2 - 2xy - y^2$
44. $d^2 - 10d + 25 - c^2 + 4c - 4$
45. $x^2 - (x - y)^2$
46. $16m^2 - 25(m - n)^2$
47. $49(m - 1)^2 - 16m^2$
48. $36(a + 2)^2 - 25a^2$
49. $(x + y)^2 + 2(x + y)z - 15z^2$
50. $(m + n)^2 + 3(m + n)p - 10p^2$
51. $6(a + b)^2 + (a + b)c - 40c^2$
52. $12(g + h)^2 - 5(g + h)j - 25j^2$
53. $(p + q)^2 - (p - q)^2$
54. $(p - q)^2 - (p + q)^2$
55. $(r + 6)^3 - 216$
56. $(b + 3)^3 - 27$
57. $27 - (m + 2n)^3$
58. $125 - (4a - b)^3$
59. $125m^6 - 8(z + x^2)^3$
60. $64p^6 - 27(p^2 - q)^3$
61. $m^{2n} - 16$
62. $p^{4n} - 49$
63. $x^{3n} - y^{6n}$
64. $a^{4p} - b^{12p}$
65. $2x^{2n} - 23x^ny^n - 39y^{2n}$
66. $3a^{2x} + 7a^xb^x + 2b^{2x}$
67. $4y^{2a} - 12y^a + 9$
68. $25x^{4c} - 20x^{2c} + 4$
69. $25q^{2r} - 30q^rt^p + 9t^{2p}$
70. $16m^{2p} + 56m^pn^q + 49n^{2q}$
71. $125r^{6k} + 216s^{12k}$
72. $216p^{12r} + 125q^{15r}$
73. $6(m + p)^{2k} + (m + p)^k - 15$
74. $8(2k + q)^{4z} - 2(2k + q)^{2z} - 3$

Factor the coefficient of smallest absolute value from each of the following.

75. $9.44x^3 + 48.144x^2 - 30.208x + 37.76$ **76.** $12.915m^2 + 2.05m - 23.575m^3 + 20.5$

77. $64.616z^4 - 23.64z^2 + 7.88z + 114.26$ **78.** $1.47y^5 - 8.526y^3 + 69.384y^2 - 36.75$

Factor the variable of smallest exponent, together with any numerical common factor, from each of the following expressions. (For example, factor $9x^{-2} - 6x^{-3}$ as $3x^{-3}(3x - 2)$.)

79. $p^{-4} - p^{-2}$ **80.** $m^{-1} + 3m^{-5}$ **81.** $12k^{-3} + 4k^{-2} - 8k^{-1}$

82. $6a^{-5} - 10a^{-4} + 18a^{-2}$ **83.** $100p^{-6} - 50p^{-2} + 75p^2$ **84.** $32y^{-3} + 48y - 64y^2$

Factor each of the following.

85. $(3z - z^2)^2 + (9 - z^2)^2 + (3 - z)^2$ **86.** $(4r^2 - r)^3 - (1 - 4r)^3 + (16r^2 - 1)^3$

87. $x^2 + y^2 + z^2 - (x + y + z)^2$ **88.** $r^2(t - s) + s^2(r - t) + t^2(s - r)$

Factor $x^4 + 4x^2 + 16$ as follows:

$$x^4 + 4x^2 + 16 = (x^4 + 8x^2 + 16) - 4x^2$$
$$= (x^2 + 4)^2 - (2x)^2$$
$$= (x^2 + 4 + 2x)(x^2 + 4 - 2x).$$

Use this procedure to factor each of the following polynomials.

89. $x^4 + 64$ **90.** $r^4 - 6r^2 + 1$ **91.** $p^4 + 9p^2 + 81$

92. $x^4 - 18x^2 + 1$ **93.** $z^4 - 11z^2 + 25$ **94.** $m^4 - 22m^2 + 9$

The expressions in Exercises 95–96 are the result of standard processes of calculus. Factor each expression.

95. $(x^3 + 5)^2(4)(5x - 11)^3(5) + (5x - 11)^4(2)(x^3 + 5)(3x^2)$

96. $(2x^2 + 9)^3(2)(x^5 - 4)(5x^4) + (x^5 - 4)^2(3)(2x^2 + 9)^2(4x)$

97. Factor $4(x^2 + 3)^{-3}(4x + 7)^{-4}$ from
$(x^2 + 3)^{-2}(-3)(4x + 7)^{-4}(4) + (4x + 7)^{-3}(-2)(x^2 + 3)^{-3}(2x).$

98. Factor $(7x - 8)^{-6}(3x + 2)^{-3}$ from
$(7x - 8)^{-5}(-2)(3x + 2)^{-3}(3) + (3x + 2)^{-2}(-5)(7x - 8)^{-6}(7).$

1.6 Fractional Expressions

The quotient of two algebraic expressions (with denominator not 0) is a **fractional expression.** The most common fractional expressions are the quotients of two polynomials; these quotients are called **rational expressions.** Since fractional expressions involve quotients, it is important to keep track of the values of the variables that satisfy the requirement that no denominator be 0. The set of such values is the **domain** of the expression. We shall make the following agreement.

Domain

> Unless otherwise stated, the **domain** of a fractional expression is the set of all real numbers for which no denominator is 0.

For example, -2 is not in the domain of the rational expression

$$\frac{x + 6}{x + 2}$$

since -2 (when substituted for x) makes the denominator equal 0. The domain of this rational expression can be written $\{x \mid x \neq -2\}$. In a similar way, the domain of

$$\frac{(x + 6)(x + 4)}{(x + 2)(x + 4)}$$

can be written $\{x \mid x \neq -2 \text{ and } x \neq -4\}$.

Just as the fraction 6/8 is written in lowest terms as 3/4, rational expressions must also be written in lowest terms. Do this with the **fundamental principle:**

Fundamental Principle

For any rational number a/b and nonzero real number c,

$$\frac{ac}{bc} = \frac{a}{b}.$$

Example 1 Write each expression in lowest terms.

(a) $\dfrac{2p^2 + 7p - 4}{5p^2 + 20p}$

Factor the numerator and denominator to get

$$\frac{2p^2 + 7p - 4}{5p^2 + 20p} = \frac{(2p - 1)(p + 4)}{5p(p + 4)}.$$

By the fundamental principle,

$$\frac{2p^2 + 7p - 4}{5p^2 + 20p} = \frac{2p - 1}{5p}.$$

The domain of the original expression is $\{p \mid p \neq 0 \text{ and } p \neq -4\}$, so this result is valid only for values of p other than 0 and -4. From now on, we shall always assume such restrictions when writing rational expressions in lowest terms.

(b) $\dfrac{6 - 3k}{k^2 - 4}$

Factor to get $\dfrac{6 - 3k}{k^2 - 4} = \dfrac{3(2 - k)}{(k + 2)(k - 2)}.$

The factors $2 - k$ and $k - 2$ have exactly opposite signs. Because of this, multiply numerator and denominator by -1, as follows.

$$\frac{6 - 3k}{k^2 - 4} = \frac{3(2 - k)(-1)}{(k + 2)(k - 2)(-1)}$$

Since $(k - 2)(-1) = -k + 2$, or $2 - k$, the denominator becomes

$$\frac{6 - 3k}{k^2 - 4} = \frac{3(2 - k)(-1)}{(k + 2)(2 - k)},$$

finally giving $\dfrac{6 - 3k}{k^2 - 4} = \dfrac{-3}{k + 2}$.

Working in an alternate way would give the equivalent result $3/(-k - 2)$. ■

To multiply or divide rational expressions, again use properties and definitions from Section 1.1.

Multiplication and Division

For rational numbers a/b and c/d,

$$\frac{a}{b} \cdot \frac{c}{d} = \frac{ac}{bd},$$

$$\frac{a}{b} \div \frac{c}{d} = \frac{a}{b} \cdot \frac{d}{c}, \qquad \frac{c}{d} \neq 0.$$

The result for division comes from the definition of division given earlier:

$$\frac{a}{b} \div \frac{c}{d} = \frac{\dfrac{a}{b}}{\dfrac{c}{d}} = \frac{a}{b} \cdot \left(\frac{c}{d}\right)^{-1} = \frac{a}{b} \cdot \frac{d}{c} = \frac{ad}{bc}.$$

As examples of these operations,

$$\frac{5}{8} \cdot \frac{3}{7} = \frac{5 \cdot 3}{8 \cdot 7} = \frac{15}{56} \quad \text{and} \quad \frac{2}{9} \div \frac{3}{7} = \frac{2}{9} \cdot \frac{7}{3} = \frac{14}{27}.$$

Example 2   Multiply or divide, as indicated.

(a) $\dfrac{3m^2 - 2m - 8}{3m^2 + 14m + 8} \cdot \dfrac{3m + 2}{3m + 4} = \dfrac{(m - 2)(3m + 4)}{(m + 4)(3m + 2)} \cdot \dfrac{3m + 2}{3m + 4}$

$$= \frac{(m - 2)(3m + 4)(3m + 2)}{(m + 4)(3m + 2)(3m + 4)} = \frac{m - 2}{m + 4}$$

(b) $\dfrac{5}{8m + 16} \div \dfrac{7}{12m + 24} = \dfrac{5}{8(m + 2)} \div \dfrac{7}{12(m + 2)}$

$$= \frac{5}{8(m + 2)} \cdot \frac{12(m + 2)}{7}$$

$$= \frac{5 \cdot 12(m + 2)}{8 \cdot 7(m + 2)}$$

$$= \frac{15}{14}$$

(c) $\dfrac{3p^2 + 11p - 4}{24p^3 - 8p^2} \div \dfrac{9p + 36}{24p^4 - 36p^3} = \dfrac{(p + 4)(3p - 1)}{8p^2(3p - 1)} \div \dfrac{9(p + 4)}{12p^3(2p - 3)}$

$$= \dfrac{(p + 4)(3p - 1)(12p^3)(2p - 3)}{8p^2(3p - 1)(9)(p + 4)}$$

$$= \dfrac{12p^3(2p - 3)}{9 \cdot 8p^2} = \dfrac{p(2p - 3)}{6} \quad \blacksquare$$

Add or subtract rational expressions with properties given earlier.

Addition and Subtraction

For rational numbers a/b, c/b, and c/d,

$$\dfrac{a}{b} + \dfrac{c}{b} = \dfrac{a + c}{b} \quad \text{or} \quad \dfrac{a}{b} + \dfrac{c}{d} = \dfrac{ad + bc}{bd}$$

$$\dfrac{a}{b} - \dfrac{c}{b} = \dfrac{a - c}{b} \quad \text{or} \quad \dfrac{a}{b} - \dfrac{c}{d} = \dfrac{ad - bc}{bd}.$$

The left hand results are for rational expressions with the same denominators. The results on the right come from the fundamental principle. For example,

$$\dfrac{a}{b} + \dfrac{c}{d} = \dfrac{a \cdot d}{b \cdot d} + \dfrac{c \cdot b}{d \cdot b} = \dfrac{ad}{bd} + \dfrac{bc}{bd} = \dfrac{ad + bc}{bd}.$$

As numerical examples,

$$\dfrac{4}{7} - \dfrac{1}{7} = \dfrac{4 - 1}{7} = \dfrac{3}{7} \quad \text{and} \quad \dfrac{2}{3} - \dfrac{1}{4} = \dfrac{2 \cdot 4 - 3 \cdot 1}{3 \cdot 4} = \dfrac{8 - 3}{12} = \dfrac{5}{12}.$$

In practice, rational expressions are normally added or subtracted after rewriting all the rational expressions so that they have the same denominator. This common denominator is found with the following steps.

Common Denominator

1. Write each denominator as a product of prime factors.
2. Form a product of all the different prime factors. Each factor should have as exponent the *highest* exponent which appears on that factor in any of the denominators.

Example 3   Add or subtract, as indicated.

(a) $\dfrac{5}{9x^2} + \dfrac{1}{6x}$

Write each denominator as a product of prime factors, as follows.

$$9x^2 = 3^2 \cdot x^2 \quad \text{and} \quad 6x = 2^1 \cdot 3^1 x^1$$

To get the common denominator, form the product of all the prime factors, with each factor having the highest exponent which appears on it in any of the denominators. Here the highest exponent on 2 is 1, while both 3 and x have a highest exponent of 2; the common denominator is

$$2^1 \cdot 3^2 \cdot x^2 = 18x^2.$$

Now write both of the given rational expressions with this denominator, giving

$$\frac{5}{9x^2} + \frac{1}{6x} = \frac{5 \cdot 2}{9x^2 \cdot 2} + \frac{1 \cdot 3x}{6x \cdot 3x} = \frac{10}{18x^2} + \frac{3x}{18x^2} = \frac{10 + 3x}{18x^2}.$$

(b) $\dfrac{y + 2}{y^2 - y} - \dfrac{3y}{2y^2 - 4y + 2}$

Factor each denominator.

$$\frac{y + 2}{y^2 - y} - \frac{3y}{2y^2 - 4y + 2} = \frac{y + 2}{y(y - 1)} - \frac{3y}{2(y - 1)^2}$$

The common denominator, by the method above, is $2y(y - 1)^2$. Write each rational expression with this denominator, as follows.

$$\frac{y + 2}{y(y - 1)} - \frac{3y}{2(y - 1)^2} = \frac{2(y - 1)(y + 2)}{2y(y - 1)^2} - \frac{y \cdot 3y}{2y(y - 1)^2}$$

$$= \frac{2(y^2 + y - 2)}{2y(y - 1)^2} - \frac{3y^2}{2y(y - 1)^2}$$

$$= \frac{2y^2 + 2y - 4 - 3y^2}{2y(y - 1)^2} = \frac{-y^2 + 2y - 4}{2y(y - 1)^2} \quad \blacksquare$$

Complex Fractions Any quotient of two rational expressions is called a **complex fraction.** Complex fractions can often be simplified by the methods shown in the following examples.

Example 4 Simplify each complex fraction.

(a) $\dfrac{6 - \dfrac{5}{k}}{1 + \dfrac{5}{k}}$

Multiply both numerator and denominator by the common denominator k.

$$\frac{k\left(6 - \dfrac{5}{k}\right)}{k\left(1 + \dfrac{5}{k}\right)} = \frac{6k - k\left(\dfrac{5}{k}\right)}{k + k\left(\dfrac{5}{k}\right)} = \frac{6k - 5}{k + 5}$$

(b) $\dfrac{\dfrac{a}{a + 1} + \dfrac{1}{a}}{\dfrac{1}{a} + \dfrac{1}{a + 1}}$

Multiply both numerator and denominator by the common denominator of all the fractions, in this case $a(a + 1)$. Doing so gives

$$\frac{\dfrac{a}{a + 1} + \dfrac{1}{a}}{\dfrac{1}{a} + \dfrac{1}{a + 1}} = \frac{\left(\dfrac{a}{a + 1} + \dfrac{1}{a}\right) a(a + 1)}{\left(\dfrac{1}{a} + \dfrac{1}{a + 1}\right) a(a + 1)} = \frac{a^2 + (a + 1)}{(a + 1) + a} = \frac{a^2 + a + 1}{2a + 1}.$$

As an alternate method of solution, first perform the indicated additions in the numerator and denominator, and then divide.

$$\frac{\dfrac{a}{a + 1} + \dfrac{1}{a}}{\dfrac{1}{a} + \dfrac{1}{a + 1}} = \frac{\dfrac{a^2 + 1(a + 1)}{a(a + 1)}}{\dfrac{1(a + 1) + 1(a)}{a(a + 1)}} = \frac{\dfrac{a^2 + a + 1}{a(a + 1)}}{\dfrac{2a + 1}{a(a + 1)}}$$

$$= \frac{a^2 + a + 1}{a(a + 1)} \cdot \frac{a(a + 1)}{2a + 1} = \frac{a^2 + a + 1}{2a + 1} \quad ■$$

The next example shows how negative exponents can lead to rational expressions.

Example 5   Simplify $\dfrac{(x + y)^{-1}}{x^{-1} + y^{-1}}$. Write the result with only positive exponents.

Use the definition of negative integer exponent to get

$$\frac{(x + y)^{-1}}{x^{-1} + y^{-1}} = \frac{\dfrac{1}{x + y}}{\dfrac{1}{x} + \dfrac{1}{y}} = \frac{\dfrac{1}{x + y}}{\dfrac{y + x}{xy}} = \frac{1}{x + y} \cdot \frac{xy}{x + y} = \frac{xy}{(x + y)^2}. \quad ■$$

1.6 Exercises

Give the domain of each of the following.

1. $\dfrac{x - 2}{x + 6}$

2. $\dfrac{x + 5}{x - 3}$

3. $\dfrac{2x}{5x - 3}$

4. $\dfrac{6x}{2x - 1}$

5. $\dfrac{-8}{x^2 + 1}$

6. $\dfrac{3x}{3x^2 + 7}$

Write each of the following in lowest terms.

7. $\dfrac{25p^3}{10p^2}$

8. $\dfrac{14z^3}{6z^2}$

9. $\dfrac{8k + 16}{9k + 18}$

10. $\dfrac{20r + 10}{30r + 15}$

11. $\dfrac{3(t + 5)}{(t + 5)(t - 3)}$

12. $\dfrac{-8(y - 4)}{(y + 2)(y - 4)}$

13. $\dfrac{8x^2 + 16x}{4x^2}$

14. $\dfrac{36y^2 + 72y}{9y}$

15. $\dfrac{m^2 - 4m + 4}{m^2 + m - 6}$

16. $\dfrac{r^2 - r - 6}{r^2 + r - 12}$

17. $\dfrac{8m^2 + 6m - 9}{16m^2 - 9}$

18. $\dfrac{6y^2 + 11y + 4}{3y^2 + 7y + 4}$

Perform each operation.

19. $\dfrac{x(y + 2)}{4} \div \dfrac{x}{2}$

20. $\dfrac{mn}{m + n} \div \dfrac{p}{m + n}$

21. $\dfrac{15p^3}{9p^2} \div \dfrac{6p}{10p^2}$

22. $\dfrac{3r^2}{9r^3} \div \dfrac{8r^3}{6r}$

23. $\dfrac{2k + 8}{6} \div \dfrac{3k + 12}{2}$

24. $\dfrac{5m + 25}{10} \cdot \dfrac{12}{6m + 30}$

25. $\dfrac{9y - 18}{6y + 12} \cdot \dfrac{3y + 6}{15y - 30}$

26. $\dfrac{12r + 24}{36r - 36} \div \dfrac{6r + 12}{8r - 8}$

27. $\dfrac{x^2 + x}{5} \cdot \dfrac{25}{xy + y}$

28. $\dfrac{3m - 15}{4m - 20} \cdot \dfrac{m^2 - 10m + 25}{12m - 60}$

29. $\dfrac{4a + 12}{2a - 10} \div \dfrac{a^2 - 9}{a^2 - a - 20}$

30. $\dfrac{6r - 18}{9r^2 + 6r - 24} \cdot \dfrac{12r - 16}{4r - 12}$

31. $\dfrac{p^2 - p - 12}{p^2 - 2p - 15} \cdot \dfrac{p^2 - 9p + 20}{p^2 - 8p + 16}$

32. $\dfrac{x^2 + 2x - 15}{x^2 + 11x + 30} \cdot \dfrac{x^2 + 2x - 24}{x^2 - 8x + 15}$

33. $\dfrac{k^2 - k - 6}{k^2 + k - 12} \cdot \dfrac{k^2 + 3k - 4}{k^2 + 2k - 3}$

34. $\dfrac{n^2 - n - 6}{n^2 - 2n - 8} \div \dfrac{n^2 - 9}{n^2 + 7n + 12}$

35. $\dfrac{m^2 + 3m + 2}{m^2 + 5m + 4} \div \dfrac{m^2 + 5m + 6}{m^2 + 10m + 24}$

36. $\dfrac{y^2 + y - 2}{y^2 + 3y - 4} \div \dfrac{y^2 + 3y + 2}{y^2 + 4y + 3}$

37. $\dfrac{2m^2 - 5m - 12}{m^2 - 10m + 24} \div \dfrac{4m^2 - 9}{m^2 - 9m + 18}$

38. $\dfrac{6n^2 - 5n - 6}{6n^2 + 5n - 6} \cdot \dfrac{12n^2 - 17n + 6}{12n^2 - n - 6}$

39. $\left(1 + \dfrac{1}{x}\right)\left(1 - \dfrac{1}{x}\right)$

40. $\left(3 + \dfrac{2}{y}\right)\left(3 - \dfrac{2}{y}\right)$

41. $\dfrac{x^3 + y^3}{x^2 - y^2} \cdot \dfrac{x + y}{x^2 - xy + y^2}$

42. $\dfrac{8y^3 - 125}{4y^2 - 20y + 25} \cdot \dfrac{2y - 5}{y}$

43. $\dfrac{x^3 + y^3}{x^3 - y^3} \cdot \dfrac{x^2 - y^2}{x^2 + 2xy + y^2}$

44. $\dfrac{x^2 - y^2}{(x - y)^2} \cdot \dfrac{x^2 - xy + y^2}{x^2 - 2xy + y^2} \div \dfrac{x^3 + y^3}{(x - y)^4}$

45. $\dfrac{8}{r} + \dfrac{6}{r}$

46. $\dfrac{3}{y} + \dfrac{4}{y}$

47. $\dfrac{3}{2k} + \dfrac{5}{3k}$

48. $\dfrac{8}{5p} + \dfrac{3}{4p}$

49. $\dfrac{2}{3y} - \dfrac{1}{4y}$

50. $\dfrac{6}{11z} - \dfrac{5}{2z}$

51. $\dfrac{a + 1}{2} - \dfrac{a - 1}{2}$

52. $\dfrac{y + 6}{5} - \dfrac{y - 6}{5}$

53. $\dfrac{3}{p} + \dfrac{1}{2}$

54. $\dfrac{9}{r} - \dfrac{2}{3}$

55. $\dfrac{2}{y} - \dfrac{1}{4}$

56. $\dfrac{6}{11} + \dfrac{3}{a}$

57. $\dfrac{1}{6m} + \dfrac{2}{5m} + \dfrac{4}{m}$

58. $\dfrac{8}{3p} + \dfrac{5}{4p} + \dfrac{9}{2p}$

59. $\dfrac{1}{y} + \dfrac{1}{y + 1}$

60. $\dfrac{2}{3(x - 1)} + \dfrac{1}{4(x - 1)}$

61. $\dfrac{2}{a + b} - \dfrac{1}{2(a + b)}$

62. $\dfrac{3}{m} - \dfrac{1}{m - 1}$

63. $\dfrac{1}{a + 1} - \dfrac{1}{a - 1}$

64. $\dfrac{1}{x + z} + \dfrac{1}{x - z}$

65. $\dfrac{m + 1}{m - 1} + \dfrac{m - 1}{m + 1}$

66. $\dfrac{2}{x - 1} + \dfrac{1}{1 - x}$

67. $\dfrac{3}{a - 2} - \dfrac{1}{2 - a}$

68. $\dfrac{q}{p - q} - \dfrac{q}{q - p}$

69. $\dfrac{x + y}{2x - y} - \dfrac{2x}{y - 2x}$

70. $\dfrac{m - 4}{3m - 4} + \dfrac{3m + 2}{4 - 3m}$

71. $\dfrac{1}{a^2 - 5a + 6} - \dfrac{1}{a^2 - 4}$

72. $\dfrac{-3}{m^2 - m - 2} - \dfrac{1}{m^2 + 3m + 2}$

73. $\dfrac{1}{x^2 + x - 12} - \dfrac{1}{x^2 - 7x + 12} + \dfrac{1}{x^2 - 16}$

74. $\dfrac{2}{2p^2 - 9p - 5} + \dfrac{p}{3p^2 - 17p + 10} - \dfrac{2p}{6p^2 - p - 2}$

75. $\dfrac{3a}{a^2 + 5a - 6} - \dfrac{2a}{a^2 + 7a + 6}$

76. $\dfrac{2k}{k^2 + 4k + 3} + \dfrac{3k}{k^2 + 5k + 6}$

77. $\dfrac{4x - 1}{x^2 + 3x - 10} + \dfrac{2x + 3}{x^2 + 4x - 5}$

78. $\dfrac{3y + 5}{y^2 - 9y + 20} + \dfrac{2y - 7}{y^2 - 2y - 8}$

79. $\left(\dfrac{3}{p - 1} - \dfrac{2}{p + 1} \right) \left(\dfrac{p - 1}{p} \right)$

80. $\left(\dfrac{y}{y^2 - 1} - \dfrac{y}{y^2 - 2y + 1} \right) \left(\dfrac{y - 1}{y + 1} \right)$

81. $\dfrac{\dfrac{1}{x + h} - \dfrac{1}{x}}{h}$

82. $\dfrac{1}{h} \left(\dfrac{1}{(x + h)^2 + 9} - \dfrac{1}{x^2 + 9} \right)$

83. $\dfrac{1 + \dfrac{1}{x}}{1 - \dfrac{1}{x}}$

84. $\dfrac{2 - \dfrac{2}{y}}{2 + \dfrac{2}{y}}$

85. $\dfrac{\dfrac{1}{x + 1} - \dfrac{1}{x}}{\dfrac{1}{x}}$

86. $\dfrac{\dfrac{1}{y + 3} - \dfrac{1}{y}}{\dfrac{1}{y}}$

87. $\dfrac{1 + \dfrac{1}{1 - b}}{1 - \dfrac{1}{1 + b}}$

88. $m - \dfrac{m}{m + \dfrac{1}{2}}$

89. $\dfrac{m - \dfrac{1}{m^2 - 4}}{\dfrac{1}{m + 2}}$

90. $\dfrac{\dfrac{3}{p^2 - 16} + p}{\dfrac{1}{p - 4}}$

91. $\dfrac{\dfrac{8}{z^2 - 25z} + 3}{\dfrac{5}{z(z + 1)}}$

92. $\dfrac{9 - \dfrac{1}{r^2 - r}}{\dfrac{3}{r(r + 2)}}$

Perform all indicated operations and write all answers with positive integer exponents.

93. $(2^{-1} - 3^{-1})^{-1}$

94. $(4^{-1} + 5^{-1})^{-1}$

95. $\dfrac{3^{-1} - 4^{-1}}{4^{-1}}$

96. $\dfrac{6^{-1} + 5^{-1}}{6^{-1}}$

97. $\dfrac{a^{-1} + b^{-1}}{(ab)^{-1}}$

98. $\dfrac{p^{-1} - q^{-1}}{(pq)^{-1}}$

99. $\dfrac{r^{-1} + q^{-1}}{r^{-1} - q^{-1}} \cdot \dfrac{r - q}{r + q}$

100. $\dfrac{xy^{-1} + yx^{-1}}{x^2 + y^2}$

101. $(a + b)^{-1}(a^{-1} + b^{-1})$

102. $(m^{-1} + n^{-1})^{-1}$

103. $(x - 9y^{-1})[(x - 3y^{-1})(x + 3y^{-1})]^{-1}$

104. $(m + n)^{-1}(m^{-2} - n^{-2})^{-1}$

Simplify each of the following, which result from standard processes of calculus.

105. $4 \left(\dfrac{x^3 + 2}{x - 5} \right)^3 \left(\dfrac{(x - 5)(3x^2) - (x^3 + 2)}{(x - 5)^2} \right)$

106. $3 \left(\dfrac{2x^2 - 9}{x^2 + 1} \right)^2 \left(\dfrac{(x^2 + 1)(4x) - (2x^2 - 9)}{(x^2 + 1)^2} \right)$

1.7 Radicals

Since $12^2 = 144$, the number 12 is the positive square root of 144. This is written

$$\sqrt{144} = 12.$$

Also, $\sqrt{400} = 20,$ $\sqrt{0.0001} = .01,$ and $\sqrt{\dfrac{25}{16}} = \dfrac{5}{4}.$

Generalizing, the symbol $\sqrt{a}$ is defined as follows.

Square Root

> For positive real numbers a and b,
> $$\sqrt{a} = b \quad \text{if and only if} \quad a = b^2.$$
> The number b is the positive or **principal square root** of a. Also, $\sqrt{0} = 0$.

If $a > 0$ and if $a = b^2$, then it is also true that $a = (-b)^2$, making two real numbers, one positive and one negative, whose square equals a given positive real number. However, as in the definition above, the **principal square root** is the positive one. Since the positive square root is used more often than the negative one, sometimes the positive or principal square root is called just **the square root.** Also, it is important to remember that

> the symbol $\sqrt{a}$ is reserved for the *positive* number whose square is a.

The symbol $-\sqrt{a}$ is used for the negative square root.

The **cube root** of a number can be defined in a manner similar to that used for square roots.

Cube Root

> For any real numbers a and b,
> $$\sqrt[3]{a} = b \quad \text{if and only if} \quad a = b^3.$$

There is a fundamental difference between the definitions of square root and cube root. While $\sqrt{a}$ was defined only for *nonnegative* real numbers a, the number $\sqrt[3]{a}$ is defined for *any* real number a.

Example 1   (a) $\sqrt{64} = 8$ since $8 > 0$ and $8^2 = 64$

(b) $-\sqrt{100} = -10$

(c) $\sqrt{-4}$ is not a real number, since there is no real number whose square is -4

(d) $\sqrt[3]{8} = 2$ since $2^3 = 8$

(e) $\sqrt[3]{-1000} = -10$ since $(-10)^3 = -1000$ ∎

If $a > 0$, then both $\sqrt{a}$ and $\sqrt[3]{a}$ are positive. If $a < 0$, then $\sqrt{a}$ is not a real number, while $\sqrt[3]{a} < 0$. Generalizing leads to the following definition.

nth Root

> If a and b are nonnegative real numbers and n is a positive integer, or if both a and b are negative and n is an odd positive integer, then
> $$\sqrt[n]{a} = b \quad \text{if and only if} \quad a = b^n.$$

The expression $\sqrt[n]{a}$ is called the **principal nth root** of a (abbreviated as the **nth root of** a); n is the **index** of the radical expression $\sqrt[n]{a}$. The number a is the **radicand**, and $\sqrt[n]{}$ is a **radical**. Abbreviate $\sqrt[2]{a}$ as just $\sqrt{a}$.

The following chart summarizes the conditions necessary for $\sqrt[n]{a}$ to exist.

If n is a positive integer and a is a real number, then $\sqrt[n]{a}$ is defined as follows.

n is	$a > 0$	$a < 0$	$a = 0$
even	$\sqrt[n]{a}$ is the positive real number such that $(\sqrt[n]{a})^n = a$	$\sqrt[n]{a}$ is not a real number	$\sqrt[n]{a}$ is 0
odd	$\sqrt[n]{a}$ is the real number such that $(\sqrt[n]{a})^n = a$		$\sqrt[n]{a}$ is 0

Example 2   (a) $\sqrt[5]{32} = 2$ since $2^5 = 32$ (c) $\sqrt[7]{-128} = -2$, since $(-2)^7 = -128$
(b) $\sqrt[4]{256} = 4$, since $4^4 = 256$ (d) $\sqrt[6]{-64}$ is not a real number   ∎

By the definition of $\sqrt[n]{a}$, for any positive integer n, if $\sqrt[n]{a}$ exists, then
$$(\sqrt[n]{a})^n = a.$$

If a is positive, or if a is negative and n is an odd positive integer, then
$$\sqrt[n]{a^n} = a.$$

Because of the conditions just given, it is *not* necessarily true that $\sqrt{x^2} = x$. For example, if $x = -5$,
$$\sqrt{x^2} = \sqrt{(-5)^2} = \sqrt{25} = 5 \neq x.$$

To take care of the fact that a negative value of x can produce a positive result for the square root, use the following rule, which involves absolute value.

> For any real number a,
> $$\sqrt{a^2} = |a|.$$
>
> *Also, if n is any even positive integer, then*
> $$\sqrt[n]{a^n} = |a|.$$

We shall prove only the first part of this statement, when $n = 2$. The statement is certainly true if a is positive. If a is negative, then $-a$ is positive. Since $(-a)^2 = a^2$,

$$\sqrt{(-a)^2} = -a.$$

Since $|a| = -a$ if a is negative, $\sqrt{a^2} = |a|$ for *all* real numbers a. For example,

$$\sqrt{(-9)^2} = |-9| = 9, \quad \text{and} \quad \sqrt{13^2} = |13| = 13.$$

To avoid difficulties when working with variable radicands, we will assume that all variables in radicands represent only nonnegative real numbers.

Three key rules for working with radicals are given below.

Rules for Radicals

For all real numbers a and b and positive integers m and n for which the indicated radicals exist,

$$\sqrt[n]{a} \cdot \sqrt[n]{b} = \sqrt[n]{ab}$$

$$\sqrt[n]{\frac{a}{b}} = \frac{\sqrt[n]{a}}{\sqrt[n]{b}} \quad b \neq 0$$

$$\sqrt[m]{\sqrt[n]{a}} = \sqrt[mn]{a}.$$

We shall prove only the first of these rules, with the other parts left for the exercises. To prove the first rule, let $x = \sqrt[n]{a}$ and let $y = \sqrt[n]{b}$. Then, by the definition of nth root, $x^n = a$ and $y^n = b$. Hence, $x^n \cdot y^n = ab$. However, $x^n \cdot y^n = (xy)^n$, giving

$$(xy)^n = ab.$$

This means xy is an nth root of ab, or

$$xy = \sqrt[n]{ab}.$$

Substituting the original values of x and y gives

$$\sqrt[n]{a} \cdot \sqrt[n]{b} = \sqrt[n]{ab}.$$

Example 3   Use the rules for radicals to simplify each of the following:

(a) $\sqrt{6} \cdot \sqrt{54} = \sqrt{6 \cdot 54} = \sqrt{324} = 18$

(b) $\sqrt{\frac{7}{64}} = \frac{\sqrt{7}}{\sqrt{64}} = \frac{\sqrt{7}}{8}$

(c) $\sqrt[3]{\frac{11}{64}} = \frac{\sqrt[3]{11}}{\sqrt[3]{64}} = \frac{\sqrt[3]{11}}{4}$

(d) $\sqrt[7]{\sqrt[3]{2}} = \sqrt[21]{2}$

(e) $\sqrt[4]{\sqrt{3}} = \sqrt[8]{3}.$ ∎

As in Example 3, the various rules for radicals can be used to help simplify expressions. For example, since $300 = 100 \cdot 3$,

$$\sqrt{300} = \sqrt{100 \cdot 3} = \sqrt{100} \cdot \sqrt{3} = 10\sqrt{3}.$$

By definition, an expression containing a radical is **simplified** when the following four conditions are satisfied.

Simplified Radicals

1. All possible factors have been removed from under the radical sign.
2. The index on the radical is as small as possible.
3. All radicals are removed from any denominators (a process called **rationalizing** the denominator).
4. All indicated operations have been performed (if possible).

Example 4 Simplify each of the following. Assume that all variables represent nonnegative real numbers.

(a) $\sqrt{175} = \sqrt{25 \cdot 7} = \sqrt{25} \cdot \sqrt{7} = 5\sqrt{7}$

(b) $\sqrt[3]{81x^5y^7z^6} = \sqrt[3]{27 \cdot 3 \cdot x^3 \cdot x^2 \cdot y^6 \cdot y \cdot z^6}$

$$= \sqrt[3]{(27x^3y^6z^6)(3x^2y)} = 3xy^2z^2\sqrt[3]{3x^2y}$$

(c) $\sqrt{98x^3y} + 3x\sqrt{32xy}$

First remove all perfect square factors from under the radical. Then use the distributive property, as follows.

$$\sqrt{98x^3y} + 3x\sqrt{32xy} = \sqrt{49 \cdot 2 \cdot x^2 \cdot x \cdot y} + 3x\sqrt{16 \cdot 2 \cdot x \cdot y}$$

$$= 7x\sqrt{2xy} + (3x)(4)\sqrt{2xy}$$

$$= 7x\sqrt{2xy} + 12x\sqrt{2xy} = 19x\sqrt{2xy}$$

(d) $\sqrt[3]{64m^4p^5} - \sqrt[3]{-27m^{10}p^{14}} = \sqrt[3]{(64m^3p^3)(mp^2)} - \sqrt[3]{(-27m^9p^{12})(mp^2)}$

$$= 4mp\sqrt[3]{mp^2} - (-3m^3p^4)\sqrt[3]{mp^2}$$

$$= 4mp\sqrt[3]{mp^2} + 3m^3p^4\sqrt[3]{mp^2}$$

$$= (4 + 3m^2p^3)mp\sqrt[3]{mp^2} \quad \blacksquare$$

Multiplying radical expressions is much like multiplying polynomials.

Example 5

$$(\sqrt{2} + 3)(\sqrt{8} - 5) = \sqrt{2}(\sqrt{8}) - \sqrt{2}(5) + 3\sqrt{8} - 3(5)$$

$$= \sqrt{16} - 5\sqrt{2} + 3(2\sqrt{2}) - 15$$

$$= 4 - 5\sqrt{2} + 6\sqrt{2} - 15$$

$$= -11 + \sqrt{2} \quad \blacksquare$$

The next example shows how to rationalize the denominator (remove any radicals from the denominator) in an expression containing radicals.

Example 6   Simplify each of the following expressions.

(a) $\dfrac{4}{\sqrt{3}}$

To rationalize the denominator, multiply by $\sqrt{3}/\sqrt{3}$ (which equals 1) so that the denominator of the product is a rational number. Work as follows.

$$\frac{4}{\sqrt{3}} \cdot \frac{\sqrt{3}}{\sqrt{3}} = \frac{4\sqrt{3}}{3}$$

(b) $\sqrt[4]{\dfrac{3}{5}}$

Start by using the fact that the radical of a quotient can be written as the quotient of radicals. To rationalize the denominator, multiply numerator and denominator by $\sqrt[4]{5^3}$. Use this number so that the denominator will be a rational number.

$$\sqrt[4]{\frac{3}{5}} = \frac{\sqrt[4]{3}}{\sqrt[4]{5}} = \frac{\sqrt[4]{3} \cdot \sqrt[4]{5^3}}{\sqrt[4]{5} \cdot \sqrt[4]{5^3}} = \frac{\sqrt[4]{3 \cdot 5^3}}{\sqrt[4]{5^4}} = \frac{\sqrt[4]{375}}{5}. \quad \blacksquare$$

In Example 7 below, the denominator is $1 - \sqrt{2}$. To rationalize this denominator, multiply numerator and denominator by $1 + \sqrt{2}$. This number is chosen because if a is rational and b is a nonnegative rational number, then

$$(a - \sqrt{b})(a + \sqrt{b}) = a^2 - (\sqrt{b})^2 = a^2 - b,$$

a rational number.

Example 7   Rationalize the denominator of $\dfrac{1}{1 - \sqrt{2}}$.

As just mentioned, multiply numerator and denominator by $1 + \sqrt{2}$.

$$\frac{1}{1 - \sqrt{2}} = \frac{1(1 + \sqrt{2})}{(1 - \sqrt{2})(1 + \sqrt{2})} = \frac{1 + \sqrt{2}}{-1} = -1 - \sqrt{2}. \quad \blacksquare$$

1.7 Exercises

Simplify each of the following. Assume that all variables represent nonnegative real numbers, and that no denominators are zero.

1. $\sqrt{50}$

2. $\sqrt{12}$

3. $\sqrt[3]{250}$

4. $\sqrt[3]{128}$

5. $-\sqrt{\dfrac{9}{5}}$

6. $\sqrt{\dfrac{3}{8}}$

7. $\sqrt{5} + \sqrt{45}$

8. $\sqrt{6} + \sqrt{54}$

9. $4\sqrt{3} - 5\sqrt{12} + 3\sqrt{75}$

10. $2\sqrt{5} - 3\sqrt{20} + 2\sqrt{45}$

11. $3\sqrt{28} - 5\sqrt{63} + \sqrt{112}$

12. $-\sqrt{12} + 2\sqrt{27} + 6\sqrt{48}$

13. $-\sqrt[3]{\dfrac{3}{2}}$

14. $-\sqrt[3]{\dfrac{4}{5}}$

15. $\sqrt[4]{\dfrac{3}{2}}$

16. $\sqrt[4]{\dfrac{32}{81}}$

17. $3\sqrt[3]{16} - 4\sqrt[3]{2}$

18. $\sqrt[3]{2} - \sqrt[3]{16} + 2\sqrt[3]{54}$

19. $2\sqrt[3]{3} + 4\sqrt[3]{24} - \sqrt[3]{81}$

20. $\sqrt[3]{32} - 5\sqrt[3]{4} + 2\sqrt[3]{108}$

21. $\dfrac{1}{\sqrt{3}} - \dfrac{2}{\sqrt{12}} + 2\sqrt{3}$

22. $\dfrac{1}{\sqrt{2}} + \dfrac{3}{\sqrt{8}} + \dfrac{3}{\sqrt{32}}$

23. $\dfrac{5}{\sqrt[3]{2}} - \dfrac{2}{\sqrt[3]{16}} + \dfrac{1}{\sqrt[3]{54}}$

24. $\dfrac{-4}{\sqrt[3]{3}} + \dfrac{1}{\sqrt[3]{24}} - \dfrac{2}{\sqrt[3]{81}}$

25. $\sqrt{2x^3y^2z^4}$

26. $\sqrt{98r^3s^4t^{10}}$

27. $\sqrt[3]{16z^5x^8y^4}$

28. $\sqrt[4]{x^8y^6z^{10}}$

29. $\sqrt{a^3b^5} - 2\sqrt{a^7b^3} + \sqrt{a^3b^9}$

30. $\sqrt{p^7q^3} - \sqrt{p^5q^9} + \sqrt{p^9q}$

31. $(\sqrt{2} + 3)(\sqrt{2} - 3)$

32. $(\sqrt{5} + \sqrt{2})(\sqrt{5} - \sqrt{2})$

33. $(\sqrt[3]{11} - 1)(\sqrt[3]{11^2} + \sqrt[3]{11} + 1)$

34. $(\sqrt[3]{7} + 3)(\sqrt[3]{7^2} - 3\sqrt[3]{7} + 9)$

35. $(\sqrt{3} + \sqrt{8})^2$

36. $(\sqrt{2} - 1)^2$

37. $(3\sqrt{2} + \sqrt{3})(2\sqrt{3} - \sqrt{2})$

38. $(4\sqrt{5} - 1)(3\sqrt{5} + 2)$

39. $(2\sqrt[3]{3} + 1)(\sqrt[3]{3} - 4)$

40. $(\sqrt[3]{4} + 3)(5\sqrt[3]{4} + 1)$

41. $\sqrt{\dfrac{2}{3x}}$

42. $\sqrt{\dfrac{5}{3p}}$

43. $\sqrt{\dfrac{x^5y^3}{z^2}}$

44. $\sqrt{\dfrac{g^3h^5}{r^3}}$

45. $\sqrt[3]{\dfrac{8}{x^2}}$

46. $\sqrt[3]{\dfrac{9}{16p^4}}$

47. $-\sqrt[3]{\dfrac{k^5m^3r^2}{r^8}}$

48. $-\sqrt[3]{\dfrac{9x^5y^6}{z^5w^2}}$

49. $\sqrt[4]{\dfrac{g^3h^5}{9r^6}}$

50. $\sqrt[4]{\dfrac{32x^6}{y^5}}$

51. $\dfrac{\sqrt[3]{mn} \cdot \sqrt[3]{m^2}}{\sqrt[3]{n^2}}$

52. $\dfrac{\sqrt[3]{8m^2n^3} \cdot \sqrt[3]{2m^2}}{\sqrt[3]{32m^4n^3}}$

53. $\dfrac{\sqrt[4]{32x^5y} \cdot \sqrt[4]{2xy^4}}{\sqrt[4]{4x^3y^2}}$

54. $\dfrac{\sqrt[4]{rs^2t^3} \cdot \sqrt[4]{r^3s^2t}}{\sqrt[4]{r^2t^3}}$

55. $\sqrt[3]{\sqrt{4}}$

56. $\sqrt[4]{\sqrt[3]{2}}$

57. $\sqrt[6]{\sqrt[3]{x}}$

58. $\sqrt[8]{\sqrt[4]{y}}$

59. $\sqrt{\dfrac{2m - 3p}{2m + 3p}} \cdot \sqrt{\dfrac{4m^2 + 12mp + 9p^2}{4m^2 - 9p^2}}$

60. $\sqrt{\dfrac{6k^2 + kr - 15r^2}{15k^2 - 34kr + 15r^2}} \cdot \sqrt{\dfrac{3k - 5r}{3k + 5r}}$

61. $\dfrac{3}{1 - \sqrt{2}}$

62. $\dfrac{2}{1 + \sqrt{5}}$

63. $\dfrac{\sqrt{3}}{4 + \sqrt{3}}$

64. $\dfrac{2\sqrt{7}}{3 - \sqrt{7}}$

65. $\dfrac{4}{2 - \sqrt{y}}$

66. $\dfrac{-1}{\sqrt{k} - 2}$

67. $\dfrac{1}{\sqrt{m} - \sqrt{p}}$

68. $\dfrac{\sqrt{z}}{1 - \sqrt{z}}$

69. $\dfrac{2}{\sqrt{5} - \sqrt{3} + 1}$

70. $\dfrac{4}{\sqrt{7} - \sqrt{2} + 3}$

71. $\dfrac{12}{\sqrt[3]{4}}$

72. $\dfrac{16}{\sqrt[3]{9}}$

73. $\dfrac{11}{\sqrt[3]{7}}$

74. $\dfrac{-45}{\sqrt[3]{10}}$

75. $\dfrac{y}{\sqrt{x} + y + z}$

76. $\dfrac{p}{p + q - \sqrt{r}}$

77. $\dfrac{2}{3 + \sqrt{1 + k}}$

78. $\dfrac{-5}{1 - \sqrt{3 - p}}$

79. $\dfrac{m}{\sqrt{p}} + \dfrac{p}{\sqrt{m}}$

80. $\dfrac{a}{\sqrt{b}} - \dfrac{b}{\sqrt{a}}$

81. $\dfrac{\sqrt{x} + \sqrt{x + 1}}{\sqrt{x} - \sqrt{x + 1}}$

82. $\dfrac{\sqrt{p} + \sqrt{p^2 - 1}}{\sqrt{p} - \sqrt{p^2 - 1}}$

83. $\dfrac{5}{\sqrt[3]{a} + \sqrt[3]{b}}$

84. $\dfrac{1}{\sqrt[3]{m} - \sqrt[3]{n}}$

85. $\dfrac{5}{(2 - \sqrt{3})(1 + \sqrt{2})}$

86. $\dfrac{-2}{(4 + \sqrt{3})(3 - \sqrt{2})}$

87. $\dfrac{6 - \sqrt{3}}{(5 - \sqrt{2})(3 + \sqrt{5})}$

88. $\dfrac{1 - \sqrt{7}}{(2 + \sqrt{10})(1 - \sqrt{5})}$

Choose the smaller in each of the following pairs of numbers. Do not use a calculator.

89. $\dfrac{1}{\sqrt{3} - \sqrt{2}}$ or $\dfrac{\sqrt{3} + \sqrt{2}}{2}$

90. $\dfrac{\sqrt{6} + \sqrt{11}}{4}$ or $\dfrac{1}{\sqrt{11} - \sqrt{6}}$

91. $\dfrac{\sqrt{7} + \sqrt{2}}{7}$ or $\dfrac{1}{\sqrt{7} - \sqrt{2}}$

92. $\dfrac{1}{\sqrt{10} - \sqrt{5}}$ or $\dfrac{\sqrt{10} + \sqrt{5}}{6}$

In advanced mathematics it is sometimes useful to write a radical expression with a rational numerator. The procedure is similar to rationalizing the denominator. Rationalize the numerator of each of the following.

93. $\dfrac{\sqrt{3}}{2}$

94. $\dfrac{\sqrt{6}}{5}$

95. $\dfrac{1 + \sqrt{2}}{2}$

96. $\dfrac{1 - \sqrt{3}}{3}$

97. $\dfrac{\sqrt{x}}{1 + \sqrt{x}}$

98. $\dfrac{\sqrt{p}}{1 - \sqrt{p}}$

99. $\dfrac{\sqrt{x} + \sqrt{x + 1}}{\sqrt{x} - \sqrt{x + 1}}$

100. $\dfrac{\sqrt{p} + \sqrt{p^2 - 1}}{\sqrt{p} - \sqrt{p^2 - 1}}$

Use a calculator to evaluate each of the following. Write all answers in scientific notation with three significant digits. For fourth roots, use the square root key twice.

101. $\sqrt{2.876 \times 10^7}$

102. $\sqrt{5.432 \times 10^9}$

103. $\sqrt[4]{3.87 \times 10^{-4}}$

104. $\sqrt[4]{5.913 \times 10^{-8}}$

105. $\dfrac{2.04 \times 10^{-3}}{\sqrt{5.97 \times 10^{-5}}}$

106. $\dfrac{3.86 \times 10^{-5}}{\sqrt{4.82 \times 10^{-5}}}$

107. $\sqrt{(4.721)^2 + (8.963)^2 - 2(4.721)(8.963)(.0468)}$

108. $\sqrt{(157.3)^2 + (184.7)^2 - 2(157.3)(184.7)(.9082)}$

Simplify each of the following.

109. $(\sqrt[3]{m + 2} - \sqrt[3]{m - 2})[(\sqrt[3]{m + 2})^2 + \sqrt[3]{m^2 - 4} + (\sqrt[3]{m - 2})^2]$

110. $(\sqrt[3]{3 - r} + \sqrt[3]{3 + r})[(\sqrt[3]{3 - r})^2 - \sqrt[3]{9 - r^2} + (\sqrt[3]{3 + r})^2]$

111. $(3\sqrt{z^2 - 2} - 2\sqrt{z^2 + 2})(3\sqrt{z^2 - 2} + 2\sqrt{z^2 + 2})(\sqrt{5z} + \sqrt{26})^{-1}$

112. $(4\sqrt{a^2 - 1} - \sqrt{a^2 + 3})(4\sqrt{a^2 - 1} + \sqrt{a^2 + 3})(\sqrt{15a} - \sqrt{19})^{-1}$

Prove each of the following (for all real numbers a and b and for any positive integers m and n for which all the following are real numbers).

113. $\dfrac{\sqrt[n]{a}}{\sqrt[n]{b}} = \sqrt[n]{\dfrac{a}{b}}$ $(b \neq 0)$

114. $\sqrt[m]{\sqrt[n]{a}} = \sqrt[mn]{a}$

115. Use a calculator to find an approximate value for $\sqrt{5 + 2\sqrt{6}}$.

116. Show that $\sqrt{5 + 2\sqrt{6}} = \sqrt{2} + \sqrt{3}$.

1.8 Rational Exponents

For a nonzero real number a, a meaning has now been given to a^n for all integer values of n. In this section, meaning is given to a^n for *rational* (and not just integer) values of the exponent. To start, let us define $a^{1/n}$ for positive integers n. Any definition of $a^{1/n}$ should be consistent with past rules of exponents. In particular, the rule $(b^m)^n = b^{mn}$ should still be valid. Replacing b^m with $a^{1/n}$ gives

$$(a^{1/n})^n = a^{n/n} = a^1 = a.$$

This means $a^{1/n}$ must be an nth root of a, as defined below.

$a^{1/n}$

> If a is a real number, n is a positive integer, and $\sqrt[n]{a}$ is a real number, then
> $$a^{1/n} = \sqrt[n]{a}.$$

Example 1 (a) $16^{1/2} = \sqrt{16} = 4$

(b) $(-32)^{1/5} = \sqrt[5]{-32} = -2$ ■

For rational exponents in general, $a^{m/n}$ must be defined so that all the rules for integer exponents still hold. For the rule $(b^m)^n = b^{mn}$ to hold, $(a^{1/n})^m$ must equal $a^{m/n}$, with $a^{m/n}$ defined as follows.

$a^{m/n}$

> For all integers m and all positive integers n such that m/n is in lowest terms, and all nonzero real numbers a for which $\sqrt[n]{a}$ is a real number,
> $$a^{m/n} = (\sqrt[n]{a})^m.$$

Example 2 (a) $125^{2/3} = (\sqrt[3]{125})^2 = 5^2 = 25$

(b) $x^{5/2} = (\sqrt{x})^5 = x^2\sqrt{x}$ $(x > 0)$

(c) $(16m)^{3/4} = (\sqrt[4]{16m})^3 = (2\sqrt[4]{m})^3 = 8(\sqrt[4]{m})^3$ $(m > 0)$ ■

We shall now show that $(a^{1/n})^m = (a^m)^{1/n}$, so that $a^{m/n}$ is also equal to $\sqrt[n]{a^m}$. To prove that $(a^{1/n})^m = (a^m)^{1/n}$, we shall use some of the rules for radicals and exponents developed in this chapter. Start with $(a^{1/n})^m$, and raise it to the nth power, to get

$$[(a^{1/n})^m]^n.$$

Now use properties of exponents, getting

$$[(a^{1/n})^m]^n = (a^{1/n})^{mn} = [(a^{1/n})^n]^m = a^m.$$

Because of this result, $(a^{1/n})^m$ must be an nth root of a^m, so that

$$(a^{1/n})^m = (a^m)^{1/n}.$$

Since $a^{m/n} = (a^{1/n})^m$ by definition, then

$$a^{m/n} = (a^m)^{1/n}.$$

Writing this result with radicals gives the next theorem.

Theorem on $a^{m/n}$

> For all integers m, and all positive integers n, such that m/n is in lowest terms, and all nonzero real numbers a for which all indicated radicals exist,
>
> $$a^{m/n} = (\sqrt[n]{a})^m \quad \text{or} \quad a^{m/n} = \sqrt[n]{a^m}.$$

This result gives two ways to evaluate $a^{m/n}$. Find $\sqrt[n]{a}$ and raise the result to the mth power, or find the nth root of a^m. In practice, it is usually easier to find $(\sqrt[n]{a})^m$. For example, $27^{4/3}$ can be evaluated in either of two ways:

$$27^{4/3} = (27^{1/3})^4 = 3^4 = 81$$

or

$$27^{4/3} = (27^4)^{1/3} = 531,441^{1/3} = 81.$$

The form $(27^{1/3})^4$ is easier to evaluate.

It can be shown that all the earlier results concerning integer exponents also apply to rational exponents. These rules are listed below.

Rules for Exponents

> Let r and s be rational numbers. The results below are valid for all real numbers a and b for which all indicated expressions exist.
>
> $$a^r \cdot a^s = a^{r+s} \qquad\qquad \left(\frac{a}{b}\right)^r = \frac{a^r}{b^r}$$
>
> $$\frac{a^r}{a^s} = a^{r-s} \qquad\qquad (a^r)^s = a^{rs}$$
>
> $$(ab)^r = a^r \cdot b^r \qquad\qquad a^{-r} = \frac{1}{a^r}$$

Example 3   Evaluate each of the following. Assume all variables represent positive real numbers.

(a) $81^{5/4} \cdot 4^{-3/2} = (81^{1/4})^5 \cdot \dfrac{1}{(4^{1/2})^3} = 3^5 \cdot \dfrac{1}{2^3} = \dfrac{243}{8}$

(b) $6y^{2/3} \cdot 2y^{1/2} = 12y^{2/3 + 1/2} = 12y^{7/6}$

(c) $\left(\dfrac{3m^{5/6}}{y^{3/4}}\right)^2 \cdot \left(\dfrac{8y^3}{m^6}\right)^{2/3} = \dfrac{9m^{5/3}}{y^{3/2}} \cdot \dfrac{4y^2}{m^4} = 36m^{5/3-4}y^{2-3/2} = \dfrac{36y^{1/2}}{m^{7/3}}$ ∎

Rational exponents can be used to simplify expressions with radicals. To do this, convert the radicals into expressions with rational exponents, simplify, and then convert back to radical form. The next example shows how this process works.

Example 4   Evaluate each of the following. Assume all variables represent positive real numbers.

(a) $\dfrac{\sqrt[4]{9}}{\sqrt{3}} = \dfrac{9^{1/4}}{3^{1/2}} = \dfrac{(3^2)^{1/4}}{3^{1/2}} = \dfrac{3^{1/2}}{3^{1/2}} = 1$

(b) $\dfrac{\sqrt[3]{m^8}}{\sqrt{m^5}} = \dfrac{m^{8/3}}{m^{5/2}} = m^{8/3 - 5/2} = m^{1/6} = \sqrt[6]{m}$

(c) $\sqrt{a^3 b^5}\,\sqrt[3]{a^4 b^2} = (a^3 b^5)^{1/2}(a^4 b^2)^{1/3}$

$\qquad = (a^{3/2} b^{5/2})(a^{4/3} b^{2/3})$

$\qquad = (a^{3/2 + 4/3})(b^{5/2 + 2/3})$

$\qquad = a^{17/6} b^{19/6} = a^2 \cdot a^{5/6} \cdot b^3 \cdot b^{1/6} = a^2 b^3 \sqrt[6]{a^5 b}$ ∎

The next examples show how to factor with rational exponents and how to simplify fractional expressions involving rational exponents.

Example 5   Factor out the lowest power of the variable. Assume all variables represent positive real numbers.

(a) $4m^{1/2} + 3m^{3/2} = m^{1/2}(4 + 3m)$

To check this result, multiply $m^{1/2}$ and $4 + 3m$.

(b) $y^{-1/3} + y^{2/3} = y^{-1/3}(1 + y)$ ∎

Example 6   Simplify $(m - 1)^{-1/2} + 3(m - 1)^{1/2}$.

Start by writing $(m - 1)^{-1/2}$ as $1/(m - 1)^{1/2}$. Then use a common denominator of $(m - 1)^{1/2}$.

$$\frac{1}{(m-1)^{1/2}} + 3(m-1)^{1/2} = \frac{1}{(m-1)^{1/2}} + \frac{3(m-1)^{1/2}(m-1)^{1/2}}{(m-1)^{1/2}}$$

$$= \frac{1}{(m-1)^{1/2}} + \frac{3(m-1)}{(m-1)^{1/2}}$$

$$= \frac{1 + 3(m-1)}{(m-1)^{1/2}}$$

$$= \frac{3m - 2}{(m-1)^{1/2}}$$

Alternatively, factor out $(m - 1)^{-1/2}$ from the original expression. ∎

1.8 Exercises

Simplify each of the following. Assume all variables represent nonnegative real numbers.

1. $4^{1/2}$

2. $25^{1/2}$

3. $8^{2/3}$

4. $81^{3/4}$

5. $27^{-2/3}$

6. $32^{-4/5}$

7. $\left(\dfrac{4}{9}\right)^{-3/2}$

8. $\left(\dfrac{1}{8}\right)^{-5/3}$

9. $(243)^{-3/5}$

10. $(121)^{-3/2}$

11. $(.0001)^{-3/2}$

12. $(.001)^{-5/3}$

13. $(16p^4)^{1/2}$ **14.** $(36r^6)^{1/2}$ **15.** $(27x^6)^{2/3}$ **16.** $(64a^{12})^{5/6}$

Write each of the following using rational exponents instead of radical signs. Assume that all variables represent positive real numbers.

17. $\sqrt[3]{x^2}$ **18.** $\sqrt[3]{y^4}$ **19.** $\sqrt[3]{z^5}$ **20.** $p\sqrt[3]{p^4}$

21. $y^3\sqrt[4]{y^6}$ **22.** $p^2\sqrt[3]{\sqrt[4]{p^8}}$ **23.** $m^{-2/3}\sqrt[4]{m^6}$ **24.** $y^{-3/2}\sqrt[4]{\sqrt{y^6}}$

Rewrite each of the following, using only positive exponents. Assume that all variables represent positive real numbers and that variables used as exponents represent rational numbers.

25. $(m^{2/3})(m^{5/3})$ **26.** $(x^{4/5})(x^{2/5})$ **27.** $(1 + n)^{1/2}(1 + n)^{3/4}$

28. $(m + 7)^{-1/6}(m + 7)^{-2/3}$ **29.** $(2y^{3/4}z)(3y^{1/4}z^{-1/3})$ **30.** $(4a^{-1/2}b^{2/3})(a^{3/2}b^{-1/3})$

31. $(r^{5/7}s^{3/5})(r^{-1/2}s^{-1/4})$ **32.** $(t^{3/2}w^{3/5})(t^{-1/3}w^{-1/5})$ **33.** $\dfrac{a^{4/3} \cdot b^{1/2}}{a^{2/3} \cdot b^{-3/2}}$

34. $\dfrac{x^{1/3} \cdot y^{2/3} \cdot z^{1/4}}{x^{5/3} \cdot y^{-1/3} \cdot z^{3/4}}$ **35.** $\dfrac{k^{-3/5} \cdot h^{-1/3} \cdot t^{2/5}}{k^{-1/5} \cdot h^{-2/3} \cdot t^{1/5}}$ **36.** $\dfrac{m^{7/3} \cdot n^{-2/5} \cdot p^{3/8}}{m^{-2/3} \cdot n^{3/5} \cdot p^{-5/8}}$

37. $\dfrac{(5x)^{-2}(x^{-3})^{-4}}{(25^{-1}x^{-3})^{-1}}$ **38.** $\dfrac{(2k)^{-3}(k^{-5})^{-1}}{(6k^{-2})^{-1}(k^3)^{-6}}$ **39.** $\left[\dfrac{(4^{-2}a)^{-1}(2a^{-3}b^{-1})^{-1}b^{-4}}{(16^{-1}a^5)(a^{-2}b^{-1})^2}\right]^2$

40. $\left[\dfrac{(7z^{-1})^2(2z^{-3})^{-2}y^{-4}}{(14^{-1}z^4y^2)(y^{-3})^{-2}}\right]^3$ **41.** $\dfrac{6m^{-2}p - 5m^{-2}}{36m^{-3}p^2 - 25m^{-3}}$ **42.** $\dfrac{2p^{-1} + 3p^{-1}q}{4p^{-2} - 9p^{-2}q^2}$

43. $\left(\dfrac{x^4y^3z}{16x^{-6}yz^5}\right)^{-1/2}$ **44.** $\left(\dfrac{p^3r^9}{27p^{-3}r^{-6}}\right)^{-1/3}$ **45.** $(r^{3/p})^{2p}(r^{1/p})^{p^2}$ **46.** $(m^{2/x})^{x/3}(m^{x/4})^{2/x}$

47. $\dfrac{m^{1-a} \cdot m^a}{m^{-1/2}}$ **48.** $\dfrac{(y^{3-b})(y^{2b-1})}{y^{1/2}}$ **49.** $\dfrac{(p^{1/n})(p^{1/m})}{p^{-m/n}}$ **50.** $\dfrac{(q^{2r/3})(q^r)^{-1/3}}{(q^{4/3})^{1/r}}$

Find each of the following products. Assume that all variables represent positive real numbers.

51. $y^{5/8}(y^{3/8} - 10y^{11/8})$ **52.** $p^{11/5}(3p^{4/5} + 9p^{19/5})$

53. $-4k(k^{7/3} - 6k^{1/3})$ **54.** $-5y(3y^{9/10} + 4y^{3/10})$

55. $(x + x^{1/2})(x - x^{1/2})$ **56.** $(2z^{1/2} + z)(z^{1/2} - z)$

57. $(r^{1/2} - r^{-1/2})^2$ **58.** $(p^{1/2} - p^{-1/2})(p^{1/2} + p^{-1/2})$

59. $[(x^{1/2} - x^{-1/2})^2 + 4]^{1/2}$ **60.** $[(x^{1/2} + x^{-1/2})^2 - 4]^{1/2}$

Write each of the following as a single radical. Assume all variables represent nonnegative real numbers.

61. $\sqrt{r^3} \cdot \sqrt[3]{r}$ **62.** $\sqrt[6]{y^3} \cdot \sqrt[4]{y^3}$ **63.** $\sqrt[10]{p^8} \cdot \sqrt{p}$ **64.** $\sqrt[8]{y^6} \cdot \sqrt[3]{y}$

65. $\sqrt[4]{p^2q^8} \cdot \sqrt[3]{p^5}$ **66.** $\sqrt[6]{q^4r^{18}} \cdot \sqrt[3]{r^5}$ **67.** $\sqrt{m^2n} \cdot \sqrt[3]{m^4n^2}$ **68.** $\sqrt[4]{y^2r^3} \cdot \sqrt[3]{y^6}$

69. $\sqrt[5]{m^3n^6} \cdot \sqrt[3]{m^4n^5}$ **70.** $\sqrt[6]{r^8s^5} \cdot \sqrt[3]{r^3s^7}$

Simplify each expression.

71. $4(p - 3)^{-1/2} + 2(p - 3)^{1/2}$ **72.** $-3(r + 1)^{1/2} - (r + 1)^{-1/2}$

73. $-(2a - 5)^{-3/2} + 3(2a - 5)^{-1/2}$ **74.** $(3k - 2)^{-1/2} + 4(3k - 2)^{-3/2}$

75. $\dfrac{\dfrac{m}{\sqrt{m-1}} - \sqrt{m-1}}{m-1}$

76. $\dfrac{r^2\sqrt{4-r} - \dfrac{1}{\sqrt{4-r}}}{4-r}$

Factor, using the given common factor. Assume all variables represent positive real numbers.

77. $4k^{7/4} + k^{3/4}$; $k^{3/4}$

78. $y^{9/2} - 3y^{5/2}$; $y^{5/2}$

79. $9z^{-1/2} + 2z^{1/2}$; $z^{-1/2}$

80. $3m^{2/3} - 4m^{-1/3}$; $m^{-1/3}$

81. $p^{-3/4} - 2p^{-1/4}$; $p^{-1/4}$

82. $6r^{-2/3} - 5r^{-5/3}$; $r^{-5/3}$

83. $(p+4)^{-3/2} + (p+4)^{-1/2} + (p+4)^{1/2}$;
$(p+4)^{-3/2}$

84. $(3r+1)^{-2/3} + (3r+1)^{1/3} + (3r+1)^{4/3}$;
$(3r+1)^{-2/3}$

Simplify by factoring each of the following expressions, which were obtained as a result of standard processes in calculus. → *bad bad word* !!! (...)

85. $\dfrac{(p+1)^{1/2} - p(\frac{1}{2})(p+1)^{-1/2}}{p+1}$

86. $\dfrac{(r-2)^{2/3} - r(\frac{2}{3})(r-2)^{-1/3}}{(r-2)^{4/3}}$

87. $\dfrac{3(2x^2+5)^{1/3} - x(2x^2+5)^{-2/3}(4x)}{(2x^2+5)^{2/3}}$

88. $\dfrac{-(m^3+m)^{2/3} + m(\frac{2}{3})(m^3+m)^{-1/3}(3m^2+1)}{(m^3+m)^{4/3}}$

89. $(x^{-1}-5)^3(2)(2-x^{-2})(2x^{-3}) + (2-x^{-2})^2(3)(x^{-1}-5)^2(-x^{-2})$

90. $(6+x^{-4})^2(3)(3x-2x^{-1})^2(3+2x^{-2}) + (3x-2x^{-1})^3(2)(6+x^{-4})(-4x^{-5})$

91. Give examples of numbers a and b where a is rational, b is rational, and a^b is irrational.

92. Give examples of numbers a and b where a is rational, b is irrational, and a^b is rational.

One important application of mathematics to business and management concerns supply and demand. Usually, as the price of an item increases, the supply increases and the demand decreases. By studying past records of supply and demand at different prices, economists can construct an equation which describes (approximately) supply and demand for a given item. The next two exercises show examples of this.

93. The price of a certain solar heater (in hundreds of dollars) is approximated by p, where

$$p = 2x^{1/2} + 3x^{2/3}$$

and x is the number of units supplied. Find the price when the supply is 64 units.

94. The demand for a certain commodity and the price are related by

$$p = 1000 - 200x^{-2/3} \qquad (x > 0),$$

where x is the number of units of the product demanded. Find the price when the demand is 27.

In our system of government, the president is elected by the electoral college and not by individual voters. Because of this, smaller states have a greater voice in the selection of a president than they would otherwise. Two political scientists have studied the problems of campaigning for president under the current system and have concluded that candidates should allot their money according to the formula

$$\text{amount for large state} = \left(\frac{E_{\text{large}}}{E_{\text{small}}}\right)^{3/2} \times \text{amount for small state}.$$

Here E_{large} represents the electoral vote of the large state, and E_{small} represents the electoral vote of the small state. Find the amount that should be spent in each of the following larger states if \$1,000,000 is spent in the small state and the following statements are true.

95. The large state has 48 electoral votes, and the small state has 3.

96. The large state has 36 electoral votes, and the small state has 4.

97. Six votes in a small state; 28 in a large.

98. Nine votes in a small state; 32 in a large.

A Delta Airlines map gives a formula for calculating the visible distance from a jet plane to the horizon. On a clear day, this distance is approximated by

$$D = 1.22x^{1/2},$$

where x is altitude in feet, and D is distance to the horizon in miles. Find D for an altitude of

99. 5000 feet **100.** 10,000 feet **101.** 30,000 feet **102.** 40,000 feet.

The Galápagos Islands are a chain of islands ranging in size from 2 to 2249 square miles. A biologist has shown that the number of different land-plant species on an island in this chain is related to the size of the island by

$$S = 28.6A^{0.32},$$

where A is the area of an island in square miles and S is the number of different plant species on that island. Estimate S (rounding to the nearest whole number) for islands of area

103. 1 square mile **104.** 25 square miles **105.** 300 square miles **106.** 2000 square miles.

1.9 Complex Numbers

So far, this book has dealt only with real numbers. However, the set of real numbers does not include all the numbers needed in algebra. For example, using real numbers alone, it is not possible to find a number whose square is -1. Finding such a number requires that the real number system be extended. To achieve this extension of the real number system, a new number i is defined as follows.

Definition of i

$$i = \sqrt{-1} \quad \text{or} \quad i^2 = -1$$

Numbers of the form $a + bi$, where a and b are real numbers, are called **complex numbers.** Each real number is a complex number, since a real number a may be thought of as the complex number $a + 0i$. A complex number of the form $0 + bi$, where b is nonzero, is called an **imaginary number** (sometimes a *pure* imaginary number). Both the set of real numbers and the set of imaginary numbers are subsets of the set of complex numbers. (See Figure 7, which is an extension of Figure 1 in Section 1.1.) A complex number that is written in the form $a + bi$ or $a + ib$ is in **standard form.** (The form $a + ib$ is used to simplify certain symbols such as $i\sqrt{5}$, since $\sqrt{5}i$ could be too easily mistaken for $\sqrt{5i}$.)

Example 1 The following statements identify different kinds of complex numbers.

(a) -8 and $\sqrt{7}$ and π are real numbers and complex numbers.

(b) $3i$ and $-11i$ and $i\sqrt{14}$ are imaginary numbers and complex numbers.

(c) $1 - 2i$ and $8 - 8i\sqrt{3}$ are complex numbers. ■

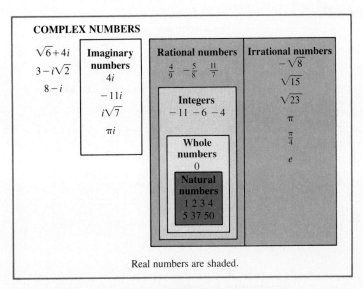

Figure 7

Example 2 The list below shows several numbers, along with the standard form of the number.

Number	Standard form
$6i$	$0 + 6i$
-9	$-9 + 0i$
0	$0 + 0i$
$9 - i$	$9 - i$
$i - 1$	$-1 + i$ ■

Equality for complex numbers is defined as follows.

Equality

> For real numbers a, b, c, and d,
>
> $$a + bi = c + di \quad \text{if and only if} \quad a = c \text{ and } b = d.$$

Example 3 Solve $2 + mi = k + 3i$ for real numbers m and k.

By the definition of equality, $2 + mi = k + 3i$ if and only if $2 = k$ and $m = 3$. ■

Simplify an expression of the form $\sqrt{-a}$, where a is a positive real number, with the following definition.

$\sqrt{-a}$

$$\text{If } a > 0, \text{ then } \sqrt{-a} = i\sqrt{a}.$$

Example 4   (a) $\sqrt{-16} = i\sqrt{16} = 4i$ (b) $\sqrt{-70} = i\sqrt{70}$ ∎

Products or quotients with square roots of negative numbers may be simplified using the fact that $\sqrt{-a} = i\sqrt{a}$ for positive numbers a. The next example shows how to do this.

Example 5   (a) $\sqrt{-7} \cdot \sqrt{-7} = i\sqrt{7} \cdot i\sqrt{7}$
$$= i^2 \cdot (\sqrt{7})^2$$
$$= (-1) \cdot 7 = -7$$

(b) $\sqrt{-6} \cdot \sqrt{-10} = i\sqrt{6} \cdot i\sqrt{10}$
$$= i^2 \cdot \sqrt{6 \cdot 10}$$
$$= -1 \cdot 2\sqrt{15} = -2\sqrt{15}$$

(c) $\dfrac{\sqrt{-20}}{\sqrt{-2}} = \dfrac{i\sqrt{20}}{i\sqrt{2}} = \sqrt{10}$

(d) $\dfrac{\sqrt{-48}}{\sqrt{24}} = \dfrac{i\sqrt{48}}{\sqrt{24}} = i\sqrt{2}$ ∎

When working with negative radicands, use the definition $\sqrt{-a} = i\sqrt{a}$ before using any of the other rules for radicals. In particular, the rule $\sqrt{c} \cdot \sqrt{d} = \sqrt{cd}$ is valid only when c and d are *not* both negative. For example,

$$\sqrt{(-4)(-9)} = \sqrt{36} = 6,$$

while
$$\sqrt{-4} \cdot \sqrt{-9} = 2i(3i) = 6i^2 = -6,$$

so
$$\sqrt{(-4)(-9)} \neq \sqrt{-4} \cdot \sqrt{-9}.$$

Operations on Complex Numbers   Complex numbers may be added, subtracted, multiplied, and divided, as shown by the following definitions and examples.

The *sum* of two complex numbers $a + bi$ and $c + di$ is defined as follows.

Sum

$$(a + bi) + (c + di) = (a + c) + (b + d)i$$

Example 6   (a) $(3 - 4i) + (-2 + 6i) = [3 + (-2)] + [-4 + 6]i$
$$= 1 + 2i$$

(b) $(-9 + 7i) + (3 - 15i) = -6 - 8i$ ∎

Since $(a + bi) + (0 + 0i) = a + bi$ for all complex numbers $a + bi$, the number $0 + 0i$ is called the **addition identity** for complex numbers. The sum of $a + bi$ and $-a - bi$ is $0 + 0i$, so the number $-a - bi$ is called the **negative** or **addition inverse** of $a + bi$.

Using this definition of addition inverse, *subtraction* of the complex numbers $a + bi$ and $c + di$ is defined as follows:

$$(a + bi) - (c + di) = (a + bi) + (-c - di).$$

This definition is often written as

Subtraction

$$(a + bi) - (c + di) = (a - c) + (b - d)i.$$

Example 7   Subtract as indicated.

(a) $(-4 + 3i) - (6 - 7i) = (-4 - 6) + [3 - (-7)]i$
$$= -10 + 10i$$

(b) $(12 - 5i) - (8 - 3i) = 4 - 2i$ ■

The product of two complex numbers can be found by multiplying as if the numbers were binomials and using the fact that $i^2 = -1$, as follows.

$$(a + bi)(c + di) = ac + adi + bic + bidi$$
$$= ac + adi + bci + bdi^2$$
$$= ac + (ad + bc)i + bd(-1)$$
$$(a + bi)(c + di) = (ac - bd) + (ad + bc)i$$

Based on this, the *product* of the complex numbers $a + bi$ and $c + di$ is defined in the following way.

Product

$$(a + bi)(c + di) = (ac - bd) + (ad + bc)i$$

This definition is hard to remember. To find a given product, it is usually better to multiply as with binomials. The next example shows this.

Example 8   Find each of the following products.

(a) $(2 - 3i)(3 + 4i) = 2(3) - 3i(3) + 2(4i) - 3i(4i)$
$$= 6 - 9i + 8i - 12i^2$$
$$= 6 - i - 12(-1) = 18 - i$$

(b) $(5 - 4i)(7 - 2i) = 5(7) - 4i(7) + 5(-2i) - 4i(-2i)$
$$= 35 - 28i - 10i + 8i^2$$
$$= 35 - 38i + 8(-1)$$
$$= 27 - 38i$$

(c) $(6 + 5i)(6 - 5i) = 6 \cdot 6 - 6 \cdot 5i + 6 \cdot 5i - (5i)^2$

$\qquad\qquad\qquad\quad = 36 - 25i^2$

$\qquad\qquad\qquad\quad = 36 - 25(-1)$

$\qquad\qquad\qquad\quad = 36 + 25$

$\qquad\qquad\qquad\quad = 61$

(d) i^{15}

Since $i^2 = -1$, the value of a power of i is found by writing the given power as a product involving i^2. For example, $i^3 = i^2 \cdot i = (-1) \cdot i = -i$. Also, $i^4 = i^2 \cdot i^2 = (-1)(-1) = 1$. Using i^4 to rewrite i^{15} gives

$$i^{15} = i^{12} \cdot i^3 = (i^4)^3 \cdot i^3 = (1)^3 \, (-i) = -i. \quad \blacksquare$$

Methods similar to those of part (d) of this last example give the following list of **powers of *i*:**

Powers of *i*

$i^1 = i$	$i^5 = i$	$i^9 = i$
$i^2 = -1$	$i^6 = -1$	$i^{10} = -1$
$i^3 = -i$	$i^7 = -i$	$i^{11} = -i$
$i^4 = 1$	$i^8 = 1$	$i^{12} = 1,$

and so on.

Example 8(c) showed that $(6 + 5i)(6 - 5i) = 61$. The numbers $6 + 5i$ and $6 - 5i$ differ only in their middle signs; these numbers are **conjugates** of each other. Conjugates are useful because the product of a complex number and its conjugate is always a real number (see Exercise 100).

Example 9 The following list shows several pairs of conjugates, along with the products of the conjugates.

Number	Conjugate	Product
$3 - i$	$3 + i$	$(3 - i)(3 + i) = 10$
$2 + 7i$	$2 - 7i$	$(2 + 7i)(2 - 7i) = 53$
$-6i$	$6i$	$(-6i)(6i) = 36 \quad \blacksquare$

The fact that the product of a complex number and its conjugate is always a real number is used to find the *quotient* of two complex numbers, as shown in the next example.

Example 10 **(a)** Find $\dfrac{3 + 2i}{5 - i}$.

Multiply numerator and denominator by the conjugate of $5 - i$.

$$\frac{3 + 2i}{5 - i} = \frac{(3 + 2i)(5 + i)}{(5 - i)(5 + i)}$$

$$= \frac{15 + 3i + 10i + 2i^2}{25 - i^2}$$

$$= \frac{13 + 13i}{26} = \frac{1}{2} + \frac{1}{2}i$$

To check this answer, show that

$$(5 - i)\left(\frac{1}{2} + \frac{1}{2}i\right) = 3 + 2i.$$

(b) $\dfrac{3}{i} = \dfrac{3(-i)}{i(-i)}$ $-i$ is the conjugate of i

$$= \frac{-3i}{-i^2}$$

$$= \frac{-3i}{1}$$ $-i^2 = -(-1) = 1$

$$= -3i$$ $0 - 3i$ in standard form   ■

1.9 Exercises

Identify each number as real, imaginary, or complex.

1. $-9i$ **2.** 6   **3.** π **4.** $-\sqrt{7}$

5. $i\sqrt{6}$ **6.** $-3i$ **7.** $2 + 5i$ **8.** $-7 - 6i$

Write each of the following in standard form.

9. $\sqrt{-100}$ **10.** $\sqrt{-169}$ **11.** $-\sqrt{-400}$ **12.** $-\sqrt{-225}$

13. $-\sqrt{-39}$ **14.** $-\sqrt{-95}$ **15.** $5 + \sqrt{-4}$ **16.** $-7 + \sqrt{-100}$

17. $-6 - \sqrt{-196}$ **18.** $13 + \sqrt{-16}$ **19.** $9 - \sqrt{-50}$ **20.** $-11 - \sqrt{-24}$

21. $\sqrt{-5} \cdot \sqrt{-5}$ **22.** $\sqrt{-20} \cdot \sqrt{-20}$ **23.** $\sqrt{-8} \cdot \sqrt{-2}$ **24.** $\sqrt{-27} \cdot \sqrt{-3}$

25. $\dfrac{\sqrt{-40}}{\sqrt{-10}}$ **26.** $\dfrac{\sqrt{-190}}{\sqrt{-19}}$ **27.** $\dfrac{\sqrt{-6} \cdot \sqrt{-2}}{\sqrt{3}}$ **28.** $\dfrac{\sqrt{-12} \cdot \sqrt{-6}}{\sqrt{8}}$

Add or subtract. Write each result in standard form.

29. $(3 + 2i) + (4 - 3i)$ **30.** $(4 - i) + (2 + 5i)$

31. $(-2 + 3i) - (-4 + 3i)$ **32.** $(-3 + 5i) - (-4 + 3i)$

33. $(2 - 5i) - (3 + 4i) - (-2 + i)$ **34.** $(-4 - i) - (2 + 3i) + (-4 + 5i)$

35. $-i - 2 - (3 - 4i) - (5 - 2i)$ **36.** $3 - (4 - i) - 4i + (-2 + 5i)$

Multiply. Write each result in standard form.

37. $(2 + i)(3 - 2i)$ **38.** $(-2 + 3i)(4 - 2i)$ **39.** $(2 + 4i)(-1 + 3i)$

40. $(1 + 3i)(2 - 5i)$ **41.** $(-3 + 2i)^2$ **42.** $(2 + i)^2$

43. $(3 + i)(-3 - i)$ **44.** $(-5 - i)(5 + i)$ **45.** $(2 + 3i)(2 - 3i)$

46. $(6 - 4i)(6 + 4i)$ **47.** $(\sqrt{6} + i)(\sqrt{6} - i)$ **48.** $(\sqrt{2} - 4i)(\sqrt{2} - 4i)$

49. $i(3 - 4i)(3 + 4i)$ **50.** $i(2 + 7i)(2 - 7i)$ **51.** $3i(2 - i)^2$

52. $-5i(4 - 3i)^2$

Divide. Write each result in standard form.

53. $\dfrac{1 + i}{1 - i}$ **54.** $\dfrac{2 - i}{2 + i}$ **55.** $\dfrac{4 - 3i}{4 + 3i}$ **56.** $\dfrac{5 - 2i}{6 - i}$

57. $\dfrac{3 - 4i}{2 - 5i}$ **58.** $\dfrac{1 - 3i}{1 + i}$ **59.** $\dfrac{-3 + 4i}{2 - i}$ **60.** $\dfrac{5 + 6i}{5 - 6i}$

61. $\dfrac{2}{i}$ **62.** $\dfrac{-7}{3i}$ **63.** $\dfrac{1 - \sqrt{-5}}{3 + \sqrt{-4}}$ **64.** $\dfrac{2 + \sqrt{-3}}{1 - \sqrt{-9}}$

Find each of the following powers of i.

65. i^5 **66.** i^8 **67.** i^9 **68.** i^{11}

69. i^{12} **70.** i^{25} **71.** i^{43} **72.** $1/i^9$

73. $1/i^{12}$ **74.** i^{-6} **75.** i^{-15} **76.** i^{-49}

Perform the indicated operations and write your answers in standard form.

77. $\dfrac{2 + i}{3 - i} \cdot \dfrac{5 + 2i}{1 + i}$ **78.** $\dfrac{1 - i}{2 + i} \cdot \dfrac{4 + 3i}{1 + i}$ **79.** $\dfrac{6 + 2i}{5 - i} \cdot \dfrac{1 - 3i}{2 + 6i}$ **80.** $\dfrac{5 - 3i}{1 + 2i} \cdot \dfrac{2 - 4i}{1 + i}$

81. $\dfrac{5 - i}{3 + i} + \dfrac{2 + 7i}{3 + i}$ **82.** $\dfrac{4 - 3i}{2 + 5i} + \dfrac{8 - i}{2 + 5i}$ **83.** $\dfrac{6 + 2i}{1 + 3i} + \dfrac{2 - i}{1 - 3i}$ **84.** $\dfrac{4 - i}{3 + 4i} - \dfrac{3 + 2i}{3 - 4i}$

85. $\dfrac{6 + 3i}{1 - i} - \dfrac{2 - i}{4 + i}$ **86.** $\dfrac{2 - 3i}{2 + i} + \dfrac{6 + i}{3 + 5i}$

Use the definition of equality for complex numbers to solve the following equations for real numbers a and b.

87. $a + bi = 23 + 5i$ **88.** $a + bi = -2 + 4i$ **89.** $a + bi = 18 - 3i$ **90.** $2 + bi = a - 4i$

91. $a + 3i = 5 + 3bi + 2a$ **92.** $4a - 2bi + 7 = 3i + 3a + 5$

93. $i(2b + 6) - 3 = 4(bi + a)$ **94.** $3i + 2(a - 1) = 4 + 2i(b + 3)$

95. Let $z = 6 - 5i$ and find $4i - 3z$. **96.** Let $z = 1 - 7i$ and find $2z - 9i$.

Let $z = a + bi$ for real numbers a and b, and let $\bar{z} = a - bi$, the conjugate of z. For example, if $z = 8 - 9i$, then $\bar{z} = 8 + 9i$. Prove each of the following properties of conjugates.

97. $\bar{\bar{z}} = z$ **98.** $\bar{z} = z$ if and only if $b = 0$

99. $\overline{-z} = -\bar{z}$ **100.** $z \cdot \bar{z}$ is a real number

Prove that the complex numbers $z_1 = a + bi$, $z_2 = c + di$, and $z_3 = e + fi$ satisfy each of the following properties.

101. commutative property for addition:
$z_1 + z_2 = z_2 + z_1$

102. commutative property for multiplication:
$z_1 z_2 = z_2 z_1$

103. associative property for addition:
$(z_1 + z_2) + z_3 = z_1 + (z_2 + z_3)$

104. associative property for multiplication:
$(z_1z_2)z_3 = z_1(z_2z_3)$

105. distributive property: $z_1(z_2 + z_3) = z_1z_2 + z_1z_3$

106. closure property of addition: $z_1 + z_2$ is a complex number

107. closure property of multiplication: z_1z_2 is a complex number

Evaluate $8z - z^2$ by replacing z with the indicated complex number.

108. $2 + i$ **109.** $4 - 3i$ **110.** $-6i$

Find all complex numbers $a + bi$ such that the square $(a + bi)^2$ is

111. real **112.** imaginary

113. Show that $\dfrac{\sqrt{2}}{2} + \dfrac{\sqrt{2}}{2}i$ is a square root of i.

114. Show that $\dfrac{\sqrt{3}}{2} - \dfrac{1}{2}i$ is a cube root of $-i$.

Chapter I Summary

Key Words

real numbers	absolute value	prime polynomial
addition inverse	exponent	common factor
multiplication inverse	scientific notation	rational expressions
reciprocal	algebraic expression	complex fraction
natural numbers	term	square root
whole numbers	coefficient	radical
integers	polynomial	nth root
rational numbers	degree of a polynomial	principal nth root
irrational numbers	trinomial	radical expression
number line	binomial	conjugate
coordinate	monomial	complex number
inequalities	difference of two squares	imaginary number

Review Exercises

Identify the property which tells why the following are true.

1. $6 \cdot 4 = 4 \cdot 6$ **2.** $8(5 + 9) = (5 + 9)8$ **3.** $3 + (-3) = 0$

4. If $x \neq 0$, then $x \cdot \dfrac{1}{x} = 1$ **5.** $4 \cdot 6 + 4 \cdot 12 = 4(6 + 12)$ **6.** $3(4 \cdot 2) = (3 \cdot 4)2$

7. Is the set of rational numbers closed for division?

8. Is the set of irrationals closed for division?

Simplify each of the following.

9. $(4 + 2 \cdot 8) \div 3$ **10.** $-7 + (-6)(-2) - 8$ **11.** $\dfrac{-9 + (2)(-8) + 7}{3(-4) - (-1 - 3)}$ **12.** $\dfrac{4 - (-8)(-2) - 7}{(-2)(-9) - 4(-4)}$

Let set $K = \{-12, -6, -9/10, -\sqrt{7}, 0, 1/8, \pi/4, 6, \sqrt{11}\}$. List the elements of K that are

13. natural numbers   **14.** whole numbers   **15.** integers

16. rational numbers   **17.** irrational numbers   **18.** real numbers.

For Exercises 19–27, choose all words from the following list which apply.

(a) natural numbers   **(b)** whole numbers   **(c)** integers   **(d)** rational numbers
(e) irrational numbers   **(f)** real numbers   **(g)** meaningless

19. 22   **20.** 0   **21.** $\sqrt{36}$ **22.** $-\sqrt{25}$

23. -1 **24.** 5/8   **25.** $\sqrt{15}$ **26.** $-6/0$

27. $3\pi/4$

Write the following numbers in numerical order, from smallest to largest.

28. $\dfrac{5}{6}, \dfrac{1}{2}, -\dfrac{2}{3}, -\dfrac{5}{4}, -\dfrac{3}{8}$ **29.** $|6 - 4|,\ -|-2|,\ |8 + 1|,\ -|3 - (-2)|$

30. $\sqrt{7},\ -\sqrt{8},\ -|\sqrt{16}|,\ |-\sqrt{12}|$

Write without absolute value bars.

31. $|3 - \sqrt{7}|$ **32.** $|\sqrt{8} - 3|$ **33.** $|m - 3|$, if $m > 3$ **34.** $|-6 - x^2|$

In each of the following exercises, the coordinates of three points are given.
Find   **(a)** $d(A, B)$ **(b)** $d(A, B) + d(B, C)$.

35. $A, -2; B, -1; C, 10$ **36.** $A, -8; B, -12; C, -15$

Under what conditions are the following statements true?

37. $|x| = x$ **38.** $|a + b| = -a - b$ **39.** $|x + y| = -|x| - |y|$
40. $|x| \le 0$ **41.** $d(A, B) = 0$ **42.** $d(A, B) + d(B, C) = d(A, C)$

Simplify each of the following. Assume all variables represent nonzero real numbers and
variables used as exponents represent rational numbers.

43. $(-6x^2 - 4x + 11) + (-2x^2 - 11x + 5)$ **44.** $(3x^3 - 9x^2 - 5) - (-4x^3 + 6x^2) + (2x^3 - 9)$
45. $(8k - 7)(3k + 2)$ **46.** $(3r - 2)(r^2 + 4r - 8)$

47. $(x + 2y - z)^2$ **48.** $(r^z - 8)(r^z + 8)$ **49.** $\dfrac{2x^5 t^2 \cdot 4x^3 t}{16x^2 t^3}$ **50.** $\dfrac{(2k^2)^2(3k^3)}{(4k^4)^2}$

Factor as completely as possible.

51. $7z^2 - 9z^3 + z$ **52.** $12p^5 - 8p^4 + 20p^3$ **53.** $6m^2 - 13m - 5$
54. $15a^2 + 7ab - 2b^2$ **55.** $30m^5 - 35m^4 n - 25m^3 n^2$ **56.** $2x^2 p^3 - 8xp^4 + 6p^5$
57. $169y^4 - 1$ **58.** $49m^8 - 9n^2$ **59.** $(x - 1)^2 - 4$
60. $a^3 - 27b^3$ **61.** $r^9 - 8(r^3 - 1)^3$ **62.** $(2p - 1)^3 + 27p^3$
63. $3(m - n) + 4k(m - n)$ **64.** $6(z - 4)^2 + 9(z - 4)^3$ **65.** $2bx - b + 6x - 3$
66. $3x(x - 2) + 9x^2(x - 2) - 12x^3(x - 2)$
67. $y^{2k} - 9$ **68.** $z^{6n} - 100$

Perform the indicated operation.

69. $\dfrac{3m - 9}{8m} \cdot \dfrac{16m + 24}{15}$

70. $\dfrac{5x^2y}{x + y} \cdot \dfrac{3x + 3y}{30xy^2}$

71. $\dfrac{x^2 + x - 2}{x^2 + 5x + 6} \div \dfrac{x^2 + 3x - 4}{x^2 + 4x + 3}$

72. $\dfrac{27m^3 - n^3}{3m - n} \div \dfrac{9m^2 + 3mn + n^2}{9m^2 - n^2}$

73. $\dfrac{p^2 - 36q^2}{(p - 6q)^2} \cdot \dfrac{p^2 - 5pq - 6q^2}{p^2 - 6pq + 36q^2} \div \dfrac{5p}{p^3 + 216q^3}$

74. $\dfrac{3r^3 - 9r^2}{r^2 - 9} \div \dfrac{8r^3}{r + 3}$

75. $\dfrac{1}{4y} + \dfrac{8}{5y}$

76. $\dfrac{m}{4 - m} + \dfrac{3m}{m - 4}$

77. $\left(1 - \dfrac{3}{p}\right)\left(1 + \dfrac{3}{p}\right)$

78. $\dfrac{3}{x^2 - 4x + 3} - \dfrac{2}{x^2 - 1}$

79. $\dfrac{\dfrac{1}{p} + \dfrac{1}{q}}{1 - \dfrac{1}{pq}}$

80. $\left(\dfrac{1}{(x + h)^2 + 16} - \dfrac{1}{x^2 + 16}\right) \div h$

81. $\dfrac{3 + \dfrac{2m}{m^2 - 4}}{\dfrac{5}{m - 2}}$

82. $\dfrac{x^{-1} - 2y^{-1}}{y^{-1} - x^{-1}}$

Simplify each of the following. Assume all variables represent positive real numbers.

83. $\dfrac{(p^4)(p^{-2})}{p^{-6}}$

84. $(-6x^2y^{-3}z^2)^{-2}$

85. $\dfrac{6^{-1}r^3s^{-2}}{6r^4s^{-3}}$

86. $\dfrac{(3m^{-2})^{-2}(m^2n^{-4})^3}{9m^{-3}n^{-5}}$

87. $\dfrac{(2x^{-3})^2(3x^2)^{-2}}{6(x^2y^3)}$

88. $\dfrac{k^{2+p} \cdot k^{-4p}}{k^{6p}}$ (p is a rational number)

Simplify. Assume that all variables represent positive real numbers.

89. $\sqrt{200}$

90. $\sqrt[4]{1250}$

91. $\sqrt{\dfrac{7}{3r}}$

92. $\sqrt{\dfrac{2^7y^8}{m^3}}$

93. $-\sqrt[3]{\dfrac{r^6m^5}{z^2}}$

94. $\sqrt[4]{\sqrt[3]{m}}$

95. $(\sqrt[3]{2} + 4)(\sqrt[3]{2^2} - 4\sqrt[3]{2} + 16)$

96. $\dfrac{\sqrt[4]{8p^2q^5} \cdot \sqrt[4]{2p^3q}}{\sqrt[4]{p^5q^2}}$

97. $\sqrt{18m^3} - 3m\sqrt{32m} + 5\sqrt{m^3}$

98. $\sqrt{75y^5} - y^2\sqrt{108y} + 2y\sqrt{27y^3}$

99. $\dfrac{3}{\sqrt{5}} - \dfrac{2}{\sqrt{45}} + \dfrac{6}{\sqrt{80}}$

100. $\dfrac{-12}{\sqrt[3]{4}}$

101. $\dfrac{z}{\sqrt{z - 1}}$

102. $\dfrac{6}{3 - \sqrt{2}}$

103. $\dfrac{1}{\sqrt{5} + 2}$

104. $\dfrac{\sqrt{x} - \sqrt{x - 2}}{\sqrt{x} + \sqrt{x - 2}}$

Simplify each of the following. Assume that all variables represent positive real numbers, and variables used as exponents are rational numbers.

105. $36^{-3/2}$

106. $(125m^6)^{-2/3}$

107. $(8r^{3/4}s^{2/3})(2r^{3/2}s^{5/3})$

108. $(7r^{1/2})(2r^{3/4})(-r^{1/6})$

109. $\dfrac{p^{-3/4} \cdot p^{5/4} \cdot p^{-1/4}}{p \cdot p^{3/4}}$

110. $\left(\dfrac{y^6x^3z^{-2}}{16x^5z^4}\right)^{-1/2}$

111. $\dfrac{m^{2+p} \cdot m^{-2}}{m^{3p}}$

112. $\dfrac{z^{-p+1} \cdot z^{-8p}}{z^{-9p}}$

Find each product. Assume that all variables represent positive real numbers.

113. $2z^{1/3}(5z^2 - 2)$

114. $-m^{3/4}(8m^{1/2} + 4m^{-3/2})$

115. $(p + p^{1/2})(3p - 5)$

116. $(m^{1/2} - 4m^{-1/2})^2$

Write in standard form.

117. $(1 - i) - (3 + 4i) + 2i$

118. $(2 - 5i) + (9 - 10i) - 3$

119. $(6 - 5i) + (2 + 7i) - (3 - 2i)$

120. $(4 - 2i) - (6 + 5i) - (3 - i)$

121. $(3 + 5i)(8 - i)$

122. $(4 - i)(5 + 2i)$

123. $(2 + 6i)^2$

124. $(6 - 3i)^2$

125. $(1 - i)^3$

126. $(2 + i)^3$

127. i^{17}

128. i^{52}

129. $\dfrac{6 + 2i}{3 - i}$

130. $\dfrac{2 - 5i}{1 + i}$

131. $\dfrac{2 + i}{1 - 5i} \cdot \dfrac{1 + i}{3 - i}$

132. $\dfrac{4 + 3i}{1 - i} \cdot \dfrac{2 - 3i}{2 + i}$

133. $\dfrac{8 - i}{2 + i} + \dfrac{3 + 2i}{4i}$

134. $\dfrac{6 + 3i}{1 + i} + \dfrac{1 - i}{2 + 2i}$

135. $\sqrt{-12}$

136. $\sqrt{-18}$

Simplify each of the following.

137. $(|a| + a)(|a| - a)$

138. $[2a^4 + (b^2 - a^2)^2 - (a^2 - b^2)^2]^{1/2}[a^6 - (-a^2)^3]^{1/2}$

Suppose a value of the number a has been given. Let $b = a + 1$ and $c = \sqrt{a + b}$. Show that

139. $b(a - c)^2 + a(b + c)^2 = c^2(ab + c^2)$

140. $b(a - c)^2 + ab(1 + c)^2 = b^2(c^2 + a)$

141. If $x + |x| = x - |x|$, what is x?

142. Prove that if $(x + 1)^2 = (|x| + 1)^2$, then $x \geq 0$.

Suppose $0 < a < 1$ and $a + b = 1$. Are the following less than, equal to, or greater than 1?

143. $a^2 + b^2$

144. $\sqrt{a} + \sqrt{b}$

What are the order relations among 1, a, a^2, $\sqrt{a}$, $1/a$, b, b^2, $\sqrt{b}$, and $1/b$ if

145. $\sqrt{b} < a < b$

146. $b < a < \sqrt{b}$

147. If $a > 0$, show that $a + 1/a \geq 2$.

148. If $a > 0$, show that $4a + 1 < (\sqrt{a} + \sqrt{a + 1})^2 < 4a + 2$.

Exercises 137–48: Reproduced from *Calculus*, 2nd edition, by Leonard Gillman and Robert H. McDowell, by permission of W. W. Norton & Company, Inc. Copyright © 1978, 1973 by W. W. Norton & Company, Inc.

2

Equations and Inequalities

For many people, the study of algebra is really the study of equations. Many applications of mathematics require the solution of one or more equations. The study of inequalities has also become important as more and more applications—especially in fields such as business—utilize inequalities.

An **equation** is a statement that two expressions are equal. Examples of equations include

$$x + 2 = 9, \qquad 11y = 5y + 6y, \qquad \text{and} \quad x^2 - 2x - 1 = 0.$$

In this chapter we discuss the solution of several different kinds of equations and inequalities.

2.1 Linear Equations

To **solve** an equation means to find the number or numbers that make the equation a true statement. A number that is a solution of an equation is said to **satisfy** the equation. The set of all solutions for an equation makes up its **solution set.** An equation which is satisfied by every number which is a meaningful replacement for the variable is called an **identity.** Examples of identities are

$$3x + 4x = 7x \qquad \text{and} \qquad x^2 - 3x + 2 = (x - 2)(x - 1).$$

Equations that are satisfied by some numbers, but not satisfied by others, are called **conditional equations.** Examples of conditional equations are

$$2m + 3 = 7 \qquad \text{and} \qquad \frac{5r}{r - 1} = 7.$$

To verify that an equation is an identity, show that the two sides are algebraically equivalent for the same set of values of x.

Example 1   Decide if the following equations are identities or conditional equations.

(a) $9p^2 - 25 = (3p + 5)(3p - 5)$

Since the product of $3p + 5$ and $3p - 5$ is $9p^2 - 25$, the equation is true for *every* value of p and is an identity.

(b) $\dfrac{(x - 3)(x + 2)}{x - 3} = x + 2$

On the left side of the equation, 3 cannot be used as a replacement for x. For this reason, the equation is an identity only for $x \neq 3$.

(c) $5y - 4 = 11$

Choosing the value 3 as a replacement for y gives

$$5 \cdot 3 - 4 = 11$$
$$11 = 11,$$

a true statement. On the other hand, $y = 4$ gives

$$5 \cdot 4 - 4 = 11$$
$$16 = 11,$$

a false statement. Since the equation $5y - 4 = 11$ is true for some values of y, but not all, it is a conditional equation. (By the way, the word *some* in mathematics means "at least one." By this usage, $5y - 4 = 11$ is true for *some* replacements of y, even though it turns out to be true only for $y = 3$.) ∎

Any two equations with the same solution set are **equivalent equations.** For example $x + 1 = 5$ and $6x + 3 = 27$ are equivalent equations since they both have the same solution set, $\{4\}$. On the other hand, the equations

$$x + 1 = 5 \qquad \text{and} \qquad (x - 4)(x + 2) = 0$$

are not equivalent. The number 4 is a solution of both equations, but the equation $(x - 4)(x + 2) = 0$ also has -2 as a solution.

One way to solve an equation is to rewrite it as successively simpler equivalent equations. These simpler equivalent equations are derived using the properties of Chapter 1 that allow the same number to be added to each side of an equation or to be multiplied on each side of an equation.

This section discusses methods of solving equations that are equivalent to linear equations.

Linear Equation

An equation that can be written in the form

$$ax + b = 0,$$

where a and b are real numbers, with $a \neq 0$, is a **linear equation.**

The equation $ax + b = 0$ can be solved by writing the following sequence of equivalent equations:

$$ax + b = 0 \qquad \text{Given equation}$$

$$ax + b + (-b) = 0 + (-b) \qquad \text{Add } -b \text{ on each side}$$

$$ax = -b \qquad \text{Inverse and identity properties}$$

$$\frac{1}{a} \cdot ax = \frac{1}{a} \cdot (-b) \qquad \text{Multiply by } 1/a \text{ on each side}$$

$$x = -\frac{b}{a} \qquad \text{Inverse and identity properties}$$

At which point in this solution was the fact that $a \neq 0$ used?

The work above shows that $-b/a$ is the only possible solution of the equation $ax + b = 0$. To show that $-b/a$ is indeed a solution, replace x with $-b/a$, getting

$$a\left(-\frac{b}{a}\right) + b = 0$$

$$-b + b = 0$$

$$0 = 0,$$

a true statement. In summary,

Solution of a Linear Equation

> the linear equation $ax + b = 0$ has exactly one solution, $-\dfrac{b}{a}$.

It is not really necessary to remember this result. To find the solution for a given linear equation, go through the steps necessary to produce a sequence of simpler equivalent equations.

Example 2 Solve $3(2x - 4) = 7 - (x + 5)$.

Using the distributive property and then collecting like terms gives the following sequence of simpler equivalent equations.

$$3(2x - 4) = 7 - (x + 5)$$

$$6x - 12 = 7 - x - 5 \qquad \text{Distributive property}$$

$$6x - 12 = 2 - x. \qquad \text{Collect like terms}$$

Now adding the same expressions to each side of the equation gives

$$x + 6x - 12 = x + 2 - x \qquad \text{Add } x \text{ to each side}$$

$$7x - 12 = 2$$

$$12 + 7x - 12 = 12 + 2 \qquad \text{Add 12 to each side}$$

$$7x = 14.$$

Finally, multiplying each side by the same number, $\frac{1}{7}$, produces

$$\frac{1}{7} \cdot 7x = \frac{1}{7} \cdot 14$$

$$x = 2.$$

To check this proposed solution, replace x with 2 in the original equation, getting

$$
\begin{array}{ll}
3(2x - 4) = 7 - (x + 5) & \text{Original equation} \\
3(2 \cdot 2 - 4) = 7 - (2 + 5) & \text{Let } x = 2 \\
3(4 - 4) = 7 - (7) & \\
0 = 0. & \text{True}
\end{array}
$$

Since replacing x with 2 results in a true statement, 2 is the solution of the given equation. The solution set is therefore $\{2\}$. ∎

Example 3   Solve $\dfrac{3p - 1}{3} - \dfrac{2p}{p - 1} = p$.

At first glance, this equation does not satisfy the definition of a linear equation given above. However, the equation does appear in proper form after algebraic simplification. To obtain a simpler equivalent equation, first multiply both sides by $3(p - 1)$, where we must assume $p \neq 1$. Doing this gives

$$3(p - 1) \left(\frac{3p - 1}{3} \right) - 3(p - 1) \left(\frac{2p}{p - 1} \right) = 3(p - 1)p$$

$$(p - 1)(3p - 1) - 3(2p) = 3p(p - 1)$$

$$3p^2 - 4p + 1 - 6p = 3p^2 - 3p.$$

An even simpler equivalent equation comes from combining terms and adding $-3p^2$ to both sides, producing

$$
\begin{array}{ll}
-10p + 1 = -3p & \\
1 = 7p & \text{Add } 10p \text{ to each side} \\
\frac{1}{7} = p. & \text{Multiply each side by } \frac{1}{7}
\end{array}
$$

Check 1/7 in the given equation to verify that the solution set is $\{1/7\}$. The restriction $p \neq 1$ does not affect the solution set here, since $1/7 \neq 1$. ∎

Example 4   Solve $\dfrac{x}{x - 2} = \dfrac{2}{x - 2} + 2$.

Multiply both sides of the equation by $x - 2$, assuming that $x - 2 \neq 0$ (or $x \neq 2$), to get

$$x = 2 + 2(x - 2)$$

$$x = 2 + 2x - 4$$

$$x = 2.$$

It was necessary to assume $x - 2 \neq 0$ in order to multiply both sides of the equation by $x - 2$. The proposed solution of 2, however, makes $x - 2 = 0$, meaning that the given equation has no solution. (To see that 2 is not a solution, substitute 2 for x in the given equation.) The solution set is $\emptyset$, the set containing no elements. (The set $\emptyset$ is called the **empty set.**) ∎

Sometimes an equation with more than one letter must be solved for a specified variable. This process is shown in the next example. (As a general rule letters from the beginning of the alphabet, such as a, b, and c are used to represent constants, while letters such as x, y, and z represent variables.)

Example 5   Solve the equation $3(2x - 5a) + 4b = 4x - 2$ for x.
Using the distributive property gives

$$6x - 15a + 4b = 4x - 2.$$

Treat x as the variable and the other letters as constants. Get all terms with x on one side and all terms without x on the other side.

$$6x - 4x = 15a - 4b - 2$$
$$2x = 15a - 4b - 2$$
$$x = \frac{15a - 4b - 2}{2} \quad ∎$$

2.1 Exercises $\overset{(H.W)}{\text{every other odd}}$

Decide whether each of the following equations is an identity or a conditional equation.

1. $x^2 + 5x = x(x + 5)$

2. $3y + 4 = 5(y - 2)$

3. $2(x - 7) = 5x + 3 - x$

4. $2x - 4 = 2(x - 2)$

5. $\dfrac{m + 3}{m} = 1 + \dfrac{3}{m}$

6. $\dfrac{p}{2 - p} = \dfrac{2}{p} - 1$

7. $4q^2 - 25 = (2q + 5)(2q - 5)$

8. $3(k + 2) - 5(k + 2) = -2k - 4$

Decide which of the following pairs of equations are equivalent.

9. $3x - 5 = 7$
$-6x + 10 = 14$

10. $-x = 2x + 3$
$-3x = 3$

11. $\dfrac{3x}{x - 1} = \dfrac{2}{x - 1}$
$3x = 2$

12. $\dfrac{x + 1}{12} = \dfrac{5}{12}$
$x + 1 = 5$

13. $\dfrac{x}{x - 2} = \dfrac{2}{x - 2}$
$x = 2$

14. $\dfrac{x + 3}{x + 1} = \dfrac{2}{x + 1}$
$x = -1$

15. $x = 4$
$x^2 = 16$

16. $z^2 = 9$
$z = 3$

Solve each of the following equations.

17. $4x - 1 = 15$

18. $-3y + 2 = 5$

19. $.2m - .5 = .1m + .7$

20. $.01p + 3.1 = 2.03p - 2.96$

21. $\dfrac{5}{6}k - 2k + \dfrac{1}{3} = \dfrac{2}{3}$

22. $\dfrac{3}{4} + \dfrac{1}{5}r - \dfrac{1}{2} = \dfrac{4}{5}r$

23. $3r + 2 - 5(r + 1) = 6r + 4$

24. $5(a + 3) + 4a - 5 = -(2a - 4)$

25. $2[m - (4 + 2m) + 3] = 2m + 2$

26. $4[2p - (3 - p) + 5] = -7p - 2$

27. $\dfrac{3x - 2}{7} = \dfrac{x + 2}{5}$

28. $\dfrac{2p + 5}{5} = \dfrac{p + 2}{3}$

29. $\dfrac{3k - 1}{4} = \dfrac{5k + 2}{8}$

30. $\dfrac{9x - 1}{6} = \dfrac{2x + 7}{3}$

31. $\dfrac{x}{3} - 7 = 6 - \dfrac{3x}{4}$

32. $\dfrac{y}{3} + 1 = \dfrac{2y}{5} - 4$

33. $\dfrac{1}{4p} + \dfrac{2}{p} = 3$

34. $\dfrac{2}{t} + 6 = \dfrac{5}{2t}$

35. $\dfrac{m}{2} - \dfrac{1}{m} = \dfrac{6m + 5}{12}$

36. $\dfrac{-3k}{2} + \dfrac{9k - 5}{6} = \dfrac{11k + 8}{k}$

37. $\dfrac{2r}{r - 1} = 5 + \dfrac{2}{r - 1}$

38. $\dfrac{3x}{x + 2} = \dfrac{1}{x + 2} - 4$

39. $\dfrac{5}{2a + 3} + \dfrac{1}{a - 6} = 0$

40. $\dfrac{2}{x + 1} = \dfrac{3}{2x - 5}$

41. $\dfrac{4}{x - 3} - \dfrac{8}{2x + 5} + \dfrac{3}{x - 3} = 0$

42. $\dfrac{5}{2p + 3} - \dfrac{3}{p - 2} = \dfrac{4}{2p + 3}$

43. $\dfrac{3}{2m + 4} = \dfrac{1}{m + 2} - 2$

44. $\dfrac{8}{3k - 9} - \dfrac{5}{k - 3} = 4$

45. $\dfrac{2p}{p - 2} = 3 + \dfrac{4}{p - 2}$

46. $\dfrac{5k}{k + 4} = 3 - \dfrac{20}{k + 4}$

47. $2(m + 1)(m - 1) = (2m + 3)(m - 2)$

48. $(2y - 1)(3y + 2) = 6(y + 2)^2$

49. $(3x - 4)^2 - 5 = 3(x + 5)(3x + 2)$

50. $(2x + 5)^2 = 3x^2 + (x + 3)^2$

Solve each of the following equations for x.

51. $2(x - a) + b = 3x + a$

52. $5x - (2a + c) = a(x + 1)$

53. $ax + b = 3(x - a)$

54. $4a - ax = 3b + bx$

55. $\dfrac{x}{a - 1} = ax + 3$

56. $\dfrac{2a}{x - 1} = a - b$

57. $a^2x + 3x = 2a^2$

58. $ax + b^2 = bx - a^2$

59. $3x = (2x - 1)(m + 4)$

60. $-x = (5x + 3)(3k + 1)$

When a loan is paid off early, a portion of the finance charge must be returned to the borrower. By one method of calculating finance charge (called the *rule of 78*), the amount of *unearned interest* (finance charge to be returned) is given by

$$u = f \cdot \dfrac{n(n + 1)}{q(q + 1)},$$

where u represents unearned interest, f is the original finance charge, n is the number of payments remaining when the loan is paid off, and q is the original number of payments. Find the amount of the unearned interest in each of the following.

61. Original finance charge = \$800, loan scheduled to run 36 months, paid off with 18 payments remaining

62. Original finance charge = \$1400, loan scheduled to run 48 months, paid off with 12 payments remaining

63. Original finance charge = \$950, loan scheduled to run 24 months, paid off with 6 payments remaining

64. Original finance charge = \$175, loan scheduled to run 12 months, paid off with 3 payments remaining

65. Find the error in the following.

$$x^2 + 2x - 15 = x^2 - 3x$$
$$(x + 5)(x - 3) = x(x - 3)$$
$$x + 5 = x$$
$$5 = 0$$

Find the value of k that will make each equation equivalent to $x = 2$.

66. $9x - 7 = k$

67. $-5x + 11x - 2 = k + 4$

68. $\dfrac{8}{k + x} = 4$

69. $\sqrt{x + k} = 0$

70. $\sqrt{3x - 2k} = 4$

Solve each of the following equations. Round to the nearest hundredth.

71. $9.06x + 3.59(8x - 5) = 12.07x + .5612$

72. $-5.74(3.1 - 2.7p) = 1.09p + 5.2588$

73. $\dfrac{2.5x - 7.8}{3.2} + \dfrac{1.2x + 11.5}{5.8} = 6$

74. $\dfrac{4.19x + 2.42}{.05} - \dfrac{5.03x - 9.74}{.02} = 1$

75. $\dfrac{2.63r - 8.99}{1.25} - \dfrac{3.90r - 1.77}{2.45} = r$

76. $\dfrac{8.19m + 2.55}{4.34} - \dfrac{8.17m - 9.94}{1.04} = 4m$

2.2 Formulas and Applications

Mathematics is an important problem-solving tool. Many times the solution of a problem depends on the use of a formula which expresses a relationship among several variables. For example, the formula

$$A = \frac{24f}{b(p + 1)} \qquad (*)$$

gives the approximate annual interest rate for a consumer loan paid off with monthly payments. Here f is the finance charge on the loan, p is the number of payments, and b is the original amount of the loan.

Suppose the number of payments, p, must be found when the other quantities are known. To do this, solve the equation for p by treating p as the variable and the other letters as constants. Begin by multiplying both sides of formula $(*)$ by $p + 1$. This gives

$$(p + 1)A = (p + 1)\frac{24f}{b(p + 1)}$$

$$(p + 1)A = \frac{24f}{b}.$$

Multiplying both sides by $1/A$ produces

$$\frac{1}{A}(p + 1)A = \frac{1}{A} \cdot \frac{24f}{b}$$

$$p + 1 = \frac{24f}{Ab}.$$

(Here we must assume $A \neq 0$. Why is this a very safe assumption?)

Finally, add -1 to both sides to get

$$p = \frac{24f}{Ab} - 1.$$

This process is called **solving for a specified variable.**

Example 1   Solve $J\left(\dfrac{x}{k} + a\right) = x$ for x.

To get all terms with x on one side of the equation and all terms without x on the other, first use the distributive property.

$$J\left(\frac{x}{k}\right) + Ja = x$$

Eliminate the denominator, k, by assuming $k \neq 0$ and multiplying both sides by k.

$$kJ\left(\frac{x}{k}\right) + kJa = kx$$

$$Jx + kJa = kx$$

Then add $-Jx$ to both sides to get the two terms with x together.

$$kJa = kx - Jx$$

$$kJa = x(k - J) \qquad \text{Factor the right side}$$

Assuming $k \neq J$ permits multiplying both sides by $1/(k - J)$ to find

$$x = \frac{kJa}{k - J}. \qquad \blacksquare$$

One of the main reasons for learning mathematics is to be able to use it in solving practical problems. However, for most students, learning how to apply mathematical skills to real situations is the most difficult task they face. In the rest of this section we give a few hints that may help you with applications.

A common difficulty with "word problems" is trying to do everything at once. It is usually best to attack the problem in stages.

Solving Word Problems

1. Read the problem very carefully.
2. Decide on an unknown, and name it with a variable that you *write down*. Many students, eager to get on with writing an equation, try to skip this step. But it is important. If you don't know what "x" represents, how can you write a meaningful equation or interpret a result?
3. Draw a sketch, if appropriate, showing the information given in the problem.
4. Decide on a variable expression to represent any other unknowns in the problem. For example, if W represents the width of a rectangle, L represents the length, and you know that the length is one more than twice the width, *write down* $L = 1 + 2W$.
5. Finally, use the results of Steps 2 and 4 to write an equation.

Notice how each of the steps listed is carried out in the following examples.

Example 2 If the length of a side of a square is increased by 3 cm, the perimeter of the new square is 40 cm more than twice the length of a side of the original square. Find the dimensions of the original square.

First, what should the variable represent? Since the length of a side of the original square is needed, let the variable represent that length. Write this down:

$$x = \text{length of side of the original square.}$$

Now draw a figure using the given information, as in Figure 1.

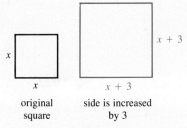

original   side is increased
square   by 3

Figure I

The length of a side of the new square is 3 cm more than the length of a side of the old square. Write a variable expression for that.

$$x + 3 = \text{length of side of the new square}$$

Now write a variable expression for the perimeter of the new square. Since the perimeter of a square is 4 times the length of a side,

$$4(x + 3) = \text{perimeter of the new square.}$$

For the next step, use the given information of the problem to write an equation. The perimeter of the new square is 40 cm more than twice the length of a side of the original square, so the equation is

the new perimeter	is	40	more than	twice the side of the original square
$4(x + 3)$	$=$	40	$+$	$2x.$

Solve the equation as follows:

$$4(x + 3) = 40 + 2x$$
$$4x + 12 = 40 + 2x$$
$$2x = 28$$
$$x = 14.$$

This solution should be checked using the words of the original problem. The length of a side of the new square would be $14 + 3 = 17$ cm; its perimeter would be $4(17) = 68$ cm. Twice the length of a side of the original square is $2(14) = 28$ cm. Since $40 + 28 = 68$, the solution satisfies the problem. ■

Example 3   Chuck travels 80 km in the same time that Mary travels 180 km. Mary travels 50 kph faster than Chuck. Find the rate of each person.

Let x represent Chuck's rate. Since Mary traveled 50 kph faster,

$$x + 50 = \text{rate for Mary}.$$

Constant velocity problems of this kind are solved with the formula

$$d = rt,$$

where d is distance traveled in t hours at a constant rate r. A chart such as the one below can be very helpful in organizing the information in the problem.

For Chuck, $d = 80$ and $r = x$. Find t by solving the formula $d = rt$ for t, getting $t = d/r$. For Chuck, $t = d/r = 80/x$. For Mary, $d = 180$, $r = x + 50$, and $t = 180/(x + 50)$. This information is summarized in the following chart.

	d	r	t
Chuck	80	x	$\dfrac{80}{x}$
Mary	180	$x + 50$	$\dfrac{180}{x + 50}$

Since they both traveled for the same time,

$$\frac{80}{x} = \frac{180}{x + 50}.$$

Solve the equation by first multiplying both sides by $x(x + 50)$, getting

$$x(x + 50) \cdot \frac{80}{x} = x(x + 50) \cdot \frac{180}{x + 50}$$

$$80(x + 50) = 180x$$

$$80x + 4000 = 180x$$

$$4000 = 100x$$

$$40 = x.$$

Since x represents Chuck's rate, Chuck went 40 kph. Mary's rate is $x + 50$, or $40 + 50 = 90$ kph.   ∎

Example 4   Elizabeth Thornton is a chemist. She needs a 20% solution of potassium permanganate. She has a 15% solution on hand, as well as a 30% solution. How many liters of the 15% solution should she add to 3 liters of the 30% solution to get her 20% solution?

Let x be the number of liters of the 15% solution to be added. See Figure 2. Arrange the information of the problem in a table.

Strength	Liters of solution	Liters of pure potassium permanganate
15%	x	$.15x$
30%	3	$.30(3)$
20%	$3 + x$	$.20(3 + x)$

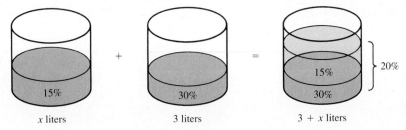

Figure 2

Since the number of liters of pure potassium permanganate in the 15% solution plus the number of liters in the 30% solution must equal the number of liters in the final 20% solution,

liters in 15%	liters in 30%	liters in 20%
$.15x$	$+$ $.30(3)$	$=$ $.20(3 + x).$

Solve this equation as follows.

$$.15x + .90 = .60 + .20x$$
$$.30 = .05x$$
$$6 = x$$

By this result, 6 liters of the 15% solution should be mixed with 3 liters of the 30% solution, giving $6 + 3 = 9$ liters of 20% solution.   ∎

Example 5   One computer can do a job twice as fast as another. Working together, both computers can do the job in 8/3 hr. How long would it take the faster computer, working alone, to do the job?

Let x be the number of hours it would take the faster computer, working alone, to do the job. Then the slower computer will do the job alone in $2x$ hr.

In 1 hr the faster computer will do $1/x$ of the job, while in 1 hr the slower computer will do $1/(2x)$ of the job. Since the total job can be done by the two machines working together in 8/3 hr, then in 1 hr the fraction $1/(8/3)$ of the job will be done. This information leads to the equation below.

portion done by faster computer in 1 hr	portion done by slower computer in 1 hr	portion done by the two computers in 1 hr
$\dfrac{1}{x}$	$+$ $\dfrac{1}{2x}$	$=$ $\dfrac{1}{\dfrac{8}{3}}$

Rewrite $1/(8/3)$ as $1 \cdot (3/8)$, or 3/8, getting

$$\frac{1}{x} + \frac{1}{2x} = \frac{3}{8}.$$

Multiply both sides of the equation by $8x$.

$$8x\left(\frac{1}{x}\right) + 8x\left(\frac{1}{2x}\right) = 8x\left(\frac{3}{8}\right)$$

$$8 + 4 = 3x$$

$$12 = 3x$$

$$4 = x$$

The faster computer could do the entire job, working alone, in 4 hr. The slower computer would need $2(4) = 8$ hr.   ∎

Example 6   John Miller receives a $14,000 bonus from his company. He invests part of the money in 6% tax-free bonds and the remainder at 15%. He earns $1335 per year in interest from the investments. Find the amount he has invested at each rate.

Let x represent the amount Miller invests at 6%, so that $14,000 - x$ is the amount invested at 15%. Since interest is the product of principal, rate, and time ($i = prt$), for one year,

interest at 6% $= x \cdot 6\% \cdot 1 = .06x$

interest at 15% $= (14,000 - x) \cdot 15\% \cdot 1 = .15(14,000 - x)$.

Since the total annual interest is $1335,

$$.06x + .15(14,000 - x) = 1335.$$

Solve this equation, getting

$$.06x + .15(14,000) - .15x = 1335$$

$$-.09x + 2100 = 1335$$

$$-.09x = -765$$

$$x = 8500.$$

Miller invested $8500 at 6%, and $14,000 - \$8500 = \$5500$ at 15%.   ∎

2.2 Exercises

Solve each of the following for the variable indicated. Assume all denominators are nonzero.

1. $PV = k$ for V

2. $F = whA$ for h

3. $V = lwh$ for l

4. $i = prt$ for p

5. $V = V_0 + gt$ for g

6. $s = s_0 + gt^2 + k$ for g

7. $s = \frac{1}{2} gt^2$ for g

8. $A = \frac{1}{2}(B + b)h$ for h

9. $A = \frac{1}{2}(B + b)h$ for B

10. $C = \frac{5}{9}(F - 32)$ for F

11. $S = 2\pi(r_1 + r_2)h$ for r_1

12. $A = P\left(1 + \frac{i}{m}\right)$ for m

13. $g = \frac{4\pi^2 l}{t^2}$ for l

14. $P = \frac{E^2 R}{r + R}$ for R

15. $S = 2\pi rh + 2\pi r^2$ for h

16. $u = f \cdot \frac{k(k + 1)}{n(n + 1)}$ for f

17. $A = \frac{24f}{b(p + 1)}$ for f

18. $A = \frac{24f}{b(p + 1)}$ for b

19. $\dfrac{1}{R} = \dfrac{1}{r_1} + \dfrac{1}{r_2}$ for R **20.** $m = \dfrac{Ft}{v_1 - v_2}$ for v_2

Solve each of the following problems that have solutions.

21. A triangle has a perimeter of 30 cm. Two sides of the triangle are both twice as long as the shortest side. Find the length of the shortest side.

22. The length of a rectangle is 3 cm less than twice the width. The perimeter is 54 cm. Find the width.

23. A box contains 11 nickels. How many quarters must be added so that the box will contain $2.30?

24. A cash drawer contains 15 twenty-dollar bills. How many five-dollar bills must be added to bring the value of the money up to $390?

25. A pharmacist wishes to strengthen a mixture which is 10% alcohol to one which is 30% alcohol. How much pure alcohol should he add to 7 liters of the 10% mixture?

26. A student needs 10% hydrochloric acid for a chemistry experiment. How much 5% acid should be mixed with 60 ml of 20% acid to get a 10% solution?

27. The Old Time Goodies Store sells mixed nuts. Cashews sell for $4 per quarter kg, hazelnuts for $3 per quarter kg, and peanuts for $1 per quarter kg. How many kg of peanuts should be added to 10 kg of cashews and 8 kg of hazelnuts to make a mixture which will sell for $2.50 per quarter kg?

28. The Old Time Goodies Store also sells candy. The manager wants to prepare 200 kg of a mixture for a special Halloween promotion to sell at $4.84 per kg. How much $4 per kg candy should be combined with $5.20 per kg candy for the required mix?

Exercises 29 and 30 depend on the idea of the octane rating of gasoline, a measure of its antiknock qualities. In one measure of octane, a standard fuel is made with only two ingredients: heptane and isooctane. For this fuel, the octane rating is the percent of isooctane. An actual gasoline blend is then compared to a standard fuel. For example, a gasoline with an octane rating of 98 has the same antiknock properties as a standard fuel that is 98% isooctane.

29. How many liters of 94-octane gasoline should be mixed with 200 liters of 99-octane gasoline to get a mixture that is 97-octane?

30. A service station has 92-octane and 98-octane gasoline. How many liters of each should be mixed to provide 12 liters of 96-octane gasoline needed for chemical research?

31. On a vacation trip, Jose averaged 50 mph traveling from Amarillo to Flagstaff. Returning by a different route which covered the same number of miles, he averaged 55 mph. What is the distance between the two cities if his total traveling time was 32 hr?

32. Cindy left by plane to visit her mother in Hartford, 420 km away. Fifteen minutes later, her mother left to meet her at the airport. She drove the 20 km to the airport at 40 kph, arriving just as the plane taxied in. What was the speed of the plane?

33. Russ and Janet are running in the Apple Hill Fun Run. Russ runs at 7 mph, Janet at 5 mph. If they start at the same time, how long will it be before they are 1/2 mi apart?

34. If the run in Exercise 33 has a staggered start, and Janet starts first, with Russ starting 10 min later, how long will it be before he catches up with her?

35. Joann took 20 min to drive her boat upstream to waterski at her favorite spot. Coming back later in the day, at the same boat speed, took her 15 min. If the current in that part of the river is 5 kph, what was her boat speed?

36. Joe traveled against the wind in a small plane for 3 hr. The return trip with the wind took 2.8 hr. Find the speed of the wind if the speed of the plane in still air is 180 mph.

37. Mark can clean the house in 9 hr, while Wendy needs 6 hr. How long will it take them to clean the house if they work together?

38. Helen can paint a room in 5 hr. Jay can paint the same room in 4 hr. (He does a sloppier job.) How long will it take them to paint the room together?

39. Two chemical plants are polluting a river. If plant A produces a predetermined maximum amount of pollution in half the time as plant B, and together they produce the maximum pollution in 26 hr, how long will it take plant B alone?

40. A sewage treatment plant has two inlet pipes to its settling pond. One can fill the pond in 10 hr, the other in 12 hr. If the first pipe is open for 5 hr and then the second pipe is opened, how long will it take to fill the pond?

41. An inlet pipe can fill Dominic's pool in 5 hr, while an outlet pipe can empty it in 8 hr. In his haste to watch television, Dominic left both pipes open. How long did it take to fill the pool?

42. Suppose Dominic discovered his error (see Exercise 41) after an hour-long program. If he then closed the outlet pipe, how much longer would be needed to fill the pool?

43. A clock radio is on sale for $49. If the sale price is 15% less than the regular price, what was the regular price?

44. A shopkeeper prices his items 20% over their wholesale price. If a lamp is marked $74, what was its wholesale price?

45. Jim Marshall invests $20,000 received from an insurance settlement in two ways, some at 13% and some at 16%. Altogether, he makes $2840 per year interest. How much is invested at each rate?

46. Susan Hessney received $52,000 profit from the sale of some land. She invested part at 7 1/2% interest and the rest at 9 1/2% interest. She earned a total of $4520 interest per year. How much did she invest at each rate?

47. Bill Cornett won $100,000 in a state lottery. He first paid income tax of 40% on the winnings. Of the rest, he invested some at 8 1/2% and some at 16%, making $5550 interest per year. How much is invested at each rate?

48. Marjorie Williams earned $48,000 from royalties on her cookbook. She paid a 40% income tax on these royalties. The balance was invested in two ways, at 7 1/2% and at 10 1/2%. The investments produced $2550 interest income per year. Find the amount invested at each rate.

49. Melissa Graves bought two plots of land for a total of $120,000. On the first plot, she made a profit at 15%. On the second, she lost 10%. Her total profit was $5500. How much did she pay for each piece of land?

50. Suppose $10,000 is invested at 6%. How much additional money must be invested at 8% to produce a yield on the entire amount invested of 7.2%?

51. Kathryn Johnson earns take-home pay of $198 a week. If her deductions for retirement, union dues, medical plan, and so on amount to 26% of her wages, what is her weekly pay before deductions?

52. Barbara Burnett gives 10% of her net income to the church. This amounts to $80 a month. In addition, her paycheck deductions are 24% of her gross monthly income. What is her gross monthly income?

53. A bank pays 7% interest on passbook accounts and 10% interest on long-term deposits. Suppose a depositor divides $20,000 among the two types of deposits. Find the amount deposited at each rate if the total annual income from interest is $2500.

54. Miriam Cross wishes to sell a piece of property for $125,000. She wishes the money to be paid off in two ways—a short-term note at 12% and a long-term note at 10%. Find the amount of each note if the total annual interest income is $2000.

55. In planning his retirement, Byron Hopkins deposits some money at 12% with twice as much deposited at 10%. The total deposit is $60,000. Find the amount deposited at each rate if the total annual interest income is $6400.

56. A church building fund has invested $75,000 in two ways: part of the money at 9% and four times as much at 12%. Find the amount invested at each rate if the total annual income from interest is $8550.

57. A cashier has some $5 bills and some $10 bills. The total number of bills is n, and the total value of the money is v. Find the number of each kind of bill that the cashier has.

58. Let $0 < a < b < c < 100$. How many liters of $a\%$ solution should be mixed with m liters of $c\%$ solution to make a $b\%$ mixture?

59. Biologists can estimate the number of individuals of a species in an area. Suppose, for example, that 100 animals of the species are caught and marked. A period of time is permitted to elapse, and then b animals are caught. If c of these ($c \le b$) are marked, show how biologists would estimate the total number of individuals in the area.

60. Suppose B dollars are invested, some at $m\%$ and the rest at $n\%$. If a total of I dollars in interest is earned per year, find the amount invested at each rate.

2.3 Quadratic Equations

An equation of the form $ax + b = 0$ is a linear equation, while an equation of the form $ax^2 + bx + c = 0$ is called quadratic. That is,

Quadratic Equation	an equation that can be written in the form $$ax^2 + bx + c = 0,$$ where a, b, and c are real numbers with $a \ne 0$, is a **quadratic equation.**

(Why is the restriction $a \ne 0$ necessary?) A quadratic equation written in the form $ax^2 + bx + c = 0$ is in **standard form.**

The simplest method of solving a quadratic equation, but one that is not always easily applied, is factoring. This method depends on the **zero-factor property:**

Zero-Factor Property	If a and b are complex numbers, with $ab = 0$, then $a = 0$ or $b = 0$ or both.

The next example shows how the zero-factor property is used to solve a quadratic equation.

Example 1   Solve $6r^2 + 7r = 3$.

First write the equation in standard form as

$$6r^2 + 7r - 3 = 0.$$

Now factor $6r^2 + 7r - 3$ to get

$$(3r - 1)(2r + 3) = 0.$$

By the zero-factor property, the product $(3r - 1)(2r + 3)$ can equal 0 only if

$$3r - 1 = 0 \quad \text{or} \quad 2r + 3 = 0.$$

Solve each of these linear equations separately to find that the solutions of the original equation are 1/3 and $-3/2$. Check these solutions by substituting back in the original equation. The solution set is $\{1/3, -3/2\}$.   ∎

A quadratic equation of the form $x^2 = k$ can be solved by factoring using the following sequence of equivalent equations.

$$x^2 = k$$
$$x^2 - k = 0$$
$$(x - \sqrt{k})(x + \sqrt{k}) = 0$$

$$x - \sqrt{k} = 0 \quad \text{or} \quad x + \sqrt{k} = 0$$
$$x = \sqrt{k} \quad \text{or} \quad x = -\sqrt{k}$$

This proves the following statement, sometimes called the **square root property.**

Square Root Property

The solution set of $x^2 = k$ is $\{\sqrt{k}, -\sqrt{k}\}$.

This solution set is often abbreviated as $\{\pm\sqrt{k}\}$. Both solutions are real if $k > 0$ and imaginary if $k < 0$. If $k = 0$, there is only one solution.

Example 2   Solve each equation.

(a) $z^2 = 17$

The solution set is $\{\pm\sqrt{17}\}$.

(b) $m^2 = -25$

Since $\sqrt{-25} = 5i$, the solution set of $m^2 = -25$ is $\{\pm 5i\}$.

(c) $(y - 4)^2 = 12$

Use a generalization of the square root property, working as follows.

$$(y - 4)^2 = 12$$
$$y - 4 = \pm\sqrt{12}$$
$$y = 4 \pm \sqrt{12}$$
$$y = 4 \pm 2\sqrt{3}$$

The solution set is $\{4 \pm 2\sqrt{3}\}$.   ∎

Completing the Square   As suggested by the equation in Example 2(c), a quadratic equation can be solved with the square root property by first writing the given equation in the form $(x + n)^2 = k$ for suitable numbers n and k. The next two examples show how to write a quadratic equation in this form. (The process explained in the next two examples, *completing the square,* is also used in later chapters when drawing the graphs of equations.)

Example 3   Solve $x^2 - 2x = 15$.

To rewrite this equation in the form $(x + n)^2 = k$, the left side must be rewritten as a perfect square. Expanding $(x + n)^2$ gives $x^2 + 2xn + n^2$. Get $x^2 - 2x$ in this form by first looking at the terms of first degree, $-2x$ and $2xn$. These terms are equal if

$$2xn = -2x$$

or

$$n = -1.$$

If $n = -1$, then $n^2 = (-1)^2 = 1$. In this way, $x^2 - 2x$ can be converted into a perfect square by adding 1, since

$$x^2 - 2x + 1 = (x - 1)^2.$$

If 1 is added to $x^2 - 2x$, the left side of the given equation, then 1 must also be added to the right side, with $x^2 - 2x = 15$ becoming

$$x^2 - 2x + 1 = 15 + 1,$$

or, after factoring on the left,

$$(x - 1)^2 = 16.$$

This equation, $(x - 1)^2 = 16$, is nothing more than the original equation, $x^2 - 2x = 15$, rewritten in an alternate form. In this new form, the equation can be solved by the square root property:

$$x - 1 = \pm 4$$
$$x = 1 \pm 4.$$

There are two solutions:

$$1 + 4 = 5 \qquad \text{and} \qquad 1 - 4 = -3.$$

The solution set for $x^2 - 2x = 15$ is $\{-3, 5\}$.   ■

In Example 3, the equation $x^2 - 2x = 15$ was rewritten as $(x - 1)^2 = 16$, a form suitable for solution by the square root property. This process of rewriting a quadratic equation in the form $(x + n)^2 = k$ is called **completing the square.** (A summary of the steps in completing the square is given after the next example.)

Example 4   Solve $9z^2 - 12z - 1 = 0$.

To rewrite this equation in the form $(z + n)^2 = k$, it is necessary that the coefficient of z^2 be 1. Get this coefficient by multiplying both sides by 1/9:

$$z^2 - \frac{4}{3}z - \frac{1}{9} = 0.$$

Now add 1/9 on both sides:

$$z^2 - \frac{4}{3}z = \frac{1}{9}.$$

The left side must be rewritten as a perfect square, $(z + n)^2$. Since $(z + n)^2 = z^2 + 2zn + n^2$, the terms of first degree of $z^2 + 2zn + n^2$ and $z^2 - (4/3)z$ must satisfy

$$2zn = -\frac{4}{3}z,$$

or

$$n = -\frac{2}{3}.$$

This number is always half the coefficient of the first degree term. If $n = -2/3$, then $n^2 = 4/9$, which should be added to both sides, giving

$$z^2 - \frac{4}{3}z + \frac{4}{9} = \frac{1}{9} + \frac{4}{9}.$$

Factoring on the left yields

$$\left(z - \frac{2}{3}\right)^2 = \frac{5}{9}.$$

Now use the square root property to get

$$z - \frac{2}{3} = \pm\sqrt{\frac{5}{9}}$$

$$z - \frac{2}{3} = \pm\frac{\sqrt{5}}{3}$$

$$z = \frac{2}{3} \pm \frac{\sqrt{5}}{3}.$$

These two solutions can be written as

$$\frac{2 \pm \sqrt{5}}{3},$$

with the solution set $\{(2 \pm \sqrt{5})/3\}$. ■

The process used in Examples 3 and 4, *completing the square,* is summarized as follows.

Completing the Square

To solve $ax^2 + bx + c = 0$, $a \neq 0$, by **completing the square:**

1. If $a \neq 1$, multiply both sides by $1/a$. Then rewrite the equation so that the constant is alone on one side of the equals sign.
2. Square half the coefficient of x, and add the square to both sides.
3. Factor, and use the square root property.

Quadratic Formula   The method of completing the square can be used to solve any quadratic equation. However, in the long run it is better to start with the general quadratic equation,

$$ax^2 + bx + c = 0,$$

and using the method of completing the square, solve this equation for x in terms of the constants a, b, and c. The result will be a general formula for solving any quadratic equation. To find this general formula, start with the fact that in a quadratic equation $a \neq 0$, and multiply both sides by $1/a$ to get

$$x^2 + \frac{b}{a} x + \frac{c}{a} = 0.$$

Adding $-c/a$ to both sides gives

$$x^2 + \frac{b}{a} x = -\frac{c}{a}.$$

Now take half of b/a, and square the result.

$$\frac{1}{2} \cdot \frac{b}{a} = \frac{b}{2a} \quad \text{and} \quad \left(\frac{b}{2a}\right)^2 = \frac{b^2}{4a^2}$$

Add the square to both sides, producing

$$x^2 + \frac{b}{a} x + \frac{b^2}{4a^2} = \frac{b^2}{4a^2} - \frac{c}{a}.$$

The expression on the left side of the equals sign can be written as the square of a binomial, while the expression on the right can be simplified. Doing all this yields

$$\left(x + \frac{b}{2a}\right)^2 = \frac{b^2 - 4ac}{4a^2}.$$

By the square root property, this last statement leads to

$$x + \frac{b}{2a} = \sqrt{\frac{b^2 - 4ac}{4a^2}} \quad \text{or} \quad x + \frac{b}{2a} = -\sqrt{\frac{b^2 - 4ac}{4a^2}}.$$

Since $4a^2 = (2a)^2$,

$$x + \frac{b}{2a} = \frac{\sqrt{b^2 - 4ac}}{|2a|} \quad \text{or} \quad x + \frac{b}{2a} = \frac{-\sqrt{b^2 - 4ac}}{|2a|}.$$

If $a > 0$, then $|2a| = 2a$, giving

$$x = \frac{-b + \sqrt{b^2 - 4ac}}{2a} \quad \text{or} \quad x = \frac{-b - \sqrt{b^2 - 4ac}}{2a}. \tag{$*$}$$

If $a < 0$, then $|2a| = -2a$, giving the same two solutions as in $(*)$, except in reversed order. In either case, the solutions can be written as

$$x = \frac{-b + \sqrt{b^2 - 4ac}}{2a} \quad \text{or} \quad x = \frac{-b - \sqrt{b^2 - 4ac}}{2a}.$$

A more compact form of this result, called the **quadratic formula,** is given below.

Quadratic Formula

The solutions of the quadratic equation $ax^2 + bx + c = 0$, where $a \neq 0$, are

$$\frac{-b \pm \sqrt{b^2 - 4ac}}{2a}.$$

Example 5   Solve $x^2 - 4x + 1 = 0$.

Here $a = 1$, $b = -4$, and $c = 1$. Substitute these values into the quadratic formula, producing

$$\begin{aligned} x &= \frac{-b \pm \sqrt{b^2 - 4ac}}{2a} \\ &= \frac{-(-4) \pm \sqrt{(-4)^2 - 4(1)(1)}}{2(1)} \\ &= \frac{4 \pm \sqrt{16 - 4}}{2} \\ &= \frac{4 \pm 2\sqrt{3}}{2} \\ &= \frac{2(2 \pm \sqrt{3})}{2} \\ x &= 2 \pm \sqrt{3}. \end{aligned}$$

The solution set is $\{2 + \sqrt{3}, 2 - \sqrt{3}\}$, abbreviated $\{2 \pm \sqrt{3}\}$.   ■

Example 6   Solve $2y^2 = y - 4$.

To find the values of a, b, and c, first rewrite the equation as $2y^2 - y + 4 = 0$. Then $a = 2$, $b = -1$, and $c = 4$.

$$\begin{aligned} y &= \frac{-(-1) \pm \sqrt{(-1)^2 - 4(2)(4)}}{2(2)} \qquad \text{By quadratic formula} \\ &= \frac{1 \pm \sqrt{1 - 32}}{4} \\ y &= \frac{1 \pm \sqrt{-31}}{4} = \frac{1 \pm i\sqrt{31}}{4} \end{aligned}$$

The solutions are complex numbers; the solution set is $\{(1 \pm i\sqrt{31})/4\}$.   ■

The next example illustrates the use of the quadratic formula when one or more of the coefficients of the quadratic equation are imaginary numbers.

Example 7 Solve $im^2 + 5m - 3i = 0$.

For this equation, $a = i$, $b = 5$, and $c = -3i$, with

$$m = \frac{-5 \pm \sqrt{5^2 - 4(i)(-3i)}}{2i}$$

$$= \frac{-5 \pm \sqrt{25 + 12i^2}}{2i} = \frac{-5 \pm \sqrt{25 - 12}}{2i}$$

$$m = \frac{-5 \pm \sqrt{13}}{2i}.$$

Simplify by multiplying numerator and denominator by $-i$ (the same result would be obtained using i). This gives

$$m = \frac{(-5 \pm \sqrt{13})(-i)}{2i(-i)}.$$

In the denominator, $2i(-i) = -2i^2 = -2(-1) = 2$, so the final result is

$$m = \frac{5i \pm i\sqrt{13}}{2}.$$

The solution set is $\{(5i \pm i\sqrt{13})/2\}$. ■

The Discriminant The quantity under the radical in the quadratic formula, $b^2 - 4ac$, is called the **discriminant.** When the numbers a, b, and c are *integers* (but not necessarily otherwise), the value of the discriminant can be used to determine whether the solutions will be rational, irrational, or complex numbers. If the discriminant is 0, there will be only one solution. (Why?)

Example 8 Use the discriminant to determine whether the solutions of $5x^2 + 2x - 4 = 0$ are rational, irrational, or complex.

The discriminant is

$$b^2 - 4ac = (2)^2 - (4)(5)(-4) = 84.$$

Since the discriminant is positive, there are two real number solutions. Since 84 is not a perfect square, the solutions will be irrational numbers. ■

The discriminant of a quadratic equation gives the following information about the solutions of the equation.

Discriminant

Discriminant	Number of solutions	Kind of solutions
Positive, perfect square	two	rational
Positive, but not a perfect square	two	irrational
Zero	one	rational
Negative	two	complex

Example 9   Find a value of k so that there is exactly one solution to the equation

$$16p^2 + kp + 25 = 0.$$

A quadratic equation with real coefficients will have exactly one solution if the discriminant is zero. Here, $a = 16$, $b = k$, and $c = 25$, giving the discriminant

$$b^2 - 4ac = k^2 - 4(16)(25) = k^2 - 1600.$$

The discriminant is 0 if $k^2 - 1600 = 0$
or if $k^2 = 1600,$

from which $k = \pm 40$.   ■

Word problems often lead to quadratic equations, as in the next example.

Example 10   Michael wants to make an exposed gravel border of uniform width around a rectangular pool in his garden. The pool is 10 ft by 6 ft. He has enough material to cover 36 sq ft. How wide will the border be?

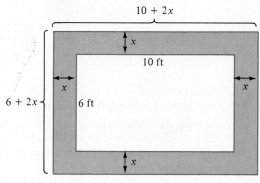

Figure 3

A sketch of the pool with border is shown in Figure 3. Let x represent the width of the border. Then the width of the large rectangle is $6 + 2x$ and its length is $10 + 2x$. The area of the large rectangle is $(6 + 2x)(10 + 2x)$. The area of the pool is $6 \cdot 10 = 60$. The area of the border is found by subtracting the area of the pool from the area of the larger rectangle. This difference should be 36 sq ft.

$$(6 + 2x)(10 + 2x) - 60 = 36$$

Now solve this equation.

$$60 + 32x + 4x^2 - 60 = 36$$
$$4x^2 + 32x - 36 = 0$$
$$x^2 + 8x - 9 = 0$$
$$(x + 9)(x - 1) = 0$$

The solutions are -9 and 1. Since -9 cannot be the width of the border, the border must be 1 ft wide.   ■

H.W
1/30/57

2.3 Exercises Fri

Solve the following equations by factoring or by using the square root property.

1. $p^2 = 16$ **2.** $k^2 = 25$ **3.** $x^2 = 27$ **4.** $y^2 = 24$

5. $(m - 3)^2 = 5$ **6.** $(p + 2)^2 = 7$ **7.** $(3k - 1)^2 = 12$ **8.** $(4t + 1)^2 = 20$

9. $p^2 - 5p + 6 = 0$ **10.** $q^2 + 2q - 8 = 0$ **11.** $6z^2 - 5z - 50 = 0$ **12.** $6r^2 + 7r = 3$

13. $8k^2 + 14k + 3 = 0$ **14.** $18r^2 - 9r - 2 = 0$

Solve the following equations by completing the square.

15. $p^2 - 8p + 15 = 0$ **16.** $m^2 + 5m = 6$ **17.** $x^2 - 2x - 4 = 0$

18. $r^2 + 8r + 13 = 0$ **19.** $2p^2 + 2p + 1 = 0$ **20.** $9z^2 - 12z + 8 = 0$

Solve the following equations.

21. $m^2 - m - 1 = 0$ **22.** $y^2 - 3y - 2 = 0$ **23.** $2s^2 + 2s = 3$

24. $t^2 - t = 3$ **25.** $x^2 - 6x + 7 = 0$ **26.** $11p^2 - 7p + 1 = 0$

27. $n^2 + 4 = 3n$ **28.** $9p^2 = 25 + 30p$ **29.** $2m^2 = m - 1$

30. $4z^2 - 12z + 11 = 0$ **31.** $x^2 = 2x - 5$ **32.** $3k^2 + 2 = k$

33. $4 - \dfrac{11}{x} - \dfrac{3}{x^2} = 0$ **34.** $3 - \dfrac{4}{p} = \dfrac{2}{p^2}$ **35.** $2 - \dfrac{5}{r} + \dfrac{3}{r^2} = 0$

36. $2 - \dfrac{5}{k} + \dfrac{2}{k^2} = 0$

Use a calculator to give the solutions of the following equations to the nearest thousandth.

37. $5n^2 - 4 = 3n$ **38.** $9p^2 - 30p = 25$ **39.** $2m^2 = m + 2$

40. $4n^2 + 12n - 9 = 0$ **41.** $x^2 - 2 = 3x$ **42.** $3k^2 - 2 = 4k$

Use the quadratic formula to solve the following equations.

43. $m^2 - \sqrt{2}m - 1 = 0$ **44.** $z^2 - \sqrt{3}z - 2 = 0$ **45.** $\sqrt{2}p^2 - 3p + \sqrt{2} = 0$

46. $-\sqrt{6}k^2 - 2k + \sqrt{6} = 0$ **47.** $x^2 + ix + 1 = 0$ **48.** $3im^2 - 2m + i = 0$

49. $ip^2 - 2p + i = 0$ **50.** $2ix^2 - 3x + 3i = 0$

51. $(1 + i)x^2 - x + (1 - i) = 0$ **52.** $(2 + i)r^2 - 3r + (2 - i) = 0$

Solve each of the following equations by factoring first and then using the quadratic formula.

53. $x^3 - 1 = 0$ **54.** $x^3 + 64 = 0$ **55.** $x^3 + 27 = 0$

56. $x^3 + 1 = 0$ **57.** $x^3 - 64 = 0$ **58.** $x^3 - 27 = 0$

59. $8p^3 + 125 = 0$ **60.** $64r^3 - 343 = 0$

Identify the values of a, b, and c for each of the following; then evaluate the discriminant $b^2 - 4ac$ and use it to predict the type of solutions. Do not solve the equations.

61. $x^2 + 8x + 16 = 0$ **62.** $x^2 - 5x + 4 = 0$ **63.** $3m^2 - 5m + 2 = 0$

64. $8y^2 = 14y - 3$ **65.** $4p^2 = 6p + 3$ **66.** $2r^2 - 4r + 1 = 0$

67. $9k^2 + 11k + 4 = 0$ **68.** $3z^2 = 4z - 5$

Solve each of the following for the indicated variable. Assume all denominators are nonzero and all variables represent positive real numbers.

69. $s = \dfrac{1}{2} gt^2$ for t **70.** $A = \pi r^2$ for r **71.** $L = \dfrac{d^4 k}{h^2}$ for h **72.** $F = \dfrac{kMv^2}{r}$ for v

73. $s = s_0 + gt^2 + k$ for t **74.** $P = \dfrac{E^2 R}{(r + R)^2}$ for R

75. $S = 2\pi rh + 2\pi r^2$ for r **76.** $pm^2 - 8qm + \dfrac{1}{r} = 0$ for m

Solve the following problems.

77. Find two consecutive even integers whose product is 288.

78. The sum of the squares of two consecutive integers is 481. Find the integers.

79. Two integers have a sum of 10. The sum of the squares of the integers is 148. Find the integers.

80. A shopping center has a rectangular area of 40,000 square yards enclosed on three sides for a parking lot. The length is 200 yards more than twice the width. What are the dimensions of the lot?

81. An ecology center wants to set up an experimental garden. It has 300 meters of fencing to enclose a rectangular area of 5000 square meters. Find the dimensions of the rectangle.

82. Alfredo went into a frame-it-yourself shop. He wanted a frame 3 centimeters longer than wide. The frame he chose extends 1.5 centimeters beyond the picture on each side. Find the outside dimensions of the frame if the area of the unframed picture is 70 square centimeters.

83. Joan wants to buy a rug for a room that is 12 feet by 15 feet. She wants to leave a uniform strip of floor around the rug. She can afford 108 square feet of carpeting. What dimensions should the rug have?

84. Max can clean the garage in 9 hours less time than his brother Paul. Working together, they can do the job in 20 hours. How long would it take each one to do the job alone?

85. Dolores drives 10 mph faster than Steve. Both start at the same time for Atlanta from Chattanooga, a distance of about 100 miles. It takes Steve 1/3 of an hour longer than Dolores to make the trip. What is Steve's average speed?

86. Amy walks 1 mph faster than her friend Lisa. In a walk for charity, both walked the full distance of 24 miles. Lisa took 2 hours longer than Amy. What was Lisa's average speed?

Find the solution for each of the following. Round to the nearest hundredth.

87. One leg of a right triangle is 4 cm longer than the other leg. The hypotenuse (longest side) is 10 cm longer than the shorter leg. Find each of the three sides of the triangle.

88. A rectangle has a diagonal of 16 m. The width of the rectangle is 3.2 m less than the length. Find the length and width of the rectangle.

Find all values for k for which the following equations have exactly one solution.

89. $9x^2 + kx + 4 = 0$ **90.** $25m^2 - 10m + k = 0$ **91.** $y^2 + 11y + k = 0$

92. $z^2 - 5z + k = 0$ **93.** $kr^2 + (2k + 6)r + 16 = 0$ **94.** $ky^2 + 2(k + 4)y + 25 = 0$

Let r_1 and r_2 be the solutions of the quadratic equation $ax^2 + bx + c = 0$. Show that

95. $r_1 + r_2 = -\dfrac{b}{a}$

96. $r_1 r_2 = \dfrac{c}{a}$

97. Let m, n, and k be real numbers. Find the solutions of the equation $(x - m)(x - n) = k^2$ and show that they are real.

98. Suppose one solution of the equation $km^2 + 10m = 8$ is -4. Find the value of k, and the other solution.

For the following equations,   **(a)** solve for x in terms of y,   **(b)** solve for y in terms of x.

99. $4x^2 - 2xy + 3y^2 = 2$

100. $3y^2 + 4xy - 9x^2 = -1$

101. This exercise gives an alternate derivation of the quadratic formula. Start with $ax^2 + bx + c = 0$, with $a \neq 0$, and give a reason for each of the following steps:
(a) $4a^2x^2 + 4abx = -4ac$
(b) $4a^2x^2 + 4abx + b^2 = b^2 - 4ac$
(c) $(2ax + b)^2 = b^2 - 4ac$
(d) $x = \dfrac{-b \pm \sqrt{b^2 - 4ac}}{2a}$

102. Show that the process of completing the square is equivalent to writing $ax^2 + bx + c$ as

$$a\left[\left(x + \frac{b}{2a} \right)^2 + \left(\frac{c}{a} - \frac{b^2}{4a^2} \right) \right].$$

2.4 Equations Reducible to Quadratics

The equation $12m^4 - 11m^2 + 2 = 0$ is not a quadratic equation because of the m^4 term. However, making the substitutions

$$x = m^2 \quad \text{and} \quad x^2 = m^4$$

changes the equation into

$$12x^2 - 11x + 2 = 0,$$

which is a quadratic equation. This quadratic equation can be solved for x, and then from $x = m^2$, the values of m, the solutions of the original equation, can be found.

An equation such as $12m^4 - 11m^2 + 2 = 0$ is said to be **quadratic in form** if it can be written as

$$au^2 + bu + c = 0,$$

where $a \neq 0$ and u is some algebraic expression.

Example 1   Solve $12m^4 - 11m^2 + 2 = 0$.

As mentioned above, this equation is quadratic in form. By making the substitution $x = m^2$, the equation becomes

$$12x^2 - 11x + 2 = 0,$$

which can be solved by factoring, as follows.

$$(3x - 2)(4x - 1) = 0$$

$$x = 2/3 \quad \text{or} \quad x = 1/4$$

To find m, use the fact that $x = m^2$ and replace x with 2/3 and with 1/4, giving

$$m^2 = \frac{2}{3} \qquad \text{or} \qquad m^2 = \frac{1}{4}$$

$$m = \pm \sqrt{\frac{2}{3}} \qquad m = \pm \sqrt{\frac{1}{4}}$$

$$m = \pm \frac{\sqrt{2}}{\sqrt{3}}$$

$$m = \frac{\pm \sqrt{6}}{3} \qquad \text{or} \qquad m = \pm \frac{1}{2}.$$

These four solutions of the given equation $12m^4 - 11m^2 + 2 = 0$ make up the solution set $\{\sqrt{6}/3, -\sqrt{6}/3, 1/2, -1/2\}$, abbreviated $\{\pm \sqrt{6}/3, \pm 1/2\}$. As before, check these solutions by substituting back into the *original* equation. ■

Example 2   Solve $6p^{-2} + p^{-1} = 2$.
  Let $u = p^{-1}$ and rearrange terms to get

$$6u^2 + u - 2 = 0.$$

Factor on the left, and then place each factor equal to 0, giving

$$(3u + 2)(2u - 1) = 0$$

$$3u + 2 = 0 \qquad \text{or} \qquad 2u - 1 = 0$$

$$u = -\frac{2}{3} \qquad\qquad u = \frac{1}{2}.$$

Since $u = p^{-1}$, $p^{-1} = -\dfrac{2}{3}$ or $p^{-1} = \dfrac{1}{2}$,

from which $p = -\dfrac{3}{2}$ or $p = 2$.

The solution set of $6p^{-2} + p^{-1} = 2$ is $\{-3/2, 2\}$. ■

Example 3   Solve $4p^4 - 16p^2 + 13 = 0$.
  Let $u = p^2$ to get

$$4u^2 - 16u + 13 = 0.$$

To solve this equation, use the quadratic formula with $a = 4$, $b = -16$, and $c = 13$. Substituting these values into the quadratic formula gives

$$u = \frac{-(-16) \pm \sqrt{(-16)^2 - 4(4)(13)}}{2(4)} = \frac{16 \pm \sqrt{48}}{8}$$

$$= \frac{16 \pm 4\sqrt{3}}{8} = \frac{4 \pm \sqrt{3}}{2}.$$

Since $p^2 = u$,

$$p^2 = \frac{4 + \sqrt{3}}{2} \qquad \text{or} \qquad p^2 = \frac{4 - \sqrt{3}}{2}.$$

Finally,

$$p = \pm \sqrt{\frac{4 + \sqrt{3}}{2}} \qquad \text{or} \qquad p = \pm \sqrt{\frac{4 - \sqrt{3}}{2}}.$$

Rationalizing the denominators gives the solution set

$$\left\{ \frac{\pm\sqrt{8 + 2\sqrt{3}}}{2}, \ \frac{\pm\sqrt{8 - 2\sqrt{3}}}{2} \right\}. \qquad \blacksquare$$

To solve equations containing radicals or rational exponents, such as $x = \sqrt{15 - 2x}$, or $(x + 1)^{1/2} = x$, use the following result.

> If P and Q are algebraic expressions, then every solution of the equation $P = Q$ is also a solution of the equation $(P)^n = (Q)^n$, for any positive integer n.

Be very careful when using this result. It does *not* say that the equations $P = Q$ and $(P)^n = (Q)^n$ are equivalent; it says only that each solution of the original equation $P = Q$ is also a solution of the new equation $(P)^n = (Q)^n$. However, the new equation may have *more* solutions than the original equation. For example, the solution set of the equation $x = -2$ is $\{-2\}$. Squaring both sides of the equation $x = -2$ gives the new equation $x^2 = 4$, which has solution set $\{-2, 2\}$. Since the solution sets are not equal, the equations are not equivalent. As this example shows, it is essential to check all proposed solutions back in the original equation.

Example 4 Solve $x = \sqrt{15 - 2x}$.

The equation $x = \sqrt{15 - 2x}$ can be solved by squaring both sides as follows:

$$x^2 = (\sqrt{15 - 2x})^2$$
$$x^2 = 15 - 2x$$
$$x^2 + 2x - 15 = 0$$
$$(x + 5)(x - 3) = 0$$
$$x = -5 \qquad \text{or} \qquad x = 3.$$

Now it is necessary to check the proposed solutions in the original equation,

$$x = \sqrt{15 - 2x}.$$

If $x = -5$, does $x = \sqrt{15 - 2x}$?   If $x = 3$, does $x = \sqrt{15 - 2x}$?

$$-5 = \sqrt{15 + 10} \qquad\qquad\qquad 3 = \sqrt{15 - 6}$$
$$-5 = 5 \quad \text{False} \qquad\qquad\qquad 3 = 3 \quad \text{True}$$

As this check shows, only 3 is a solution, giving the solution set $\{3\}$. $\blacksquare$

Example 5 Solve $\sqrt{2x + 3} - \sqrt{x + 1} = 1$.

Separate the radicals by writing the equation as

$$\sqrt{2x + 3} = 1 + \sqrt{x + 1}.$$

Now square both sides. Be very careful when squaring on the right side of this equation. Recall that $(a + b)^2 = a^2 + 2ab + b^2$; replace a with 1 and b with $\sqrt{x + 1}$ to get the next equation, the result of squaring both sides of $\sqrt{2x + 3} = 1 + \sqrt{x + 1}$.

$$2x + 3 = 1 + 2\sqrt{x + 1} + x + 1$$
$$x + 1 = 2\sqrt{x + 1}$$

One side of the equation still contains a radical; to eliminate it, square both sides again. This gives

$$x^2 + 2x + 1 = 4(x + 1)$$
$$x^2 - 2x - 3 = 0$$
$$(x - 3)(x + 1) = 0$$
$$x = 3 \quad \text{or} \quad x = -1.$$

Check these proposed solutions in the original equation.

$$\text{If } x = 3, \text{ does } \sqrt{2x + 3} - \sqrt{x + 1} = 1?$$
$$\sqrt{9} - \sqrt{4} = 1$$
$$3 - 2 = 1 \qquad \text{True}$$
$$\text{If } x = -1, \text{ does } \sqrt{2x + 3} - \sqrt{x + 1} = 1?$$
$$\sqrt{1} - \sqrt{0} = 1$$
$$1 - 0 = 1 \qquad \text{True}$$

Both proposed solutions 3 and -1 are solutions of the original equation, giving $\{3, -1\}$ as the solution set. ■

Example 6 Solve $(5x^2 - 6)^{1/4} = x$.

Since the equation involves a fourth root, begin by raising both sides to the fourth power.

$$[(5x^2 - 6)^{1/4}]^4 = x^4$$
$$5x^2 - 6 = x^4$$
$$x^4 - 5x^2 + 6 = 0$$

Now substitute y for x^2, yielding

$$y^2 - 5y + 6 = 0$$
$$(y - 3)(y - 2) = 0$$
$$y = 3 \quad \text{or} \quad y = 2.$$

Since $y = x^2$,
$$x^2 = 3 \quad \text{or} \quad x^2 = 2$$
$$x = \pm\sqrt{3} \quad \text{or} \quad x = \pm\sqrt{2}.$$

Checking the four proposed solutions, $\sqrt{3}$, $-\sqrt{3}$, $\sqrt{2}$, and $-\sqrt{2}$, in the original equation shows that only $\sqrt{3}$ and $\sqrt{2}$ are solutions, so the solution set is $\{\sqrt{3}, \sqrt{2}\}$. ∎

2.4 Exercises

Find all real solutions for each of the following equations by first rewriting in quadratic form.

1. $m^4 - 8m^2 + 15 = 0$

2. $3k^4 + 10k^2 - 25 = 0$

3. $2r^4 - 7r^2 + 5 = 0$

4. $4x^4 - 8x^2 + 3 = 0$

5. $(g - 2)^2 - 6(g - 2) + 8 = 0$

6. $(p + 2)^2 - 2(p + 2) - 15 = 0$

7. $-(r + 1)^2 - 3(r + 1) + 3 = 0$

8. $-2(z - 4)^2 + 2(z - 4) + 3 = 0$

9. $6(k + 2)^4 - 11(k + 2)^2 + 4 = 0$

10. $8(m - 4)^4 - 10(m - 4)^2 + 3 = 0$

11. $7p^{-2} + 19p^{-1} = 6$

12. $5k^{-2} - 43k^{-1} = 18$

13. $(r - 1)^{2/3} + (r - 1)^{1/3} = 12$

14. $(y + 3)^{2/3} - 2(y + 3)^{1/3} - 3 = 0$

15. $5r = r^{-1}$

16. $a^{-1} = a^{-2}$

17. $\dfrac{17}{p^2 + 1} - \dfrac{22}{(p^2 + 1)^2} = 3$

18. $6 = \dfrac{7}{2y - 3} + \dfrac{3}{(2y - 3)^2}$

19. $r^3 - 13r^{3/2} + 40 = 0$

20. $a^3 - 8a^{3/2} + 7 = 0$

21. $1 + 3(r^2 - 1)^{-1} = 28(r^2 - 1)^{-2}$

22. $5(m^2 + 1)^{-2} = 4(m^2 + 1)^{-1} + 1$

23. $2(1 + 2\sqrt{x})^2 - (1 + 2\sqrt{x}) = 21$

24. $20(2 - \sqrt{m})^2 + 11(2 - \sqrt{m}) = 3$

25. $\left(\dfrac{k^2 + 3k}{4}\right)^2 - \dfrac{11}{6}\left(\dfrac{k^2 + 3k}{4}\right) + \dfrac{1}{2} = 0$

26. $\left(\dfrac{z^2 + 5z}{6}\right)^2 + \dfrac{4}{3}\left(\dfrac{z^2 + 5z}{6}\right) = \dfrac{7}{3}$

27. $12\left(m - \dfrac{2}{m}\right)^2 - 13m + \dfrac{26}{m} = 14$

28. $6\left(2k - \dfrac{1}{k}\right)^2 - 22k + \dfrac{11}{k} = 35$

29. $\sqrt{2m + 1} = 2\sqrt{m}$

30. $3\sqrt{p} = \sqrt{8p + 16}$

31. $\sqrt{3z + 7} = 3z + 5$

32. $\sqrt{4r + 13} = 2r - 1$

33. $\sqrt{4k + 5} - 2 = 2k - 7$

34. $\sqrt{6m + 7} - 1 = m + 1$

35. $\sqrt{4x + 3} = x$

36. $\sqrt{2t + 4} = t$

37. $\sqrt{y} = \sqrt{y - 5} + 1$

38. $\sqrt{2m} = \sqrt{m + 7} - 1$

39. $\sqrt{r + 5} - 2 = \sqrt{r - 1}$

40. $\sqrt{m + 7} + 3 = \sqrt{m - 4}$

41. $\sqrt{y + 2} = \sqrt{2y + 5} - 1$

42. $\sqrt{4x + 1} = \sqrt{x - 1} + 2$

43. $\sqrt{2\sqrt{7x + 2}} = \sqrt{3x + 2}$

44. $\sqrt{3\sqrt{2m + 3}} = \sqrt{5m - 6}$

45. $\sqrt{x + 2} = \sqrt{4 + 7\sqrt{x}}$

46. $3 - \sqrt{x} = \sqrt{2\sqrt{x} - 3}$

47. $\sqrt[3]{4n + 3} = \sqrt[3]{2n - 1}$

48. $\sqrt[3]{2z} = \sqrt[3]{5z + 2}$

49. $\sqrt[3]{t^2 + 2t - 1} = \sqrt[3]{t^2 + 3}$

50. $\sqrt[3]{2x^2 - 5x + 4} = \sqrt[3]{2x^2}$

51. $(2r + 5)^{1/3} - (6r - 1)^{1/3} = 0$

52. $(3m + 7)^{1/3} - (4m + 2)^{1/3} = 0$

53. $\sqrt[4]{q - 15} = 2$

54. $\sqrt[4]{3x + 1} = 1$

55. $\sqrt[4]{y^2 + 2y} - \sqrt[4]{3} = 0$

56. $\sqrt[4]{k^2 + 6k} - 2 = 0$

57. $(z^2 + 24z)^{1/4} - 3 = 0$

58. $(3t^2 + 52t)^{1/4} - 4 = 0$

59. $(2r - 1)^{2/3} = r^{1/3}$

60. $(z - 3)^{2/5} = (4z)^{1/5}$

61. $k^{2/3} = 2k^{1/3}$

62. $3m^{3/4} = m^{1/2}$

63. $p(2 + p)^{-1/2} + (2 + p)^{1/2} = 0$

64. $4(3 + 2m)^{1/2} - (3 + 2m)^{-1/2} = 0$

65. $(1 + 5r)^{1/3} - 2(1 + 5r)^{-2/3} = 0$

66. $(2k - 9)^{-2/3} + 4(2k - 9)^{1/3} = 0$

Find all solutions for each of the following, rounded to the nearest thousandth.

67. $6p^4 - 41p^2 + 63 = 0$

68. $20z^4 - 67z^2 + 56 = 0$

69. $3k^2 - \sqrt{4k^2 + 3} = 0$

70. $2r^2 - \sqrt{r^2 + 1} = 0$

Solve each equation for the indicated variable. Assume all denominators are nonzero and all variables represent positive real numbers.

71. $d = k\sqrt{h}$ for h

72. $v = \dfrac{k}{\sqrt{d}}$ for d

73. $P = 2\sqrt{\dfrac{L}{g}}$ for L

74. $c = \sqrt{a^2 + b^2}$ for a

75. $x^{2/3} + y^{2/3} = a^{2/3}$ for y

76. $m^{3/4} + n^{3/4} = 1$ for m

2.5   Inequalities

An equation says that two expressions are equal, while an **inequality** says that one expression is greater than, greater than or equal to, less than, or less than or equal to, another. As with equations, a value of the variable for which the inequality is true is a *solution* of the inequality; the set of all such solutions makes up the **solution set** of the inequality. Two inequalities with the same solution set are **equivalent.**

Inequalities are solved with the following properties of real numbers.

Properties of Inequalities

For real numbers a, b, and c,

(a) if $a < b$, then $a + c < b + c$

(b) if $a < b$, and if $c > 0$, then $ac < bc$

(c) if $a < b$, and if $c < 0$, then $ac > bc$.

Similar properties are valid if $<$ is replaced with $>$, $\leq$, or $\geq$. Pay careful attention to part (c): if both sides of an inequality are multiplied by a negative number, the direction of the inequality symbol must be reversed. For example, starting with the true statement $-3 < 5$, multiplying both sides by the positive number 2 gives

$$-3 \cdot 2 < 5 \cdot 2$$
$$-6 < 10,$$

still a true statement. On the other hand, starting with $-3 < 5$ and multiplying both sides by the *negative* number -2 gives a true result only if the direction of the inequality symbol is reversed:

$$-3(-2) > 5(-2)$$
$$6 > -10.$$

To prove the properties of inequalities listed above, first recall from Chapter 1 that $a < b$ means that $b - a$ is positive. The proofs also depend on the fact that the sum or product of two positive numbers is positive.

To prove part (a), use the fact that $a < b$ to see that $b - a$ is positive. Rewrite $b - a$ as

$$b - a = (b + c) - (a + c).$$

Since $b - c$ is positive, $(b + c) - (a + c)$ must also be positive, so that

$$a + c < b + c.$$

To prove part (b), again the assumption $a < b$ makes $b - a$ positive. Since c is assumed positive, the product $(b - a)c$ is positive. By the distributive property,

$$(b - a)c = bc - ac,$$

and $bc - ac$ must be positive, giving

$$ac < bc.$$

For part (c), again we are given $a < b$, so that $b - a$ is positive. We are also told that $c < 0$. Since c is less than 0, then $0 - c$, or $-c$, must be positive. Thus,

$$(b - a)(-c) = -bc + ac = ac - bc$$

is positive. If $ac - bc$ is positive,

$$bc < ac.$$

Similar proofs could be given if $<$ is replaced with $>$, $\leq$, or $\geq$.

Example 1 Solve the inequality $-3x + 5 > -7$.

Use the properties of inequalities. First, adding -5 on both sides gives

$$-3x + 5 + (-5) > -7 + (-5)$$
$$-3x > -12.$$

Now multiply both sides by $-1/3$. Since $-1/3 < 0$, reverse the direction of the inequality symbol to get

$$-\frac{1}{3}(-3x) < -\frac{1}{3}(-12)$$

$$x < 4$$

The original inequality is satisfied by any real number less than 4. The solution set can be written $\{x | x < 4\}$. A graph of the solution set is shown in Figure 4, where the parenthesis is used to show that 4 itself does not belong to the solution set. ■

Figure 4

The set $\{x|x < 4\}$, the solution set for the inequality in Example 1, is an example of an **interval.** A simplified notation, called **interval notation,** is used for writing intervals. With this notation, the interval in Example 1 can be written as just $(-\infty, 4)$. The symbol $-\infty$ is not a real number; the symbol is used as a convenience to show that the interval includes all real numbers less than 4. The interval $(-\infty, 4)$ is an example of an **open interval,** since the endpoint, 4, is not part of the interval. Examples of other sets written in interval notation are shown below. Square brackets are used to show that the given number *is* part of the graph. Whenever two real numbers a and b are used to write an interval, it is assumed that $a < b$.

Type of interval	Set	Interval notation	Graph
Open interval	$\{x|a < x\}$	$(a, +\infty)$	
	$\{x|a < x < b\}$	(a, b)	
	$\{x|x < b\}$	$(-\infty, b)$	
Half-open interval	$\{x|a \leq x\}$	$[a, +\infty)$	
	$\{x|a < x \leq b\}$	$(a, b]$	
	$\{x|a \leq x < b\}$	$[a, b)$	
	$\{x|x \leq b\}$	$(-\infty, b]$	
Closed interval	$\{x|a \leq x \leq b\}$	$[a, b]$	

Example 2 Solve $4 - 3y \leq 7 + 2y$. Write the solution in interval notation and graph the solution on a number line.

Write the following series of equivalent inequalities.

$$4 - 3y \leq 7 + 2y$$
$$-4 - 2y + 4 - 3y \leq -4 - 2y + 7 + 2y$$
$$-5y \leq 3$$
$$(-1/5)(-5y) \geq (-1/5)(3)$$
$$y \geq -3/5$$

In set-builder notation, the solution set is $\{y|y \geq -3/5\}$, while in interval notation the solution set is $[-3/5, +\infty)$. See Figure 5 for the graph of the solution set. ■

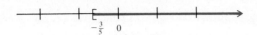

Figure 5

Figure 6

From now on, we shall write the solutions of all inequalities with interval notation.

The inequality $-2 < 5 + 3m < 20$ says that $5 + 3m$ is between -2 and 20. Solve this inequality using an extension of the properties of inequalities given above.

Example 3   Solve $-2 < 5 + 3m < 20$.

Write equivalent inequalities as follows.

$$-2 < 5 + 3m < 20$$
$$-7 < 3m < 15$$
$$-7/3 < m < 5$$

The solution, graphed in Figure 6, is the interval $(-7/3, 5)$. ■

Quadratic Inequalities   Section 2.3 introduced quadratic equations. Now we will look at *quadratic inequalities*.

Quadratic Inequality	A **quadratic inequality** is an inequality that can be written in the form $$ax^2 + bx + c < 0,$$ for real numbers $a \neq 0$, b, and c. (The symbol $<$ can be replaced with $>$, $\leq$, or $\geq$.)

Example 4   Solve the quadratic inequality $x^2 - x - 12 < 0$.

Begin by finding the values of x which satisfy $x^2 - x - 12 = 0$.

$$x^2 - x - 12 = 0$$
$$(x + 3)(x - 4) = 0$$
$$x = -3 \quad \text{or} \quad x = 4$$

These two points, -3 and 4, divide a number line into the three regions shown in Figure 7. If a point in region A, for example, leads to negative values for the polynomial $x^2 - x - 12$, then all points in region A will lead to negative values.

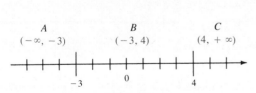

Figure 7

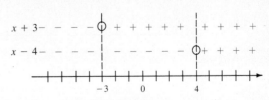

Figure 8

To find the regions that make $x^2 - x - 12$ negative (< 0), sketch the **sign graph** shown in Figure 8. A sign graph shows where factors are positive or negative. First decide on the sign of the factor $x + 3$ in each of the three regions. Then do the same thing for the factor $x - 4$.

Now consider the sign of the product of the two factors in each region. As the sign graph shows, both factors are negative in the interval $(-\infty, -3)$; therefore their product is positive in that interval. For the interval $(-3, 4)$, one factor is positive, while the other is negative, giving a negative product. In the last region, $(4, +\infty)$, both factors are positive so their product is positive. The polynomial $x^2 - x - 12$ takes on negatives values (what the original inequality calls for) when the product of its factors is negative, that is, for the interval $(-3, 4)$. The graph of this solution set is shown in Figure 9.   ■

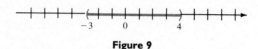

Figure 9

Example 5   Solve the inequality $2x^2 + 5x - 12 \geq 0$.

Begin by finding the values of x which satisfy $2x^2 + 5x - 12 = 0$.

$$2x^2 + 5x - 12 = 0$$
$$(2x - 3)(x + 4) = 0$$
$$x = 3/2 \quad \text{or} \quad x = -4$$

These two points divide the number line into the three regions shown in the sign graph of Figure 10. Since both factors are negative in the first interval, their product, $2x^2 + 5x - 12$, is positive there. In the second interval, the factors have opposite signs, and therefore their product is negative. Both factors are positive in the third interval with their product also positive there. Thus, the polynomial $2x^2 + 5x - 12$ is positive or zero in the interval $(-\infty, -4]$ and also in the interval $[3/2, +\infty)$. Since both of these intervals belong to the solution set, the result can be written as the *union** of the two intervals, or

$$(-\infty, -4] \cup [3/2, +\infty).$$

The graph of the solution set is shown in Figure 11.   ■

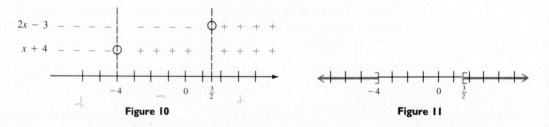

Figure 10   **Figure 11**

*The **union** of sets A and B, written $A \cup B$, is defined as $A \cup B = \{x | x$ is an element of A or x is an element of $B\}$.

Rational Inequalities   The inequalities discussed in the remainder of this section involve quotients of algebraic expressions, and for this reason they are called **rational inequalities.** These inequalities can be solved with a sign graph in much the same way that we solved quadratic inequalities.

Example 6   Solve the inequality $\dfrac{5}{x + 4} \geq 1$.

It is tempting to begin the solution by multiplying both sides of the inequality by $x + 4$, but to do this it would be necessary to consider whether $x + 4$ is positive or negative. Instead, subtract 1 from both sides of the inequality, getting

$$\frac{5}{x + 4} - 1 \geq 0.$$

Writing the left side as a single fraction yields

$$\frac{5 - (x + 4)}{x + 4} \geq 0$$

or

$$\frac{1 - x}{x + 4} \geq 0.$$

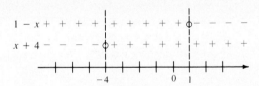

get 1 fraction

$$\frac{5}{x+4} - \frac{1(x+4)}{1(x+4)}$$

$$\frac{5-x-4}{x+4}$$

$$\frac{1-x}{x+4} \geq 0$$

The quotient can change sign only when the denominator is 0 or when the numerator is 0. This means

$$1 - x = 0 \qquad \text{or} \qquad x + 4 = 0$$
$$x = 1 \qquad\qquad\qquad x = -4.$$

Make a sign graph as before. This time consider the sign of the *quotient* of the two quantities rather than their product. See Figure 12.

$$1 - x + \ + \ + \ + \ \ |\ + \ + \ + \ + \ + \ \phi - \ - \ - \ -$$
$$x + 4 - \ - \ - \ - \ \phi + \ + \ + \ + \ + \ |\ + \ + \ + \ +$$

$$\underset{-4 \qquad\qquad\qquad 0 \ \ 1}{\xrightarrow{\hspace{6cm}}}$$

Figure 12

The quotient of two numbers is positive if both numbers are positive or if both numbers are negative. On the other hand, the quotient is negative if the two numbers have opposite signs. The sign graph in Figure 12 shows that values in the interval $(-4, 1)$ give a positive quotient and are part of the solution. With a quotient, the endpoints must be considered separately to make sure that no denominator is 0. With this inequality, -4 gives a 0 denominator while 1 satisfies the given inequality. In interval notation, the solution set is $(-4, 1]$.   ∎

As suggested by Example 6, be very careful with the endpoints of the intervals in the solution of rational inequalities.

Example 7   Solve $\dfrac{2x - 1}{3x + 4} < 5$.

Begin by subtracting 5 on both sides and combining the terms on the left into a single fraction. This process gives

$$\frac{2x - 1}{3x + 4} < 5$$

$$\frac{2x - 1}{3x + 4} - 5 < 0$$

$$\frac{2x - 1 - 5(3x + 4)}{3x + 4} < 0$$

$$\frac{-13x - 21}{3x + 4} < 0.$$

To draw a sign graph, first solve the equations

$$-13x - 21 = 0 \quad \text{and} \quad 3x + 4 = 0,$$

getting the solutions

$$x = -\frac{21}{13} \quad \text{and} \quad x = -\frac{4}{3}.$$

Use the values $-21/13$ and $-4/3$ to divide the number line into three intervals. Now complete a sign graph and find the intervals where the quotient is negative. See Figure 13.

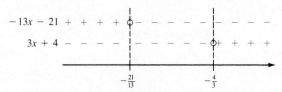

Figure 13

From the sign graph, values of x in the two intervals $(-\infty, -21/13)$ and $(-4/3, +\infty)$ make the quotient negative, as required. Neither endpoint satisfies the given inequality, so the solution set should be written $(-\infty, -21/13) \cup (-4/3, +\infty)$.   ∎

2.5 Exercises

Write each of the following in interval notation. Graph each interval.

1. $-1 < x < 4$ **2.** $x \geq -3$ **3.** $x < 0$ **4.** $8 > x > 3$

5. $2 > x \geq 1$ **6.** $-4 \geq x > -5$ **7.** $-9 > x$ **8.** $6 \leq x$

Using the variable x, write each of the following intervals as an inequality.

9. $(-4, 3)$ **10.** $[2, 7)$ **11.** $(-\infty, -1]$ **12.** $(3, +\infty)$

13.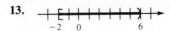

14.

15.

16.

Solve the following inequalities. Write the solutions in interval notation.

17. $2x + 1 \leq 9$

18. $3y - 2 \leq 10$

19. $-3p - 2 \leq 1$

20. $-5r + 3 \geq -2$

21. $2(m + 5) - 3m + 1 \geq 5$

22. $6m - (2m + 3) \geq 4m - 5$

23. $8k - 3k + 2 < 2(k + 7)$

24. $2 - 4x + 5(x - 1) < -6(x - 2)$

25. $\dfrac{4x + 7}{-3} \leq 2x + 5$

26. $\dfrac{2z - 5}{-8} \leq 1 - z$

27. $2 \leq y + 1 \leq 5$

28. $-3 \leq 2t \leq 6$

29. $-10 > 3r + 2 > -16$

30. $4 > 6a + 5 > -1$

31. $-3 \leq \dfrac{x - 4}{-5} < 4$

32. $1 < \dfrac{4m - 5}{-2} < 9$

33. $-4 < 1 - 3x < 2$

34. $-6 \leq 2 - 5y \leq -2$

35. $2 \geq 5 - \dfrac{3}{4}m > -5$

36. $9 > 4 - \dfrac{1}{2}k \geq -1$

very other
odd
37-65

37. $y^2 - 10y + 25 < 25$

38. $m^2 + 6m + 9 < 9$

39. $x^2 - x \leq 6$

40. $r^2 + r < 12$

41. $2k^2 - 9k > -4$

42. $3n^2 < -10 - 13n$

43. $x^2 + 5x - 2 < 0$

44. $4x^2 + 3x + 1 \leq 0$

45. $x^3 - 4x \leq 0$

46. $r^3 - 9r \geq 0$

47. $4m^3 + 7m^2 - 2m > 0$

48. $6p^3 - 11p^2 + 3p > 0$

67,69

Solve the following inequalities. Give the answer in interval notation.

49. $\dfrac{m - 3}{m + 5} \leq 0$

50. $\dfrac{r + 1}{r - 4} > 0$

51. $\dfrac{k - 1}{k + 2} > 1$

52. $\dfrac{a - 6}{a + 2} < -1$

53. $\dfrac{3}{x - 6} \leq 2$

54. $\dfrac{1}{k - 2} < \dfrac{1}{3}$

55. $\dfrac{1}{m - 1} < \dfrac{5}{4}$

56. $\dfrac{6}{5 - 3x} \leq 2$

57. $\dfrac{10}{3 + 2x} \leq 5$

58. $\dfrac{1}{x + 2} \geq 3$

59. $\dfrac{7}{k + 2} \geq \dfrac{1}{k + 2}$

60. $\dfrac{5}{p + 1} > \dfrac{12}{p + 1}$

61. $\dfrac{3}{2r - 1} > \dfrac{-4}{r}$

62. $\dfrac{-5}{3h + 2} \geq \dfrac{5}{h}$

63. $\dfrac{4}{y - 2} \leq \dfrac{3}{y - 1}$

64. $\dfrac{4}{n + 1} < \dfrac{2}{n + 3}$

65. $\dfrac{y + 3}{y - 5} \leq 1$

66. $\dfrac{a + 2}{3 + 2a} \leq 5$

Solve the following rational inequalities using methods similar to those used above.

67. $\dfrac{2x - 3}{x^2 + 1} \geq 0$

68. $\dfrac{9x - 8}{4x^2 + 25} < 0$

69. $\dfrac{(3x - 5)^2}{(2x - 5)^3} > 0$

70. $\dfrac{(5x - 3)^3}{(8x - 25)^2} \leq 0$

71. $\dfrac{(2x - 3)(3x + 8)}{(x - 6)^3} \geq 0$

72. $\dfrac{(9x - 11)(2x + 7)}{(3x - 8)^3} > 0$

Use the discriminant to find the values of k where the following equations have real solutions.

73. $x^2 - kx + 8 = 0$

74. $x^2 + kx - 5 = 0$

75. $x^2 + kx + 2k = 0$

76. $kx^2 + 4x + k = 0$

A product will break even or produce a profit only if the revenue from selling the product at least equals the cost of producing it. Find all intervals when the following products will at least break even.

77. The cost to produce x units of wire is $C = 50x + 5000$, while the revenue is $R = 60x$.

78. The cost to produce x units of squash is $C = 100x + 6000$, while the revenue is $R = 500x$.

79. $C = 70x + 500$; $R = 60x$.

80. $C = 1000x + 5000$; $R = 900x$.

81. The commodity market is very unstable; money can be made or lost quickly when investing in soybeans, wheat, and so on. Suppose that an investor kept track of her total profit, P, at time t, measured in months, after she began investing, and found that

$$P = 4t^2 - 29t + 30.$$

Find the time intervals where she has been ahead. (Hint: $t > 0$ in this case.)

82. Suppose the velocity of an object is given by

$$v = 2t^2 - 5t - 12,$$

where t is time in seconds. (Here, t can be positive or negative.) Find the intervals where the velocity is negative.

83. An analyst has found that his company's profits, in hundred thousands of dollars, are given by

$$P = 3x^2 - 35x + 50,$$

where x is the amount, in hundreds, spent on advertising. For what values of x does the company make a profit?

84. The manager of a large apartment complex has found that the amount of profit he makes is given by

$$P = -x^2 + 250x - 15,000,$$

where x is the number of units rented. For what values of x does the complex produce a profit?

85. The formula for converting from Celsius to Fahrenheit temperature is $F = 9C/5 + 32$. What temperature range in °F corresponds to 0° to 30°C?

86. A projectile is fired from ground level. After t seconds its height above the ground is $220t - 16t^2$ feet. For what time period is the projectile at least 624 feet above the ground?

Solve each inequality.

87. $\dfrac{1}{2} < \dfrac{3}{q} < \dfrac{2}{3}$

88. $\dfrac{5}{8} \leq \dfrac{2}{r} \leq \dfrac{11}{12}$

89. $3b < \dfrac{1}{5bx} < \dfrac{2}{b}$, if b is a positive constant

90. $6c > \dfrac{3}{5cx} \geq \dfrac{1}{c}$, if c is a positive constant

91. If $a > b > 0$, show that $1/a < 1/b$.

92. If $a > b$, is it always true that $1/a < 1/b$?

93. Suppose $a > b > 0$. Show that $a^2 > b^2$.

94. Suppose $a > b > 0$. Show that $(\sqrt{a} - \sqrt{b})^2 > 0$ and that $b < \sqrt{ab} < \dfrac{a + b}{2}$.

95. Let $b > 0$. When is $b^2 > b$?

96. If $a < b$ and $c < d$, show that $a + c < b + d$.

Suppose a, b, c, and d are positive real numbers with $ad < bc$. Show that

97. $\dfrac{a}{b} < \dfrac{a + c}{b + d}$

98. $\dfrac{a + c}{b + d} < \dfrac{c}{d}$

Find all positive solutions of each of the following, given that n is a positive integer greater than 1.

99. $\dfrac{n - 1}{x} \leq \dfrac{n}{x + 1}$

100. $\dfrac{(n - 1)^2}{x^2} \leq \dfrac{n^2}{(x + 1)^2}$

2.6 Absolute Value Equations and Inequalities

This section discusses methods of solving equations and inequalities involving absolute value. Recall from Chapter 1 that the absolute value of a number a, written $|a|$, gives the distance from a to 0 on a number line. By this definition, the absolute value equation $|x| = 3$ can be solved by finding all real numbers at a distance of 3 units from 0. As shown in the graph of Figure 14, there are two numbers satisfying this condition, 3 and -3, making $\{3, -3\}$ the solution set of the equation $|x| = 3$.

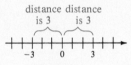

distance distance
is 3   is 3

Figure 14

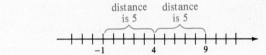

distance   distance
is 5   is 5

Figure 15

Example 1   Solve $|p - 4| = 5$.

The expression $|p - 4|$ represents the distance between p and 4. The equation $|p - 4| = 5$ can be solved by finding all real numbers that are 5 units from 4. As shown in Figure 15, these numbers are -1 and 9. The solution set is $\{-1, 9\}$. ■

The definition of absolute value leads to the following property of absolute value.

> If b is positive, then
>
> $$|a| = b \quad \text{if and only if} \quad a = b \quad \text{or} \quad a = -b.$$

Example 2   Solve $|4m - 3| = |m + 6|$.

By a generalization of the property just given, this equation will be true if either

$$4m - 3 = m + 6 \qquad \text{or} \qquad 4m - 3 = -(m + 6).$$

Solve each of these equations separately. Starting with $4m - 3 = m + 6$ gives

$$4m - 3 = m + 6$$
$$3m = 9$$
$$m = 3.$$

If $4m - 3 = -(m + 6)$, then

$$4m - 3 = -m - 6$$
$$5m = -3$$
$$m = -\frac{3}{5}.$$

The solution set of $|4m - 3| = |m + 6|$ is $\{3, -3/5\}$.   ■

Absolute Value Inequalities   The method used to solve absolute value equations can be generalized to solve inequalities with absolute value.

Example 3   Solve: **(a)** $|x| < 5$ **(b)** $|x| > 5$.

(a) Since absolute value gives the distance from a number to 0, the inequality $|x| < 5$ will be satisfied by all real numbers whose distance from 0 is less than 5. As shown in Figure 16, the solution includes all numbers from -5 to 5, or $-5 < x < 5$. In interval notation, the solution is written as the open interval $(-5, 5)$. A graph of the solution set is shown in Figure 16.

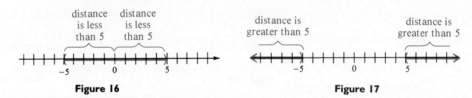

Figure 16   **Figure 17**

(b) In a similar way, the solution of $|x| > 5$ is made up of all real numbers whose distance from 0 is greater than 5. This includes those numbers greater than 5 or those less than -5, or

$$x < -5 \quad \text{or} \quad x > 5.$$

In interval notation, the solution is written $(-\infty, -5) \cup (5, +\infty)$. A graph of the solution set is shown in Figure 17.   ■

The following properties of absolute value can be obtained from the definition of absolute value.

Properties of Absolute Value

If b is a positive number,

(a) $|a| < b$ if and only if $-b < a < b$;

(b) $|a| > b$ if and only if $a < -b$ or $a > b$.

Example 4   Solve $|x - 2| < 5$.

To solve this inequality, find all real numbers whose distance from 2 is less than 5. As shown in Figure 18, the solution set is the interval $(-3, 7)$.

The solution of this inequality can also be found using property (a) above. Let $a = x - 2$ and $b = 5$, so that $|x - 2| < 5$ if and only if

$$-5 < x - 2 < 5.$$

Adding 2 to each portion of this inequality produces

$$-3 < x < 7,$$

again giving the interval solution $(-3, 7)$.   ■

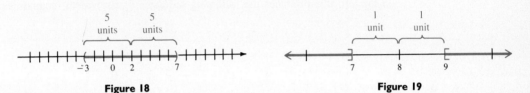

Figure 18   Figure 19

Example 5   Solve $|x - 8| \geq 1$.

We need to find all numbers whose distance from 8 is greater than or equal to 1. As shown in Figure 19 the solution set is $(-\infty, 7] \cup [9, +\infty)$. To find the solution using property (b) above, let $a = x - 8$ and $b = 1$ so that $|x - 8| \geq 1$ if and only if

$$x - 8 \leq -1 \quad \text{or} \quad x - 8 \geq 1.$$

Solve each inequality separately to get the same solution set, $(-\infty, 7] \cup [9, +\infty)$, as mentioned above.   ■

Example 6   Solve $|2 - 7m| - 1 > 4$.

In order to use the properties of absolute value given above, first add 1 to both sides; this gives

$$|2 - 7m| > 5.$$

Now use property (b) above. By this property, $|2 - 7m| > 5$ if and only if

$$2 - 7m < -5 \quad \text{or} \quad 2 - 7m > 5.$$

Solve each of these inequalities separately to get the solution set $(-\infty, -3/7) \cup (1, +\infty)$.   ■

Example 7   Solve $|2 - 5x| \geq -4$.

Since the absolute value of a number is always nonnegative, $|2 - 5x| \geq -4$ is always true. The solution set includes all real numbers. In interval notation, the solution set is $(-\infty, +\infty)$.   ■

The following examples show a method for solving absolute value inequalities with variable denominators.

Example 8 Solve $\left| \dfrac{3k + 1}{k - 1} \right| < 2.$

By property (a) of absolute values, this inequality is satisfied if and only if

$$-2 < \frac{3k + 1}{k - 1} < 2,$$

which is equivalent to

$$-2 < \frac{3k + 1}{k - 1} \quad \text{and} \quad \frac{3k + 1}{k - 1} < 2. \qquad (*)$$

Now solve each of these inequalities separately. Start by rewriting the inequality on the left; this produces the following sequence of equivalent inequalities.

$$\frac{3k + 1}{k - 1} > -2$$

$$\frac{3k + 1}{k - 1} + 2 > 0$$

$$\frac{3k + 1 + 2(k - 1)}{k - 1} > 0$$

$$\frac{5k - 1}{k - 1} > 0$$

By the methods presented in the previous section, the solution set of this inequality is

$$(-\infty, \ 1/5) \cup (1, \ +\infty).$$

Solving the second inequality from $(*)$ above gives

$$\frac{3k + 1}{k - 1} < 2$$

$$\frac{3k + 1}{k - 1} - 2 < 0$$

$$\frac{3k + 1 - 2(k - 1)}{k - 1} < 0$$

$$\frac{k + 3}{k - 1} < 0.$$

Again using the methods of the previous section, the solution set here is the interval $(-3, 1)$.

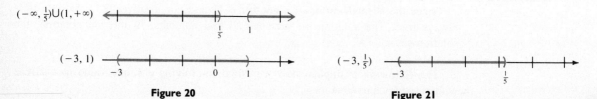

$(-\infty, \frac{1}{5}) \cup (1, +\infty)$

$(-3, 1)$

Figure 20

$(-3, \frac{1}{5})$

Figure 21

Statement ($*$) says that the given inequality is true if and only if *both* the separate inequalities are true. To find where both are true, draw graphs of the separate solution sets, as in Figure 20.

Inspecting these graphs shows that the only values that make both inequalities true at the same time are those in the interval $(-3, 1/5)$. So this interval, shown in Figure 21, is the solution set of the original inequality.   ∎

Example 9   Solve $\left| \dfrac{y}{2y + 1} \right| > 3$.

By property (b) above, this inequality will be true if and only if

$$\frac{y}{2y + 1} < -3 \quad \text{or} \quad \frac{y}{2y + 1} > 3.$$

Again, solve each inequality separately. Starting with the one on the left gives

$$\frac{y}{2y + 1} + 3 < 0$$

$$\frac{y + 3(2y + 1)}{2y + 1} < 0$$

$$\frac{7y + 3}{2y + 1} < 0.$$

The solution set here can be shown to equal $(-1/2, -3/7)$. Now solve the inequality on the right above, getting

$$\frac{y}{2y + 1} > 3$$

$$\frac{y}{2y + 1} - 3 > 0$$

$$\frac{y - 3(2y + 1)}{2y + 1} > 0$$

$$\frac{-5y - 3}{2y + 1} > 0.$$

The solution set here is $(-3/5, -1/2)$. In the first step of this example, we said that the given inequality is true if either of our separate inequalities is true. Thus, the solution of the given inequality is the union of the solution sets of the separate inequalities. As shown in Figure 22, this union is

$$(-3/5, -1/2) \cup (-1/2, -3/7).$$

The solution set is graphed in Figure 23.   ∎

(a) $\left(-\frac{3}{5}, -\frac{1}{2}\right)$

$-\frac{3}{5}$ $-\frac{1}{2}$

(b) $\left(-\frac{1}{2}, -\frac{3}{7}\right)$

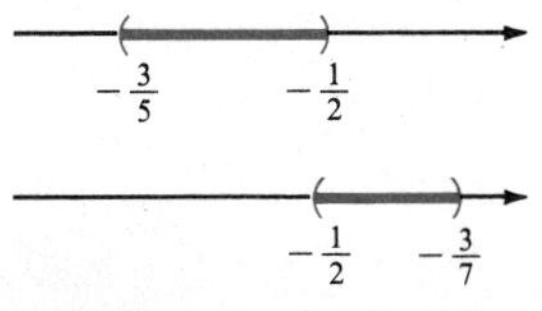

$-\frac{1}{2}$ $-\frac{3}{7}$

$\left(-\frac{3}{5}, -\frac{1}{2}\right) \cup \left(-\frac{1}{2}, -\frac{3}{7}\right)$

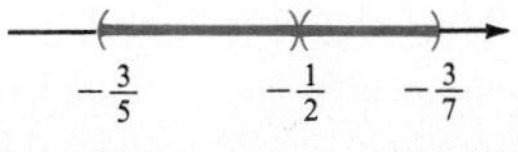

$-\frac{3}{5}$ $-\frac{1}{2}$ $-\frac{3}{7}$

Figure 22   **Figure 23**

2.6 Exercises

Solve each of the following equations.

1. $|a - 2| = 1$ **2.** $|x - 3| = 2$ **3.** $|3m - 1| = 2$ **4.** $|4p + 2| = 5$

5. $|5 - 3x| = 3$ **6.** $|-3a + 7| = 3$ **7.** $\left|\dfrac{z - 4}{2}\right| = 5$ **8.** $\left|\dfrac{m + 2}{2}\right| = 7$

9. $\left|\dfrac{5}{r - 3}\right| = 10$ **10.** $\left|\dfrac{3}{2h - 1}\right| = 4$ **11.** $\left|\dfrac{6y + 1}{y - 1}\right| = 3$ **12.** $\left|\dfrac{3a - 4}{2a + 3}\right| = 1$

13. $|2k - 3| = |5k + 4|$ **14.** $|p + 1| = |3p - 1|$ **15.** $|4 - 3y| = |7 + 2y|$ **16.** $|2 + 5a| = |4 - 6a|$

17. $|x + 2| = |x - 1|$ **18.** $|y - 5| = |y + 3|$

Solve each of the following inequalities. Give the solution in interval notation.

19. $|x| \leq 3$ **20.** $|y| \leq 10$ **21.** $|m| > 1$ **22.** $|z| > 4$

23. $|a| < -2$ **24.** $|b| > -5$ **25.** $|x| - 3 \leq 7$ **26.** $|r| + 3 \leq 10$

27. $|2x + 5| < 3$ **28.** $\left|x - \dfrac{1}{2}\right| < 2$ **29.** $|3m - 2| > 4$ **30.** $|4x - 6| > 10$

31. $|3z + 1| \geq 7$ **32.** $|8b + 5| \geq 7$ **33.** $\left|5x + \dfrac{1}{2}\right| - 2 < 5$ **34.** $\left|x + \dfrac{2}{3}\right| + 1 < 4$

35. $\left|\dfrac{2x + 3}{x}\right| < 1$ **36.** $\left|\dfrac{7 - 4y}{y}\right| < 3$ **37.** $\left|\dfrac{6 + 2y}{y - 5}\right| > 2$ **38.** $\left|\dfrac{3 - 3p}{p + 4}\right| > 5$

39. $\left|\dfrac{2}{q - 2}\right| \leq 4$ **40.** $\left|\dfrac{5}{t - 1}\right| \leq 3$ **41.** $\left|\dfrac{x - 1}{x - 2}\right| \geq 3$ **42.** $\left|\dfrac{2r + 1}{r}\right| \geq 5$

43. $\left|\dfrac{3z - 5}{2z}\right| > 4$ **44.** $\left|\dfrac{m + 4}{3m}\right| \geq 2$ **45.** $\left|\dfrac{6y - 5/2}{2y - 1}\right| > 1$ **46.** $\left|\dfrac{2k + 1}{3 - k}\right| < \dfrac{5}{2}$

Solve each of the following equations.

47. $|x + 1| = 2x$ **48.** $|m - 3| = 4m$

49. $|2k - 1| = k + 2$ **50.** $|3r + 5| = r + 3$

51. $|p| = |p|^2$ **52.** $5|m| = |m|^2$

53. $|1 - 6q|^2 - 4|1 - 6q| - 45 = 0$ **54.** $|6a - 5|^2 + 4|6a - 5| - 12 = 0$

55. $6|3r + 4|^2 + |3r + 4| = 2$ **56.** $5|3 - 4m|^2 - 6|3 - 4m| = 8$

Solve each inequality.

57. $|r^2 - 3| < 1$ **58.** $|m^2 - 7| \leq 2$ **59.** $|z^2 - 8| \geq 7$ **60.** $|y^2 - 6| > 5$

61. Solve $|m + 2| \leq |4m - 1|$ by first dividing each side by $|4m - 1|$.

62. Solve $|3x - 1| < 2|2x + 1|$.

Write each of the following using absolute value statements. For example, write "k is at least 4 units from 1" as $|k - 1| \geq 4$.

63. x is within 4 units of 2   **64.** m is no more than 8 units from 9   **65.** z is no less than 2 units from 12

66. p is at least 5 units from 9   **67.** k is 6 units from 1   **68.** r is 5 units from 3

69. If x is within .0004 units of 2, then y is within .00001 units of 7.

70. y is within 10^{-6} units of 10 whenever x is within 2×10^{-4} units of 5.

71. If $|x - 2| < 3$, find the values of m and n such that $m < 3x + 5 < n$.

72. If $|x + 8| < 16$, find the values of p and q so that $p < 2x - 1 < q$.

73. If $|x - 1| < 10^{-6}$, show that $|7x - 7| < 10^{-5}$.

74. Suppose $|x - k| < 10^{-9}$ and show that $|9x - 9k| < 10^{-8}$.

Solve each inequality.

75. $|4 + p| + |4 - p| \leq 8$

76. $|3 - r| + |3 + r| \leq 6$

Chapter 2 Summary

Key Words

identity

conditional equation

equivalent equations

linear equation

quadratic equation

square root property

completing the square

quadratic formula

discriminant

inequality

quadratic inequality

Review Exercises

Solve each of the following equations.

1. $2m + 7 = 3m + 1$

2. $4k - 2(k - 1) = 12$

3. $5y - 2(y + 4) = 3(2y + 1)$

4. $\dfrac{x - 3}{2} = \dfrac{2x + 1}{3}$

5. $\dfrac{p}{2} - \dfrac{3p}{4} = 8 + \dfrac{p}{3}$

6. $\dfrac{2r}{5} - \dfrac{r - 3}{10} = \dfrac{3r}{5}$

7. $\dfrac{2z}{5} - \dfrac{4z - 3}{10} = \dfrac{1 - z}{10}$

8. $\dfrac{p}{p + 2} - \dfrac{3}{4} = \dfrac{2}{p + 2}$

9. $(x - 3)(2x + 1) = 2(x + 2)(x - 4)$

10. $(3k + 1)^2 = 6k^2 + 3(k - 1)^2$

Solve for x.

11. $3(x + 2b) + a = 2x - 6$

12. $9x - 11(k + p) = x(a - 1)$

13. $\dfrac{x}{m - 2} = kx - 3$

14. $r^2x - 5x = 3r^2$

Solve each of the following for the indicated variable.

15. $2a + ay = 4y - 4a$ for y

16. $\dfrac{3m}{m - x} = 2m + x$ for x

17. $F = \dfrac{9}{5}C + 32$ for C

18. $A = P + Pi$ for P

19. $A = I\left(1 - \dfrac{j}{n}\right)$ for j

20. $A = \dfrac{24f}{b(p + 1)}$ for f

21. $\dfrac{1}{k} = \dfrac{1}{r_1} + \dfrac{1}{r_2}$ for r_1

22. $m = \dfrac{Ft}{\sqrt{I} - \sqrt{2}}$ for t

23. $V = \pi r^2 L$ for L

24. $P(r + R)^2 = E^2R$ for P

25. $\dfrac{xy^2 - 5xy + 4}{3x} = 2p$ for x

26. $\dfrac{zx^2 - 5x + z}{z + 1} = 9$ for z

Solve each of the following problems.

27. A stereo is on sale for 15% off. The sale price is $425. What was the original price?

28. To make a special mix for Valentine's Day, the owner of a candy store wants to combine chocolate hearts which sell for $5 per pound with candy kisses which sell for $3.50 per pound. How many pounds of each should be used to get 30 pounds of a mix which can be sold for $4.50 per pound?

29. Two people are stuffing envelopes for a political campaign. Working together, they can stuff 5000 envelopes in 4 hours. If the first person worked alone, it would take 6 hours to stuff the envelopes. How long would it take the second person, working alone, to stuff the envelopes?

30. Maria can ride her bike to the university library in 20 minutes. The trip home, which is all uphill, takes her half an hour. If her rate is 8 mph slower on the return trip, how far does she live from the library?

Solve each equation.

31. $(b + 7)^2 = 5$ **32.** $(3y - 2)^2 = 8$ **33.** $2a^2 + a - 15 = 0$ **34.** $12x^2 = 8x - 1$

35. $2q^2 - 11q = 21$ **36.** $3x^2 + 2x = 16$ **37.** $2 - \dfrac{5}{p} = \dfrac{3}{p^2}$ **38.** $\dfrac{4}{m^2} = 2 + \dfrac{7}{m}$

39. $ix^2 - 4x + i = 0$ **40.** $4p^2 - ip + 2 = 0$

Evaluate the discriminant for each of the following, and use it to predict the type of solutions for the equation.

41. $8y^2 = 2y - 6$ **42.** $6k^2 - 2k = 3$ **43.** $16r^2 + 3 = 26r$ **44.** $8p^2 + 10p = 7$

45. $25z^2 - 110z + 121 = 0$ **46.** $4y^2 - 8y + 17 = 0$

Solve each word problem.

47. Calvin wants to fence off a rectangular playground next to an apartment building. Since the building forms one boundary, he needs to fence only the other three sides. The area of the playground is to be 11,250 square meters. He has enough material to build 325 meters of fence. Find the length and width of the playground.

48. Steve and Paula sell pies. It takes Paula one hour longer than Steve to bake a day's supply of pies. Working together, it takes them 6/5 hours to bake the pies. How long would it take Steve working alone?

Solve each equation.

49. $4a^4 + 3a^2 - 1 = 0$ **50.** $2x^4 = x^2$ **51.** $(r + 1)^2 - 3(r + 1) = 4$

52. $4(y - 2)^2 - 9(y - 2) + 2 = 0$ **53.** $(2z + 3)^{2/3} + (2z + 3)^{1/3} = 6$

54. $\sqrt{x - 7} = 10$ **55.** $\sqrt{2p + 1} = 8$ **56.** $5\sqrt{m} = \sqrt{3m + 2}$

57. $\sqrt{4y - 2} = \sqrt{3y + 1}$ **58.** $\sqrt{2x + 3} = x + 2$ **59.** $\sqrt{p + 2} = 2 + p$

60. $\sqrt{k} = \sqrt{k + 3} - 1$ **61.** $\sqrt{x^2 + 3x} - 2 = 0$ **62.** $\sqrt[3]{2r} = \sqrt[3]{3r + 2}$

63. $\sqrt[3]{6y + 2} = \sqrt[3]{4y}$ **64.** $(x - 2)^{2/3} = x^{1/3}$ **65.** $\sqrt{3 + x} = \sqrt{3x + 7} - 2$

66. $\sqrt{4 + 3y} = \sqrt{y + 5} + 1$ **67.** $\sqrt{8 - a} + 1 = \sqrt{10 - 6a}$ **68.** $2\sqrt{6 - r} = 2 + \sqrt{7 - 3r}$

Solve each of the following inequalities. Write solutions in interval notation.

69. $-9x < 4x + 7$

70. $11y \geq 2y - 8$

71. $-5z - 4 \geq 3(2z - 5)$

72. $-(4a + 5) < 3a - 2$

73. $3r - 4 + r > 2(r - 1)$

74. $7p - 2(p - 3) \leq 5(2 - p)$

75. $5 \leq 2x - 3 \leq 7$

76. $-8 < 3a - 5 < -1$

77. $-5 < \dfrac{2p - 1}{-3} \leq 2$

78. $3 < \dfrac{6z + 5}{-2} < 7$

79. $x^2 + 3x - 4 \leq 0$

80. $p^2 + 4p > 21$

81. $6m^2 - 11m < 10$

82. $k^2 - 3k - 5 \geq 0$

83. $z^3 - 16z \leq 0$

84. $2r^3 - 3r^2 - 5r < 0$

85. $\dfrac{3a - 2}{a} > 4$

86. $\dfrac{5p + 2}{p} < -1$

87. $\dfrac{3}{r - 1} \leq \dfrac{5}{r + 3}$

88. $\dfrac{3}{x + 2} > \dfrac{2}{x - 4}$

Work the following word problems.

89. Steve and Paula (from Exercise 48 above) have found that the profit from their pie shop is given by

$$P = -x^2 + 28x + 60,$$

where x is the number of units of pies sold daily. For what values of x is the profit positive?

90. A projectile is thrown upward. Its height in feet above the ground after t seconds is $320t - 16t^2$.   **(a)** After how many seconds in the air will it hit the ground?   **(b)** During what time interval is the projectile more than 576 feet above the ground?

Solve each equation.

91. $|a + 4| = 7$

92. $|3 - 2m| = 5$

93. $|2 - y| = 3$

94. $\left|\dfrac{r - 5}{3}\right| = 6$

95. $\left|\dfrac{7}{2 - 3a}\right| = 9$

96. $\left|\dfrac{8r - 1}{3r + 2}\right| = 7$

97. $|5r - 1| = |2r + 3|$

98. $|k + 7| = |k - 8|$

Solve each inequality. Write solutions in interval notation.

99. $|m| \leq 7$

100. $|z| > -1$

101. $|b| \leq -1$

102. $|5m - 8| \leq 2$

103. $|7k - 3| < 5$

104. $|2p - 1| > 2$

105. $|3r + 7| > 5$

106. $\left|\dfrac{1}{k + 3}\right| > 3$

107. $\left|\dfrac{3r}{r - 1}\right| \geq 3$

108. $\left|\dfrac{2p - 1}{p + 2}\right| \leq 1$

109. Let

$$f = 2x\left[\frac{1}{2}(x^2 + 1)^{-1/2}(2x)\right] + 2(x^2 + 1)^{1/2}.$$

Find all intervals where   **(a)** $f > 0$; and   **(b)** $f < 0$.

110. Let

$$g = \frac{(x^2 + 5)^{1/2} - x[(1/2)(x^2 + 5)^{-1/2}(2x)]}{x^2 + 5}$$

Find all intervals where   **(a)** $g > 0$; and   **(b)** $g < 0$.

Set up Exercises 111 and 112.

111. A book is to contain 36 square inches of printed material per page, with margins of 1 inch along the sides, and $1\frac{1}{2}$ inches along the top and bottom. Let x represent the width of the printed area, and write an expression for the area of the entire page.

112. A hunter is at a point on a riverbank. He wants to get to his cabin, located 3 miles north and 8 miles west (see the figure). He can travel 5 miles per hour on the river but only 2 miles per hour on this very rocky ground. If he travels $8 - x$ miles along the river and then walks in a straight line to the cabin, find an expression for the total time that he travels.

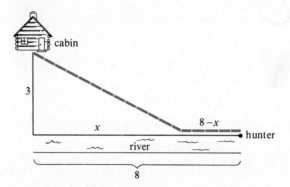

113. If $y = 2x + |2 - x|$, express x in terms of y.

114. Show that $(s + t + |s - t|)/2$ equals the larger of s and t.

115. Show that $(s + t - |s - t|)/2$ equals the smaller of s and t.

116. **(a)** Prove that $|A - B| \le |A - W| + |W - B|$ for all real numbers A, B, and W.
(b) Describe those situations in which the preceding "less-than-or-equal-to" statement is actually an equality.

Solve for x and express the solution in terms of intervals.

117. $\dfrac{1}{|x - 4|} < \dfrac{1}{|x + 7|}$

118. $\dfrac{1}{|x - 3|} - \dfrac{1}{|x + 4|} \ge 0$

119. Find the smallest value of M such that $\left|\dfrac{1}{x}\right| \le M$ for all x in the interval $[2, 7]$.

120. Find the smallest value of M such that $\left|\dfrac{1}{x + 7}\right| \le M$ for all x in the interval $(-4, 2)$.

121. Use the triangle inequality [of Section 1.2] to find a value of M such that

$$|x^3 - 2x + 1| \le M$$

for all x in the interval $(-2, 3)$.

Exercise 113: George B. Thomas, Jr. and Ross L. Finney, *Calculus and Analytic Geometry*. Copyright © 1979, Addison-Wesley, Reading, Massachusetts, p. 57. Reprinted with permission. Exercises 114–16: From *Calculus*, 2nd edition by Stanley I. Grossman. Copyright © 1981 by Academic Press, Inc. Reprinted by permission of Academic Press, Inc. and the author. Exercises 117–22: From *Calculus with Analytic Geometry* by Howard Anton, pp. 21, 22. Copyright © 1984 by Anton Textbooks, Inc. Reprinted by permission of John Wiley & Sons, Inc.

122. Find a value of M such that $\left|\dfrac{x+3}{x-3}\right| \le M$ for all x in the interval $[-3/4, 1/4]$.

Cumulative Review Exercises, Chapters I and 2

[1.1]   Choose all words from the following list that describe the number given: (a) whole number   (b) integer   (c) rational number   (d) irrational number   (e) real number.

1. $-11/3$ **2.** -5 **3.** $\sqrt{49}$ **4.** $\sqrt{3}$

Which of the following sets are closed with respect to the indicated operations?

5. Integers, subtraction   **6.** Integers, division

7. Odd integers, addition   **8.** Even integers, multiplication

[1.2]   Write the following numbers, in numerical order, from smallest to largest.

9. $-|-7|,\;\; |-3|,\;\; -|2|$

10. $|-14|,\;\; -|-1|,\;\; |0|,\;\; -|-4|$

11. $|-8+2|,\;\; -|3|,\;\; -|-2|,\;\; -|-2|+(-3),\;\; -|-8|-|-6|$

12. $-2-|-4|,\;\; -3+|2|,\;\; -4,\;\; -5+|-3|$

Write each of the following without absolute value bars.

13. $-|0|$ **14.** $|-5|-(-2)$ **15.** $|7-\sqrt{5}|$ **16.** $|2-\sqrt{11}|$

17. $|m-3y|$, if $m/3 > y$ **18.** $|5+y^2|$

[1.3]   Simplify each of the following. Write the results with only positive exponents. Assume all variables represent positive real numbers.

19. $(-5)^{-3}$ **20.** $2^{-4}+3^{-2}$ **21.** $(r^{-3})(r^{-2})(r^5)$ **22.** $\dfrac{p^3z^2}{p^{-1}z^{-3}}$

23. $[(3^{-2})^2]^{-1}$ **24.** $\dfrac{(2x^3)^{-2}(2^2x^5)^{-1}}{(2x^4)^{-3}}$

25. $\dfrac{(5p^2q)^{-1}(5p^3q^{-2})^2}{5(pq)^{-3}(p^4q^{-2})^{-1}}$ **26.** $\dfrac{.0000015 \times 40{,}000{,}000{,}000}{.00000000000003}$

[1.4]   Find each of the following.

27. $(-9m^2+11m-2)+(4m^2-8m+7)$ **28.** $(-7z^2+8z-1)-(-4z^2-7z-9)$

29. $(k-7)(3k-8)$ **30.** $(9w+5)^2$

31. $(3k-5)^3$ **32.** $(4k+3)(2k^2+5k+6)$

33. $(y+z+2)(3y-2z+5)$ **34.** $(2r+3s+3)(3r-s+2)$

35. $\dfrac{15x^4+30x^3+12x^2-9}{3x}$ **36.** $\dfrac{16a^6+24a^5-48a^4+12a}{8a^2}$

[1.5]   Factor as completely as possible.

37. $3m^3+9m+15m^5$ **38.** $r^2+15r+54$

39. $6q^2-q-12$ **40.** $10b^2-19b-15$

41. $8a^3 + 125$

42. $64p^6 - 27q^9$

43. $rs + rt - ps - pt$

44. $2m + 6 - am - 3a$

45. $(z - 4)^2 - (z + 4)^2$

46. $6(r + s)^2 + 13(r + s) - 5$

[1.6]   Perform each of the following operations.

47. $\dfrac{2x - 2}{3} \cdot \dfrac{6x - 6}{(x - 1)^3}$

48. $\dfrac{3m - 15}{4m - 20} \cdot \dfrac{m^2 - 10m + 25}{12m - 60}$

49. $\dfrac{3z^2 + z - 2}{4z^2 - z - 5} \div \dfrac{3z^2 + 11z + 6}{4z^2 + 7z - 15}$

50. $\dfrac{1}{a + 1} - \dfrac{1}{a - 1}$

51. $\dfrac{m + 3}{m - 7} + \dfrac{m + 5}{2m - 14}$

52. $\dfrac{1}{x^2 + x - 12} - \dfrac{1}{x^2 - 7x + 12}$

53. $\dfrac{m^{-1} - n^{-1}}{(mn)^{-1}}$

54. $\dfrac{z^{-1} - y^{-1}}{z^{-1} + y^{-1}} \cdot \dfrac{z + y}{z - y}$

[1.7]   Simplify each of the following. Assume that all variables represent positive real numbers.

55. $\sqrt{1000}$

56. $\sqrt[3]{54}$

57. $-\sqrt[4]{32}$

58. $\sqrt{24 \cdot 3^2 \cdot 2^4}$

59. $\sqrt[3]{25 \cdot 3^4 \cdot 5^3}$

60. $\sqrt{50p^7q^8}$

61. $\sqrt[4]{1875h^5y^6q^9}$

62. $\dfrac{\sqrt[3]{a^3b^7c^7} \cdot \sqrt[3]{a^6b^8c^9}}{\sqrt[3]{a^7b^3c^5}}$

63. $\sqrt{8} + 5\sqrt{32} - 7\sqrt{128}$

64. $\dfrac{15}{\sqrt{3}} - \dfrac{2}{\sqrt{27}} + \dfrac{4}{\sqrt{12}}$

65. $\dfrac{1}{2 - \sqrt{7}}$

66. $\dfrac{\sqrt{p} + \sqrt{p + 1}}{\sqrt{p} - \sqrt{p + 1}}$

[1.8]   Rewrite each of the following, using only positive exponents. Assume that all variables represent positive real numbers, and that variables used as exponents represent rational numbers.

67. $32^{-6/5}$

68. $(625z^8)^{1/2}$

69. $(a - b)^{2/3} \cdot (a - b)^{-5/3}$ $(a > b)$

70. $(7k^{3/4}x^{1/8})(9k^{7/4}x^{3/8})$

71. $(3a^{-2/3}b^{5/3})(8a^2b^{-10/3})$

72. $z^{1+r} \cdot z^{3-2r}$

73. $\dfrac{r^{1/3} \cdot s^{5/3} \cdot t^{1/2}}{r^{-2/3} \cdot s^2 \cdot t^{-3/2}}$

74. $\dfrac{q^r \cdot q^{-5r}}{q^{-3r}}$

[1.9]   Perform the following operations.

75. $(-3 + 5i) - (-9 + 3i)$

76. $(-6 + 2i) + (-1 + 7i)$

77. $(1 + 3i)(2 - 5i)$

78. $(-2 + 5i)^2$

79. $i(2 + 3i)^2$

80. $(1 - 2i)^3$

81. $\dfrac{4 + 3i}{1 + i}$

82. $\dfrac{3 + 7i}{5 - 3i}$

83. $\dfrac{i}{3 + 2i}$

84. $\dfrac{5 - 2i}{12 + i} - \dfrac{2 - 9i}{4 + 3i}$

Simplify each of the following.

85. $\sqrt{-400}$

86. $-\sqrt{-39}$

87. $\sqrt{-3} \cdot \sqrt{-7}$

88. $\sqrt{-11} \cdot \sqrt{-6}$

89. Find i^{15}

90. Find i^{245}.

[2.1]   Solve each equation.

91. $5(a + 3) + 4a - 5 = -(2a - 4)$

92. $\dfrac{x}{3} - 7 = 6 - \dfrac{3x}{4}$

93. $\dfrac{-7r}{2} + \dfrac{3r - 5}{4} = \dfrac{2r + 5}{4}$

94. $\dfrac{1}{z - 5} = 2 - \dfrac{3}{z - 5}$

95. $(3x - 4)^2 - 5 = 3(x + 5)(3x + 2)$

96. $\dfrac{x}{b - 1} = 5x + 3b$ (Solve for x.)

Solve each of the following for the indicated variable.

97. $v = v_0 + gt$ for t

98. $s = s_0 + gt^2 + k$ for t^2

99. $A = \dfrac{1}{2}(b + B)h$ for B

100. $S = 2\pi(r_1 + r_2)h$ for r_2

[2.2]   Solve each of the following.

101. A triangle has a perimeter of 54 cm. Two of the sides of the triangle are equal in length, with the third side 6 cm shorter than either of the two equal sides. Find the lengths of the three sides of the triangle.

102. After a lottery win, John has $90,000 to invest. He puts part of the money in a certificate of deposit at 10%, with $10,000 less than this amount put into a real estate scheme paying 12%. The total annual income from the investments is $9800. How much does he have invested at each rate?

103. Suppose $40,000 is invested at 7%. How much additional money would have to be invested at 11% to make the yield on the entire amount equal to 9.4%?

104. A student needs 25% acid for an experiment. How many ml of 50% acid should be mixed with 40 ml of 15% acid to get the necessary 25% acid?

105. How many pounds of coffee worth $6 per pound should be mixed with 20 pounds of coffee selling for $4.50 per pound to get a mixture that can be sold for $5 per pound?

106. Tom can run 6 miles per hour, while Roy runs 4 miles per hour. If they start running at the same time, how long would it take them to be 3/4 miles apart?

107. A boat can go 15 km upstream in the same time that it takes to go 27 km downstream. The speed of the current is 2 km per hour. Find the speed of the boat in still water.

108. An inlet pipe can fill a swimming pool in 1 day. An outlet can empty the pool in 36 hours. Suppose that both the inlet and the outlet are opened. How long would it take to fill the pool 5/8 full?

[2.3]   Solve each equation.

109. $(5r - 3)^2 = 7$

110. $(7q + 2)^2 = 40$

111. $8k^2 + 14k + 3 = 0$

112. $2s^2 + 2s = 3$

113. $4z^2 - 4z - 5 = 0$

114. $12r^2 = 4r$

115. $x^2 - 2x + 2 = 0$

116. $9k^2 - 12k + 8 = 0$

117. $12y^2 - 4y + 3 = 0$

118. $25z^2 + 30z + 11 = 0$

119. $im^2 + 2m + i = 0$

120. $w^2 - 7iw + 4 = 0$

Solve for the indicated variable.

121. $Z = 10a^2yb$ for a

122. $9k = 12qw + 5w^2$ for w

Solve each of the following problems.

123. Find two consecutive odd integers whose product is -1.

124. The area of a field is 9600 m². One side is 40 m longer than the other side. Find the length and width of the field.

125. Person A can do a job in 5 hours. Working with person B, the job takes 3 hours. How long would it take B working alone to do the job?

126. One leg of a right triangle is 3 cm longer than three times the length of the shorter leg. The hypotenuse is 1 cm longer than the longer leg. Find the lengths of the sides of the triangle.

[2.4]   Solve each equation.

127. $2z^4 - 7z^2 + 3 = 0$

128. $-(r + 1)^2 - 3(r + 1) + 3 = 0$

129. $(m - 1)^{2/3} + 3(m - 1)^{1/3} = 10$

130. $6z^{-2} + 7z^{-1} + 2 = 0$

131. $\sqrt{3s} - s = -6$

132. $\sqrt{r + 3} = \sqrt{2r - 1} - 1$

133. $(z^2 - 18z)^{1/4} = 0$

134. $p^{2/3} = 9p^{1/3}$

135. $(m^2 - 1)^2 - 5m^2 = 1$

136. $(y^2 + 2)^2 + y^2 = 10$

[2.5–6]   Solve each inequality. Write all solutions in interval notation.

137. $12m - 17 \geq 8m + 7$

138. $-h \leq 6h + 30$

139. $-15 < -2y + 3 < -1$

140. $z^2 + 6z + 16 < 8$

141. $y^2 + 6y \geq 0$

142. $p^2 < -1$

143. $2t^3 - 2t^2 - 12t \leq 0$

144. $\dfrac{a - 6}{a + 2} < -1$

145. $\dfrac{3}{y + 6} \geq \dfrac{1}{y - 2}$

146. $|z| > 8$

147. $|b| < 3$

148. $|x - 1/2| < 2$

149. $|2x + 5| > 3$

150. $|3 - 5k| > 2$

151. $\left|\dfrac{3x + 1}{x - 1}\right| \leq 2$

152. $\left|\dfrac{m + 4}{m - 3}\right| > 2$

Solve each of the following equations.

153. $|a - 2| = 1$

154. $\left|\dfrac{6y + 1}{y - 1}\right| = 3$

155. $|3z - 1| = |2z + 5|$

156. $|m + 2| = |m + 5|$

3

Functions and Graphs

All the equations and inequalities in the previous chapter involved only *one* variable. However, it is very common for a practical application to use *two* variables, with the value of one variable depending on the value of the other variable. As an example, the graph on the left in Figure 1, from *Road & Track* magazine, shows how the speed in miles per hour of a Porsche 928 depends on the time *t* in seconds after the car has started from a dead stop.

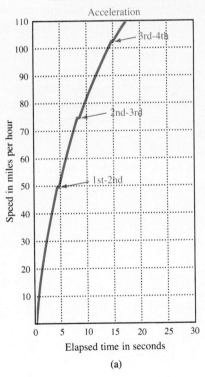

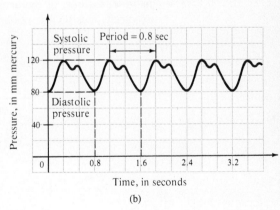

Figure 1

Graph on right: from *Calculus for the Life Sciences* by Rodolfo De Sapio, 1978. Reprinted by permission of W. H. Freeman and Company. Graph on left: from "Road Tests," *Road & Track*, April 1985. Copyright © 1985, CBS Consumer Publishing Division. Reprinted by permission.

The graph on the right in Figure 1 shows how a person's blood pressure depends on time. (Systolic and diastolic pressures are the upper and lower limits in the periodic changes in pressure that produce the pulse. The length of time between peaks is called the *period* of the pulse.)

Each of the graphs in Figure 1 is the graph of a **function,** a rule or procedure giving just one value of one variable from a given value of another variable. In each of these examples, a given value of time can be used to find just one value of the other variable, speed or blood pressure, respectively. Before getting to a complete discussion of functions later in the chapter, we need to look at the topic of graphing, since functions are often studied by looking at their graphs.

3.1 | A Two-Dimensional Coordinate System

Chapter 1 showed a correspondence between real numbers and points on a number line, a correspondence set up by establishing a coordinate system for the line. This idea can be extended to two dimensions: in two dimensions the correspondence is between *pairs* of real numbers and points on a plane. One way to get this correspondence is by drawing two perpendicular lines, one horizontal and one vertical. These lines intersect at a point O called the **origin.** The horizontal line is the **x-axis,** and the vertical line is the **y-axis.**

Starting at the origin, the *x*-axis can be made into a number line by placing positive numbers to the right and negative numbers to the left. The *y*-axis can be made into a number line with positive numbers going up and negative numbers down.

The *x*-axis and *y*-axis set up a **rectangular coordinate system,** or **Cartesian coordinate system** (named for one of its co-inventors, René Descartes; the other co-inventor was Pierre de Fermat). The plane into which the coordinate system is introduced is the **coordinate plane,** or **xy-plane.** The *x*-axis and *y*-axis divide the plane into four regions, or **quadrants,** labeled as shown in Figure 2. The points on the *x*-axis and *y*-axis themselves belong to no quadrant.

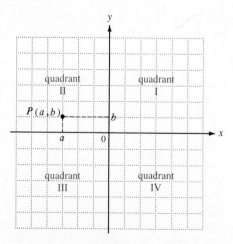

Figure 2

Find a pair of numbers corresponding to a given point P as follows. Start at P (see Figure 2), and draw a vertical line cutting the x-axis at a. Draw a horizontal line cutting the y-axis at b. Then point P has **coordinates** (a, b), where (a, b) is an *ordered pair* of numbers. An **ordered pair** of numbers consists of two numbers, written in parentheses, in which the sequence of the numbers is important. For example, $(4, 2)$ and $(2, 4)$ are not the same ordered pair since the sequence of the numbers is different.

As an example, Figure 3 shows the point A which corresponds to the ordered pair $(3, 4)$. Also in Figure 3, point B corresponds to the ordered pair $(-5, 6)$, C to $(-2, -4)$, D to $(4, -3)$, and E to $(-3, 0)$.

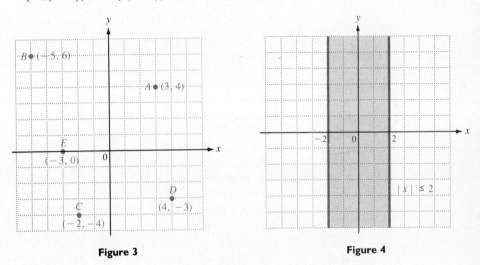

Figure 3   Figure 4

Example 1   Graph the set of all points (x, y) satisfying the inequality $|x| \le 2$.
As shown in the last chapter, $|x| \le 2$ means

$$-2 \le x \le 2.$$

The graph of the set of all ordered pairs (x, y) whose x-coordinate satisfies the inequality $-2 \le x \le 2$ is bounded by vertical lines through $(2, 0)$ and $(-2, 0)$. Graph the region satisfying $|x| \le 2$ as shaded in Figure 4.   ■

Distance Formula   The Pythagorean theorem (in a right triangle with shorter sides a and b, and longest side c, $a^2 + b^2 = c^2$) may be used to obtain a formula for the distance between any two points in the plane. To get this formula, start with two points on a horizontal line, as in Figure 5(a) on the next page. Use the symbol $P(x_1, y_1)$ to represent point P having coordinates (x_1, y_1). The distance between points $P(x_1, y_1)$ and $Q(x_2, y_1)$ can be found by subtracting the x-coordinates. (Absolute value is used to make sure that the distance is not negative—recall the work with distance in Chapter 1.) From this, the distance between points P and Q is $|x_1 - x_2|$. If $d(P, Q)$ represents the distance between P and Q, then

$$d(P, Q) = |x_1 - x_2|.$$

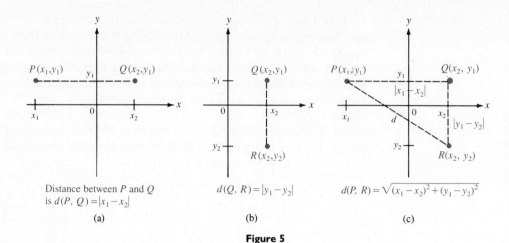

Distance between P and Q
is $d(P, Q) = |x_1 - x_2|$

(a)

$d(Q, R) = |y_1 - y_2|$

(b)

$d(P, R) = \sqrt{(x_1 - x_2)^2 + (y_1 - y_2)^2}$

(c)

Figure 5

Figure 5(b) shows points $Q(x_2, y_1)$ and $R(x_2, y_2)$ on a vertical line. To find the distance between Q and R, subtract the y-coordinates, finding

$$d(Q, R) = |y_1 - y_2|.$$

Finally, Figure 5(c) shows two points, $P(x_1, y_1)$ and $R(x_2, y_2)$, which are *not* on a horizontal or vertical line. To find $d(P, R)$, construct the right triangle shown in the figure. One side of this triangle is horizontal and has length $|x_1 - x_2|$. The other side is vertical and has length $|y_1 - y_2|$. By the Pythagorean theorem,

$$[d(P, R)]^2 = |x_1 - x_2|^2 + |y_1 - y_2|^2.$$

Writing $|x_1 - x_2|^2$ as the equal expression $(x_1 - x_2)^2$ and $|y_1 - y_2|^2$ as $(y_1 - y_2)^2$ gives the following result, called the **distance formula.**

Distance Formula

Suppose $P(x_1, y_1)$ and $R(x_2, y_2)$ are two points in a coordinate plane. Then the distance between P and R, written $d(P, R)$, is

$$d(P, R) = \sqrt{(x_1 - x_2)^2 + (y_1 - y_2)^2}.$$

Example 2   Find the distance between $P(-8, 4)$ and $Q(3, -2)$.
According to the distance formula,

$$d(P, Q) = \sqrt{(-8 - 3)^2 + [4 - (-2)]^2}$$
$$= \sqrt{(-11)^2 + 6^2}$$
$$= \sqrt{121 + 36}$$
$$= \sqrt{157}.$$

Using a calculator with a square root key gives 12.530 as an approximate value for $\sqrt{157}$. ∎

Example 3   Are the points $M(-2, 5)$, $N(12, 3)$, and $Q(10, -11)$ the vertices of a right triangle?

To decide whether or not the triangle determined by these three points is a right triangle, use the converse of the Pythagorean theorem: if the sides a, b, and c of a triangle satisfy $a^2 + b^2 = c^2$, then the triangle is a right triangle. A triangle with the three given points as vertices is shown in Figure 6. This triangle is a right triangle if the square of the length of the longest side equals the sum of the squares of the lengths of the other two sides. Use the distance formula to find the length of each side of the triangle.

$$d(M, N) = \sqrt{[12 - (-2)]^2 + (3 - 5)^2} = \sqrt{196 + 4} = \sqrt{200}$$
$$d(M, Q) = \sqrt{[10 - (-2)]^2 + (-11 - 5)^2} = \sqrt{144 + 256} = \sqrt{400} = 20$$
$$d(N, Q) = \sqrt{(10 - 12)^2 + (-11 - 3)^2} = \sqrt{4 + 196} = \sqrt{200}$$

By these results,

$$[d(M, Q)]^2 = [d(M, N)]^2 + [d(N, Q)]^2,$$

proving that the triangle is a right triangle with hypotenuse connecting M and Q. ∎

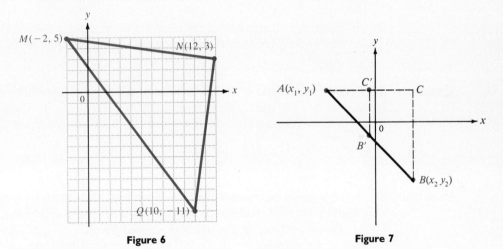

Figure 6   Figure 7

Midpoint Formula   The distance formula is used to find the distance between any two points in a plane, while the **midpoint formula** is used to find the coordinates of the midpoint of a line segment.

To develop this formula, let $A(x_1, y_1)$ and $B(x_2, y_2)$ be two different points in a plane (see Figure 7). Assume that A and B are not on a horizontal or vertical line. Let C be the intersection of the horizontal line through A and the vertical line through B. Let B' (read "B-prime") be the midpoint of segment AB. Draw a line through B' and parallel to segment BC. Let C' be the point where this line cuts segment AC. If the coordinates of B' are (x', y'), then C' has coordinates (x', y_1). Since B' is the midpoint of AB, point C' must be the midpoint of segment AC (why?), and

$$d(C, C') = d(C', A),$$

or

$$|x_2 - x'| = |x' - x_1|.$$

Because $d(C, C')$ and $d(C', A)$ must be positive the only solutions for this equation are found if

$$x_2 - x' = x' - x_1,$$

or

$$x_2 + x_1 = 2x'.$$

Finally

$$x' = \frac{x_1 + x_2}{2},$$

so that the x-coordinate of the midpoint is the average of the x-coordinates of the endpoints of the segment. In a similar manner, the y-coordinate of the midpoint is $(y_1 + y_2)/2$, proving the following result.

Midpoint Formula	The coordinates of the midpoint of the line segment with endpoints having coordinates (x_1, y_1) and (x_2, y_2) is $$\left(\frac{x_1 + x_2}{2}, \frac{y_1 + y_2}{2}\right).$$

In words: the coordinates of the midpoint of a segment are found by finding the *average* of the x-coordinates and the *average* of the y-coordinates of the endpoints of the segment.

Example 4   Find the coordinates of the midpoint M of the segment with endpoints having coordinates $(8, -4)$ and $(-9, 6)$.

Use the midpoint formula to find that the coordinates of M are

$$\left(\frac{8 + (-9)}{2}, \frac{-4 + 6}{2}\right) = \left(-\frac{1}{2}, 1\right). \quad\blacksquare$$

Example 5   A line segment has an endpoint with coordinates $(2, -8)$ and a midpoint with coordinates $(-1, -3)$. Find the coordinates of the other endpoint of the segment.

The x-coordinate of the midpoint is found from $(x_1 + x_2)/2$. Here, the x-coordinate of the midpoint is -1. To find x_2, use this formula, with $x_1 = 2$, getting

$$-1 = \frac{2 + x_2}{2}$$

$$-2 = 2 + x_2$$

$$-4 = x_2.$$

In the same way, $y_2 = 2$; the endpoint has coordinates $(-4, 2)$. $\quad\blacksquare$

The two formulas derived in this section can be used to prove various results from geometry.

Example 6   Prove that the diagonals of a parallelogram bisect each other.

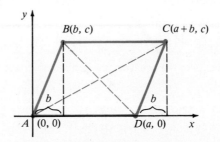

Figure 8

Figure 8 shows parallelogram $ABCD$ having diagonals AC and BD. The figure has been placed on a coordinate system with A at the origin and side AD along the x-axis. Assign coordinates (b, c) to B and $(a, 0)$ to D. Since DC is parallel to AB and is the same length as AB, use results from congruent triangles of geometry and write the coordinates of C as $(a + b, c)$. Show that the diagonals bisect each other by showing that they have the same midpoint. By the midpoint formula,

$$\text{midpoint of } AC = \left(\frac{a + b + 0}{2}, \frac{c + 0}{2}\right) = \left(\frac{a + b}{2}, \frac{c}{2}\right)$$

$$\text{midpoint of } BC = \left(\frac{a + b}{2}, \frac{c + 0}{2}\right) = \left(\frac{a + b}{2}, \frac{c}{2}\right).$$

The midpoints are the same, so AC and BD must bisect each other.   ■

3.1 Exercises

Plot the following points in the xy-plane. Identify the quadrant for each.

1. $A(6, -5)$ **2.** $B(8, 3)$ **3.** $C(-4, 7)$

4. $D(-9, -8)$ **5.** $E(0, -5)$ **6.** $F(-8, 0)$

Graph the set of all points satisfying the following conditions for ordered pairs (x, y).

7. $x = 0$ **8.** $x > 0$ **9.** $y \leq 0$ **10.** $y = 0$

11. $xy < 0$ **12.** $\dfrac{x}{y} > 0$ **13.** $|x| = 4, y \geq 2$ **14.** $|y| = 3, x \geq 4$

15. $|y| < 2, x > 1$ **16.** $|x| < 3, y < -2$ **17.** $2 \leq |x| \leq 3, y \geq 2$ **18.** $1 \leq |y| \leq 4, x < 3$

Find the distance $d(P, Q)$ and the coordinates of the midpoint of segment PQ.

19. $P(5, 7), Q(13, -1)$ **20.** $P(-2, 5), Q(4, -3)$

21. $P(-8, -2), Q(-3, -5)$ **22.** $P(-6, -10), Q(6, 5)$

23. $P(\sqrt{2}, -\sqrt{5}), Q(3\sqrt{2}, 4\sqrt{5})$ **24.** $P(5\sqrt{7}, -\sqrt{3}), Q(-7, 8\sqrt{3})$

Give the distance between the following points rounded to the nearest thousandth.

25. $(5, 7), (2, 14)$ **26.** $(-4, 6), (8, -5)$ **27.** $(3, -7), (-5, 19)$ **28.** $(-9, -2), (-1, -15)$

Find the coordinates of the other endpoint of the segments with endpoints and midpoints having coordinates as given.

29. endpoint $(-3, 6)$, midpoint $(5, 8)$

30. endpoint $(2, -8)$, midpoint $(3, -5)$

31. endpoint $(6, -1)$, midpoint $(-2, 5)$

32. endpoint $(-5, 3)$, midpoint $(-7, 6)$

Decide whether or not the following points are the vertices of a right triangle.

33. $(-2, 5)$, $(1, 5)$, $(1, 9)$

34. $(-9, -2)$, $(-1, -2)$, $(-9, 11)$

35. $(-4, 0)$, $(1, 3)$, $(-6, -2)$

36. $(-8, 2)$, $(5, -7)$, $(3, -9)$

37. $(\sqrt{3}, 2\sqrt{3} + 3)$, $(\sqrt{3} + 4, -\sqrt{3} + 3)$, $(2\sqrt{3}, 2\sqrt{3} + 4)$

38. $(4 - \sqrt{3}, -2\sqrt{3})$, $(2 - \sqrt{3}, -\sqrt{3})$, $(3 - \sqrt{3}, -2\sqrt{3})$

Use the distance formula to decide whether or not the following points lie on a straight line.

39. $(0, 7)$, $(3, -5)$, $(-2, 15)$

40. $(1, -4)$, $(2, 1)$, $(-1, -14)$

41. $(0, -9)$, $(3, 7)$, $(-2, -19)$

42. $(1, 3)$, $(5, -12)$, $(-1, 11)$

Find all values of x or y such that the distance between the given points is as indicated.

43. $(x, 7)$ and $(2, 3)$ is 5

44. $(5, y)$ and $(8, -1)$ is 5

45. $(3, y)$ and $(-2, 9)$ is 12

46. $(x, 11)$ and $(5, -4)$ is 17

47. (x, x) and $(2x, 0)$ is 4

48. (y, y) and $(0, 4y)$ is 6

49. Show that the points $(-2, 2)$, $(13, 10)$, $(21, -5)$, and $(6, -13)$ are the vertices of a square.

50. Are the points $A(1, 1)$, $B(5, 2)$, $C(3, 4)$, $D(-1, 3)$ the vertices of a parallelogram? Of a rhombus (all sides equal in length)?

51. Use the distance formula and write an equation for all points that are 5 units from $(0, 0)$. Sketch a graph showing these points.

52. Write an equation for all points 3 units from $(-5, 6)$. Sketch a graph showing these points.

53. Find all points (x, y) with $x = y$ that are 4 units from $(1, 3)$.

54. Find all points satisfying $x + y = 0$ that are 8 units from $(-2, 3)$.

55. Write an equation for the points on the perpendicular bisector of the line segment with endpoints at $(0, 0)$ and $(-8, -10)$.

56. Let point A be $(-3, 0)$ and point B be $(3, 0)$. Write an expression for all points (x, y) such that the sum of the distances from A to (x, y) and from (x, y) to B is 8. Simplify the result so that no radicals are involved.

57. Let a be a positive number. Show that the distance between the points (ax_1, ay_1) and (ax_2, ay_2) is a times the distance between (x_1, y_1) and (x_2, y_2).

Use the midpoint formula and distance formula, as necessary, to prove each of the following.

58. The midpoint of the hypotenuse of a right triangle is equally distant from all three vertices.

59. The diagonals of a rectangle are equal in length.

60. The line segment connecting the midpoints of two adjacent sides of any quadrilateral is the same length as the line segment connecting the midpoints of the other two sides.

61. The diagonals of an isosceles trapezoid are equal.

62. If the diagonals of a parallelogram are equal in length, then the parallelogram is a rectangle.

63. Find the length of the hypotenuse in each right triangle of the figure.

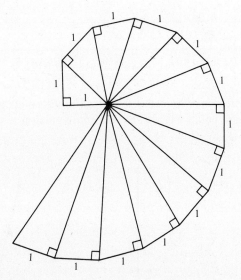

Find the value of x in each right triangle.

64.

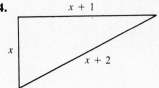

$x + 1$

x

$x + 2$

65.

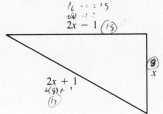

$2x - 1$

$2x + 1$

66.

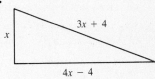

$3x + 4$

x

$4x - 4$

3.2 Graphs

For any set of ordered pairs of real numbers, there is a corresponding set of points in the coordinate plane. For each ordered pair (x, y) in the set, there is a corresponding point (x, y) in the coordinate plane. The set of all points in the plane corresponding to the set of ordered pairs is the **graph** of the set of ordered pairs. For now, we can find graphs only by identifying a reasonable number of ordered pairs. The points corresponding to these ordered pairs are then located, and used to try to decide on the shape of the entire graph. Later, more useful methods of identifying particular graphs are developed.

Example 1   Draw the graph of $S = \{(x, y)|y = -4x + 3\}$.

Use the equation $y = -4x + 3$ to obtain several sample ordered pairs belonging to S. Find these ordered pairs by selecting a number of values for x (or y) and then finding the corresponding values for the other variable. For example, if $x = -3$, then $y = -4(-3) + 3 = 15$, producing the ordered pair $(-3, 15)$. Additional ordered pairs found in this way are given in the following table.

x	-3	-2	-1	0	1	2	3
y	15	11	7	3	-1	-5	-9
ordered pair	$(-3, 15)$	$(-2, 11)$	$(-1, 7)$	$(0, 3)$	$(1, -1)$	$(2, -5)$	$(3, -9)$

The ordered pairs from this table lead to the points that have been plotted in Figure 9(a). These points suggest that the entire graph is a straight line, as drawn in Figure 9(b). ■

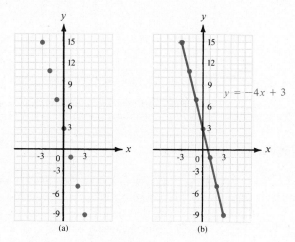

(a)   (b)

Figure 9

The set of ordered pairs in Example 1, $S = \{(x, y)|y = -4x + 3\}$ involves the equation $y = -4x + 3$. For each value of x that might be chosen, this equation can be used to find a corresponding value of y. By using all possible such values of x, the equation $y = -4x + 3$ leads to a set of ordered pairs (a, b) such that $b = -4a + 3$. The ordered pairs (a, b) are **solutions** of the equation $y = -4x + 3$. The set of all these solutions has a graph, the **graph of the equation.** Using this definition, the graph of the equation $y = -4x + 3$ is the same as the graph of the set $\{(x, y)|y = -4x + 3\}$.

Example 2 Graph the equation $y = x^2 + 2$.

As in the previous example, choose several values of x and find the corresponding values of y.

x	-4	-3	-2	-1	0	1	2	3	4
y	18	11	6	3	2	3	6	11	18

The ordered pairs obtained from this table were plotted in Figure 10(a), and a smooth curve was then drawn through the points as in Figure 10(b). This graph, called a **parabola,** will be studied in more detail later. The lowest point on this parabola, the point (0, 2), is called the **vertex** of the parabola. In $y = x^2 + 2$, any value at all may be chosen for x. However, any real value of x makes $x^2 \geq 0$, so that $x^2 + 2 \geq 2$. Since $y = x^2 + 2$, the value of y will always be at least equal to 2, or $y \geq 2$. ■

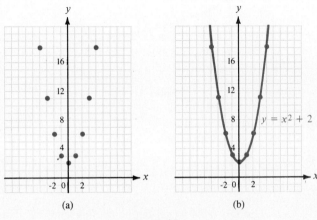

(a)   (b)

Figure 10

There is a danger in the method used in Examples 1 and 2—we might choose a few values for x, find the corresponding values of y, begin to sketch a graph through these few points, and then make a completely wrong guess as to the shape of the graph. For example, choosing only -1, 0, and 1 as values of x in Example 2 above would produce only the three points $(-1, 3)$, $(0, 2)$, and $(1, 3)$. These three points would not be nearly enough to give sufficient information to determine the proper graph for $y = x^2 + 2$. However, this section involves only elementary graphs, and when more complicated graphs are presented later, more accurate methods of working with them will be developed.

Example 3   Graph $x = y^2$.

Since y is squared, it is probably easier to choose values of y and then to find the corresponding values of x. For example, choosing the value 2 for y gives $x = 2^2 = 4$. Choosing -2 for y gives $x = (-2)^2 = 4$, the same result. The following table shows the values of x corresponding to various values of y.

y	0	1	-1	2	-2	3	-3
x	0	1	1	4	4	9	9

The ordered pairs from this table were used to get the points plotted in Figure 11. (Don't forget that x always goes first in the ordered pair.) A smooth curve was then drawn through the resulting points. This curve is a parabola with vertex (0, 0), opening to the right. Here, y can take on any value. Since $x = y^2$, then $x \geq 0$. ■

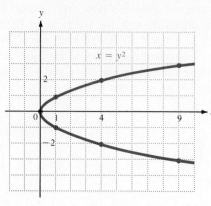

Figure 11

Figure 12

Example 4 Graph $y = |x|$.

Start with a table.

x	-4	-3	-2	-1	0	1	2	3	4
y	4	3	2	1	0	1	2	3	4

Use this table to get the points in Figure 12. The graph drawn through these points is made up of portions of two straight lines. Here x may represent any real number, while $y \geq 0$. ∎

Example 5 Graph $xy = -4$.

Make a table of values. Since this graph is more complicated than the ones above, it is a good idea to use more points. Also, neither x nor y can be equal to 0 so it is a good idea to make several choices for x that are close to 0.

x	-8	-4	-2	-1	$-\dfrac{1}{2}$	$-\dfrac{1}{4}$	$-\dfrac{1}{16}$	$-\dfrac{1}{64}$
y	$\dfrac{1}{2}$	1	2	4	8	16	64	256

x	$\dfrac{1}{64}$	$\dfrac{1}{16}$	$\dfrac{1}{4}$	$\dfrac{1}{2}$	1	2	4	8
y	-256	-64	-16	-8	-4	-2	-1	$-\dfrac{1}{2}$

As x approaches 0 from the left, y gets larger and larger. As x approaches 0 from the right, y gets more and more negative. If $x = 0$, there is no value of y; therefore the graph cannot cross the y-axis. Also, if $y = 0$, there is no value of x, so the graph cannot cross the x-axis either. The variables x and y can take on any value except 0; here $x \neq 0$ and $y \neq 0$. The table was used to get enough points to draw the curve in Figure 13. ∎

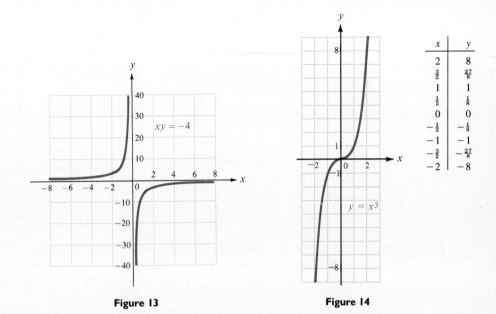

x	y
2	8
$\frac{3}{2}$	$\frac{27}{8}$
1	1
$\frac{1}{2}$	$\frac{1}{8}$
0	0
$-\frac{1}{2}$	$-\frac{1}{8}$
-1	-1
$-\frac{3}{2}$	$-\frac{27}{8}$
-2	-8

Figure 13　　　　　　　　**Figure 14**

Example 6　Graph $y = x^3$.

A table of values and the graph are shown in Figure 14. Both the variables x and y can take on any values at all.　■

Circles　A **circle** is the set of all points in a plane which lie a given distance from a given point. The given distance is the **radius** of the circle and the given point is the **center.** The equation of a circle can be found by using the distance formula discussed in Section 1 of this chapter.

For example, Figure 15 shows a circle of radius 3 with center at the origin. To find the equation of this circle, let (x, y) be any point on the circle. The distance between (x, y) and the center of the circle, $(0, 0)$, is given by

$$\sqrt{(x - 0)^2 + (y - 0)^2}.$$

Since this distance equals the radius, 3,

$$\sqrt{(x - 0)^2 + (y - 0)^2} = 3$$
$$\sqrt{x^2 + y^2} = 3$$
$$x^2 + y^2 = 9.$$

As suggested by the graph, the possible values of x are $-3 \le x \le 3$, while the possible values of y are $-3 \le y \le 3$.

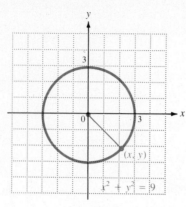

Figure 15   Figure 16

Example 7   Find an equation for the circle having radius 6 and center at $(-3, 4)$.

This circle is shown in Figure 16. Its equation can be found by using the distance formula. Start by letting (x, y) be any point on the circle. The distance from (x, y) to $(-3, 4)$ is given by

$$\sqrt{[x - (-3)]^2 + (y - 4)^2} = \sqrt{(x + 3)^2 + (y - 4)^2}.$$

This same distance is given by the radius, 6. Therefore,

$$\sqrt{(x + 3)^2 + (y - 4)^2} = 6$$

or
$$(x + 3)^2 + (y - 4)^2 = 36.$$

The graph in Figure 16 suggests that the possible values of x are $-9 \leq x \leq 3$ while the possible values of y are $-2 \leq y \leq 10$. ∎

Generalizing the work of Example 7 to a circle of radius r and center (h, k) would give the following result.

Center-Radius Form of the Equation of a Circle

> The circle with center (h, k) and radius r has equation
>
> $$(x - h)^2 + (y - k)^2 = r^2,$$
>
> the **center-radius form** of the equation of a circle. As a special case,
>
> $$x^2 + y^2 = r^2$$
>
> is the equation of a circle with radius r and center at the origin.

Starting with the center-radius form of the equation of a circle, $(x - h)^2 + (y - k)^2 = r^2$, and squaring $x - h$ and $y - k$ gives a result of the form

$$x^2 + y^2 + Cx + Dy + E = 0, \qquad (*)$$

where C, D, and E are real numbers, the **general form** of the equation of a circle.

Also, starting with an equation similar to (∗), the process of *completing the square* discussed in Section 2.3 can be used to get an equation of the form

$$(x - h)^2 + (y - k)^2 = m$$

for some number m. If $m > 0$, then $r^2 = m$, and the graph is that of a circle with radius $\sqrt{m}$. If $m = 0$, the graph is the single point (h, k), while there is no graph if $m < 0$.

Example 8 Is $x^2 - 6x + y^2 + 10y + 25 = 0$ the equation of a circle?

Since this equation has the form of equation (∗) above, it either represents a circle, a single point, or no points at all. To decide which, complete the square on x and y separately, as explained in Section 2.3. Start with

$$(x^2 - 6x \quad) + (y^2 + 10y \quad) = -25.$$

Half of -6 is -3, and $(-3)^2 = 9$. Also, half of 10 is 5, and $5^2 = 25$. Add 9 and 25 on the left, and to compensate, add 9 and 25 on the right:

$$(x^2 - 6x + 9) + (y^2 + 10y + 25) = -25 + 9 + 25$$
$$(x - 3)^2 + (y + 5)^2 = 9.$$

Since $9 > 0$, the equation represents a circle with center at $(3, -5)$ and radius 3. ■

Example 9 Is $x^2 + 10x + y^2 - 4y + 33 = 0$ the equation of a circle?

Completing the square as above gives

$$(x^2 + 10x + 25) + (y^2 - 4y + 4) = -33 + 25 + 4$$
$$(x + 5)^2 + (y - 2)^2 = -4.$$

Since $-4 < 0$, there are no ordered pairs (x, y), with x and y real numbers, satisfying the equation. The graph of the given equation would contain no points. ■

3.2 Exercises

Graph each of the following.

1. $y = 8x - 3$ 2. $y = 2x + 7$ 3. $y = 3x$ 4. $y = -2x$

5. $3y + 4x = 12$ 6. $5y - 3x = 15$ 7. $y = 3x^2$ 8. $y = 5x^2$

9. $y = -x^2$ 10. $y = -2x^2$ 11. $y = x^2 - 8$ 12. $y = x^2 + 6$

13. $y = 4 - x^2$ 14. $y = -2 - x^2$ 15. $xy = -9$ 16. $xy = 25$

17. $4x = y^2$ 18. $9x = y^2$ 19. $16y^2 = -x$ 20. $4y^2 = -x$

21. $y^2 = x + 2$ 22. $y^2 = -5 + x$ 23. $y = x^3 - 3$ 24. $y = x^3 + 4$

25. $y = 1 - x^3$ 26. $y = -5 - x^3$ 27. $2y = x^4$ 28. $y = -x^4$

29. $y = |x| + 4$ 30. $y = |x| - 3$ 31. $y = |x| - 2$ 32. $y = -|x| + 1$

33. $y = 3 - |x|$ 34. $y = -4 + |x|$ 35. $y = |x + 3|$ 36. $y = -|x - 2|$

37. $x^2 + y^2 = 36$ 38. $x^2 + y^2 = 81$ 39. $(x - 2)^2 + y^2 = 36$

40. $x^2 + (y + 3)^2 = 49$

41. $(x - 4)^2 + (y + 3)^2 = 4$

42. $(x + 3)^2 + (y - 2)^2 = 16$

43. $(x + 2)^2 + (y - 5)^2 = 12$

44. $(x - 4)^2 + (y - 3)^2 = 8$

Find equations for each of the following circles.

45. center $(1, 4)$, radius 3

46. center $(-2, 5)$, radius 4

47. center $(-8, 6)$, radius 5

48. center $(3, -2)$, radius 2

49. center $(-1, 2)$, passing through $(2, 6)$

50. center $(2, -7)$, passing through $(-2, -4)$

51. center $(-3, -2)$, tangent to the x-axis

52. center $(5, -1)$, tangent to the y-axis

For each of the following that are equations of circles, give the center and radius of the circle.

53. $x^2 + 6x + y^2 + 8y = -9$

54. $x^2 - 4x + y^2 + 12y + 4 = 0$

55. $x^2 - 12x + y^2 + 10y + 25 = 0$

56. $x^2 + 8x + y^2 - 6y = -16$

57. $x^2 + 8x + y^2 - 14y + 65 = 0$

58. $x^2 - 2x + y^2 + 1 = 0$

59. $x^2 + y^2 = 2y + 48$

60. $x^2 + 4x + y^2 = 21$

61. $x^2 - 2.84x + y^2 + 1.4y + 1.8664 = 0$

62. $x^2 + 7.4x + y^2 - 3.8y + 16.09 = 0$

63. Find the equation of the circle having the points $(3, -5)$ and $(-7, 2)$ as endpoints of a diameter.

64. Suppose a circle is tangent to both axes, has its center in the third quadrant, and has a radius of $\sqrt{2}$. Find an equation for the circle.

65. One circle has center at $(3, 4)$ and radius 5. A second circle has center at $(-1, -3)$ and radius 4. Do the circles cross?

66. Does the circle with radius 6 and center at $(0, 5)$ cross the circle with center at $(-5, -4)$ with radius 4?

Use the appropriate formula to find the circumference and area of each of the circles having equations as follows.

67. $(x - 2)^2 + (y + 4)^2 = 25$

68. $(x + 3)^2 + (y - 1)^2 = 9$

69. $x^2 + 2x + y^2 - 4y + 1 = 0$

70. $x^2 - 6x + y^2 + 8y = -16$

The **unit circle** is a circle centered at the origin and having radius 1. Show that each of the following points lies on the unit circle. Only an approximation is possible in Exercises 75 and 76.

71. $(-1/2, \sqrt{3}/2)$

72. $(-\sqrt{2}/2, -\sqrt{2}/2)$

73. $(-\sqrt{39}/8, 5/8)$

74. $(11/13, 4\sqrt{3}/13)$

75. $(-.47691285, .87895059)$

76. $(.21514257, -.97658265)$

77. Decide if each of the following points is *inside, on,* or *outside* the circle with center $(1, -4)$ and radius 6:
(a) $(3, -2)$ (b) $(9, 1)$ (c) $(7, -4)$ (d) $(0, 9)$.

78. Find any points of intersection of the circles having equations $x^2 - 4x + y^2 - 10y = -4$ and $x^2 + 8x + y^2 - 10y = -40$, given that $x = -3$ for one point of intersection.

3.3 Symmetry and Translations

The graph of $y = x^2 + 2$, shown in Figure 10 in the previous section, could be cut in half by the y-axis with each half the mirror image of the other half. A graph with this property is said to be **symmetric with respect to the y-axis.** As we shall see in this section, symmetry is helpful in drawing graphs.

Figure 17(a) shows another graph which is symmetric to the y-axis. As this graph suggests, a graph is symmetric with respect to the y-axis if the point $(-x, y)$ is on the graph whenever (x, y) is on the graph.

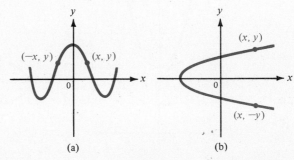

Figure 17

If the graph in Figure 17(b) were folded in half along the x-axis, the portion at the top would exactly match the portion at the bottom. Such a graph is **symmetric with respect to the x-axis:** the point $(x, -y)$ is on the graph whenever the point (x, y) is on the graph.

The following test tells when a graph is symmetric with respect to the x-axis or y-axis.

Symmetry with Respect to an Axis

> The graph of an equation is **symmetric with respect to the y-axis** if the replacement of x with $-x$ results in an equivalent equation.
>
> The graph of an equation is **symmetric with respect to the x-axis** if the replacement of y with $-y$ results in an equivalent equation.

Example 1 Test for symmetry with respect to the x-axis or y-axis.

(a) $y = x^2 + 4$

Replace x with $-x$:

$$y = x^2 + 4 \text{ becomes } y = (-x)^2 + 4 = x^2 + 4.$$

The result is the same as the original equation so that the graph, shown in Figure 18 on the next page, is symmetric with respect to the y-axis. Check that the graph is *not* symmetric with respect to the x-axis.

(b) $x = y^2 - 3$

Replace y with $-y$ to get $x = (-y)^2 - 3 = y^2 - 3$, the same as the original equation. The graph is symmetric with respect to the x-axis as shown in Figure 19. Is the graph symmetric with respect to the y-axis?

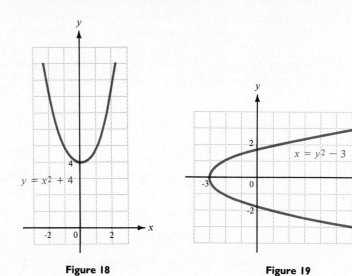

$y = x^2 + 4$

$x = y^2 - 3$

Figure 18   **Figure 19**

(c) $x^2 + y^2 = 16$

First replace x with $-x$ to see that the graph is symmetric with respect to the y-axis. Then replace y with $-y$, to see that the graph, a circle of radius 4 centered at the origin, is also symmetric with respect to the x-axis.

(d) $2x + y = 4$

Replace x with $-x$ and y with $-y$; in neither case does an equivalent equation result. This graph is symmetric with respect to neither the x-axis nor the y-axis. ■

Another kind of symmetry is found when a graph can be rotated 180° about the origin, with the result coinciding exactly with the original graph. Symmetry of this type is called **symmetry with respect to the origin.** It turns out that rotating a graph 180° is equivalent to saying that the point $(-x, -y)$ is on the graph whenever (x, y) is on the graph. Figure 20 shows two graphs that are symmetric with respect to the origin.

Symmetry with Respect to the Origin	The graph of an equation is **symmetric with respect to the origin** if the replacement of both x with $-x$ and y with $-y$ results in an equivalent equation.

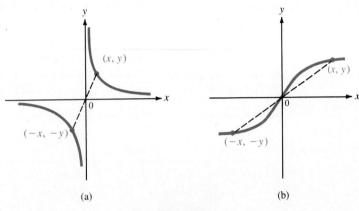

Figure 20

Example 2   Are the following graphs symmetric with respect to the origin?

(a) $x^2 + y^2 = 16$

Replace x with $-x$ and y with $-y$ to get

$$(-x)^2 + (-y)^2 = 16 \qquad \text{or} \qquad x^2 + y^2 = 16,$$

an equivalent equation. This graph, shown in Figure 21, is symmetric with respect to the origin.

(b) $y = x^3$

Replace x with $-x$ and y with $-y$ to get

$$-y = (-x)^3, \qquad \text{or} \qquad -y = -x^3, \qquad \text{or} \qquad y = x^3,$$

an equivalent equation. The graph, symmetric with respect to the origin, is shown in Figure 22.   ∎

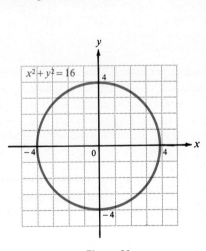

Figure 21

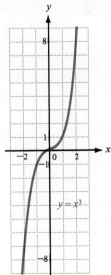

Figure 22

A graph symmetric with respect to both the x- and y-axes is automatically symmetric with respect to the origin. (See Exercise 47.) However, a graph symmetric to the origin need not be symmetric to either axis. (See Figure 22 above.) Of the three types of symmetry—with respect to the x-axis, the y-axis, and the origin—a graph possessing any two must have the third type also. (See Exercise 48.)

The various tests for symmetry are summarized below.

Tests for Symmetry

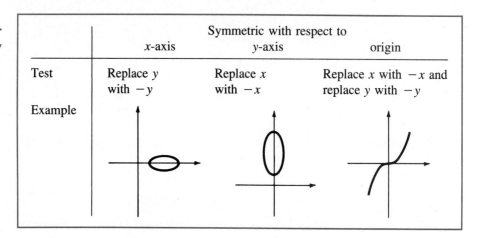

	Symmetric with respect to		
	x-axis	y-axis	origin
Test	Replace y with $-y$	Replace x with $-x$	Replace x with $-x$ and replace y with $-y$
Example			

Translations   The graph of $y = x^2$ was found earlier by choosing several values of x and finding the corresponding values for y. See Figure 23. As mentioned earlier, the graph of $y = x^2$ is called a parabola. Once this graph has been found, how could the graph of $y = x^2 - 4$ be found? For each point (x, y) on the graph of $y = x^2$, there will be a corresponding point $(x, y - 4)$ on the graph of $y = x^2 - 4$, so that each point on the graph of $y = x^2$ is moved, or *translated,* 4 units downward to get the graph of $y = x^2 - 4$. (See Figure 23 again.) Such a vertical shift is called a **vertical translation.** As an example, Figure 24 shows a graph along with two different vertical translations of it.

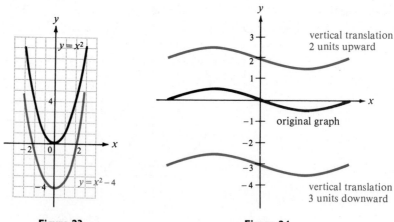

Figure 23

Figure 24

Figure 25 shows the graph of $y = x^2$ along with the graph of $y = (x - 3)^2$; this second graph is obtained from that of $y = x^2$ by a **horizontal translation.** Figure 26 shows a graph along with horizontal translations of the graph.

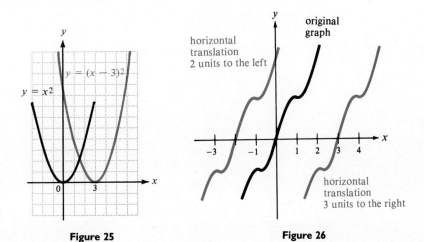

Figure 25   **Figure 26**

The following statements tell how to identify a translation.

Translations

Replacing y in an equation by $y - a$, where a is a constant, produces a **vertical translation** of the graph of $|a|$ units. The translation is upward if $a > 0$ and downward if $a < 0$.

Replacing x in an equation by $x - b$, where b is a constant, produces a **horizontal translation** of the graph of $|b|$ units. The translation is to the right if $b > 0$ and to the left if $b < 0$.

Example 3 Graph each of the following.

(a) $y = |x - 4|$.

Here x was replaced with $x - 4$ in the equation $y = |x|$. This results in a shift (or horizontal translation) 4 units to the right. See Figure 27.

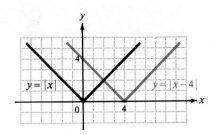

Figure 27

(b) $y = |x| - 5$

This equation can be written as the equivalent equation $y + 5 = |x|$. Here y was replaced with $y + 5$ in the equation $y = |x|$. Since $y + 5 = y - (-5)$, this replacement produces a shift, or vertical translation, of 5 units downward. See Figure 28.

(c) $y = |x + 3| + 2$

Rewrite $y = |x + 3| + 2$ as $y - 2 = |x + 3|$, to see that the graph involves both a horizontal translation of 3 units to the left and a vertical translation of 2 units upward. See Figure 29. ■

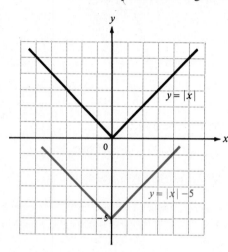

Figure 28   Figure 29

3.3 Exercises

Plot the following points, and then plot the points that are symmetric to the given point with respect to the (a) x-axis, (b) y-axis, (c) origin.

1. $(5, -3)$ **2.** $(-6, 1)$ **3.** $(-4, -2)$

4. $(-8, 3)$ **5.** $(-8, 0)$ **6.** $(0, -3)$

Use symmetry, point plotting, and translation to help graph each of the following.

7. $x^2 + y^2 = 5$ **8.** $y^2 = 4 - x^2$ **9.** $y = 4 - x^2$ **10.** $y = x^2 - 2$

11. $y = x^2 - 8x$ **12.** $y = 4x - x^2$ **13.** $x^2 + 2y^2 = 10$ **14.** $y^2 + x^2 = 8$

15. $3x^2 - y^2 = 8$ **16.** $5y^2 - x^2 = 6$ **17.** $y = 3|x|$ **18.** $y = -2|x|$

19. $|y| = 2x$ **20.** $-3|y| = 4x$ **21.** $|y| = |x + 2|$ **22.** $|x| = |y - 1|$

23. $xy = 4$ **24.** $xy = -6$ **25.** $\dfrac{x}{y} = -1$ **26.** $\dfrac{x}{y} = 2$

27. $y = x^3$ **28.** $y = -x^3$ **29.** $y = \dfrac{1}{1 + x^2}$ **30.** $y = \dfrac{-1}{x^2 + 9}$

31. $(x + 1)^2 + (y - 3)^2 = 9$ **32.** $(y + 2)^2 = 25 - (x - 3)^2$

A graph of $y = x^2$ is shown in the figure. Explain how each of the following graphs could be obtained from the graph shown. Sketch each graph.

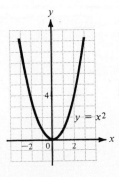

33. $y = x^2 + 2$ **34.** $y = x^2 - 7$ **35.** $y = -x^2$ **36.** $y = -x^2 + 8$

37. $y = -(x - 4)^2$ **38.** $y = -(x + 5)^2$ **39.** $y = (x - 2)^2 - 3$ **40.** $y = (x + 1)^2 + 5$

41. $y = -(x + 4)^2 + 2$ **42.** $y = -(x - 3)^2 - 1$

In Exercises 43 and 44, let F be some algebraic expression involving x as the only variable.

43. Suppose the equation $y = F$ is changed to $y = -F$. What is the relationship between the graphs of the two equations?

44. Suppose the equation $y = F$ is changed to $y = c \cdot F$, for some constant c. What is the effect on the graph of $y = F$? Discuss the effect depending on whether $c > 0$ or $c < 0$, and $|c| > 1$ or $|c| < 1$.

Sketch examples of graphs which are as follows.

45. Symmetric to the x-axis but not the y-axis

46. Symmetric to the origin but to neither the x-axis nor the y-axis

Prove each statement below.

47. A graph symmetric with respect to both the x-axis and y-axis is also symmetric to the origin.

48. A graph possessing two of the three types of symmetry, with respect to the x-axis, y-axis, and origin, must possess the third type of symmetry also.

3.4 Functions

Suppose X is the set of all students studying this book every Monday evening at the local pizza parlor. Let Y be the set of integers between 0 and 100. To each student in set X can be associated a number from Y which represents the score the student received on the last test. Typical associations between students in set X and scores in set Y are shown in Figure 30.

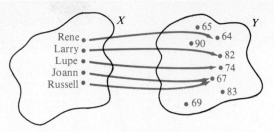

Figure 30

A correspondence such as the one shown in Figure 30 is called a **function** or a **mapping,** defined as follows.

Function

A **function** from a set X to a set Y is a correspondence that assigns to each element of X exactly one element of Y.

The set X in this definition is called the **domain** of the function.

Three things should be noticed about the function shown in Figure 30 above.

First, there is a single score associated with each student. That is, each element in X corresponds to exactly one element in Y.

Second, the same score may correspond to more than one student. In the example above, two students (Joann and Russell) have a score of 67.

Third, not every element of Y need be used; for example, none of the students had a score of 83 or 65.

While the sets X and Y above were different, in many cases X and Y have many elements in common. In fact, it is not unusual for X and Y to be equal.

Example 1   Decide whether or not the following diagrams represent functions from X to Y.

(a)

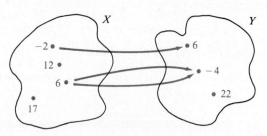

Figure 31

The diagram in Figure 31 does not represent a function from X to Y: there is no arrow leading from 17 in the domain, so that there is no element in Y corresponding to the element 17 in X.

(b)

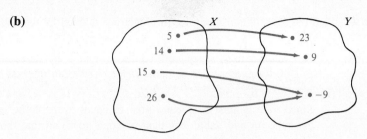

Figure 32

This diagram (Figure 32) does represent a function from X to Y since there is a single element in Y corresponding to each element in X.

(c)

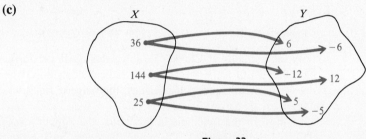

Figure 33

Since each element of set X corresponds to *two* elements in set Y, the diagram in Figure 33 does not represent a function from X to Y. ■

It is common to use the letters f, g, and h to name functions. If f is a function and x is an element in the domain X, then $f(x)$ represents the element in Y that corresponds to x in X. For example, if f is used to name the function in Figure 32 above, then

$$f(5) = 23, \quad f(14) = 9, \quad f(15) = -9, \quad \text{and} \quad f(26) = -9.$$

For a given element x in set X, the corresponding element $f(x)$ in set Y is called the **value** or **image** of f at x. The set of all possible values of $f(x)$ makes up the **range** of the function. Throughout this book, if the domain for a function specified by a formula is not given, it will be assumed to be the largest possible set of real numbers for which the formula is meaningful, unless otherwise specified. For example, suppose function f is defined by

$$f(x) = \frac{-4x}{2x - 3}.$$

With this formula, any real number can be used for x except $x = 3/2$, which makes the denominator equal 0. Assuming that the domain is the largest possible set of real numbers for which the formula is meaningful makes the domain $\{x \mid x \neq 3/2\}$.

Example 2   Find the domain and range for the functions defined by the following rules.

(a) $f(x) = x^2$

Any number may be squared, so the domain is the set of all real numbers. Since $x^2 \geq 0$ for every value of x, the range, written in interval notation, is $[0, +\infty)$.

(b) $f(x) = \sqrt{6 - x}$

Since $6 - x$ must be greater than or equal to 0 for the square root to exist, x can take on any value less than or equal to 6, so that the domain is $(-\infty, 6]$. The radical indicates the nonnegative square root, so the interval $[0, +\infty)$ is the range.

(c) $f(x) = \sqrt{x^2 + x - 6}$

The domain includes those values of x that make

$$x^2 + x - 6 \geq 0.$$

Factor to get

$$(x + 3)(x - 2) \geq 0.$$

By the methods of Chapter 2, solve this quadratic inequality to get the domain $(-\infty, -3] \cup [2, +\infty)$. Since the radical exists only when the radicand, $x^2 + x - 6$, takes on nonnegative values, the range is $[0, +\infty)$.   ■

For most of the functions in this book, the domain can be found with algebraic methods already discussed. The range, however, must often be found with graphing (see Figure 35 below), complicated algebra, or calculus.

Based on the definition of a function, functions f and g are **equal** if and only if f and g have exactly the same domains and $f(x) = g(x)$ for every value of x in the domain.

Suppose a function is defined by $f(x) = -3x + 2$. To emphasize that this statement is used to find values in the range of f, it is common to write

$$y = -3x + 2.$$

When a function is written in the form $y = f(x)$, x is called the **independent variable,** and y the **dependent variable.** There is no reason to restrict the variables to x or y—different areas of study use different variables. For example, it is common to use t for time in physics, or p for price in management.

By the definition, a function f is a rule that assigns to each element of one set exactly one element of a second set. For a particular value of x in the first set, the corresponding element in the second set is written $f(x)$. There is a distinction between f and $f(x)$: f is the function or rule, while $f(x)$ is the value obtained by applying the rule to an element x. However, it is very common to abbreviate

"the function defined by the rule $y = f(x)$"

as simply

"the function $y = f(x)$."

Example 3   Let $g(x) = 3\sqrt{x}$ and $h(x) = 1 + 4x$. Find each of the following.

(a) $g(16)$

To find $g(16)$, replace x in $g(x) = 3\sqrt{x}$ with 16, getting

$$g(16) = 3\sqrt{16} = 3 \cdot 4 = 12.$$

(b) $h(-3) = 1 + 4(-3) = -11$

(c) $g(-4)$ does not exist; -4 is not in the domain of g since $\sqrt{-4}$ is not a real number

(d) $h(\pi) = 1 + 4\pi$

(e) $g(m) = 3\sqrt{m}$, if m represents a nonnegative real number

(f) $g[h(3)]$

First find $h(3)$, as follows.

$$h(3) = 1 + 4 \cdot 3 = 1 + 12 = 13$$

Now, $g[h(3)] = g(13) = 3\sqrt{13}.$ ∎

Example 4 Let $f(x) = 2x^2 - 3x$, and find the quotient

$$\frac{f(x + h) - f(x)}{h}, \quad h \neq 0,$$

which is important in calculus.

To find $f(x + h)$, replace x with $x + h$, to get

$$\frac{f(x + h) - f(x)}{h} = \frac{2(x + h)^2 - 3(x + h) - (2x^2 - 3x)}{h}$$

$$= \frac{2(x^2 + 2xh + h^2) - 3x - 3h - 2x^2 + 3x}{h}$$

$$= \frac{2x^2 + 4xh + 2h^2 - 3x - 3h - 2x^2 + 3x}{h}$$

$$= \frac{4xh + 2h^2 - 3h}{h} = 4x + 2h - 3.$$ ∎

We have been finding the graphs of various sets of ordered pairs throughout this chapter. A function can be thought of as a set of ordered pairs; in fact, a function f produces the set of ordered pairs $\{(x, f(x)) | x$ is in the domain of $f\}$. By the definition of function, for each x which appears in the first position of an ordered pair, there will be exactly one value of $f(x)$ in the second position.

In addition, a set of ordered pairs in which any two ordered pairs that have the same first entry x also have equal values for the second entry must be a function. The following *alternate definition of a function* is based on this.

Alternate Definition of a Function

> A **function** is a set of ordered pairs in which two ordered pairs cannot have the same first entries and different second entries.

The idea of a function as a set of ordered pairs can be used to define the **graph of a function:** the graph of a function f is the set of all points in the plane of the form $(x, f(x))$, where x is in the domain of f. The graph of a function f is the same as the graph of the equation $y = f(x)$.

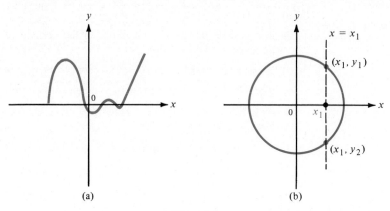

Figure 34

There is a quick way to tell if a given graph is the graph of a function or not. Figure 34 shows two graphs. In the graph of part (a), for any x that might be chosen, exactly one value of $f(x)$ or y could be found, showing that the graph is the graph of a function. On the other hand, the graph in part (b) is not the graph of a function. For example, the vertical line through x_1 leads to two different values of y, namely y_1 and y_2. This example suggests the **vertical line test** for a function.

Vertical Line Test

> If each vertical line cuts a graph in no more than one point, the graph is the graph of a function.

The graph of a function can be used to find the domain and range of the function, as shown in Figure 35.

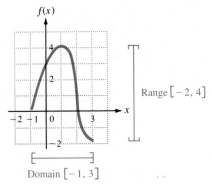

Figure 35

Odd and Even Functions As we have seen, the idea of symmetry is very useful when drawing graphs. A function whose graph is symmetric with respect to the y-axis is an *even function,* while a function having a graph symmetric to the origin is an *odd function.* (Exercises 109 and 110 call for the use of the following definition to prove this previous statement.)

Odd and Even Functions

> Suppose x and $-x$ are both in the domain of a function f. Then
>
> f is an **odd function** if $f(-x) = -f(x)$ for every x in the domain of f,
>
> f is an **even function** if $f(-x) = f(x)$ for every x in the domain of f.

Example 5   Decide if the functions defined as follows are odd, even, or neither.

(a) $f(x) = 8x^4 - 3x^2$

Replacing x with $-x$ gives

$$f(-x) = 8(-x)^4 - 3(-x)^2 = 8x^4 - 3x^2 = f(x).$$

Since $f(-x) = f(x)$ for each x in the domain of the function, f is an even function.

(b) $f(x) = 6x^3 - 9x$

Here

$$f(-x) = 6(-x)^3 - 9(-x) = -6x^3 + 9x = -f(x).$$

This function is odd.

(c) $f(x) = \dfrac{1}{x - 3}$

Replacing x with $-x$ produces

$$f(-x) = \frac{1}{-x - 3}$$

which equals neither $f(x)$ nor $-f(x)$. This function is neither odd nor even.   ■

The exponents on the terms in parts (a) and (b) of Example 5 show the origin of the names *odd* and *even*.

Increasing and Decreasing Functions   Intuitively, a function is *increasing* if its graph moves up as it goes to the right. The functions graphed in Figure 36(a) and (b) are increasing functions. On the other hand, a function is *decreasing* if its graph goes down as it goes to the right. The function graphed in Figure 36(c) is a decreasing function.

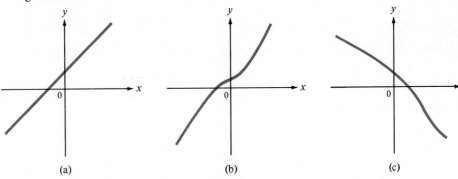

(a)   (b)   (c)

Figure 36

The function of Figure 37 is neither an increasing nor a decreasing function. However, this function is increasing on both the interval $(-\infty, -2]$ and the interval $[4, +\infty)$; also, f is decreasing on the interval $[-2, 4]$.

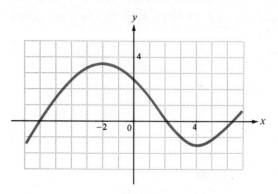

Figure 37

The idea of a function increasing and decreasing on an interval is defined below, where I represents any interval of real numbers.

Increasing and Decreasing Functions

Let f be a function, with the interval I a subset of the domain of f. Let x_1 and x_2 be in I. Then

f is **increasing** on I if $f(x_1) < f(x_2)$ whenever $x_1 < x_2$, and

f is **decreasing** on I if $f(x_1) > f(x_2)$ whenever $x_1 < x_2$.

Example 6   Find where the following functions are increasing or decreasing.

(a) The function graphed in Figure 38(a) is increasing on the interval $[-2, 6]$. It is decreasing on $(-\infty, -2]$ and $[6, +\infty)$.

(b) The function in Figure 38(b) is never increasing or decreasing. This function is an example of a **constant function,** a function defined by $f(x) = k$, for k a real number.   ■

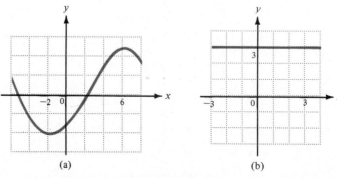

(a)   (b)

Figure 38

The work of Section 3.3 on translations also applies to the graphs of functions, as summarized below.

Translations of the Graph of a Function

Let f be a function, and let c be a positive number.

To graph:	shift the graph of $y = f(x)$ by c units:
$y = f(x) + c$	upward
$y = f(x) - c$	downward
$y = f(x + c)$	left
$y = f(x - c)$	right

Example 7   A graph of a function $y = f(x)$ is shown in Figure 39. Using this graph, find each of the following graphs.

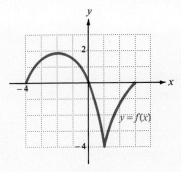

Figure 39

(a) $y = f(x) + 3$

This graph is the same as the graph in Figure 39, translated 3 units upward. See Figure 40(a).

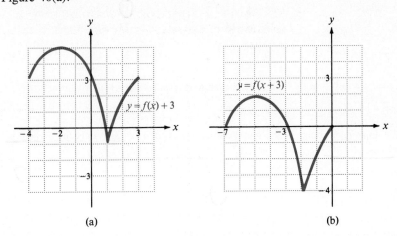

(a)   (b)

Figure 40

(b) $y = f(x + 3)$

To get the graph of $y = f(x + 3)$, translate the graph of $y = f(x)$ to the left 3 units. See Figure 40(b).   ■

3.4 Exercises

For each of the following, find the indicated function values: **(a)** $f(-2)$,   **(b)** $f(0)$,
(c) $f(1)$,   **(d)** $f(4)$.

 1.
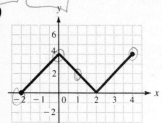

2.
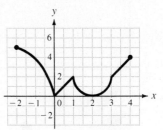

 3.
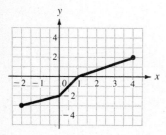

4.
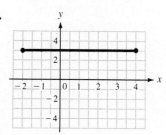

Let $f(x) = 3x - 1$ and $g(x) = |x^2 - 8|$. Find each of the following.

5. $f(0)$ **6.** $f(-1)$ **7.** $f(-3) + 2$ **8.** $f(4) - 5$

9. $g(2)$ **10.** $g(0)$ **11.** $f(a^2)$ **12.** $g(b^3)$

13. $f(1) + g(1)$ **14.** $f(-2) - g(-2)$ **15.** $[g(5)]^2$ **16.** $[f(-1)]^2$

17. $f(-2m)$ **18.** $f(-11p)$ **19.** $f(5a - 2)$ **20.** $f(3 + 2k)$

21. $g(5p - 2)$ **22.** $g(-6k + 1)$ **23.** $f(p) + g(2p)$ **24.** $g(1 - r) - f(1 - r)$

25. $g(m) \cdot f(m)$ **26.** $\dfrac{f(5r)}{g(r)}$ **27.** $f\left(\dfrac{1}{m + p}\right)$ **28.** $f\left(\dfrac{-4 + r}{4 + r}\right)$

Let $f(x) = -4.6x^2 - 8.9x + 1.3$. Find each of the following function values.

29. $f(3)$ **30.** $f(-5)$ **31.** $f(-4.2)$ **32.** $f(-1.8)$

Give the domain and range of the functions defined as follows. <u>Give only the domain in</u>
<u>Exercises 45 and 46.</u>

33. $f(x) = 2x - 1$ **34.** $g(x) = 3x + 5$ **35.** $g(x) = x^4$

36. $h(x) = (x - 2)^2$ **37.** $f(x) = \sqrt{8 + x}$ **38.** $f(x) = -\sqrt{x + 6}$

39. $h(x) = \sqrt{16 - x^2}$ **40.** $m(x) = \sqrt{x^2 - 25}$ **41.** $k(x) = (x - 4)^{1/2}$

$[-4, 4]; [0, 4]$

42. $f(x) = (3x + 2)^{1/2}$

43. $g(x) = \sqrt{\dfrac{1}{x^2 + 25}}$

44. $z(x) = -\sqrt{\dfrac{4}{x^2 + 1}}$

in class

45. $g(x) = \dfrac{2}{x^2 - 3x + 2}$

46. $h(x) = \dfrac{-4}{x^2 + 5x + 4}$

47. $r(x) = \sqrt{x^2 - 4x - 5}$

48. $r(x) = \sqrt{x^2 + 7x + 10}$

49. $f(x) = |x - 4|$

50. $k(x) = -|2x - 7|$

51. $g(x) = -\sqrt{2x + 5}$

52. $h(x) = \sqrt{1 - x}$

53. $f(x) = \sqrt{36 - x^2}$ $(-6,6)$ D : R $(0,6)$

54. $g(x) = -\sqrt{25 - x^2}$

55. $k(x) = -\sqrt{x^2 - 9}$

56. $z(x) = \sqrt{x^2 - 100}$

57.
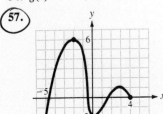

58.
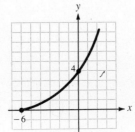

59.
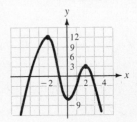

60.
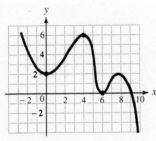

61.
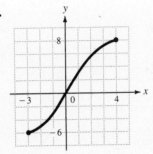

62.
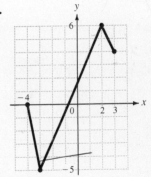

For the functions defined as follows, find **(a)** $f(x + h)$ **(b)** $f(x + h) - f(x)$, and

(c) $\dfrac{f(x + h) - f(x)}{h}$ (Assume $h \neq 0$).

63. $f(x) = x^2 - 4$

64. $f(x) = 8 - 3x^2$

65. $f(x) = 6x + 2$

66. $f(x) = 4x + 11$

67. $f(x) = 2x^3 + x^2$

68. $f(x) = -4x^3 - 8x$

Find where the functions graphed or defined as follows are increasing or decreasing. In Exercises 75–82, first graph the functions.

69.
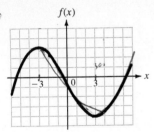

70.
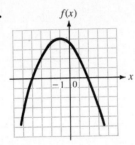

71.
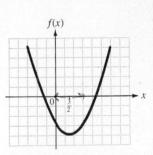

72.
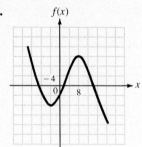

73.
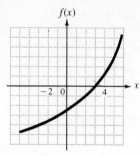

74.
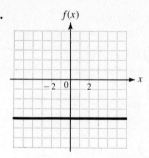

75. $f(x) = -4x + 2$ **76.** $f(x) = 5x - 1$ **77.** $f(x) = x^2 + 4$ **78.** $f(x) = -x^2 - 3$

79. $f(x) = -|x + 2|$ **80.** $f(x) = |3 - x|$ **81.** $f(x) = x + |x|$ **82.** $f(x) = |x| - x$

Decide whether the functions defined as follows are even, odd, or neither.

83. $f(x) = x^2$ **84.** $f(x) = x^3$ **85.** $f(x) = x^4 + x^2 + 5$ **86.** $f(x) = x^3 - x + 1$

87. $f(x) = 2x + 3$ **88.** $f(x) = |x|$ **89.** $f(x) = \dfrac{2}{x - 6}$ **90.** $f(x) = \dfrac{8}{x}$

Graph on the same coordinate axes the graphs of $y = f(x)$ for the given values of c.

91. $f(x) = x^2 + c;\ c = -1,\ c = 2$ **92.** $f(x) = (x + c)^2;\ c = -1,\ c = 2$

93. $f(x) = |x - c|;\ c = -2,\ c = 1$ **94.** $f(x) = c - |x|;\ c = 2,\ c = -3$

Let the graph of a function $y = f(x)$ be as shown. Sketch the graph of each of the following.

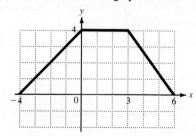

95. $y = f(x) + 4$ **96.** $y = 3 - f(x)$ **97.** $y = f(x - 1)$

98. $y = f(x + 2)$ **99.** $y = f(x + 3) - 2$ **100.** $y = f(x - 1) - 4$

101. A box is made from a piece of metal 12 by 16 inches by cutting squares of side x from each corner. (See the sketch.) Give the volume of the box as a function of x.

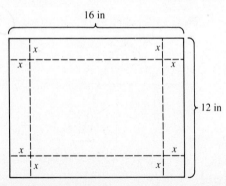

102. A cone has a radius of 6 inches and a height of 9 inches. (See the figure.) The cone is filled with water having a depth of h inches. The radius of the surface of the water is r inches. Use similar triangles to give r as a function of h.

Suppose $f(x) = x^2 + 5x$. Find all values of t so that

103. $f(t + 1) = f(2t)$.

104. $f(2t + 3) = f(t + 2)$.

Let $f(x) = x^2 - 3x$. Decide which of the following results hold for all values of x, as long as no denominator is 0.

105. $f\left(\dfrac{1}{x}\right) = \dfrac{1}{f(x)}$

106. $f(x) \cdot \dfrac{1}{f(x)} = 1$

107. $f(2x) = 2 \cdot f(x)$

108. $f(x + 5) = f(x) + f(5)$

109. Prove that an even function has a graph which is symmetric to the y-axis.

110. Prove that an odd function has a graph which is symmetric to the origin.

Suppose a function is increasing for the interval $(-\infty, +\infty)$. Could the graph of the function be symmetric to the

111. x-axis?

112. y-axis?

113. Give the area of a circle as a function of its radius; also, give the circumference as a function of the radius.

114. Write the area of a circle as a function of the diameter of the circle; then write the circumference as a function of the diameter.

115. A rectangle is inscribed in a circle of radius r. Let x represent the length of one side of the rectangle. Give the area of the rectangle as a function of r.

116. The height of a cone is half the radius of the base. Give the volume of the cone as a function of the radius of the base.

3.5 Algebra of Functions

When a company accountant sits down to estimate the firm's overhead, the first step might be to find functions representing the cost of materials, labor charges, equipment maintenance, and so on. The sum of these various functions could then be used to find the total overhead for the company. Methods of combining functions are discussed in this section.

Given two functions f and g, their *sum*, written $f + g$, is defined by

$$(f + g)(x) = f(x) + g(x),$$

for all x such that both $f(x)$ and $g(x)$ exist. Similar definitions can be given for the difference, $f - g$, product, $f \cdot g$, and quotient, f/g, of functions; however, the quotient,

$$\left(\frac{f}{g}\right)(x) = \frac{f(x)}{g(x)},$$

is defined only for values of x where both $f(x)$ and $g(x)$ exist, and $g(x) \neq 0$.

Example 1   Let $f(x) = x^2 + 1$ and $g(x) = 3x + 5$. Find each of the following.

(a) $(f + g)(1)$

Since $f(1) = 2$ and $g(1) = 8$, use the definition above to get

$$(f + g)(1) = f(1) + g(1) = 2 + 8 = 10.$$

(b) $(f - g)(-3) = f(-3) - g(-3) = 10 - (-4) = 14$

(c) $(f \cdot g)(5) = f(5) \cdot g(5) = 26 \cdot 20 = 520$

(d) $\left(\dfrac{f}{g}\right)(0) = \dfrac{f(0)}{g(0)} = \dfrac{1}{5}$ ■

The various operations on functions are defined below.

Operations on Functions

Given two functions f and g, then for all values of x for which both $f(x)$ and $g(x)$ exist, the functions $f + g$, $f - g$, $f \cdot g$, and f/g are defined as follows.

Sum	$(f + g)(x) = f(x) + g(x)$
Difference	$(f - g)(x) = f(x) - g(x)$
Product	$(f \cdot g)(x) = f(x) \cdot g(x)$
Quotient	$\left(\dfrac{f}{g}\right)(x) = \dfrac{f(x)}{g(x)}, \quad g(x) \neq 0$

Example 2   Let $f(x) = 8x - 9$ and $g(x) = \sqrt{2x - 1}$.

(a) $(f + g)(x) = f(x) + g(x) = 8x - 9 + \sqrt{2x - 1}$

(b) $(f - g)(x) = f(x) - g(x) = 8x - 9 - \sqrt{2x - 1}$

(c) $(f \cdot g)(x) = f(x) \cdot g(x) = (8x - 9)\sqrt{2x - 1}$

(d) $\left(\dfrac{f}{g}\right)(x) = \dfrac{f(x)}{g(x)} = \dfrac{8x - 9}{\sqrt{2x - 1}}$ ■

In Example 2, the domain of f is the set of all real numbers, while the domain of g, where $g(x) = \sqrt{2x - 1}$ includes just those real numbers that make $2x - 1 \geq 0$; the domain of g is the interval $[1/2, +\infty)$. The domain of $f + g$, $f - g$, and $f \cdot g$ is thus $[1/2, +\infty)$. With f/g, the denominator cannot be zero, so the value $1/2$ is excluded from the domain. The domain of f/g is $(1/2, +\infty)$.

The domains of $f + g$, $f - g$, $f \cdot g$, and f/g are summarized below. (Recall that the intersection of two sets is the set of all elements belonging to *both* sets.)

Domains

For functions f and g, the domains of $f + g$, $f - g$, and $f \cdot g$ include all real numbers in the intersection of the domains of f and g, while the domain of f/g includes those real numbers in the intersection of the domains of f and g for which $g(x) \neq 0$.

Composition of Functions   The sketch in Figure 41 shows a function f which assigns to each element x of set X some element y of set Y. Suppose also that a function g takes each element of set Y and assigns a value z of set Z. Using both f and g, then, an element x in X is assigned to an element z in Z. The result of this process is a new function h, which takes an element x in X and assigns an element z in Z.

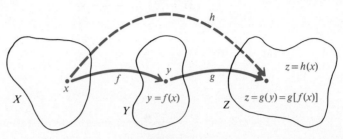

Figure 41

This function h is called the **composition** of functions g and f, written $g \circ f$, and defined as follows.

Composition of Functions

> Let f and g be functions. The **composite function,** or **composition,** of g and f, written $g \circ f$, is defined by
>
> $$(g \circ f)(x) = g[f(x)],$$
>
> for all x in the domain of f such that $f(x)$ is in the domain of g.

Example 3   Let $f(x) = 4x + 1$ and $g(x) = 2x^2 + 5x$. Find each of the following.

(a) $(g \circ f)(x)$

By definition, $(g \circ f)(x) = g[f(x)]$. Using the given functions,

$$\begin{aligned}(g \circ f)(x) &= g[f(x)] \\ &= g(4x + 1) \\ &= 2(4x + 1)^2 + 5(4x + 1) \\ &= 2(16x^2 + 8x + 1) + 20x + 5 \\ &= 32x^2 + 16x + 2 + 20x + 5 \\ &= 32x^2 + 36x + 7.\end{aligned}$$

(b) $(f \circ g)(x)$

Use the definition above with f and g interchanged, so that $(f \circ g)(x)$ becomes $f[g(x)]$. Then

$$\begin{aligned}(f \circ g)(x) &= f[g(x)] \\ &= f(2x^2 + 5x) \\ &= 4(2x^2 + 5x) + 1 \\ &= 8x^2 + 20x + 1. \quad \blacksquare\end{aligned}$$

As this example shows, it is not always true that $f \circ g = g \circ f$. In fact, $f \circ g$ is very rarely equal to $g \circ f$ (see one example in Exercise 65 below). In Example 3, the domain of both composite functions is the set of all real numbers. (We might also mention that in general, the function $f \circ g$ is not the same as the product $f \cdot g$.)

Example 4 Let $f(x) = 1/x$ and $g(x) = \sqrt{3 - x}$. Find $f \circ g$ and $g \circ f$. Give the domain of each. First find $f \circ g$.

$$(f \circ g)(x) = f[g(x)]$$
$$= f(\sqrt{3 - x})$$
$$= \frac{1}{\sqrt{3 - x}}$$

The radicand $\sqrt{3 - x}$ represents a nonzero real number only when $3 - x > 0$, or $x < 3$, so that the domain of $f \circ g$ is the interval $(-\infty, 3)$.

Use the same functions to find $g \circ f$, as follows.

$$(g \circ f)(x) = g[f(x)]$$
$$= g\left(\frac{1}{x}\right)$$
$$= \sqrt{3 - \frac{1}{x}}$$
$$= \sqrt{\frac{3x - 1}{x}}$$

The domain of $g \circ f$ is the set of all real numbers x such that $(3x - 1)/x \geq 0$. By the methods in Section 2.6, the set $(-\infty, 0) \cup [1/3, +\infty)$ is the domain of $g \circ f$. ∎

Example 5 Find functions f and g such that

$$(f \circ g)(x) = (x^2 - 5)^3 - 4(x^2 - 5) + 3.$$

One pair of functions that will work is

$$f(x) = x^3 - 4x + 3 \text{ and } g(x) = x^2 - 5.$$

Then

$$(f \circ g)(x) = f[g(x)]$$
$$= f[x^2 - 5]$$
$$= (x^2 - 5)^3 - 4(x^2 - 5) + 3.$$

Other pairs of functions f and g might also work. ∎

3.5 Exercises

For each of the pairs of functions defined as follows, find $f + g$, $f - g$, $f \cdot g$, and f/g. Give the domain of each.

1. $f(x) = 4x - 1$, $g(x) = 6x + 3$ **2.** $f(x) = 9 - 2x$, $g(x) = -5x + 2$

3. $f(x) = 3x^2 - 2x$, $g(x) = x^2 - 2x + 1$ **4.** $f(x) = 6x^2 - 11x$, $g(x) = x^2 - 4x - 5$

5. $f(x) = \sqrt{2x + 5}$, $g(x) = \sqrt{4x - 9}$ **6.** $f(x) = \sqrt{11x - 3}$, $g(x) = \sqrt{2x - 15}$

7. $f(x) = 4x^2 - 11x + 2$, $g(x) = x^2 + 5$ **8.** $f(x) = 15x^2 - 2x + 1$, $g(x) = 16 + x^2$

Let $f(x) = 4x^2 - 2x$ and let $g(x) = 8x + 1$. Find each of the following.

9. $(f + g)(3)$ **10.** $(f + g)(-5)$ **11.** $(f \cdot g)(4)$ **12.** $(f \cdot g)(-3)$

13. $\left(\dfrac{f}{g}\right)(-1)$ **14.** $\left(\dfrac{f}{g}\right)(4)$ **15.** $(f + g)(m)$ **16.** $(f - g)(2k)$

17. $(f \circ g)(2)$ **18.** $(f \circ g)(-5)$ **19.** $(g \circ f)(2)$ **20.** $(g \circ f)(-5)$

21. $(f \circ g)(k)$ **22.** $(g \circ f)(5z)$

Find $f \circ g$ and $g \circ f$ for each of the pairs of functions defined as follows.

23. $f(x) = 8x + 12$, $g(x) = 3x - 1$ **24.** $f(x) = -6x + 9$, $g(x) = 5x + 7$

25. $f(x) = 5x + 3$, $g(x) = -x^2 + 4x + 3$ **26.** $f(x) = 4x^2 + 2x + 8$, $g(x) = x + 5$

27. $f(x) = -x^3 + 2$, $g(x) = 4x$ **28.** $f(x) = 2x$, $g(x) = 6x^2 - x^3$

29. $f(x) = \dfrac{1}{x}$, $g(x) = x^2$ **30.** $f(x) = \dfrac{2}{x^4}$, $g(x) = 2 - x$

31. $f(x) = \sqrt{x + 2}$, $g(x) = 8x^2 - 6$ **32.** $f(x) = 9x^2 - 11x$, $g(x) = 2\sqrt{x + 2}$

33. $f(x) = \dfrac{1}{x - 5}$, $g(x) = \dfrac{2}{x}$ **34.** $f(x) = \dfrac{8}{x - 6}$, $g(x) = \dfrac{4}{3x}$

35. $f(x) = \sqrt{x + 1}$, $g(x) = \dfrac{-1}{x}$ **36.** $f(x) = \dfrac{8}{x}$, $g(x) = \sqrt{3 - x}$

The graphs of functions f and g are shown. Use these graphs to find each of the following values.

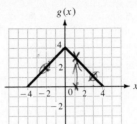
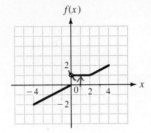

37. $f(1) + g(1)$ **38.** $f(4) - g(3)$ **39.** $f(-2) \cdot g(4)$ **40.** $\dfrac{f(4)}{g(2)}$

41. $(f \circ g)(2)$ **42.** $(g \circ f)(2)$ **43.** $(g \circ f)(-4)$ **44.** $(f \circ g)(-2)$

For each of the pairs of functions defined as follows, show that $(f \circ g)(x) = x$ and $(g \circ f)(x) = x$.

45. $f(x) = 3x, \quad g(x) = \frac{1}{3}x$

46. $f(x) = \frac{3}{4}x, \quad g(x) = \frac{4}{3}x$

47. $f(x) = 8x - 11, \quad g(x) = \frac{x + 11}{8}$

48. $f(x) = \frac{x - 3}{4}, \quad g(x) = 4x + 3$

49. $f(x) = x^3 + 6, \quad g(x) = \sqrt[3]{x - 6}$

50. $f(x) = \sqrt[5]{x - 9}, \quad g(x) = x^5 + 9$

In each of the following exercises, a function h is defined. Find functions f and g such that $h(x) = (f \circ g)(x)$. Many such pairs of functions exist.

51. $h(x) = (6x - 1)^2$

52. $h(x) = (11x^2 + 12x)^2$

53. $h(x) = \sqrt{x^2 - 1}$

54. $h(x) = \frac{1}{x^2 + 2}$

55. $h(x) = \frac{(x - 2)^2 + 1}{5 - (x - 2)^2}$

56. $h(x) = (x + 2)^3 - 3(x + 2)^2$

57. Suppose the population P of a certain species of fish depends on the number x (in hundreds) of a smaller kind of fish which serves as its food supply, so that

$$P(x) = 2x^2 + 1.$$

Suppose, also, that the number x (in hundreds) of the smaller species of fish depends upon the amount a (in appropriate units) of its food supply, a kind of plankton. Suppose

$$x = f(a) = 3a + 2.$$

Find $(P \circ f)(a)$, the relationship between the population P of the large fish and the amount a of plankton available.

58. Suppose the demand for a certain brand of vacuum cleaner is given by

$$D(p) = \frac{-p^2}{100} + 500,$$

where p is the price in dollars. If the price, in terms of the cost, c, is expressed as

$$p(c) = 2c - 10,$$

find the demand in terms of the cost.

59. An oil well off the Gulf Coast is leaking, with the leak spreading oil over the surface as a circle. At any time t, in minutes, after the beginning of the leak, the radius of the circular oil slick on the surface is $r(t) = 4t$ ft. Let $A(r) = \pi r^2$ represent the area of a circle of radius r. Find and interpret $(A \circ r)(t)$.

60. When a thermal inversion layer is over a city (such as happens often in Los Angeles), pollutants cannot rise vertically but are trapped below the layer and must disperse horizontally. Assume that a factory smokestack begins emitting a pollutant at 8 A.M. Assume that the pollutant disperses horizontally, forming a circle. If t represents the time, in hours, since the factory began emitting pollutants ($t = 0$ represents 8 A.M.), assume that the radius of the circle of pollution is $r(t) = 2t$ mi. Let $A(r) = \pi r^2$ represent the area of a circle of radius r. Find and interpret $(A \circ r)(t)$.

Let f and g be increasing functions on the same interval. Show that the following two functions are also increasing on that same interval.

61. $f + g$

62. $f \circ g$

63. What can you say about $f - g$?

64. Let $f(x) = x/(x - 1)$ for $x \neq 1$. Show that $f \circ f = x$.

65. Let $f(x) = ax + b$ and $g(x) = cx + d$. Find conditions for c and d so that $f \circ g = g \circ f$.

Recall that even and odd functions were defined in the previous section.

66. Let f be any function. Prove that the function $\quad g(x) = \dfrac{1}{2}[f(x) + f(-x)]\quad$ is even.

67. Let f be any function. Prove that the function $\quad h(x) = \dfrac{1}{2}[f(x) - f(-x)]\quad$ is odd.

68. Use the results of the previous two exercises to show that any function may be expressed as the sum of an odd function and an even function.

Prove that the

69. sum of two even functions is even.

70. product of two even functions is even.

71. sum of two odd functions is odd.

72. product of two odd functions is even.

73. product of an odd function and an even function is odd.

74. What can you say about the sum of an odd and an even function? Give examples.

3.6 Inverse Functions

Addition and subtraction are inverse operations: starting with a number x, adding 5, and then subtracting 5 gives x back as a result. Also, some functions are inverses of each other. For example, it turns out that the functions

$$f(x) = 8x \quad \text{and} \quad g(x) = \frac{1}{8}x$$

are inverses of each other. To see why, choose a value of x such as $x = 12$ and find $f(12)$.

$$f(12) = 8 \cdot 12 = 96$$

Calculating $g(96)$ gives

$$g(96) = \frac{1}{8} \cdot 96 = 12,$$

so that $(g \circ f)(12) = g[f(12)] = 12$. Also, finding $g(12)$ and then $f[g(12)]$ would show that $(f \circ g)(12) = 12$. For these two functions f and g, it turns out that for *any* value of x,

$$(f \circ g)(x) = 8\left(\frac{1}{8}x\right) = x \quad \text{and} \quad (g \circ f)(x) = \frac{1}{8}(8x) = x.$$

As we shall see, this condition makes f and g inverses of each other.

This section shows how to start with a function such as $f(x) = 8x$ and use it to obtain the inverse function $g(x) = (1/8)x$, if an inverse exists. Not all functions have inverses: the only functions with an inverse function are *one-to-one functions*.

One-to-one Functions   Given the function $y = x^2$, it is possible for two different values of x to lead to the same value of y. For example, the value $y = 4$ is obtained from either of two values of x: both $2^2 = 4$ and $(-2)^2 = 4$. On the other hand, for the function $y = 6x$, a given value of y can be found from exactly one value of x. For this function, if $y = 30$, then $x = 5$; there is no other value of x that will produce a value of 30 for y.

This second function, $y = 6x$, is an example of a **one-to-one function:** a function is one-to-one if each element in the range is obtained from exactly one element in the domain, or

One-to-one Function	a function f is **one-to-one** if $f(a) = f(b)$ implies that $a = b$.

Example I   Decide whether the functions defined as follows are one-to-one.

(a) $f(x) = -4x + 12$

Suppose $f(a) = f(b)$, or

$$-4a + 12 = -4b + 12.$$

Then
$$-4a = -4b,$$

or
$$a = b.$$

The fact that $f(a) = f(b)$ implies $a = b$, which makes the function one-to-one.

(b) $g(x) = \sqrt{25 - x^2}$

Start with $g(a) = g(b)$, or

$$\sqrt{25 - a^2} = \sqrt{25 - b^2}.$$

Squaring both sides gives
$$25 - a^2 = 25 - b^2,$$

or
$$a^2 = b^2.$$

From $a^2 = b^2$, it is *not* possible to conclude that $a = b$, since a might equal $-b$ instead. This means the function g is not one-to-one. As a numerical example, $g(3) = g(-3)$, but $3 \neq -3$. ■

There is a useful graphical test for deciding whether or not a function is one-to-one. Figure 42(a) shows the graph of a function defined by $y = f(x)$ cut by a horizontal line. As the graph suggests, $f(x_1) = f(x_2) = f(x_3)$, even though x_1, x_2, and x_3 are all distinct points. Since one value of y can be obtained from more than one value of x, the function is not one-to-one. On the other hand, drawing horizontal lines on the graph of Figure 42(b) shows that a given value of y can be obtained from only one value of x, making the function one-to-one.

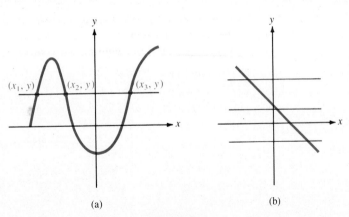

(a)   (b)

Figure 42

The examples suggest the **horizontal line test** for one-to-one functions.

Horizontal Line Test	If no horizontal line cuts the graph of a function in more than one point, then the function is one-to-one.

Inverse Functions We saw above that the functions defined as $f(x) = 8x$ and $g(x) = (1/8)x$ have the property that

$$(f \circ g)(x) = x \qquad \text{and} \qquad (g \circ f)(x) = x.$$

In other words, if $y = f(x)$, then $g(y) = x$. For example, $f(5) = 40$ and $g(40) = 5$. See the sketch in Figure 43.

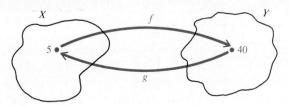

Figure 43

Functions f and g have the following property: starting with a value of x in the domain of f and finding $f(x)$, then evaluating g at $f(x)$, the result is x. Because of this property, f and g are called *inverse functions* of each other.

Inverse Functions	Let f be a one-to-one function with domain X and range Y. Let g be a function with domain Y and range X. Then g is the **inverse function** of f if $$(f \circ g)(x) = x \quad \text{for every } x \text{ in } Y,$$ and $$(g \circ f)(x) = x \quad \text{for every } x \text{ in } X.$$

A special notation is often used for inverse functions: if g is the inverse of a function f, then g is written as f^{-1} (read "f-inverse.") In the example above, $f(x) = 8x$, and $g(x) = f^{-1}(x) = (1/8)x$. Do not confuse the -1 in f^{-1} with a negative exponent: the symbol f^{-1} does not represent $1/f$; it is used for the inverse of function f.

By the definition of inverse function, the domain of f equals the range of f^{-1}, while the range of f equals the domain of f^{-1}. See Figure 44.

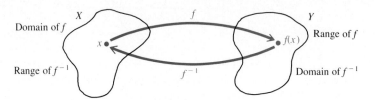

Figure 44

Example 2   Let functions f and g be defined by $f(x) = x^3 - 1$ and $g(x) = \sqrt[3]{x + 1}$, respectively. Is g the inverse function of f?

First check that f is one-to-one. Since it is, now find $(f \circ g)(x)$ and $(g \circ f)(x)$.

$$(f \circ g)(x) = f[g(x)] = (\sqrt[3]{x + 1})^3 - 1 = x + 1 - 1 = x$$
$$(g \circ f)(x) = g[f(x)] = \sqrt[3]{(x^3 - 1) + 1} = \sqrt[3]{x^3} = x.$$

Since both $(f \circ g)(x) = x$ and $(g \circ f)(x) = x$, and since the domain of f equals the range of g, and the range of f equals the domain of g, function g is indeed the inverse of function f, so that f^{-1} is given by

$$f^{-1}(x) = \sqrt[3]{x + 1}. \quad \blacksquare$$

Keep in mind that a function f can have an inverse function f^{-1} if and only if f is one-to-one. Since increasing and decreasing functions are one-to-one, they must have inverse functions.

Given a one-to-one function f and an x in the domain of f, the corresponding value in the range of f is found by means of the equation $y = f(x)$. With the inverse function f^{-1}, the value of y can be used to produce x, since $x = f^{-1}(y)$. Therefore, the equation for f^{-1} can be found by solving $y = f(x)$ for x.

For example, let $f(x) = 7x - 2$. Then $y = 7x - 2$. The function f is one-to-one, so that f^{-1} exists. Solve $y = 7x - 2$ for x, as follows:

$$y = 7x - 2$$
$$7x = y + 2$$
$$x = \frac{y + 2}{7}$$

or, since $x = f^{-1}(y)$,

$$f^{-1}(y) = \frac{y + 2}{7}.$$

For a given x in X (the domain of f), the equation $y = 7x - 2$ produces a value of $f(x)$ in Y (the range of f). The function $f^{-1}(y) = (y + 2)/7$ takes a value of y in Y and produces a value x in X. Since it is customary to use x for the domain element of a function, replace y with x in f^{-1} to get

$$f^{-1}(x) = \frac{x + 2}{7}.$$

Check that $(f \circ f^{-1})(x) = x$ and $(f^{-1} \circ f)(x) = x$, so that f^{-1} is indeed the inverse of f.

In summary, find the equation of an inverse function with the following steps.

Finding an Equation for f^{-1}

> 1. Check that the function f defined by $y = f(x)$ is a one-to-one function.
> 2. Solve for x. Let $x = f^{-1}(y)$.
> 3. Exchange x and y to get $y = f^{-1}(x)$.
> 4. Check the domains and ranges: the domain of f and the range of f^{-1} should be equal, as should the domain of f^{-1} and the range of f.

Example 3 For each of the functions defined as follows, find any inverse functions.

(a) $f(x) = \dfrac{4x + 6}{5}$

This function is one-to-one and so has an inverse. Let $y = f(x)$, and solve for x.

$$y = \frac{4x + 6}{5}$$

$$5y = 4x + 6$$

$$5y - 6 = 4x$$

$$\frac{5y - 6}{4} = x$$

Finally, exchange x and y, and let $y = f^{-1}(x)$, to get

$$\frac{5x - 6}{4} = y,$$

or

$$f^{-1}(x) = \frac{5x - 6}{4}.$$

Verify that the domain of f is the range of f^{-1} and the range of f is the domain of f^{-1}.

(b) $f(x) = x^3 - 1$

Two different values of $f(x)$ come from two different values of x, so the function is one-to-one and has an inverse. To find the inverse, first solve $y = x^3 - 1$ for x, as follows.

$$y = x^3 - 1$$

$$y + 1 = x^3$$

$$\sqrt[3]{y + 1} = x$$

Exchange x and y, to get

$$\sqrt[3]{x + 1} = y,$$

or

$$f^{-1}(x) = \sqrt[3]{x + 1}.$$

Check that the domain of f^{-1} is the range of f and the range of f^{-1} is the domain of f.

(c) $f(x) = x^2$

The two different x-values 4 and -4 give the same value of y, namely 16, showing that the function is not one-to-one and has no inverse function. ■

Suppose f and f^{-1} are inverse functions of each other. Suppose that $f(a) = b$, for real numbers a and b. Then by the definition of inverse, $f^{-1}(b) = a$. This shows that if a point (a, b) is on the graph of f, then (b, a) will belong to the graph of f^{-1}. As shown in Figure 45, the points (a, b) and (b, a) are **symmetric with respect to the line $y = x$.** Thus, the graph of f^{-1} can be obtained from the graph of f by reflecting the graph of f about the line $y = x$.

For example, Figure 46 shows the graph of $f(x) = x^3 - 1$ as a solid line and the graph of $f^{-1}(x) = \sqrt[3]{x + 1}$ as a dashed line. These graphs are symmetric with respect to the line $y = x$.

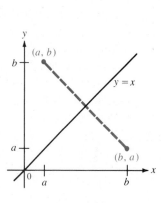

Figure 45

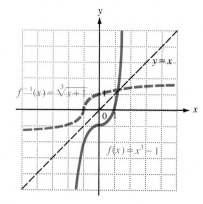

Figure 46

Example 4   Let $f(x) = \sqrt{x + 5}$ with domain $[-5, +\infty)$. Find $f^{-1}(x)$.

The function f is one-to-one and has an inverse function. To find this inverse function, start with

$$y = \sqrt{x + 5}$$

and solve for x, to get

$$y = \sqrt{x + 5}$$
$$y^2 = x + 5$$
$$y^2 - 5 = x.$$

Exchanging x and y gives

$$x^2 - 5 = y.$$

We cannot just give $x^2 - 5$ as $f^{-1}(x)$. In the definition of f above, the domain was given as $[-5, +\infty)$. The range of f is $[0, +\infty)$. Since the range of f equals the domain of f^{-1}, the function f^{-1} must be given as

$$f^{-1}(x) = x^2 - 5, \quad \text{domain } [0, +\infty).$$

As a check, the range of f^{-1}, $[-5, +\infty)$ equals the domain of f. Graphs of f and f^{-1} are shown in Figure 47. The line $y = x$ is included on the graph to show that the graphs of f and f^{-1} are mirror images with respect to this line. ■

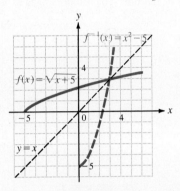

Figure 47

3.6 Exercises

Which of the functions graphed or defined as follows are one-to-one?

1.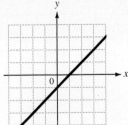

2.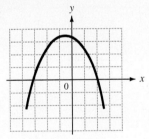

3.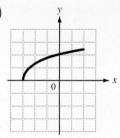

4.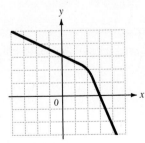

5.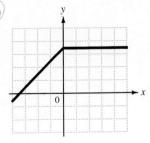

6.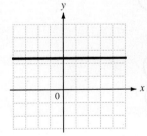

7. $y = 4x - 5$

8. $y = -x$

9. $y = 6 - x$

10. $y = -x^2$

11. $y = (x - 2)^2$

12. $y = -(x + 3)^2 - 8$

13. $y = \sqrt{36 - x^2}$

14. $y = -\sqrt{100 - x^2}$

15. $y = |25 - x^2|$

16. $y = -|16 - x^2|$

17. $y = x^3 - 1$

18. $y = -\sqrt[3]{x + 5}$

19. $y = (\sqrt{x} + 1)^2$

20. $y = (3 - 2\sqrt{x})^2$

21. $y = \dfrac{1}{x + 2}$

22. $y = \dfrac{-4}{x - 8}$

23. $y = 9$

24. $y = -4$

Which of the following pairs of functions graphed or defined as follows are inverses of each other?

25.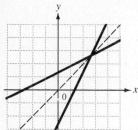

26.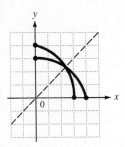

27.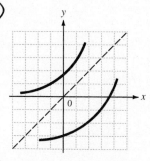

28.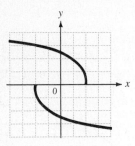

29.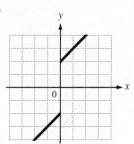

30.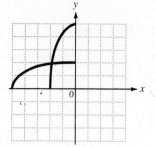

31. $f(x) = -\dfrac{3}{11}x, \quad g(x) = -\dfrac{11}{3}x$

32. $f(x) = 2x + 4, \quad g(x) = \dfrac{1}{2}x - 2$

33. $f(x) = 5x - 5, \quad g(x) = \dfrac{1}{5}x + 1$

34. $f(x) = 8x - 7, \quad g(x) = \dfrac{x + 8}{7}$

35. $f(x) = \dfrac{1}{x + 1}, \quad g(x) = \dfrac{x - 9}{12}$

36. $f(x) = \dfrac{1}{x + 1}, \quad g(x) = \dfrac{1 - x}{x}$

37. $f(x) = \dfrac{2}{x + 6}, \quad g(x) = \dfrac{6x + 2}{x}$

38. $f(x) = \dfrac{1}{x}, \quad g(x) = \dfrac{1}{x}$

39. $f(x) = x^2 + 3$, domain $[0, +\infty)$, and $g(x) = \sqrt{x - 3}$, domain $[3, +\infty)$

40. $f(x) = \sqrt{x + 8}$, domain $[-8, +\infty)$, and $g(x) = x^2 - 8$, domain $[0, +\infty)$

41. $f(x) = -|x + 5|$, domain $[-5, +\infty)$, and $g(x) = |x - 5|$, domain $[5, +\infty)$

42. $f(x) = |x - 1|$, domain $[-1, +\infty)$, and $g(x) = |x + 1|$, domain $[1, +\infty)$

Graph the inverse of each one-to-one function.

43.

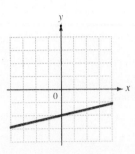

44.

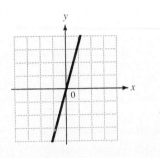

45.

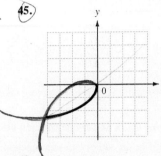

46.

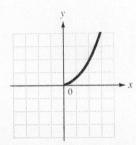

47.

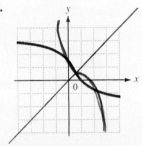

48.

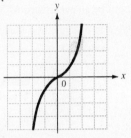

For each of the functions defined as follows that is one-to-one, write an equation for the inverse function in the form of $y = f^{-1}(x)$, and then graph f and f^{-1}.

49. $y = 4x - 5$

50. $y = 3x - 4$

51. $y = -\dfrac{2}{5}x$

52. $y = \dfrac{1}{3}x$

53. $y = -x^3 - 2$

54. $y = x^3 + 1$

55. $3x + y = 9$

56. $x + 4y = 12$

57. $y = -x^2 + 2$

58. $y = x^2$

59. $xy = 4$

60. $y = \dfrac{1}{x}$

61. $y = \dfrac{-6x + 5}{3x - 1}$

62. $y = \dfrac{8x + 3}{4x - 1}$

63. $f(x) = \sqrt{6 + x}$, domain $[-6, +\infty)$

64. $f(x) = 4 - x^2$, domain $(-\infty, 0]$

The graph of a function f is shown in the figure. Use the graph to find each of the following values.

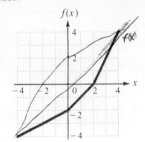

65. $f^{-1}(4)$

66. $f^{-1}(2)$

67. $f^{-1}(0)$

68. $f^{-1}(-2)$

69. $f^{-1}(-3)$

70. $f^{-1}(-4)$

Let $f(x) = x^2 + 5x$ for $x \geq -5/2$. Find each of the following, rounding to the nearest hundredth.

71. $f^{-1}(7)$ **72.** $f^{-1}(-3)$

Let $f(x) = -4x + 3$, while $g(x) = 2x^3 - 4$. Find each of the following.

73. $(f \circ g)^{-1}(x)$ **74.** $(f^{-1} \circ g^{-1})(x)$ **75.** $(g^{-1} \circ f^{-1})(x)$

76. Let f and g be functions having inverses f^{-1} and g^{-1} respectively. Let $f \circ g$ exist. Show that $(f \circ g)^{-1}$ is $(g^{-1} \circ f^{-1})$.

77. Show that a one-to-one function has exactly one inverse function.

78. Let f be an odd one-to-one function. What can you say about f^{-1}?

79. Let f be an even function. What can you say about f^{-1}?

80. Give an example of a function f such that $f = f^{-1}$.

work, but

3.7   Variation Omit

In many applications of mathematics, it is necessary to write relationships between variables. For example, in chemistry the ideal gas law shows how temperature, pressure, and volume are related. In physics, various formulas in optics show how the focal length of a lens and the size of an image are related.

If the quotient of two variables is constant, then one variable *varies directly* or is *directly proportional* to the other. This can be stated differently as follows.

Directly Proportional

> y **varies directly** as x, or y **is directly proportional** to x, means that a nonzero real number k (called the **constant of variation**) exists such that
>
> $$y = kx.$$

Here y is a function of x.

Example 1   Suppose the value of y varies directly as the value of x, and that $y = 12$ when $x = 5$. Find y when $x = 21$.

If y varies directly as x, then a real number k exists such that

$$y = kx.$$

To find the value of k, use the given information: $y = 12$ when $x = 5$. Replacing y with 12 and x with 5 gives

$$12 = k \cdot 5$$

or $$k = \frac{12}{5}.$$

In this example, then, the relationship between y and x is given by

$$y = \frac{12}{5}x.$$

Now find y when x is 21.

$$y = \frac{12}{5} \cdot 21 = \frac{252}{5}. \quad \blacksquare$$

Sometimes y varies as a power of x.

Varies as nth Power

> Let n be a positive real number. Then y **varies directly as the nth power** of x, or y is **directly proportional to the nth power** of x, if there exists a real number k such that
>
> $$y = kx^n.$$

The phrase "directly proportional" is sometimes abbreviated to just "proportional."

For example, the area of a square of side x is given by the formula $A = x^2$, so that the area *varies directly as the square* of the length of a side. Here $k = 1$.

The case where y increases as x decreases is an example of *inverse variation*.

Inverse Variation

> Let n be a positive real number. Then y **varies inversely as the nth power** of x means that there exists a real number k such that
>
> $$y = \frac{k}{x^n}.$$
>
> If $n = 1$, then $y = k/x$, and y **varies inversely** as x.

Example 2 In a certain manufacturing process, the cost of producing a single item varies inversely as the square of the number of items produced. If 100 items are produced, each costs $2. Find the cost per item if 400 items are produced.

We can let x represent the number of items produced and y the cost per item, and write

$$y = \frac{k}{x^2}$$

for some nonzero constant k. Since $y = 2$ when $x = 100$,

$$2 = \frac{k}{100^2} \quad \text{or} \quad k = 20{,}000.$$

Thus, the relationship between x and y is given by

$$y = \frac{20{,}000}{x^2}.$$

When 400 items are produced, the cost per item is given by

$$y = \frac{20{,}000}{400^2} = .125, \text{ or } 12.5\text{¢}. \quad \blacksquare$$

One variable may depend on more than one other variable. Such variation is called *combined variation*. If a variable depends on the product of two or more other variables, we refer to that as *joint variation*.

Joint Variation

> *y* **varies jointly** as the *n*th power of *x* and the *m*th power of *z* if there exists a real number *k* such that
>
> $$y = kx^n z^m.$$

The next example shows combined variation.

Example 3   Suppose that *m* varies directly as the square of *p* and inversely as *q*. Suppose also that $m = 8$ when $p = 2$ and $q = 6$. Find *m* if $p = 6$ and $q = 10$.

Here *m* depends on two variables:

$$m = \frac{kp^2}{q}.$$

Since $m = 8$ when $p = 2$ and $q = 6$,

$$8 = \frac{k \cdot 2^2}{6}$$

$$k = 12$$

and it follows that

$$m = \frac{12p^2}{q}.$$

For $p = 6$ and $q = 10$,

$$m = \frac{12 \cdot 6^2}{10}$$

$$m = \frac{216}{5}. \quad \blacksquare$$

The steps involved in solving a problem in variation can be summarized as follows:

Solving Variation Problems

> 1. Write, in an algebraic form, the general relationship among the variables. Use the constant *k*.
> 2. Substitute given values of the variables and find the value of *k*.
> 3. Substitute this value of *k* into the formula of Step 1, thus obtaining a specific formula.
> 4. Solve for the required unknown.

Omit
For Test

3.7 Exercises

Express each of the following as an equation.

1. *a* varies directly as *b*.

2. *m* is proportional to *n*.

3. *x* is inversely proportional to *y*.

4. *p* varies inversely as *y*.

5. *r* varies jointly as *s* and *t*.

6. *R* is proportional to *m* and *p*.

7. *w* is proportional to x^2 and inversely proportional to *y*.

8. *c* varies directly as *d* and inversely as f^2 and *g*.

Solve each of the following.

9. If *m* varies directly as *x* and *y*, and *m* = 10 when *x* = 4 and *y* = 7; find *m* when *x* = 11 and *y* = 8.

10. Suppose *m* varies directly as *z* and *p*. If *m* = 10 when *z* = 3 and *p* = 5, find *m* when *z* = 5 and *p* = 7.

11. Suppose *r* varies directly as the square of *m*, and inversely as *s*. If *r* = 12 when *m* = 6 and *s* = 4, find *r* when *m* = 4 and *s* = 10.

12. Suppose *p* varies directly as the square of *z*, and inversely as *r*. If *p* = 32/5 when *z* = 4 and *r* = 10, find *p* when *z* = 2 and *r* = 16.

13. Let *a* be proportional to *m* and n^2, and inversely proportional to y^3. If *a* = 9 when *m* = 4, *n* = 9, and *y* = 3, find *a* if *m* = 6, *n* = 2, and *y* = 5.

14. If *y* varies directly as *x*, and inversely as m^2 and r^2, and *y* = 5/3 when *x* = 1, *m* = 2, and *r* = 3, find *y* if *x* = 3, *m* = 1, and *r* = 8.

15. Let *r* vary directly as p^2 and q^3, and inversely as z^2 and *x*. If *r* = 334.6 when *p* = 1.9, *q* = 2.8, *z* = .4, and *x* = 3.7, find *r* when *p* = 2.1, *q* = 4.8, *z* = .9, and *x* = 1.4.

16. Let *p* vary directly as *x* and z^2, and inversely as *m*, *n*, and *q*. If *p* = 9.25 when *x* = 3.2, *z* = 10.1, *m* = 2, *n* = 1.7, and *q* = 8.3, find *p* when *x* = 8.1, *z* = 4.3, *m* = 4, *n* = 2.1, and *q* = 6.2.

17. Suppose *m* varies directly as p^2 and r^4. If *p* doubles and *r* triples, how does *m* change?

18. Let *z* vary directly as y^3 and inversely as x^2. If *y* doubles and *x* is halved, how does *z* change?

19. The distance a body falls from rest varies directly as the square of the time it falls (disregarding air resistance). If an object falls 1024 feet in 8 seconds, how far will it fall in 12 seconds?

20. Hooke's law for an elastic spring states that the distance a spring stretches varies directly as the force applied. If a force of 15 pounds stretches a certain spring 8 inches, how much will a force of 30 pounds stretch the spring?

21. In electric current flow, it is found that the resistance (measured in units called ohms) offered by a fixed length of wire of a given material varies inversely as the square of the diameter of the wire. If a wire .01 inches in diameter has a resistance of .4 ohm, what is the resistance of a wire of the same length and material but .03 inches in diameter?

22. The illumination produced by a light source varies inversely as the square of the distance from the source. The illumination of a light source at 5 meters is 70 candela. What is the illumination 12 meters from the source?

23. The pressure exerted by a certain liquid at a given point is proportional to the depth of the point below the surface of the liquid. If the pressure 20 meters below the surface is 70 kilograms per square centimeter, what pressure is exerted 40 meters below the surface?

24. The distance that a person can see to the horizon from a point above the surface of the earth varies directly as the square root of the height. A person on a hill 121 meters high can see for 15 kilometers to the horizon. How far is the horizon from a hill 900 meters high?

25. Simple interest varies jointly as principal and time. If $1000 left at interest for 2 years earned $110, find the amount of interest earned by $5000 for 5 years.

26. The volume of a right circular cylinder is jointly proportional to the square of the radius of the circular base and to the height. If the volume is 300 cubic centimeters when the height is 10.62 centimeters and the radius is 3 centimeters, find the volume for a cylinder with a radius of 4 centimeters and a height of 15.92 centimeters.

27. The Downtown Construction Company is designing a building whose roof rests on round concrete pillars. The company's engineers know that the maximum load a cylindrical column of circular cross section can hold varies directly as the fourth power of the diameter and inversely as the square of the height. If a 9 meter column 1 meter in diameter will support a load of 8 metric tons, how many metric tons will be supported by a column 12 meters high and 2/3 meter in diameter?

28. The company's engineers also know that the maximum load of a horizontal beam which is supported at both ends varies directly as the width and square of the height and inversely as the length between supports. If a beam 8 meters long, 12 centimeters wide, and 15 centimeters high can support a maximum of 400 kilograms, what will they find to be the maximum load of a beam of the same material 16 meters long, 24 centimeters wide, and 8 centimeters high?

29. The force needed to keep a car from skidding on a curve varies inversely as the radius of the curve and jointly as the weight of the car and the square of the speed. It takes 3000 pounds of force to keep a 2000 pound car from skidding on a curve of radius 500 feet at 30 mph. What force is needed to keep the same car from skidding on a curve of radius 800 feet at 60 mph?

30. The period of a pendulum varies directly as the square root of the length of the pendulum and inversely as the square root of the acceleration due to gravity. Find the period when the length is 121 centimeters and the acceleration due to gravity is 980 centimeters per second per second, if the period is 6π seconds when the length is 289 centimeters and the acceleration due to gravity is 980 centimeters per second per second.

31. The pressure on a point in a liquid is directly proportional to the distance from the surface to the point. In a certain liquid the pressure at a depth of 4 m is 60 kg per m^2. Find the pressure at a depth of 10 m.

32. The volume V of a gas varies directly as the temperature T and inversely as the pressure P. If V is 10 when T is 280 and P is 6, find V if T is 300 and P is 10.

33. A sociologist has decided to rank the cities in a nation by population, with population varying inversely as the ranking. If a city with a population of 1,000,000 ranks 8th, find the population of a city that ranks 2nd.

34. Under certain conditions, the length of time that it takes for fruit to ripen during the growing season varies inversely as the average maximum temperature during the season.

If it takes 25 days for fruit to ripen with an average maximum temperature of 80°, find the number of days it would take at 75°.

35. The number of long distance phone calls between two cities in a certain time period varies directly as the populations p_1 and p_2 of the cities, and inversely as the distance between them. If 10,000 calls are made between two cities 500 miles apart, having populations of 50,000 and 125,000, find the number of calls between two cities 800 miles apart and having populations of 20,000 and 80,000.

36. The horsepower needed to run a boat through water varies as the cube of the speed. If 80 horsepower are needed to go 15 kilometers per hour in a certain boat, how many horsepower would be needed to go 30 kilometers per hour?

37. According to Poiseuille's Law, the resistance to flow of a blood vessel, R, is directly proportional to the length, l, and inversely proportional to the fourth power of the radius, r. If $R = 25$ when $l = 12$ and $r = 0.2$, find R as r increases to 0.3, while l is unchanged.

38. The Stefan-Boltzmann Law says that the radiation of heat from an object is directly proportional to the fourth power of the Kelvin temperature of the object. For a certain object, $R = 213.73$ at room temperature (293° Kelvin). Find R if the temperature increases to 335° Kelvin.

39. Suppose a nuclear bomb is detonated at a certain site. The effects of the bomb will be felt over a distance from the point of detonation that is directly proportional to the cube root of the yield of the bomb. Suppose a 100-kiloton bomb has certain effects to a radius of 3 km from the point of detonation. Find the distance that the effects would be felt for a 1500-kiloton bomb.

40. The maximum speed possible on a length of railroad track is directly proportional to the cube root of the amount of money spent on maintaining the track. Suppose that a maximum speed of 25 km per hour is possible on a stretch of track for which $450,000 was spent on maintenance. Find the maximum speed if the amount spent on maintenance is increased to $1,750,000.

41. Assume that a person's weight increases directly as the cube of their height. Find the weight of a 20 inch, 7 pound baby who grows up to be an adult 67 inches tall (How reasonable does our assumption about weight and height seem?)

42. The cost of a pizza varies directly as the square of its radius. If a pizza with a radius of 15 cm costs $7, find the cost of a pizza having a radius of 22.5 cm. (You might want to do some research at a nearby pizza establishment and see if this assumption is reasonable.)

Chapter 3 Summary

Key Words

distance formula	domain	composition of functions
midpoint formula	range	one-to-one function
parabola	independent variable	horizontal line test
vertex	dependent variable	inverse functions
vertical translation	vertical line test for a	varies directly
horizontal translation	function	directly proportional
function	increasing function	constant of variation
mapping	decreasing function	varies inversely

Review Exercises

Graph the set of points satisfying each condition below.

1. $x < 0$ **2.** $y \geq 0$ **3.** $xy > 0$ **4.** $x = 2$

Find the distance $d(P, Q)$ and the midpoint of segment PQ.

5. $P(3, -1)$ and $Q(-4, 5)$ **6.** $P(-8, 2)$ and $Q(3, -7)$

7. Find the other endpoint of a line segment having one end at $(-5, 7)$ and having midpoint at $(1, -3)$.

8. Are the points $(5, 7)$, $(3, 9)$, $(6, 8)$ the vertices of a right triangle?

9. Find all possible values of k so that $(-1, 2)$, $(-10, 5)$, and $(-4, k)$ are the vertices of a right triangle.

10. Find all possible values of x so that the distance between $(x, -9)$ and $(3, -5)$ is 6.

11. Find all points (x, y) with $x = 6$ so that (x, y) is 4 units from $(1, 3)$.

12. Find all points (x, y) with $x + y = 0$ so that (x, y) is 6 units from $(-2, 3)$.

Prove each of the following.

13. The medians to the two equal sides of an isosceles triangle are equal in length.

14. The line segment connecting midpoints of two sides of a triangle is half as long as the third side.

Graph each of the following. Give the domain for those graphs that are graphs of functions.

15. $x + y = 4$ **16.** $3x - 5y = 20$ **17.** $y = \frac{1}{2}x^2$ **18.** $y = 3 - x^2$

19. $y = \frac{-8}{x}$ **20.** $y = -2x^3$ **21.** $y = \sqrt{x - 7}$ **22.** $(x - 3)^2 + y^2 = 16$

Find equations for each of the circles below.

23. Center $(-2, 3)$, radius 5 **24.** Center $(\sqrt{5}, -\sqrt{7})$, radius $\sqrt{3}$

25. Center $(-8, 1)$, passing through $(0, 16)$ **26.** Center $(3, -6)$, tangent to the x-axis

Find the center and radius of each of the following that are circles.

27. $x^2 - 4x + y^2 + 6y + 12 = 0$ **28.** $x^2 - 6x + y^2 - 10y + 30 = 0$

29. $x^2 + 7x + y^2 + 3y + 1 = 0$ **30.** $x^2 + 11x + y^2 - 5y + 46 = 0$

Decide whether the equations below have graphs that are symmetric to the x-axis or the y-axis.

31. $3y^2 - 5x^2 = 15$ **32.** $x + y^2 = 8$ **33.** $y^3 = x + 1$ **34.** $x^2 = y^3$

35. $|y| = -x$ **36.** $|x + 2| = |y - 3|$ **37.** $|x| = |y|$ **38.** $xy = 8$

39. Graph $y = |x|$, and use the result in Exercises 40–44.

Using your graph from Exercise 39, explain how each of the following graphs could be obtained. Sketch each graph.

40. $y = |x| - 3$ **41.** $y = -|x|$ **42.** $y = -|x| - 2$

43. $y = -|x + 1| + 3$

44. $y = 2|x - 3| - 4$

Give the domain of the functions defined as follows.

45. $y = -4 + |x|$

46. $y = 3x^2 - 1$

47. $y = (x - 4)^2$

48. $y = 8 - x$

49. $y = \dfrac{8 + x}{8 - x}$

50. $y = -\sqrt{\dfrac{5}{x^2 + 9}}$

51. $y = \sqrt{49 - x^2}$

52. $y = -\sqrt{x^2 - 4}$

The graph of a function f is shown in the figure below. Sketch the graph of each of the functions defined as follows.

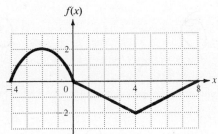

53. $y = f(x) + 3$

54. $y = f(x) - 4$

55. $y = f(x - 2)$

56. $y = f(x + 4)$

57. $y = f(x + 3) - 2$

58. $y = f(x - 1) + 4$

59. $y = |f(x)|$

For each of the following, find $\dfrac{f(x + h) - f(x)}{h}$, if $h \neq 0$.

60. $f(x) = -2x^2 + 4x - 3$

61. $f(x) = -x^3 + 2x^2$

62. $f(x) = 5$

63. $f(x) = \sqrt{x}$

64. $f(x) = -1/x$

Decide whether the functions defined as follows are odd, even, or neither.

65. $g(x) = \dfrac{x}{|x|} \ (x \neq 0)$

66. $h(x) = 9x^6 + 5|x|$

67. $f(x) = x$

68. $f(x) = -x^4 + 9x^2 - 3x^6$

Let $f(x) = 3x^2 - 4$ and $g(x) = x^2 - 3x - 4$. Find each of the following.

69. $(f + g)(x)$

70. $(f \cdot g)(x)$

71. $(f - g)(4)$

72. $(f + g)(-4)$

73. $(f + g)(2k)$

74. $(f \cdot g)(1 + r)$

75. $(f/g)(3)$

76. $(f/g)(-1)$

77. Give the domain of $(f \cdot g)(x)$.

78. Give the domain of $(f/g)(x)$.

Let $f(x) = \sqrt{x - 2}$ and $g(x) = x^2$. Find each of the following.

79. $(f \circ g)(x)$

80. $(g \circ f)(x)$

81. $(f \circ g)(-6)$

82. $(f \circ g)(2)$

83. $(g \circ f)(3)$

84. $(g \circ f)(24)$

Find all intervals where the following functions are increasing or decreasing.

85. $y = -(x + 1)^2$

86. $y = -3 + |x|$

87. $y = \dfrac{|x|}{x}$

88. $y = |x| + x^2$

Which of the functions graphed or defined as follows are one-to-one?

89. $f(x)$

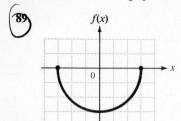

90. $f(x)$

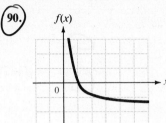

91. $f(x)$

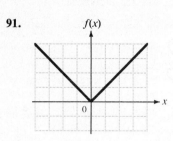

92. $y = \dfrac{8x - 9}{5}$

93. $y = -x^2 + 11$

94. $y = \sqrt{5 - x}$

95. $y = \sqrt{100 - x^2}$

96. $y = -\sqrt{1 - \dfrac{x^2}{100}}, \; x \geq 0$

For each of the functions defined as follows that is one-to-one, write an equation for the inverse function in the form $y = f^{-1}(x)$ and then graph f and f^{-1}.

97. $f(x) = 12x + 3$

98. $f(x) = \dfrac{2}{x - 9}$

99. $f(x) = x^3 - 3$

100. $f(x) = x^2 - 6$

101. $f(x) = \sqrt{25 - x^2}$, domain $[0, 5]$

102. $f(x) = -\sqrt{x - 3}$

Write each of the statements below as an equation.

103. m varies directly as the square of z

104. y varies inversely as r and directly as the cube of p

105. Y varies jointly as M and the square of N and inversely as the cube of X

106. A varies jointly as the third power of t and the fourth power of s, and inversely as p and the square of h.

Solve each problem below.

107. Suppose r varies directly as x and inversely as the square of y. If r is 10 when x is 5 and y is 3, find r when x is 12 and y is 4.

108. Suppose m varies jointly as n and the square of p, and inversely as q. If m is 20 when n is 5, p is 6, and q is 18, find m when n is 7, p is 11, and q is 2.

109. Suppose Z varies jointly as the square of J and the cube of M, and inversely as the fourth power of W. If Z is 125 when J is 3, M is 5 and W is 1, find Z if J is 2, M is 7, and W is 3.

110. The power a windmill obtains from the wind varies directly as the cube of the wind velocity. If a 10 kph wind produces 10,000 units of power, how much power is produced by a wind of 15 kph?

111. Hooke's Law for an elastic spring states that the distance a spring stretches varies directly as the force applied. If a force of 32 pounds stretches a certain spring 48 inches, how much will a force of 24 pounds stretch the spring?

112. A baseball diamond is a square, 90 feet on a side. Casey runs at a constant velocity of 30 ft/sec whether he hits a ground ball or a home run. Today, in his first time at bat, he hit a home run. Write an expression for the function that measures his line-of-sight distance from second base as a function of the time t, in sec, after he left home plate.

113. Alec, on vacation in Canada, found that he got a 12% premium on his U.S. money. When he returned, he discovered there was a 12% discount on converting his Canadian money back into U.S. currency. Describe each conversion function and then show that one is not the inverse of the other. In other words, show that after converting both ways, Alec lost money.

114. Let $f(x)$ be the fifth decimal place of the decimal expansion of x. For example, $f(1/64)$ $= f(0.015625) = 2$, $f(98.7865433210) = 4$, $f(-78.90123456) = 3$, etc. Find the domain and range of f.

115. Prove that a graph that is symmetric with respect to any two perpendicular lines is also symmetric with respect to their point of intersection.

116. If the point $P(x, y)$ is on the line through $P_1(x_1, y_1)$ and $P_2(x_2, y_2)$ such that $d(P_1, P)/d(P_1, P_2) = k$, prove that the coordinates of P are given by $x = x_1 + k(x_2 - x_1)$ and $y = y_1 + k(y_2 - y_1)$, where

$0 < k < 1$ if P is between P_1 and P_2

$k > 1$ if P is not between P_1 and P_2 and is closer to P_2

$k < 0$ if P is not between P_1 and P_2 and is closer to P_1

117. Find formulas for $(f \circ g)(x)$ if

$$f(x) = \begin{cases} 0 & \text{if } x < 0 \\ 2x & \text{if } 0 \le x \le 1 \\ 0 & \text{if } x > 1 \end{cases} \quad \text{and} \quad g(x) = \begin{cases} 1 & \text{if } x < 0 \\ x/2 & \text{if } 0 \le x \le 1 \\ 1 & \text{if } x > 1. \end{cases}$$

118. Find formulas for $(g \circ f)(x)$ for the functions in Exercise 117.

119. Find formulas for $(f \circ g)(x)$ if

$$f(x) = \begin{cases} 0 & \text{if } x < 0 \\ x^2 & \text{if } 0 \le x \le 1 \\ 0 & \text{if } x > 1 \end{cases} \quad \text{and} \quad g(x) = \begin{cases} 1 & \text{if } x < 0 \\ 2x & \text{if } 0 \le x \le 1 \\ 1 & \text{if } x > 1. \end{cases}$$

120. From the origin, chords (lines cutting the circle in two points and perpendicular to a diameter) are drawn. Prove that the set of midpoints of these chords is a circle.

121. Given $$f(x) = \begin{cases} x & \text{if } x < 1 \\ x^2 & \text{if } 1 \le x \le 9 \\ 27\sqrt{x} & \text{if } 9 < x, \end{cases}$$

prove that f has an inverse function and find $f^{-1}(x)$.

122. Determine the value of the constant k so that the function defined by

$$f(x) = \frac{x + 5}{x + k}$$

will be its own inverse.

Each of the following statements is false in general. While particular functions satisfying the statements might be found, the statements are not true for all functions. Find examples of functions showing each statement false.

123. $f(x^2) = [f(x)]^2$

124. $f(x + y) = f(x) + f(y)$

125. $f(xy) = f(x) \cdot f(y)$

126. $f\left(\dfrac{x}{y}\right) = \dfrac{f(x)}{f(y)}$

794-2178

4

Polynomial and Rational Functions

This chapter begins the study of polynomial functions.

Polynomial Function

A **polynomial function of degree n,** where n is a nonnegative integer, is a function defined by an expression of the form

$$f(x) = a_n x^n + a_{n-1} x^{n-1} + \ldots + a_1 x + a_0,$$

where $a_n, a_{n-1}, \ldots, a_1,$ and a_0 are real numbers, with $a_n \neq 0$.

The first polynomial functions studied in this chapter are the *linear functions,* which are polynomial functions of degree 1. *Quadratic functions,* polynomial functions of degree 2, are presented next. Then, after a discussion of other graphs with second-degree equations, polynomial functions of degree more than 2 are discussed. The chapter ends with a section on rational functions, which involve quotients of polynomials.

4.1 Linear Functions

As mentioned above, a linear function is a polynomial function of degree 1.

Linear Function

A function f is a **linear function** if

$$f(x) = ax + b,$$

for real numbers a and b, with $a \neq 0$.

It is customary to write the general linear function with $f(x) = ax + b$ rather than with $f(x) = a_1x + a_0$, as used with more general polynomial functions. As shown later, the name "linear" comes from the fact that every linear function has a graph that is a straight line.

A straight line is determined by two different points on the line. Two points that are useful for sketching the graph of a linear function are the *x-intercept* and the *y-intercept*. An **x-intercept** is an x-value (if any) at which a graph crosses the x-axis, while a **y-intercept** is a y-value (if any) at which a graph crosses the y-axis. The graph in Figure 1 has an x-intercept of x_1 and a y-intercept of y_1. The intercepts can't give more than two points on the line, and while only two different points are needed to graph a line, it is a good idea to find three points, using the third point as a check.

As suggested by the graph in Figure 1, x-intercepts can be found by letting $y = 0$, with y-intercepts found by letting $x = 0$. The first example shows how the intercepts can be helpful in graphing a line.

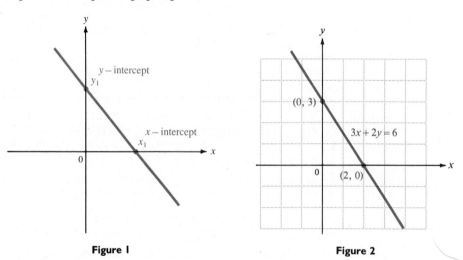

Figure I   Figure 2

Example I   Use the intercepts to graph $3x + 2y = 6$.

While $3x + 2y = 6$ is not in the form $f(x) = ax + b$, this equation actually is that of a linear function. To see why, rewrite $3x + 2y = 6$ as $y = (-3/2)x + 3$, or $f(x) = (-3/2)x + 3$. Find the y-intercept by letting $x = 0$.

$$3 \cdot 0 + 2y = 6$$
$$2y = 6$$
$$y = 3$$

For the x-intercept, let $y = 0$.

$$3x + 2 \cdot 0 = 6$$
$$3x = 6$$
$$x = 2$$

Plotting (0, 3) and (2, 0) and drawing a straight line through the two points gives the graph in Figure 2. A third point could be found as a check if desired.   ∎

Example 2   Graph $y = -3$.

Think of $y = -3$ as $y = -3 + 0 \cdot x$. This alternate form shows that for any value of x that might be chosen, y is $-3 + 0 \cdot x = -3 + 0 = -3$. Since y always equals -3, the value of y can never be 0. This means that the graph has no x-intercept. The only way a straight line can have no x-intercept is by being parallel to the x-axis, as shown in Figure 3.   ■

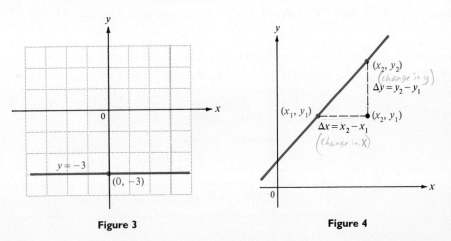

Figure 3   **Figure 4**

Slope   An important characteristic of a straight line is its *slope*, a numerical measure of the steepness of the line. To find this measure, start with the line through the two distinct points (x_1, y_1) and (x_2, y_2), as shown in Figure 4. (Since the points are distinct, $x_1 \neq x_2$). The difference

$$x_2 - x_1$$

is called the **change in x** and denoted by Δx (read "delta x"), where Δ is the Greek letter *delta*. In the same way, the **change in y** can be written

$$\Delta y = y_2 - y_1.$$

The slope of a nonvertical line, usually symbolized with the letter m, is defined as the quotient of the change in y and the change in x.

Slope

The **slope** m of the line through (x_1, y_1) and (x_2, y_2) is

$$m = \frac{\Delta y}{\Delta x} = \frac{y_2 - y_1}{x_2 - x_1}, \quad \Delta x \neq 0.$$

The slope of a line can be found only if the line is nonvertical. This guarantees that $x_2 \neq x_1$ so that the denominator $x_2 - x_1 \neq 0$. It is not possible to define the slope of a vertical line.

The slope of a vertical line is undefined.

Example 3 Find the slope of the line through each of the following pairs of points.

(a) $(-4, 8)$, $(2, -3)$

Let $x_1 = -4$, $y_1 = 8$, $x_2 = 2$, and $y_2 = -3$. Then $\Delta y = -3 - 8 = -11$ and $\Delta x = 2 - (-4) = 6$. The slope $m = \Delta y/(\Delta x) = -11/6$.

(b) $(2, 7)$, $(2, -4)$

A sketch shows that the line through $(2, 7)$ and $(2, -4)$ is vertical. As mentioned above, the slope of a vertical line is not defined. (An attempt to use the definition of slope would produce a zero denominator.)

(c) $(5, -3)$ and $(-2, -3)$

By the definition of slope,

$$m = \frac{-3 - (-3)}{-2 - 5} = \frac{0}{-7} = 0. \quad \blacksquare$$

The line through the two points in Example 3(c) is a horizontal line; this example suggests the following generalization.

The slope of a horizontal line is 0.

Figure 5 shows lines of various slopes. As the figure shows, a line with a positive slope goes up from left to right, but a line with a negative slope goes down.

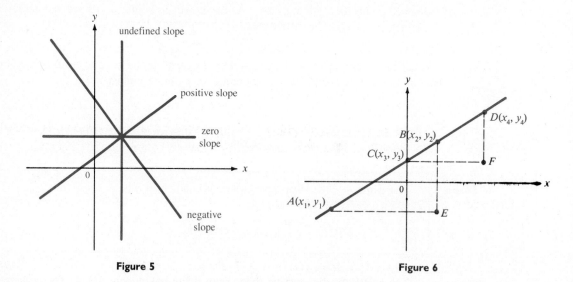

Figure 5

Figure 6

Figure 6 shows four points $A(x_1, y_1)$, $B(x_2, y_2)$, $C(x_3, y_3)$, and $D(x_4, y_4)$, all on the same straight line. Right triangles ABE and CDF have been completed. The lines

AE and *CF* are parallel, as are the lines *EB* and *FD*, making triangles *ABE* and *CDF* similar. Since similar triangles have corresponding sides proportional,

$$\frac{y_2 - y_1}{x_2 - x_1} = \frac{y_4 - y_3}{x_4 - x_3}.$$

The quotient on the left is the one used in the definition of slope, and it equals the quotient on the right for any choice of four distinct points on the line, showing that the slope of a line is the same regardless of the pair of distinct points chosen.

Example 4   Graph the line passing through $(-1, 5)$ and having slope $-5/3$.

First locate the point $(-1, 5)$, as shown in Figure 7. The slope of this line is $-5/3$, so a change of 3 units horizontally produces a change of -5 units vertically. Starting at $(-1, 5)$ and going 5 units down and 3 units across gives a second point, $(2, 0)$, which can then be used to complete the graph.   ■

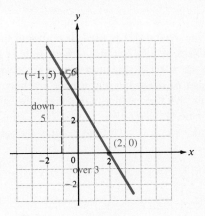

Figure 7

Some applications require a function whose graph is made up of parts of two or more lines. The next examples show such functions, called **functions defined piecewise.**

Example 5   Graph the function defined by

$$f(x) = \begin{cases} x + 1 \text{ if } x > 2 \\ -2x + 5 \text{ if } x \le 2. \end{cases}$$

For $x > 2$, graph $f(x) = x + 1$. For $x \le 2$, graph $f(x) = -2x + 5$, as shown in Figure 8(a) on the next page. A single parenthesis is used at $(2, 3)$ to show that this point is not part of the graph, while the square bracket at $(2, 1)$ shows that the point is part of the graph.   ■

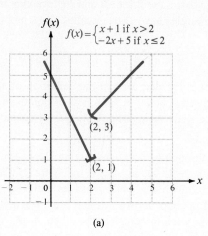

$$f(x)=\begin{cases}x+1 \text{ if } x>2 \\ -2x+5 \text{ if } x\leq 2\end{cases}$$

(a)

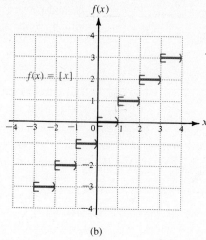

$f(x) = [x]$

(b)

Figure 8

Example 6 The symbol $[x]$ is used to represent the greatest integer less than or equal to x. For example, $[8.4] = 8$, $[-5] = -5$, $[\pi] = 3$, $[-6.9] = -7$, and so on. A graph of $f(x) = [x]$ is shown in Figure 8(b), with parentheses used for endpoints that are not part of the graph and square brackets used for endpoints that are part of the graph. As the vertical line test shows, this graph is that of a function. Also, there are no intervals where the function is increasing or decreasing. The function defined by $f(x) = [x]$ is called the **greatest-integer function.** ∎

4.1 Exercises

Graph each of the following.

1. $x - y = 4$ **2.** $x + y = 4$ **3.** $y = 3x + 1$ **4.** $2x + y = 4$

5. $3x - y = 6$ **6.** $2x - 3y = 6$ **7.** $2x + 5y = 10$ **8.** $3x + 4y = 12$

9. $4x - 3y = 9$ **10.** $3y - 5x = 15$ **11.** $x = 2$ **12.** $y = -3$

13. $y = -5$ **14.** $x = 6$ **15.** $y = 3x$ **16.** $x = -2y$

17. $x = -5y$ **18.** $y = x$

Graph the line passing through the given point and having the indicated slope.

19. Through $(-1, 3)$, $m = 3/2$ **20.** Through $(-2, 8)$, $m = -1$

21. Through $(3, -4)$, $m = -1/3$ **22.** Through $(-2, -3)$, $m = -3/4$

23. Through $(-1, 4)$, $m = 0$ **24.** Through $(2, -5)$, $m = 0$

25. Through $(3, 2/3)$, undefined slope   **26.** Through $(9/4, 2)$, undefined slope

Find the slope of each of the following lines. *Hint:* in Exercises 35–44, find two points on the line.

27. Through $(-2, 1)$ and $(3, 2)$ **28.** Through $(-2, 3)$ and $(-1, 2)$ **29.** Through $(8, 4)$ and $(-1, -3)$

30. Through $(-4, -3)$ and $(5, 0)$ **31.** Through $(4, 5)$ and $(-1, 2)$ **32.** Through $(1, 5)$ and $(-2, 5)$

33. Through (2, 3) and (2, 7)

34. Through (10, −2) and (2, −1)

35. $y = 2x$

36. $y = 3x − 2$

37. $y = 4x − 5$

38. $y = −2x + 1$

39. $3x + 4y = 6$

40. $2x + y = 8$

41. $y = 4$

42. $x = −6$

43. The x-axis

44. The y-axis

45. Through (−1.978, 4.806) and (3.759, 8.125)

46. Through (11.72, 9.811) and (−12.67, −5.009)

Use slopes to decide whether or not the following points lie on a straight line. (*Hint:* in Exercise 47, find the slopes of the lines through M and N and through M and P. If these slopes are the same, then the points lie on a straight line.)

47. $M(1, −2), N(3, −18), P(−2, 22)$

48. $A(0, −2), B(3, 7), C(−4, −14)$

49. $X(1, 3), Y(−4, 73), Z(5, −50)$

50. $R(0, −7), S(8, −11), T(−9, −25)$

Graph the functions defined as follows.

51. $f(x) = \begin{cases} x − 1 & \text{if } x \le 3 \\ 2 & \text{if } x > 3 \end{cases}$

52. $f(x) = \begin{cases} 6 − x & \text{if } x \le 3 \\ 3x − 6 & \text{if } x > 3 \end{cases}$

53. $g(x) = \begin{cases} 4 − x & \text{if } x < 2 \\ 1 + 2x & \text{if } x \ge 2 \end{cases}$

54. $h(x) = \begin{cases} −2 & \text{if } x \ge 1 \\ 2 & \text{if } x < 1 \end{cases}$

55. $k(x) = \begin{cases} 2x + 1 & \text{if } x \ge 0 \\ x & \text{if } x < 0 \end{cases}$

56. $g(x) = \begin{cases} 5x − 4 & \text{if } x \ge 1 \\ x & \text{if } x < 1 \end{cases}$

57. $h(x) = \begin{cases} 2 + x & \text{if } x < −4 \\ −x & \text{if } −4 \le x \le 5 \\ 3x & \text{if } x > 5 \end{cases}$

58. $f(x) = \begin{cases} −2x & \text{if } x < −3 \\ 3x + 1 & \text{if } −3 \le x < 2 \\ −4x & \text{if } x \ge 2 \end{cases}$

59. $p(x) = \begin{cases} |x| & \text{if } x > −2 \\ x & \text{if } x \le −2 \end{cases}$

60. $r(x) = \begin{cases} |x| − 1 & \text{if } x > −1 \\ x − 1 & \text{if } x \le −1 \end{cases}$

61. When a diabetic takes long-acting insulin, the insulin reaches its peak effect on the blood sugar level in about three hours. This effect remains fairly constant for five hours, then declines, and is very low until the next injection. In a typical patient, the level of insulin might be given by the following function.

$$i(t) = \begin{cases} 40t + 100 & \text{if } 0 \le t \le 3 \\ 220 & \text{if } 3 < t \le 8 \\ −80t + 860 & \text{if } 8 < t \le 10 \\ 60 & \text{if } 10 < t \le 24 \end{cases}$$

Here $i(t)$ is the blood sugar level, in appropriate units, at time t measured in hours from the time of the injection. Suppose a patient takes insulin at 6 A.M. Find the blood sugar level at each of the following times: **(a)** 7 A.M. **(b)** 9 A.M. **(c)** 10 A.M. **(d)** noon **(e)** 2 P.M. **(f)** 5 P.M. **(g)** midnight. **(h)** Graph $y = i(t)$.

62. To rent a midsized car from Avis costs $27 per day or fraction of a day. If you pick up the car in Lansing, and drop it in West Lafayette, there is a fixed $25 dropoff charge. Let $C(x)$ represent the cost of renting the car for x days, taking it from Lansing to West Lafayette. Find each of the following: **(a)** $C(3/4)$ **(b)** $C(9/10)$ **(c)** $C(1)$ **(d)** $C\left(1\frac{5}{8}\right)$ **(e)** $C(2.4)$. **(f)** Graph $y = C(x)$. **(g)** Is C a function? **(h)** Is C a linear function?

Graph the functions defined as follows.

63. $f(x) = [−x]$

64. $f(x) = [2x]$

65. $g(x) = [2x − 1]$

66. $h(x) = [3x + 1]$

67. $k(x) = [3x]$

68. $r(x) = [3x] + 1$

69. $h(x) = [3x] − 1$

70. $f(x) = x − [x]$

Solve each of the following problems.

71. Suppose a chain-saw rental firm charges a fixed $4 sharpening fee plus $7 per day or fraction of a day. Let $S(x)$ represent the cost of renting a saw for x days. Find the value of **(a)** $S(1)$ **(b)** $S(1.25)$ **(c)** $S(3.5)$. **(d)** Graph $y = S(x)$. **(e)** Give the domain and range of S.

72. Assume that it costs 30¢ to mail a letter weighing one ounce or less, and then 27¢ for each additional ounce or fraction of an ounce. Let $L(x)$ be the cost of mailing a letter weighing x ounces. Find **(a)** $L(.75)$ **(b)** $L(1.6)$ **(c)** $L(4)$. **(d)** Graph $y = L(x)$. **(e)** Give the domain and range of L.

73. Use the greatest integer function and write an expression for the number of ounces for which postage will be charged on a letter weighing x ounces (See Exercise 72).

74. In a recent year Washington, D.C. taxi rates were 90¢ for the first 1/9 of a mile and 10¢ for each additional 1/9 mile or fraction of 1/9. Let $C(x)$ be the cost for a taxi ride of x 1/9's of a mile. Find the following: **(a)** $C(1)$ **(b)** $C(2.3)$ **(c)** $C(8)$. **(d)** Graph $y = C(x)$. **(e)** Give the domain and range of C.

75. For a lift truck rental of no more than three days, the charge is $300. An additional charge of $75 is made for each day or portion of a day after three. Graph the ordered pairs (number of days, cost).

76. A car rental cost $37 for one day, which includes 50 free miles. Each additional 25 miles or portion costs $10. Graph the ordered pairs (miles, cost) for a one-day rental.

77. Straight line graphs are often good choices for *supply-demand curves*. In most supply-demand curves, the supply of an item goes up as the price goes up, while the demand goes down as the price goes up. Suppose that the demand and price for a certain model of electric can opener are related by

$$p = 16 - \frac{5}{4}x,$$

where p is price and x is demand, in appropriate units. Find the price when the demand is at the following levels: **(a)** 0 units **(b)** 4 units **(c)** 8 units.
Find the demand for the electric can opener at the following prices: **(d)** $6 **(e)** $11 **(f)** $16. **(g)** Graph $p = 16 - (5/4)x$.
Suppose that the price and supply of the item above are related by

$$p = \frac{3}{4}x,$$

where x represents the supply and p the price. Find the supply at the following prices: **(h)** $0 **(i)** $10 **(j)** $20. **(k)** Graph $p = (3/4)x$ on the same axes used for part (g). **(l)** Find the equilibrium supply (the supply at the point where the supply and demand curves cross). **(m)** Find the equilibrium price (the price at the equilibrium supply).

78. Let the supply and demand equations for strawberry-flavored licorice be

$$\text{supply: } p = \frac{3}{2}x \quad \text{and} \quad \text{demand: } p = 81 - \frac{3}{4}x.$$

(a) Graph these on the same axes. **(b)** Find the equilibrium demand. (See Exercise 77.) **(c)** Find the equilibrium price.

4.2 Equations of a Line

The previous section showed that the graph of a linear function is a straight line. This section shows various forms for the equation of a line. The first equation given is for the only line which is not the graph of a function, a *vertical line*. The vertical line through the point $(k, 0)$ goes through all points of the form (k, y); this means

Equation of a Vertical Line

> an equation of the vertical line through $(k, 0)$ is
> $$x = k.$$

For example, the vertical line through $(-6, 0)$ has equation $x = -6$, or $x + 6 = 0$. (Horizontal lines are the graphs of functions, as discussed later in this section.)

Point-Slope Form Suppose now that a line has slope m and goes through the fixed point (x_1, y_1), as in Figure 9. Let (x, y) be any other point on this line. By the definition of slope, the slope of this line is

$$\frac{y - y_1}{x - x_1}.$$

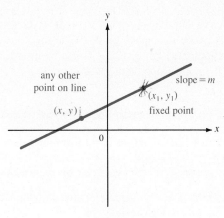

Figure 9

Since the slope of the line is m,

$$\frac{y - y_1}{x - x_1} = m.$$

Multiplying both sides by $x - x_1$ gives

$$y - y_1 = m(x - x_1) \tag{*}$$

Substituting x_1 for x and y_1 for y shows that this equation also holds for the point (x_1, y_1), showing that any point (x, y) whose coordinates satisfy equation $(*)$ above lies on the given line, and also that any point on the line has coordinates satisfying the equation. The equation of this discussion, called the *point-slope form* of the equation of a line, is summarized as follows.

Point-Slope Form

The line with slope m passing through the point (x_1, y_1) has an equation

$$y - y_1 = m(x - x_1),$$

$$\Delta y = m \, \Delta x$$

the **point-slope form** of the equation of a line.

Example 1 Write an equation of the line through $(-4, 1)$ with slope -3.

Here $x_1 = -4$, $y_1 = 1$, and $m = -3$. Use the point-slope form of the equation of a line to get

$$y - 1 = -3[x - (-4)]$$
$$y - 1 = -3(x + 4)$$
$$y - 1 = -3x - 12$$
$$3x + y = -11. \quad \blacksquare$$

Example 2 Find an equation of the line through $(-3, 2)$ and $(2, -4)$.

First find the slope, using the definition.

$$m = \frac{-4 - 2}{2 - (-3)} = -\frac{6}{5}.$$

Either $(-3, 2)$ or $(2, -4)$ can be used for (x_1, y_1). Choosing $x_1 = -3$ and $y_1 = 2$, the point-slope form gives

$$y - 2 = -\frac{6}{5}[x - (-3)]$$

$$5(y - 2) = -6(x + 3)$$
$$5y - 10 = -6x - 18$$
$$6x + 5y = -8.$$

Verify that this same equation is obtained if $(2, -4)$ is used instead of $(-3, 2)$ in the point-slope form. $\quad \blacksquare$

Slope-Intercept Form As a special case of the point-slope form of the equation of a line, suppose that a line passes through the point $(0, b)$, so the line has y-intercept b. If the line has slope m, then using the point-slope form with $x_1 = 0$ and $y_1 = b$ gives

$$y - y_1 = m(x - x_1)$$
$$y - b = m(x - 0)$$
$$y = mx + b,$$

as an equation of the line. This result is called the *slope-intercept form* of the equation of a line since it shows the slope and the y-intercept.

Slope-Intercept Form	The line with slope m and y-intercept b has an equation $$y = mx + b,$$ the **slope-intercept form** of the equation of a line.

Example 3 Find the slope and y-intercept of $3x - y = 2$. Graph the line.

First write $3x - y = 2$ in the slope-intercept form, $y = mx + b$. Do this by solving for y, getting $y = 3x - 2$. From this form of the equation, the slope is $m = 3$ and the y-intercept is $b = -2$. Draw the graph by first locating the y-intercept, as in Figure 10. Then, as in the previous section, use the slope of 3, or 3/1, to get a second point on the graph. The line through these two points is the graph of $3x - y = 2$. ■

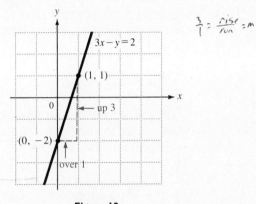

Figure 10

As mentioned in the previous section, a horizontal line has a slope of 0. Letting $m = 0$ in the slope-intercept form of the equation of a line gives $y = 0x + b$, or $y = b$, an equation of a horizontal line.

Horizontal Line	An equation of the horizontal line through $(0, b)$ is $$y = b.$$

The slope-intercept form, together with the vertical-line equation $x = k$, show that every line has an equation of the form $ax + by + c = 0$, where a and b are not both 0. Conversely, assuming $b \neq 0$, solving $ax + by + c = 0$ for y gives $y = (-a/b)x - c/b$, the equation of a line with slope $-a/b$ and y-intercept $-c/b$. If $b = 0$, solving for x gives $x = -c/a$, a vertical line. In any case, the equation $ax + by + c = 0$ has a line for its graph.

If a and b are not both 0, then the equation $ax + by + c = 0$ has a line for its graph. Also, any line has an equation of the form $ax + by + c = 0$.

Parallel and Perpendicular Lines One application of slope involves deciding whether or not two lines are parallel. Since two parallel lines are equally "steep," they should have the same slope. Also, two lines with the same "steepness" are parallel. Thus,

Parallel Lines

two nonvertical lines are parallel if and only if they have the same slope.

A proof of this statement is requested in Exercise 43.

Example 4 Find the equation of the line through the point $(3, 5)$ and parallel to the line $2x + 5y = 4$.

The point-slope form of the equation of a line requires a point that the line goes through and the slope of the line. The line here goes through $(3, 5)$. Find the slope by writing the given equation in slope-intercept form (that is, solve for y).

$$2x + 5y = 4$$

$$y = -\frac{2}{5}x + \frac{4}{5}$$

The slope is $-2/5$. Since the lines are parallel, the slope of the line whose equation is needed must also be $-2/5$. Substituting $m = -2/5$, $x_1 = 3$, and $y_1 = 5$ into the point-slope form gives

$$y - y_1 = m(x - x_1)$$

$$y - 5 = -\frac{2}{5}(x - 3)$$

$$5(y - 5) = -2(x - 3)$$

$$5y - 25 = -2x + 6$$

$$2x + 5y = 31. \quad \blacksquare$$

Two lines with the same slope are parallel. As stated below, two lines having slopes with a product of -1 are perpendicular.

Perpendicular Lines

Two lines, neither of which is vertical, are perpendicular if and only if their slopes have a product of -1.

Exercises 47–50 outline a proof of this statement.

Example 5 Find the equation of the line L through $(3, 0)$ and perpendicular to $5x - y = 4$.

Use the point-slope form to get the equation of line L. To find the slope of line L, first find the slope of $5x - y = 4$ by solving for y.

$$5x - y = 4$$

$$y = 5x - 4$$

This line has a slope of 5. Since the lines are perpendicular, if line L has slope m, then $5m = -1$ and $m = -1/5$. Now use the point-slope form to find the required equation.

$$y - 0 = -\frac{1}{5}(x - 3)$$

$$5y = -x + 3$$

$$x + 5y = 3. \quad \blacksquare$$

All the lines above have equations that could be placed in the form $ax + by = c$ for real numbers a, b, and c. The equation $ax + by = c$ is called the *standard form of the equation of a line*. The various forms of the equation of a line can be summarized as follows.

Equations of a Line

General equation	Description of line
$ax + by = c$	**Standard form** x-intercept c/a, y-intercept c/b, slope $-a/b$ (if $a \neq 0$ and $b \neq 0$)
$x = k$	**Vertical line** x-intercept k, no y-intercept, line has undefined slope
$y = k$	**Horizontal line** No x-intercept, y-intercept k, line has slope 0
$y = mx + b$	**Slope-intercept form** Slope is m, y-intercept is b
$y - y_1 = m(x - x_1)$	**Point-slope form** Slope is m, line passes through (x_1, y_1)

4.2 Exercises

For each of the following lines, write an equation in standard form.

1. Through $(1, 3)$, $m = -2$

2. Through $(2, 4)$, $m = -1$

3. Through $(-5, 4)$, $m = -3/2$

4. Through $(-4, 3)$, $m = 3/4$

5. Through $(2, 0)$, $m = -3/4$

6. Through $(0, -1)$, $m = 3/5$

7. Through $(-3, 2)$, $m = 0$

8. Through $(6, 1)$, $m = 0$

9. Through $(-8, 1)$, undefined slope

10. Through $(0, 4)$, undefined slope

11. Through $(4, 2)$ and $(1, 3)$

12. Through $(8, -1)$ and $(4, 3)$

13. Through $(-1, 3)$ and $(3, 4)$

14. Through $(6, 0)$ and $(3, 2)$

15. Through $(0, 3)$ and $(4, 0)$

16. Through $(-3, 0)$ and $(0, -5)$

17. x-intercept 3, y-intercept -2

18. x-intercept -2, y-intercept 4

19. x-intercept -5, no y-intercept

20. y-intercept 3, no x-intercept

21. Vertical, through $(-6, 5)$

22. Horizontal, through $(8, 7)$

23. Through $(-1.76, 4.25)$, slope -5.081

24. Through $(5.469, 11.08)$, slope 4.723

Write equations in standard form for each of the following lines.

25. Through $(-1, 4)$, parallel to $x + 3y = 5$

26. Through $(3, -2)$, parallel to $2x - y = 5$

27. Through $(2, -5)$, parallel to $y - 4 = 2x$

28. Through $(0, 5)$, parallel to $3x - 7y = 8$

29. Through $(3, -4)$, perpendicular to $x + y = 4$

30. Through $(-2, 6)$, perpendicular to $2x - 3y = 5$

31. Through $(1, 6)$, perpendicular to $3x + 5y = 1$

32. Through $(-2, 0)$, perpendicular to $8x - 3y = 7$

33. x-intercept -2, parallel to $y = 2x$

34. y-intercept 3, parallel to $x + y = 4$

35. Through $(-5, 7)$, perpendicular to $y = -2$

36. Through $(1, -4)$, perpendicular to $x = 4$

37. Do the points $(4, 3)$, $(2, 0)$, and $(-18, -12)$ lie on the same line? (*Hint:* Find the equation of the line through two of the points.)

38. Do the points $(4, -5)$, $(3, -5/2)$, and $(-6, 18)$ lie on a line?

39. Find k so that the line through $(4, -1)$ and $(k, 2)$ is
    (a) parallel to $3y + 2x = 6$
    (b) perpendicular to $2y - 5x = 1$.

40. Find r so that the line through $(2, 6)$ and $(-4, r)$ is
    (a) parallel to $2x - 3y = 4$
    (b) perpendicular to $x + 2y = 1$.

41. Use slopes to show that the quadrilateral with vertices at $(1, 3)$, $(-5/2, 2)$, $(-7/2, 4)$, and $(2, 1)$ is a parallelogram (has opposite sides parallel).

42. Use slopes to show that the square with vertices at $(-2, 5)$, $(4, 5)$, $(4, -1)$, and $(-2, -1)$ has diagonals that are perpendicular.

Prove each of the following statements.

43. (a) Two nonvertical parallel lines have the same slope.
    (b) Two lines with the same slope are parallel.

44. The linear function defined by $f(x) = ax + b$ is increasing if $a > 0$ and decreasing if $a < 0$.

45. The line $y = x$ is the perpendicular bisector of the segment connecting (a, b) and (b, a), where $a \neq b$.

46. The line $ax + by = 0$, where $ab \neq 0$, goes through the origin. If $b \neq 0$, the slope of the line is $-a/b$.

To prove that two perpendicular lines, neither of which is vertical, have slopes with a product of -1, go through the following steps. Let line L_1 have equation $y = m_1x + b_1$, and let line L_2 have equation $y = m_2x + b_2$. Assume that L_1 and L_2 are perpendicular, and complete right triangle MPN as shown in the figure.

47. Show that MQ has length m_1.

48. Show that QN has length $-m_2$.

49. Show that triangles MPQ and PQN are similar.

50. Show that $m_1/1 = 1/-m_2$ and that $m_1m_2 = -1$.

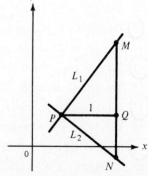

Find and simplify the slope of the lines through the points $(x, f(x))$ and $(x + h, f(x + h))$ for the functions f defined by the following rules.

51. $f(x) = x^2$ **52.** $f(x) = \sqrt{x}$ **53.** $f(x) = 1/x$ **54.** $f(x) = x^3$

Let two nonvertical lines have equations $y = m_1x + b_1$ and $y = m_2x + b_2$, respectively.

55. Set $m_1x + b_1$ equal to $m_2x + b_2$ and solve for x to find the x-coordinate of the point of intersection of the lines.

56. Show that the lines are parallel if and only if $m_1 = m_2$.

Many real situations can be described approximately by a straight-line graph. One way to find the equation of such a straight line is to use two typical data points from the graph and the point-slope form of the equation of a line. In each of the following problems, assume that the data can be approximated fairly closely by a straight line. Use the given information to find the equation of the line. Find the slope of each line.

57. A company finds that it can make a total of 20 solar heaters for $13,900, while 10 heaters cost $7500. Let y be the total cost to produce x solar heaters.

58. The sales of a small company were $27,000 in its second year of operation and $63,000 in its fifth year. Let y represent sales in year x.

59. When a certain industrial pollutant is introduced into a river, the reproduction of catfish declines. In a given period of time, three tons of the pollutant results in a fish population of 37,000. Also, 12 tons of pollutant produce a fish population of 28,000. Let y be the fish population when x tons of pollutant are introduced into the river.

60. In the snake *Lampropelbis polyzona*, total length y is related to tail length x in the domain 30 mm $\leq x \leq$ 200 mm by a linear function. Find such a linear function, if a snake 455 mm long has a 60 mm tail, and a 1050 mm snake has a 140 mm tail.

61. According to research done by the political scientist James March, if the Democrats win 45% of the two-party vote for the House of Representatives, they win 42.5% of the seats. If the Democrats win 55% of the vote, they win 67.5% of the seats. Let y be the percent of seats won, and x the percent of the two-party vote.

62. If the Republicans win 45% of the two-party vote, they win 32.5% of the seats (see Exercise 61). If they win 60% of the vote, they get 70% of the seats. Let y represent the percent of the seats, and x the percent of the vote.

4.3 Quadratic Functions

Recall that polynomial functions were defined on the first page of this chapter. The previous sections discussed linear functions, which are polynomial functions of degree 1. This section looks at polynomial functions of degree 2, called *quadratic functions*.

Quadratic Function

A function f is a **quadratic function** if

$$f(x) = ax^2 + bx + c,$$

where a, b, and c are real numbers with $a \neq 0$.

The simplest quadratic function is given by $f(x) = x^2$ with $a = 1$, $b = 0$, and $c = 0$. To find some points on the graph of this function, choose some values for x and find the corresponding values for $f(x)$, as in the chart with Figure 11. Then plot these points, and draw a smooth curve through them. (The reason for drawing a smooth curve depends on ideas from calculus.) As mentioned in Chapter 3, this graph is called a **parabola.** Every quadratic function has a graph that is a parabola.

Parabolas have symmetry about a line (the y-axis in Figure 11). The line of symmetry for a parabola is called the **axis** of the parabola. The point where the axis intersects the parabola is the **vertex** of the parabola.

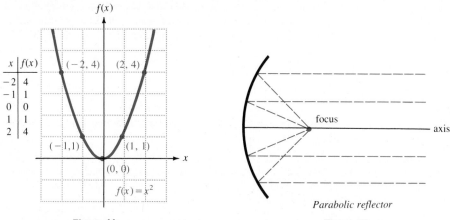

x	$f(x)$
-2	4
-1	1
0	0
1	1
2	4

Figure 11

Parabolic reflector

Figure 12

Parabolas have many practical applications. For example, if a light source is placed at the focus of a parabolic reflector, as in Figure 12, light rays reflect parallel to the axis, making a spotlight or flashlight.

The process also works in reverse. Light rays from a distant source come in parallel to the axis and are reflected to a point at the focus. (If such a reflector is aimed at the sun, a temperature of several thousand degrees may be obtained.)

Starting with $f(x) = x^2$, there are several possible ways to get a more general expression:

$f(x) = ax^2$	Multiply by a positive or negative coefficient
$f(x) = x^2 + k$	Add a positive or negative constant
$f(x) = (x - h)^2$	Replace x with $x - h$, where h is a constant
$f(x) = a(x - h)^2 + k$	Do all of the above

The graph of each of these quadratic functions is still a parabola, but modified from that of $f(x) = x^2$. The next few examples show how these changes modify the graphs of the functions. The first example shows the result of changing $f(x) = x^2$ to $f(x) = ax^2$, for a nonzero constant a.

Example 1 Graph the functions defined as follows.

(a) $f(x) = -x^2$

For a given value of x, the corresponding value of $f(x)$ will be the negative of

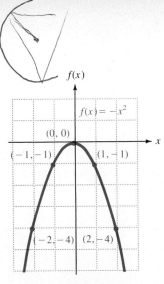

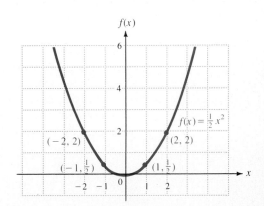

Figure 13

Figure 14

what it was for $f(x) = x^2$. Because of this, the graph of $f(x) = -x^2$ is the same shape as that of $f(x) = x^2$, but opens downward. See Figure 13.

(b) $f(x) = \dfrac{1}{2}x^2$

Choose a value of x, and then find $f(x)$. The coefficient of 1/2 will cause the resulting value of $f(x)$ to be smaller than for $f(x) = x^2$, making the parabola "broader" than $f(x) = x^2$. See Figure 14. In both parabolas of this example, the axis is the vertical line $x = 0$. ■

The next few examples show the results of horizontal and vertical translations of the graph of $f(x) = x^2$.

Example 2 Graph $f(x) = x^2 - 4$.

As shown in Chapter 3, replacing $y = f(x)$ with $y = f(x) - c$ produces a downward translation of c units. Applying this to the graph of $f(x) = x^2$ shows that the graph of $f(x) = x^2 - 4$ is the same as that of $f(x) = x^2$, but translated 4 units down. See Figure 15. The vertex of this parabola (on this parabola, the lowest point) is at $(0, -4)$. The axis of the parabola is the vertical line $x = 0$. ■

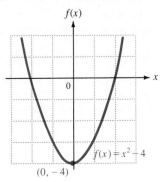

$(0, -4)$

Figure 15

Example 3   Graph $f(x) = (x - 4)^2$.

Replacing x in $y = f(x)$ with $x - c$ produces a horizontal translation of c units. This means that the graph of $f(x) = (x - 4)^2$ is the same as that of $f(x) = x^2$, but translated 4 units right. The vertex is at (4, 0). The axis of this parabola is the vertical line $x = 4$. See Figure 16.   ■

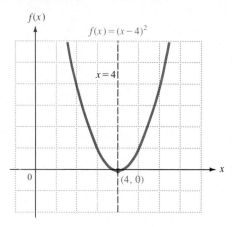

Figure 16   Figure 17

Example 4   Graph $f(x) = -(x + 3)^2 + 1$.

This parabola is translated 3 units to the left and 1 unit up. It opens downward. The vertex, the point $(-3, 1)$, is the *highest* point on the graph. The axis is the line $x = -3$. See Figure 17.   ■

Generalizing from these examples leads to the following summary.

Graph of a Quadratic Function

The graph of the quadratic function defined by

$$y = f(x) = a(x - h)^2 + k, \qquad a \neq 0,$$

1. is a parabola with vertex (h, k), and the vertical line $x = h$ as axis;
2. opens upward if $a > 0$ and downward if $a < 0$;
3. is broader than $y = x^2$ if $0 < |a| < 1$ and narrower than $y = x^2$ if $|a| > 1$.

The parabola $f(x) = ax^2 + bx + c$ has as its axis the vertical line

$$x = -\frac{b}{2a}$$

and as its vertex the point

$$\left(-\frac{b}{2a}, f\left(-\frac{b}{2a} \right) \right) = \left(-\frac{b}{2a}, \frac{-b^2 + 4ac}{4a} \right).$$

This method for finding the axis and the vertex (h, k) is derived in Exercises 57–60 below.

Example 5 Find the axis and the vertex of the parabola having equation $f(x) = 2x^2 + 4x + 5$.
Here $a = 2$, $b = 4$, and $c = 5$. From the results given above, the axis of the parabola is the vertical line

$$x = -\frac{b}{2a}$$

$$= -\frac{4}{2(2)}$$

$$x = -1.$$

The vertex is the point $(-1, f(-1))$, or, using the result above,

$$\left(-1, \frac{-4^2 + 4(2)(5)}{4(2)}\right) = \left(-1, \frac{-16 + 40}{8}\right)$$

$$= (-1, 3). \quad \blacksquare$$

Given the expression $f(x) = ax^2 + bx + c$, where a, b, and c are real numbers and $a \neq 0$, the process of *completing the square* (discussed in Chapter 2 and used with circles in Chapter 3) can be used to change $f(x) = ax^2 + bx + c$ to the form $f(x) = a(x - h)^2 + k$, so that the vertex and axis may be identified. Follow the steps given in the next two examples.

Example 6 Graph $f(x) = x^2 - 6x + 7$.
To graph this parabola, $x^2 - 6x + 7$ must be rewritten in the form $(x - h)^2 + k$. Start as follows.

$$f(x) = (x^2 - 6x \qquad) + 7$$

As shown earlier, a number must be added inside the parentheses to get a perfect square trinomial. To find this number, take half the coefficient of x and then square the result. Half of -6 is -3, and $(-3)^2$ is 9. Now add and subtract 9 inside the parentheses.

$$f(x) = (x^2 - 6x + 9 - 9) + 7$$

Factor as follows.

$$f(x) = (x^2 - 6x + 9) - 9 + 7$$
$$f(x) = (x - 3)^2 - 2$$

This result shows that the vertex of the parabola is $(3, -2)$, with axis $x = 3$. A graph is shown in Figure 18 on the next page. $\quad \blacksquare$

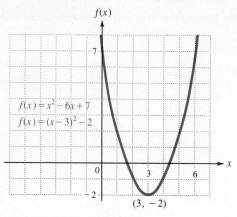

$f(x) = x^2 - 6x + 7$

$f(x) = (x - 3)^2 - 2$

$(3, -2)$

Figure 18

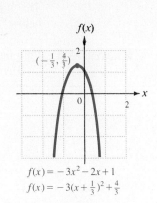

$\left(-\frac{1}{3}, \frac{4}{3}\right)$

$f(x) = -3x^2 - 2x + 1$

$f(x) = -3(x + \frac{1}{3})^2 + \frac{4}{3}$

Figure 19

Example 7   Graph $f(x) = -3x^2 - 2x + 1$.

To complete the square, first factor out -3 to get

$$f(x) = -3\left(x^2 + \frac{2}{3}x \qquad\right) + 1.$$

(This is necessary to make the coefficient of x^2 equal 1.) Half the coefficient of x is $1/3$, and $(1/3)^2 = 1/9$. Add and subtract $1/9$ inside the parentheses as follows:

$$f(x) = -3\left(x^2 + \frac{2}{3}x + \frac{1}{9} - \frac{1}{9}\right) + 1.$$

Using the distributive property and simplifying gives

$$f(x) = -3\left(x^2 + \frac{2}{3}x + \frac{1}{9}\right) - 3\left(-\frac{1}{9}\right) + 1$$

$$f(x) = -3\left(x^2 + \frac{2}{3}x + \frac{1}{9}\right) + \frac{1}{3} + 1$$

$$f(x) = -3\left(x^2 + \frac{2}{3}x + \frac{1}{9}\right) + \frac{4}{3}.$$

Factor to get

$$f(x) = -3\left(x + \frac{1}{3}\right)^2 + \frac{4}{3}.$$

Now the equation of the parabola is written in the form $f(x) = a(x - h)^2 + k$. This rewritten equation shows that the axis of the parabola is the vertical line

$$x + \frac{1}{3} = 0 \qquad \text{or} \qquad x = -\frac{1}{3}$$

and that the vertex is $(-1/3, 4/3)$. Use these results and additional ordered pairs as needed to get the graph in Figure 19. ∎

The fact that the vertex of a parabola of the form $f(x) = ax^2 + bx + c$ is the highest or lowest point on the graph can be used in applications to find a maximum or a minimum value.

Example 8 Ms. Whitney owns and operates Aunt Emma's Pie Shop. She has hired a consultant to analyze her business operations. The consultant tells her that her profit $P(x)$ is given by

$$P(x) = 120x - x^2,$$

where x is the number of units of pies that she makes. How many units of pies should be made in order to maximize the profit? What is the maximum possible profit?

The profit function can be rewritten as $P(x) = -x^2 + 120x + 0$, a quadratic function with $a = -1$, $b = 120$, and $c = 0$. Complete the square to find that the vertex of the parabola is (60, 3600). Since $a < 0$ the vertex is the highest point on the graph and produces a *maximum* rather than a minimum. Figure 20 shows that portion of the profit function in quadrant I. (Why is quadrant I the only one of interest here?) The maximum profit of $3600 is made when 60 units of pies are made. In this case, profit increases as more and more pies are made up to 60 units and then decreases as more and more pies are made past this point. ■

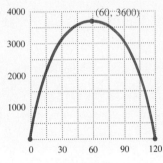

Figure 20

4.3 Exercises

1. Graph the functions defined as follows on the same coordinate system.
   (a) $f(x) = 2x^2$ **(b)** $f(x) = 3x^2$ **(c)** $f(x) = \dfrac{1}{2}x^2$ **(d)** $f(x) = \dfrac{1}{3}x^2$
   (e) How does the coefficient affect the shape of the graph?

2. Graph the functions defined as follows on the same coordinate system.
   (a) $f(x) = x^2 + 2$ **(b)** $f(x) = x^2 - 1$ **(c)** $f(x) = x^2 + 1$ **(d)** $f(x) = x^2 - 2$
   (e) How do these graphs differ from the graph of $f(x) = x^2$?

3. Graph the functions defined as follows on the same coordinate system.
   (a) $f(x) = (x - 2)^2$ **(b)** $f(x) = (x + 1)^2$ **(c)** $f(x) = (x + 3)^2$ **(d)** $f(x) = (x - 4)^2$
   (e) How do these graphs differ from the graph of $f(x) = x^2$?

4. Give the range of each of the functions defined as follows.
   (a) $f(x) = (x - 1)^2 + 5$ **(b)** $f(x) = -(x + 3)^2 - 1$
   (c) $g(x) = -2(x + 4)^2 - 3$ **(d)** $h(x) = -\dfrac{1}{2}(x - 5)^2 + 2$

Graph the functions defined as follows. Give the vertex and axis of each.

5. $f(x) = (x - 2)^2$

6. $f(x) = (x + 4)^2$

7. $g(x) = (x + 3)^2 - 4$

8. $h(x) = (x - 5)^2 - 4$

9. $k(x) = -2(x + 3)^2 + 2$

10. $F(x) = -3(x - 2)^2 + 1$

11. $H(x) = -\frac{1}{2}(x + 1)^2 - 3$

12. $G(x) = \frac{2}{3}(x - 2)^2 - 1$

13. $f(x) = x^2 - 2x + 3$

14. $f(x) = x^2 + 6x + 5$

15. $g(x) = -x^2 - 4x + 2$

16. $k(x) = -x^2 + 6x - 6$

17. $f(x) = 2x^2 - 4x + 5$

18. $g(x) = -3x^2 + 24x - 46$

Find several points satisfying each of the following equations, and then sketch the graphs.

19. $y = .14x^2 + .56x - .3$

20. $y = .82x^2 + 3.24x - .4$

21. $y = -.09x^2 - 1.8x + .5$

22. $y = -.35x^2 + 2.8x - .3$

Give the intervals where the quadratic functions defined as follows are increasing.

23. $f(x) = (x - 2)^2$

24. $f(x) = (x + 3)^2$

25. $g(x) = -x^2 + 4$

26. $h(x) = -x^2 - 6$

27. $k(x) = -2x^2 - 5x + 3$

28. $g(x) = -3x^2 - 9x + 2$

Use the quadratic formula to find the values of x where $f(x) = 0$ for each of the functions defined as follows. Use the two values you get to locate the x-value of the vertex. Find the axis of each parabola.

29. $f(x) = x^2 + 8x + 13$

30. $f(x) = x^2 - 12x + 30$

31. $f(x) = 3x^2 - 2x - 6$

32. $f(x) = 5x^2 + 6x - 3$

Use the results of Exercises 29–32 to help solve the following inequalities.

33. $x^2 + 8x + 13 < 0$ **34.** $x^2 - 12x + 30 \geq 0$ **35.** $3x^2 - 2x - 6 \geq 0$ **36.** $5x^2 + 6x - 3 < 0$

Find the distance from the origin to the vertex of each of the following parabolas.

37. $f(x) = -(x - 1)^2 - 7$

38. $f(x) = 2(x + 3)^2 + 4$

39. $f(x) = 4(x + 2)^2 + 3$

40. $f(x) = -\frac{1}{2}(x - 3)^2 + 1$

41. $f(x) = x^2 - 5x + 4$

42. $f(x) = x^2 + x - 2$

43. Glenview College wants to construct a rectangular parking lot on land bordered on one side by a highway. It has 320 feet of fencing which it will use to fence off the other three sides. What should be the dimensions of the lot if the enclosed area is to be a maximum? (*Hint:* Let x represent the width of the lot and let $320 - 2x$ represent the length. Graph the area parabola, $A = x(320 - 2x)$, and investigate the vertex.)

44. The revenue of a charter bus company depends on the number of unsold seats. If the revenue $R(x)$, is given by

$$R(x) = 5000 + 50x - x^2,$$

where x is the number of unsold seats, find the maximum revenue and the number of unsold seats which produce maximum revenue.

45. The number of mosquitoes, $M(x)$, in millions, in a certain area of Kentucky depends on the June rainfall, x, in inches, approximately as follows.

$$M(x) = 10x - x^2$$

Find the rainfall that will produce the maximum number of mosquitoes.

46. If an object is thrown upward with an initial velocity of 32 feet per second, then its height after t seconds is given by

$$h = 32t - 16t^2.$$

Find the maximum height attained by the object. Find the number of seconds it takes the object to hit the ground.

47. Find two numbers whose sum is 20 and whose product is a maximum. (*Hint:* Let x and $20 - x$ be the two numbers, and write an equation for the product.)

48. A charter flight charges a fare of $200 per person plus $4 per person for each unsold seat on the plane. If the plane holds 100 passengers, and if x represents the number of unsold seats, find the following.
  (a) An expression for the total revenue received for the flight (*Hint:* Multiply the number of people flying, $100 - x$, by the price per ticket.)
  (b) The graph for the expression in part (a)
  (c) The number of unsold seats that will produce the maximum revenue
  (d) The maximum revenue

49. The demand for a certain type of cosmetic is given by

$$p = 500 - x,$$

where p is the price per unit when x units are demanded.
  (a) Find the revenue, $R(x)$, obtained when x units are demanded. (*Hint:* Revenue = number of units demanded $\times$ price per unit.)
  (b) Graph the revenue function defined by $y = R(x)$.
  (c) From the graph of the revenue function, estimate the price that will produce maximum revenue.
  (d) What is the maximum revenue?

50. Between the months of June and October, the percent of maximum possible chlorophyll production in a leaf is approximated by $C(x)$, where

$$C(x) = 10x + 50.$$

Here x is time in months with $x = 1$ representing June. From October through December, $C(x)$ is approximated by

$$C(x) = -20(x - 5)^2 + 100,$$

with x as above. Find the percent of maximum possible chlorophyll production in each of the following months: **(a)** June **(b)** July **(c)** September **(d)** October
  (e) November **(f)** December.
  (g) Sketch a graph of $y = C(x)$, from June through December. In what month is chlorophyll production a maximum?

51. An arch is shaped like a parabola. It is 30 m wide at the base and 15 m high. How wide is the arch 10 m from the ground?

52. A culvert is shaped like a parabola, 18 cm across the top and 12 cm deep. How wide is the culvert 8 cm from the top?

53. Find a value of c so that $y = x^2 - 10x + c$ has exactly one x-intercept.

54. Find b so that $y = x^2 + bx + 9$ has exactly one x-intercept.

55. Let x be in the interval $[0, 1]$. Show that the product $x(1 - x)$ is always less than or equal to $1/4$. For what values of x does the product equal $1/4$?

56. Let $f(x) = a(x - h)^2 + k$, and show that $f(h + x) = f(h - x)$. Why does this show that the parabola is symmetric to its axis?

In Exercises 57–60, let a parabola have equation $y = ax^2 + bx + c$.

57. Let $y = 0$ and use the quadratic formula to find any x-intercepts of the graph.

58. From the discriminant of Section 2.3, when does the graph of the parabola not cross the x-axis?

59. When does the graph cross the x-axis in two distinct points?

60. The axis of the parabola is a vertical line halfway between the x-intercepts. Use the results of Exercise 57 and find the equation of the axis. Now show that the vertex of the parabola is

$$\left(-\frac{b}{2a}, \frac{-b^2 + 4ac}{4a} \right).$$

(Use the fact that the vertex is on the axis.)

61. A parabola can be defined as the set of all points in the plane equally distant from a given point (called the *focus*) and a given line (called the *directrix*) not containing the point. See the figure.

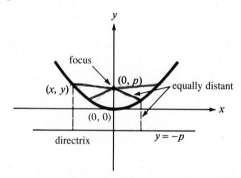

(a) Suppose (x, y) is any point on the parabola. Suppose the line mentioned in the definition is $y = -p$. Find the distance between (x, y) and the line $y = -p$. (The distance from a point to a line is the length of the perpendicular from the point to the line.)

(b) If $y = -p$ is the line mentioned in the definition, why is it reasonable to use $(0, p)$ as the given point? (See the figure.) Find the distance from (x, y) to $(0, p)$.

(c) Find an equation for the parabola in the figure.

Use the geometric definition of parabola given in Exercise 61 to find the equation of each parabola having the given point as focus.

62. $(0, 3)$, directrix $y = -3$ **63.** $(0, -5)$, directrix $y = 5$ **64.** $(-2, 0)$, directrix $x = 2$

65. $(8, 0)$, directrix $x = -8$ **66.** $(3, 6)$, vertex $(3, 4)$ **67.** $(-5, 2)$, vertex $(-5, 5)$

68. $(-4, 1)$, directrix $x = -7$ **69.** $(3, -4)$, directrix $x = 9$

70. Use the results of Exercise 61 to find the equation of a parabola with vertex at (h, k) and axis $x = h$.

71. If $b \neq 0$, show that the graph of $f(x) = x^2 + bx + b^2$ does not cross the x-axis.

72. Find a value of b so that the x-intercepts of $f(x) = x^2 + bx - 6$ are 5 units apart.

73. Find the largest possible value of y if $y = -(x - 2)^2 + 9$. Then find
    (a) the largest possible value of $\sqrt{-(x - 2)^2 + 9}$
    (b) the smallest possible value of $1/[-(x - 2)^2 + 9]$.

74. Find the smallest possible value of y if $y = 3 + (x + 5)^2$. Then find
    (a) the smallest possible value of $\sqrt{3 + (x + 5)^2}$
    (b) the largest possible value of $1/[3 + (x + 5)^2]$.

4.4 Ellipses and Hyperbolas (Optional)

As the earth travels around the sun over a year's time, it traces out a curve called an *ellipse*. There are many applications of ellipses. In one interesting application, patients with kidney stones are treated by being placed in a water bath in a tub with an elliptical cross section. Several hundred spark discharges are produced at one focus of the ellipse (see Figure 21), with the kidney stone at the other focus. The discharges go through the water, causing the stone to break up into small pieces which can be readily excreted from the body.

An ellipse is defined as follows.

Ellipse

> An **ellipse** is the set of all points in a plane the sum of whose distances from two fixed points is constant. The two fixed points are called the **foci** of the ellipse.

For example, the ellipse in Figure 21 has foci at points F and F'. By the definition, the ellipse is made up of all points P such that the sum $d(P, F) + d(P, F')$ is constant. The ellipse in Figure 21 has its **center** at the origin. Points V and V' are the **vertices** of the ellipse, the line segment connecting V and V' is the **major axis,** and the line segment connecting B and B' is the **minor axis.**

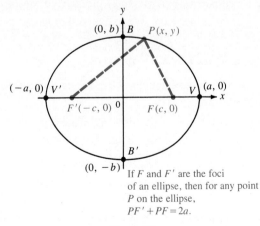

If F and F' are the foci
of an ellipse, then for any point
P on the ellipse,
$PF' + PF = 2a.$

Figure 21

As the vertical line test shows, the graph of Figure 21 is not the graph of a function, since one value of x can lead to two values of y.

To obtain an equation for an ellipse centered at the origin, let the two foci have coordinates $(-c, 0)$ and $(c, 0)$, respectively. Let the sum of the distances from any point $P(x, y)$ on the ellipse to the two foci be $2a$. By the distance formula, segment PF has length

$$d(P, F) = \sqrt{(x - c)^2 + y^2},$$

while segment PF' has length

$$d(P, F') = \sqrt{[x - (-c)]^2 + y^2} = \sqrt{(x + c)^2 + y^2}.$$

The sum of the lengths $d(P, F)$ and $d(P, F')$ must be $2a$, or

$$\sqrt{(x - c)^2 + y^2} + \sqrt{(x + c)^2 + y^2} = 2a.$$

Using algebra (see Exercise 49) and letting $a^2 - c^2 = b^2$, gives

$$\frac{x^2}{a^2} + \frac{y^2}{b^2} = 1,$$

the **standard form** of the equation of an ellipse centered at the origin with foci on the x-axis.

Letting $y = 0$ in the standard form gives

$$\frac{x^2}{a^2} + \frac{0^2}{b^2} = 1$$

$$\frac{x^2}{a^2} = 1$$

$$x^2 = a^2$$

$$x = \pm a$$

as the x-intercepts of the ellipse. The points $V'\ (-a, 0)$ and $V(a, 0)$ are the vertices of the ellipse; the segment VV' is the major axis. In a similar manner, letting $x = 0$ shows that the y-intercepts are $\pm b$; the segment connecting $(0, b)$ and $(0, -b)$ is the minor axis. We assumed throughout the work above that the foci were on the x-axis. If the foci were on the y-axis, an almost identical proof could be used to get the standard form

$$\frac{y^2}{a^2} + \frac{x^2}{b^2} = 1.$$

Do not be confused by the two standard forms—in one case a^2 is associated with x^2; in the other case a^2 is associated with y^2. However, in practice it is necessary only to find the intercepts of the graph—if the positive x-intercept is larger than the positive y-intercept, the major axis is horizontal, and otherwise it is vertical. When using the relationship $a^2 - c^2 = b^2$, or $a^2 - b^2 = c^2$, choose a^2 and b^2 so that $a^2 - b^2 > 0$. A summary of this work with ellipses follows.

Equations for Ellipses

The ellipse with center at the origin and major axis along the x-axis has equation

$$\frac{x^2}{a^2} + \frac{y^2}{b^2} = 1 \quad (a > b),$$

while the ellipse centered at the origin with major axis along the y-axis has equation

$$\frac{y^2}{a^2} + \frac{x^2}{b^2} = 1 \quad (a > b).$$

Notice that, as shown in the summary, an ellipse always has $a > b$. Also, an ellipse is symmetric with respect to its major axis and its minor axis.

Example 1   Graph $4x^2 + 9y^2 = 36$.

To obtain the standard form for the equation of an ellipse, divide each side by 36 to get

$$\frac{x^2}{9} + \frac{y^2}{4} = 1.$$

The x-intercepts of this ellipse are ± 3, and the y-intercepts ± 2. Additional ordered pairs satisfying the equation of the ellipse may be found if desired. The graph of the ellipse is shown in Figure 22.

Since $9 > 4$, find the foci by letting $c^2 = 9 - 4 = 5$ so that $c = \pm\sqrt{5}$. The major axis is along the x-axis so the foci are at $(-\sqrt{5}, 0)$ and $(\sqrt{5}, 0)$.   ■

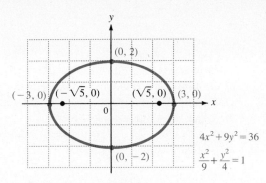

Figure 22

Example 2   Find the equation of the ellipse having center at the origin, foci at $(0, 3)$ and $(0, -3)$, and major axis of length 8 units.

Since the major axis is 8 units long,

$$2a = 8$$

or

$$a = 4.$$

To find b^2, use the relationship $a^2 - b^2 = c^2$. Here $a = 4$ and $c = 3$. Substituting for a and c gives

$$a^2 - b^2 = c^2$$
$$4^2 - b^2 = 3^2$$
$$16 - b^2 = 9$$
$$b^2 = 7.$$

Since the foci are on the y-axis, the larger intercept, a, is used to find the denominator for y^2, giving the equation in standard form as

$$\frac{y^2}{16} + \frac{x^2}{7} = 1.$$

A graph of this ellipse is shown in Figure 23. ■

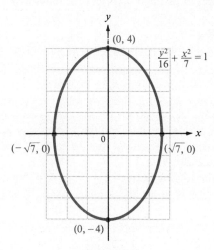

Figure 23

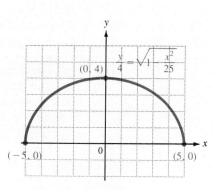

Figure 24

Example 3   Graph $\dfrac{y}{4} = \sqrt{1 - \dfrac{x^2}{25}}$.

Square both sides to get

$$\frac{y^2}{16} = 1 - \frac{x^2}{25}$$

or

$$\frac{x^2}{25} + \frac{y^2}{16} = 1,$$

the equation of an ellipse with x-intercepts ± 5 and y-intercepts ± 4. Since $\sqrt{1 - x^2/25} \geq 0$, the only possible values of y are those making $y/4 \geq 0$, giving the half-ellipse shown in Figure 24. While the graph of the ellipse $x^2/25 + y^2/16 = 1$ is not the graph of a function, the half-ellipse in Figure 24 *is* the graph of a function. The domain is the interval $[-5, 5]$ and the range is $[0, 4]$. ■

Just as a circle need not have its center at the origin, an ellipse may also have its center translated away from the origin.

Ellipse Centered at (h, k)

An ellipse centered at (h, k) with horizontal major axis of length $2a$ has equation

$$\frac{(x - h)^2}{a^2} + \frac{(y - k)^2}{b^2} = 1.$$

There is a similar result for ellipses having a vertical major axis.

This result can be proven from the definition of an ellipse.

Example 4 Graph $\dfrac{(x - 2)^2}{9} + \dfrac{(y + 1)^2}{16} = 1.$

The graph of this equation is an ellipse centered at $(2, -1)$. As mentioned earlier, ellipses always have $a > b$. For this ellipse, then, $a = 4$ and $b = 3$. Since $a = 4$ is associated with y^2, the vertices of the ellipse are on the vertical line through $(2, -1)$. Find the vertices by locating two points on the vertical line through $(2, -1)$, one 4 units up from $(2, -1)$ and one 4 units down. The vertices are $(2, 3)$ and $(2, -5)$. Locate two other points on the ellipse by locating points on a horizontal line through $(2, -1)$, one 3 units to the right and one 3 units to the left. Find additional points as needed. The final graph is in Figure 25. ■

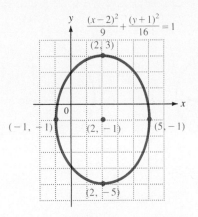

Figure 25

Hyperbolas An ellipse was defined as the set of all points in a plane with the sum of the distances from two fixed points as a constant. A **hyperbola** is the set of all points in a plane for which the *difference* of the distances from two fixed points (called **foci**) is constant.

Hyperbola

Let $F'(-c, 0)$ and $F(c, 0)$ be two points on the x-axis. A **hyperbola** is the set of all points $P(x, y)$ in a plane such that the difference of the distances $d(P, F')$ and $d(P, F)$ is a constant.

The midpoint of the segment $F'F$ is the **center** of the hyperbola. See Figure 26.

Suppose a hyperbola has center at the origin and foci at $F'(-c, 0)$ and $F(c, 0)$. Choosing $2a$ as the constant in the definition above gives

$$d(P, F') - d(P, F) = 2a.$$

The distance formula and algebraic manipulation (see Exercise 48) produce the result

$$\frac{x^2}{a^2} - \frac{y^2}{b^2} = 1,$$

where $b^2 = c^2 - a^2$. Letting $y = 0$ shows that the x-intercepts are $\pm a$. If $x = 0$ the equation becomes

$$\frac{0^2}{a^2} - \frac{y^2}{b^2} = 1$$

$$-\frac{y^2}{b^2} = 1$$

$$y^2 = -b^2,$$

which has no real number solutions, showing that this hyperbola has no y-intercepts.

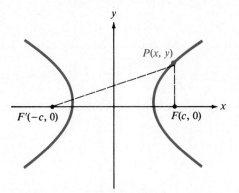

Figure 26

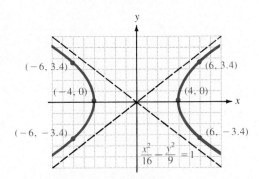

Figure 27

Example 5 Graph $\dfrac{x^2}{16} - \dfrac{y^2}{9} = 1$.

This hyperbola has x-intercepts 4 and -4 and no y-intercepts. To sketch the graph, we can find some other points that lie on the graph. For example, letting $x = 6$ gives

$$\frac{6^2}{16} - \frac{y^2}{9} = 1$$

$$-\frac{y^2}{9} = 1 - \frac{6^2}{16}$$

$$\frac{y^2}{9} = \frac{20}{16}$$

$$y^2 = \frac{180}{16} = \frac{45}{4}$$

$$y = \frac{\pm 3\sqrt{5}}{2} \approx \pm 3.4.$$

The graph includes the points $(6, 3.4)$ and $(6, -3.4)$. Also, letting $x = -6$ would still give $y \approx \pm 3.4$ with the points $(-6, 3.4)$ and $(-6, -3.4)$ also on the graph. These points, along with other points on the graph, were used to help sketch the final graph shown in Figure 27. ■

This basic information on hyperbolas is summarized as follows.

Equations for Hyperbolas

A hyperbola centered at the origin and having x-intercepts at a and $-a$ has an equation of the form

$$\frac{x^2}{a^2} - \frac{y^2}{b^2} = 1,$$

while a hyperbola centered at the origin and having y-intercepts at a and $-a$ has an equation of the form

$$\frac{y^2}{a^2} - \frac{x^2}{b^2} = 1.$$

Starting with the equation for a hyperbola $(x^2/a^2) - (y^2/b^2) = 1$ and solving for y gives

$$\frac{x^2}{a^2} - 1 = \frac{y^2}{b^2}$$

$$\frac{x^2 - a^2}{a^2} = \frac{y^2}{b^2}$$

or
$$y = \pm \frac{b}{a}\sqrt{x^2 - a^2}. \qquad (*)$$

If x^2 is very large in comparison to a^2, the difference $x^2 - a^2$ would be very close to x^2. If this happens, then the points satisfying equation $(*)$ above would be very close to one of the lines

$$y = \pm \frac{b}{a} x.$$

Thus, as $|x|$ gets larger and larger, the points of the hyperbola $x^2/a^2 - y^2/b^2 = 1$ come closer and closer to the lines $y = (\pm b/a)x$. These lines, called the **asymptotes** of the hyperbola, are very helpful when graphing the hyperbola.

Example 6   Graph $\dfrac{x^2}{25} - \dfrac{y^2}{49} = 1$.

For this hyperbola, $a = 5$ and $b = 7$. With these values, $y = (\pm b/a)x$ becomes $y = (\pm 7/5)x$. If $x = 5$, then $y = (\pm 7/5)(5) = \pm 7$, while $x = -5$ also gives $y = \pm 7$. These four points, $(5, 7)$, $(5, -7)$, $(-5, 7)$, and $(-5, -7)$, lead to the rectangle shown in Figure 28. The extended diagonals of this rectangle are the asymptotes of the hyperbola. The hyperbola crosses the x-axis at 5 and -5. The final graph is shown in Figure 28.   ■

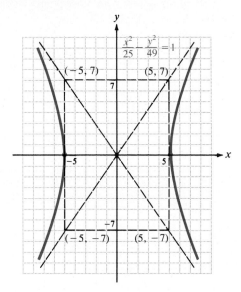

Figure 28

The rectangle used to graph the hyperbola in Example 6 is called the **fundamental rectangle.** While $a > b$ for an ellipse, the examples above show that for hyperbolas, it is possible that $a > b$ or $a < b$; other examples would show that a might equal b, also. If the foci of a hyperbola are on the y-axis, the equation of the hyperbola is of the form

$$\frac{y^2}{a^2} - \frac{x^2}{b^2} = 1, \quad \text{with asymptotes} \quad y = \pm \frac{a}{b}x.$$

If the foci of the hyperbola are on the x-axis, we found that the asymptotes have equations $y = \pm(b/a)x$, while foci on the y-axis lead to asymptotes $y = \pm(a/b)x$. There is an obvious chance for confusion here; to avoid mistakes write the equation of the hyperbola in either the form

$$\frac{x^2}{a^2} - \frac{y^2}{b^2} = 1 \quad \text{or} \quad \frac{y^2}{a^2} - \frac{x^2}{b^2} = 1,$$

and replace 1 with 0. Solving the resulting equation for y produces the proper equations for the asymptotes. (The reason why this process works is explained in more advanced courses.)

Example 7   Graph $25y^2 - 4x^2 = 100$.

Divide each side by 100 to get

$$\frac{y^2}{4} - \frac{x^2}{25} = 1.$$

This hyperbola is centered at the origin, has foci on the y-axis, and has y-intercepts 2 and -2. To find the asymptotes, replace 1 with 0, getting

$$\frac{y^2}{4} - \frac{x^2}{25} = 0$$

$$\frac{y^2}{4} = \frac{x^2}{25}$$

$$y^2 = \frac{4x^2}{25}$$

$$y = \pm\frac{2}{5}x.$$

Use the points $(5, 2)$, $(5, -2)$, $(-5, 2)$, and $(-5, -2)$ to get the fundamental rectangle shown in Figure 29. Use the diagonals of this rectangle to determine the asymptotes for the graph, as shown in Figure 29. ■

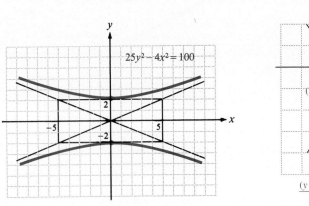

Figure 29

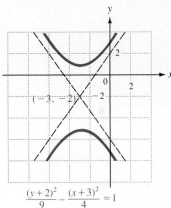

$$\frac{(y+2)^2}{9} - \frac{(x+3)^2}{4} = 1$$

Figure 30

The next example shows a hyperbola with its center translated away from the origin.

Example 8   Graph $\dfrac{(y + 2)^2}{9} - \dfrac{(x + 3)^2}{4} = 1$.

This equation represents a hyperbola centered at $(-3, -2)$. See Figure 30 above. ■

Our final example shows an equation whose graph is only half of a hyperbola.

Example 9   Graph $x = -\sqrt{1 + 4y^2}$.

Squaring both sides gives

$$x^2 = 1 + 4y^2$$

or

$$x^2 - 4y^2 = 1.$$

To find the asymptotes, rewrite 4 as $1/(1/4)$ to change this equation into

$$x^2 - \frac{y^2}{1/4} = 0$$

or

$$\frac{1}{4}x^2 = y^2,$$

giving

$$y = \pm\frac{1}{2}x.$$

Since the given equation $x = -\sqrt{1 + 4y^2}$ restricts x to nonpositive values, the graph is the left branch of the hyperbola, as shown in Figure 31.   ■

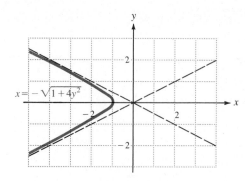

Figure 31

4.4 Exercises

Sketch the graph of each of the following. Give the endpoints of the major axis for ellipses and the equation of the asymptotes for hyperbolas. Give the center of each figure.

1. $\dfrac{x^2}{9} + \dfrac{y^2}{4} = 1$

2. $\dfrac{x^2}{16} + \dfrac{y^2}{36} = 1$

3. $\dfrac{x^2}{9} + y^2 = 1$

4. $\dfrac{y^2}{16} - \dfrac{x^2}{9} = 1$

5. $\dfrac{x^2}{6} + \dfrac{y^2}{9} = 1$

6. $\dfrac{x^2}{8} - \dfrac{y^2}{12} = 1$

7. $x^2 + 4y^2 = 16$

8. $25x^2 + 9y^2 = 225$

9. $x^2 = 9 + y^2$

10. $y^2 = 16 + x^2$

11. $2x^2 + y^2 = 8$

12. $9x^2 - 25y^2 = 225$

13. $25x^2 - 4y^2 = -100$

14. $4x^2 - y^2 = -16$

15. $\dfrac{x^2}{1/9} + \dfrac{y^2}{1/16} = 1$

16. $\dfrac{x^2}{4/25} + \dfrac{y^2}{9/49} = 1$

17. $\dfrac{64x^2}{9} + \dfrac{25y^2}{36} = 1$

18. $\dfrac{121x^2}{25} + \dfrac{16y^2}{9} = 1$

19. $\dfrac{(x-1)^2}{9} + \dfrac{(y+3)^2}{25} = 1$

20. $\dfrac{(x+3)^2}{16} + \dfrac{(y-2)^2}{36} = 1$

in class

21. $\dfrac{(x-3)^2}{16} - \dfrac{(y+2)^2}{49} = 1$

22. $\dfrac{(y-5)^2}{4} - \dfrac{(x+1)^2}{9} = 1$

23. $\dfrac{(y+1)^2}{25} - \dfrac{(x-3)^2}{36} = 1$

24. $\dfrac{(x+2)^2}{16} - \dfrac{(y+2)^2}{25} = 1$

Sketch the graph of each of the following. Identify any that are the graphs of functions.

25. $\dfrac{x}{4} = \sqrt{1 - \dfrac{y^2}{9}}$

26. $\dfrac{y}{2} = \sqrt{1 - \dfrac{x^2}{25}}$

27. $\dfrac{y}{3} = \sqrt{1 + \dfrac{x^2}{16}}$

28. $x = \sqrt{1 + \dfrac{y^2}{36}}$

in class

29. $x = -\sqrt{1 - \dfrac{y^2}{64}}$

30. $y = \sqrt{1 - \dfrac{x^2}{100}}$

31. $y = -\sqrt{1 + \dfrac{x^2}{25}}$

32. $x = -\sqrt{1 + \dfrac{y^2}{9}}$

Find equations for each of the following ellipses.

33. x-intercepts ± 4; foci at $(-2, 0)$ and $(2, 0)$

34. y-intercepts ± 3; foci at $(0, \sqrt{3})$, $(0, -\sqrt{3})$

35. Endpoints of major axis at $(6, 0)$, $(-6, 0)$; $c = 4$

36. Vertices $(0, 5)$, $(0, -5)$; $b = 2$

37. Center $(3, -2)$, $a = 5$, $c = 3$, major axis vertical

38. Center $(2, 0)$, minor axis of length 6, major axis horizontal, and of length 9

Find equations for each of the following hyperbolas.

39. x-intercepts ± 3, foci at $(-4, 0)$, $(4, 0)$

40. y-intercepts ± 5, foci at $(0, 3\sqrt{3})$, $(0, -3\sqrt{3})$

41. Asymptotes $y = \pm(3/5)x$, y-intercepts $(0, 3)$, $(0, -3)$

42. Center at the origin, passing through $(5, 3)$ and $(-10, 2\sqrt{21})$, no y-intercepts

The orbit of Mars is an ellipse, with the sun at one focus. An approximate equation for the orbit is

$$\frac{x^2}{5013} + \frac{y^2}{4970} = 1,$$

where x and y are measured in millions of miles.

43. Find the length of the major axis.

44. Find the length of the minor axis.

45. Draftspeople often use the method shown on the sketch below to draw an ellipse. Explain why the method works.

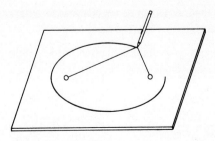

46. Ships and planes often use a location finding system called LORAN. With this system, a radio transmitter at M on the figure below left sends out a series of pulses. When each pulse is received at transmitter S, it then sends out a pulse. A ship at P receives pulses from both M and S. A receiver on the ship measures the difference in the arrival times of the pulses. The navigator then consults a special map, showing certain curves according to the differences in arrival times. In this way, the ship can be located as lying on a portion of which curve? (This method requires three transmitters acting as two pairs.)

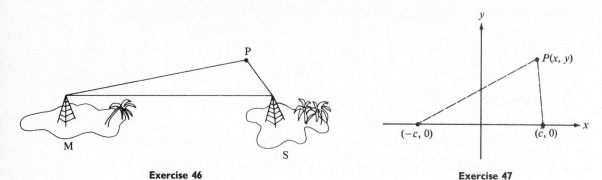

Exercise 46   Exercise 47

47. Microphones are placed at points $(-c, 0)$ and $(c, 0)$. An explosion occurs at point $P(x, y)$ having positive x-coordinate. (See the figure above right.) The sound is detected at the closer microphone t seconds before being detected at the farther microphone. Assume that sound travels at a speed of 330 m per second, and show that P must be on the hyperbola

$$\frac{x^2}{330^2 t^2} - \frac{y^2}{4c^2 - 330^2 t^2} = \frac{1}{4}.$$

48. Suppose a hyperbola has center at the origin, foci at $F'(-c, 0)$ and $F(c, 0)$, and the value $d(P, F') - d(P, F) = 2a$. Let $b^2 = c^2 - a^2$, and show that an equation of the hyperbola is

$$\frac{x^2}{a^2} - \frac{y^2}{b^2} = 1.$$

49. Derive the standard form of the equation of an ellipse centered at the origin.

50. A rod of fixed length in the xy-coordinate plane is moved so that one end is always on the x-axis and the other end is always on the y-axis. Let P be any fixed point on the rod. Show that the path of P is an ellipse.

4.5   Conic Sections (Optional)

Before giving a general classification of the graphs of this chapter, we need to discuss the graph of an equation of the form

$$x = ay^2 + by + c,$$

where a, b, and c are real numbers and $a \neq 0$. This equation is the same as the quadratic function defined by $y = ax^2 + bx + c$, with x and y interchanged. As with the graphs of inverse functions, this makes the graph of the new equation $x =$

$ay^2 + by + c$ symmetric to that of $y = ax^2 + bx + c$ with respect to the line $y = x$. Reflecting a graph about the line $y = x$ would change a vertical axis into a horizontal line, so that the graph of $x = ay^2 + by + c$ is a parabola with *horizontal axis*. That is,

Parabola with Horizontal Axis

$$x = a(y - k)^2 + h$$

is the equation of a parabola with vertex at (h, k) and axis the horizontal line $y = k$. The parabola opens to the right if $a > 0$ and to the left if $a < 0$.

Example 1   Graph $x = y^2$.

Choosing values of y and finding the corresponding values of x gives the parabola in Figure 32. Since the equation $x = y^2$ can be obtained from $y = x^2$ by exchanging x and y, the graph of $y = x^2$ is shown for comparison as a dashed line. These graphs are symmetric to each other with respect to the line $y = x$. ■

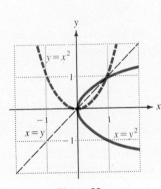

Figure 32

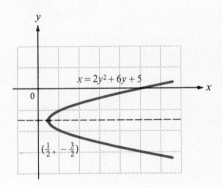

Figure 33

Example 2   Graph $x = 2y^2 + 6y + 5$.

To write this equation in the form $x = a(y - k)^2 + h$, complete the square on y as follows.

$$x = 2(y^2 + 3y \qquad) + 5 = 2\left(y^2 + 3y + \frac{9}{4} - \frac{9}{4}\right) + 5$$

$$= 2\left(y^2 + 3y + \frac{9}{4}\right) + 2\left(-\frac{9}{4}\right) + 5 = 2\left(y + \frac{3}{2}\right)^2 + \frac{1}{2}$$

As this result shows, the vertex of the parabola is the point $(1/2, -3/2)$. The axis is the horizontal line

$$y + \frac{3}{2} = 0 \quad \text{or} \quad y = -\frac{3}{2}.$$

Using the vertex and the axis and plotting a few additional points gives the graph in Figure 33. ■

The graphs of parabolas, circles, hyperbolas, and ellipses are called **conic sections** since each graph can be obtained by cutting a cone with a plane as suggested by Figure 34.

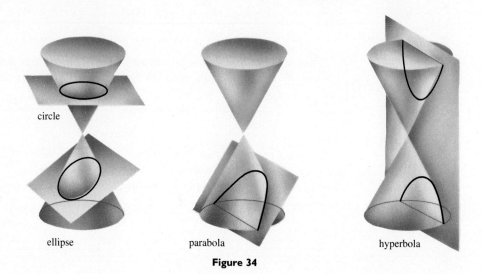

circle

ellipse   parabola   hyperbola

Figure 34

It turns out that all conic sections of the types presented in this chapter have equations of the form

$$Ax^2 + Bx + Cy^2 + Dy + E = 0,$$

where either A or C must be nonzero. The special characteristics of the equations of each of the conic sections are summarized as follows.

Equations of Conic Sections

Conic section	Characteristic	Example
Parabola	Either $A = 0$ or $C = 0$, but not both	$x^2 = y + 4$ $(y - 2)^2 = -(x + 3)$
Circle	$A = C \neq 0$	$x^2 + y^2 = 16$
Ellipse	$A \neq C, AC > 0$	$\dfrac{x^2}{16} + \dfrac{y^2}{25} = 1$
Hyperbola	$AC < 0$	$x^2 - y^2 = 1$

The graphs of the conic sections are summarized in Figure 35. Ellipses and hyperbolas having centers not at the origin can be shown in much the same way as we show circles and parabolas.

Equation	Graph	Description	Identification
$y = a(x - h)^2 + k$	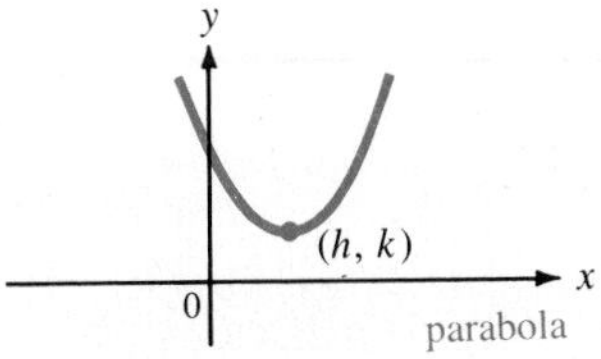 parabola	Opens upward if $a > 0$, downward if $a < 0$. Vertex is at (h, k).	x^2 term y is not squared.
$x = a(y - k)^2 + h$	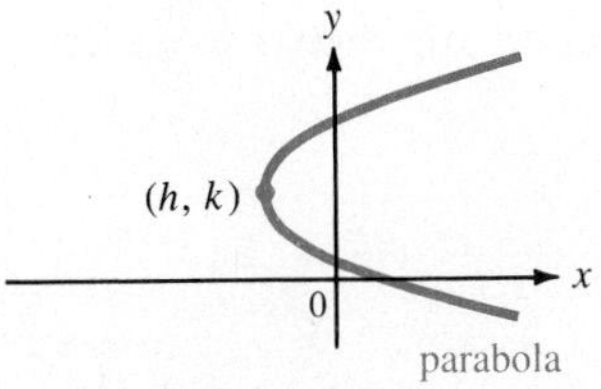 parabola	Opens to right if $a > 0$, to left if $a < 0$. Vertex is at (h, k).	y^2 term x is not squared.
$(x - h)^2 + (y - k)^2 = r^2$	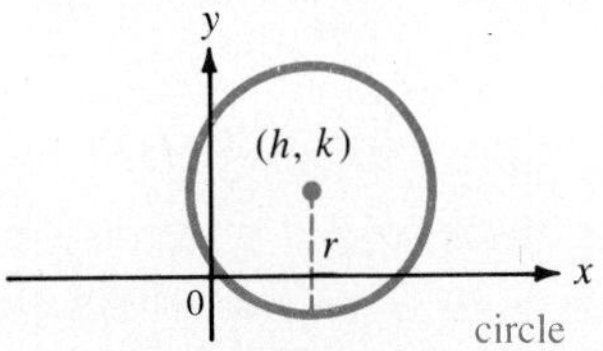 circle	Center is at (h, k), radius is r.	x^2 and y^2 terms have the same positive coefficient.
$\dfrac{x^2}{a^2} + \dfrac{y^2}{b^2} = 1$	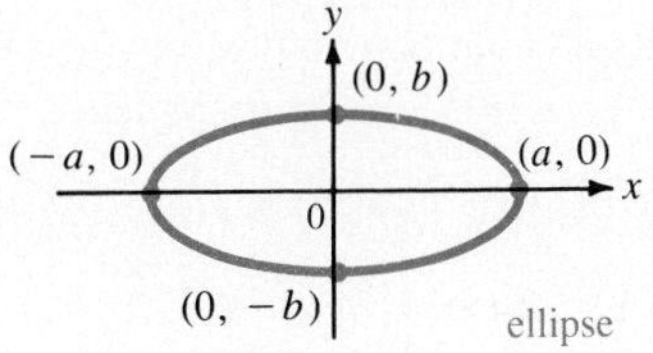 ellipse	x-intercepts are a and $-a$. y-intercepts are b and $-b$.	x^2 and y^2 terms have different positive coefficients.
$\dfrac{x^2}{a^2} - \dfrac{y^2}{b^2} = 1$	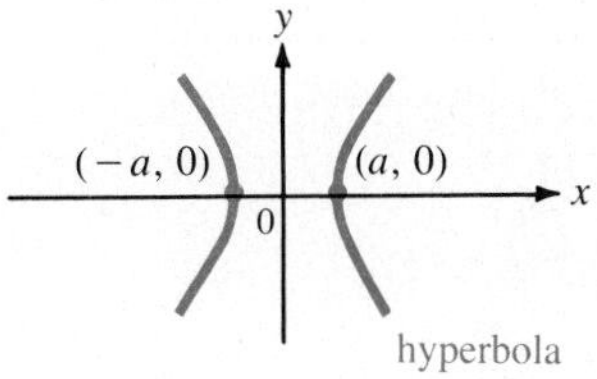 hyperbola	x-intercepts are a and $-a$. Asymptotes found from (a, b), $(a, -b)$, $(-a, -b)$, and $(-a, b)$.	x^2 has a positive coefficient. y^2 has a negative coefficient.
$\dfrac{y^2}{a^2} - \dfrac{x^2}{b^2} = 1$	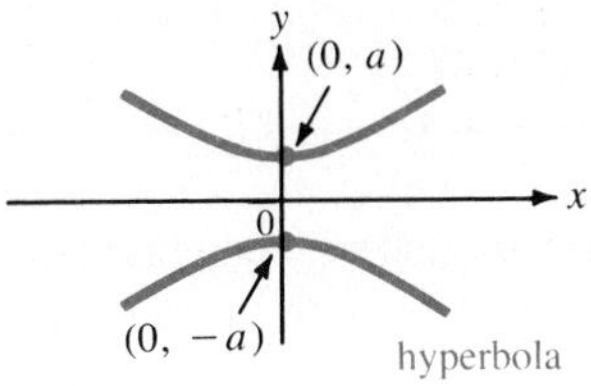 hyperbola	y-intercepts are a and $-a$. Asymptotes found from (b, a), $(b, -a)$, $(-b, a)$, and $(-b, -a)$.	y^2 has a positive coefficient. x^2 has a negative coefficient.

Figure 35

Example 3 Decide on the type of conic section represented by each of the following equations, and sketch each graph.

(a) $x^2 = 25 + 5y^2$

Rewriting the equation as

$$x^2 - 5y^2 = 25$$

or

$$\frac{x^2}{25} - \frac{y^2}{5} = 1$$

shows that the equation represents a hyperbola centered at the origin, with asymptotes

$$\frac{x^2}{25} - \frac{y^2}{5} = 0,$$

or

$$y = \frac{\pm\sqrt{5}}{5}x.$$

The x-intercepts are ± 5; the graph is shown in Figure 36.

(b) $4x^2 - 16x + 9y^2 + 54y = -61$

Since the coefficients of the x^2 and y^2 terms are unequal and both positive, this equation might represent an ellipse. (It might also represent a single point or no points at all.) To find out, complete the square on x and y.

$$4(x^2 - 4x \quad\quad) + 9(y^2 + 6y \quad\quad) = -61$$
$$4(x^2 - 4x + 4 - 4) + 9(y^2 + 6y + 9 - 9) = -61$$
$$4(x^2 - 4x + 4) - 16 + 9(y^2 + 6y + 9) - 81 = -61$$
$$4(x - 2)^2 + 9(y + 3)^2 = 36$$
$$\frac{(x - 2)^2}{9} + \frac{(y + 3)^2}{4} = 1$$

This equation represents an ellipse having center at $(2, -3)$ and graph as shown in Figure 37.

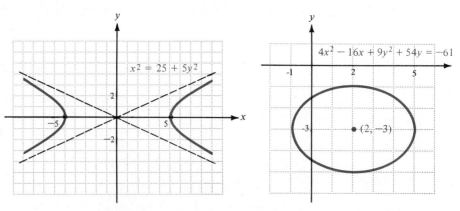

Figure 36   Figure 37

(c) $x^2 - 8x + y^2 + 10y = -41$

Complete the square on both x and y, as follows:

$$(x^2 - 8x + 16 - 16) + (y^2 + 10y + 25 - 25) = -41$$
$$(x - 4)^2 + (y + 5)^2 = 16 + 25 - 41$$
$$(x - 4)^2 + (y + 5)^2 = 0.$$

This result shows that the equation is that of a circle of radius 0; that is, the point $(4, -5)$. Had a negative number been obtained on the right (instead of 0), the equation would have represented no points at all, and there would be no graph.

(d) $x^2 - 6x + 8y - 7 = 0$

Since only one variable is squared (x, and not y), the equation represents a parabola. Complete the square.

$$(x^2 - 6x + 9 - 9) + 8y - 7 = 0$$
$$(x - 3)^2 = 16 - 8y$$
$$(x - 3)^2 = -8(y - 2)$$

The parabola has vertex at $(3, 2)$, and opens downward, as shown in Figure 38. ■

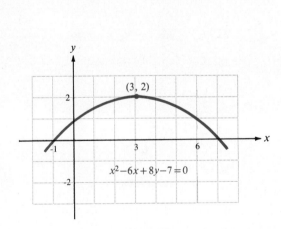

Figure 38

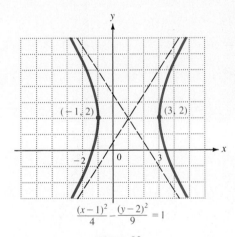

Figure 39

The next example, with Figure 39, is designed to serve as a warning about a very common error.

Example 4 Graph $4y^2 - 16y - 9x^2 + 18x = -43$.

Complete the square on x and on y.

$$4(y^2 - 4y \quad) - 9(x^2 - 2x \quad) = -43$$
$$4(y^2 - 4y + 4) - 9(x^2 - 2x + 1) = -43 + 16 - 9$$
$$4(y - 2)^2 - 9(x - 1)^2 = -36$$

Because of the -36, it is very tempting to say that this equation does not have a graph. However, the minus sign in the middle on the left shows that the graph is that of a hyperbola. Dividing through by -36 and rearranging terms gives

$$\frac{(x - 1)^2}{4} - \frac{(y - 2)^2}{9} = 1,$$

the hyperbola centered at $(1, 2)$, with graph as shown in Figure 39. ■

4.5 Exercises

Graph each horizontal parabola.

1. $x = y^2 + 2$ **2.** $x = -y^2$ **3.** $x = (y + 1)^2$ **4.** $x = (y - 3)^2$

5. $x = (y + 2)^2 - 1$ **6.** $x = (y - 4)^2 + 2$ **7.** $x = -2(y + 3)^2$ **8.** $x = -3(y - 1)^2 + 2$

9. $x = \frac{1}{2}(y + 1)^2 + 3$ **10.** $x = \frac{2}{3}(y - 3)^2 + 2$ **11.** $x = y^2 + 2y - 8$ **12.** $x = y^2 - 6y + 7$

13. $x = -2y^2 + 2y - 3$ **14.** $x = -4y^2 - 4y - 3$

Identify each equation of Exercises 15–40. Draw a graph of each that has a graph.

15. $\frac{x^2}{4} + \frac{y^2}{9} = 1$ **16.** $\frac{x^2}{4} - \frac{y^2}{9} = 1$ **17.** $\frac{x^2}{4} - \frac{y^2}{4} = 1$ **18.** $\frac{x^2}{4} + \frac{y^2}{4} = 1$

19. $x^2 + 2x = x^2 + y - 6$ **20.** $y^2 - 4y = y^2 + 3 - x$

21. $x^2 = 25 + y^2$ **22.** $x^2 = 25 - y^2$ **23.** $9x^2 + 36y^2 = 36$ **24.** $x^2 = 4y - 8$

25. $\frac{(x + 3)^2}{16} + \frac{(y - 2)^2}{16} = 1$ **26.** $\frac{(x - 4)}{8} - \frac{(y + 1)^2}{2} = 0$

27. $y^2 - 4y = x + 4$ **28.** $11 - 3x = 2y^2 - 8y$

29. $(x + 7)^2 + (y - 5)^2 + 4 = 0$ **30.** $4(x - 3)^2 + 3(y + 4)^2 = 0$

31. $3x^2 + 6x + 3y^2 - 12y = 12$ **32.** $2x^2 - 8x + 5y^2 + 20y = 12$

33. $x^2 - 6x + y = 0$ **34.** $x - 4y^2 - 8y = 0$

35. $4x^2 - 8x - y^2 - 6y = 6$ **36.** $x^2 + 2x = y^2 - 4y - 2$

37. $4x^2 - 8x + 9y^2 + 54y = -84$ **38.** $3x^2 + 12x + 3y^2 = -11$

39. $6x^2 - 12x + 6y^2 - 18y + 25 = 0$ **40.** $4x^2 - 24x + 5y^2 + 10y + 41 = 0$

4.6 Graphs of Polynomial Functions

As mentioned earlier, any function defined by an expression of the form

$$f(x) = a_n x^n + a_{n-1} x^{n-1} + \cdots + a_1 x + a_0$$

is a polynomial function of degree n, for real numbers $a_n \neq 0$, $a_{n-1}, \ldots, a_1, a_0$, and any nonnegative integer n.

We have already discussed examples of polynomial functions; a linear function defined by $f(x) = ax + b$, with $a \neq 0$, is a polynomial function of degree 1, while a quadratic function defined by $f(x) = ax^2 + bx + c$ is a polynomial function of degree 2.

In this section we shall graph polynomial functions of degree 3 or more. (This section introduces the idea of graphing these functions, and a later chapter provides more detail.) We start by graphing polynomial functions of the form $f(x) = x^n$.

Example 1   Graph the functions defined as follows.

(a) $f(x) = x^3$

Choose several values for x, and find the corresponding values of $f(x)$, or y, as shown in the table in Figure 40. Plot the resulting ordered pairs and connect the points with a smooth curve. The graph of $f(x) = x^3$ is shown as a solid line in Figure 40.

(b) $f(x) = x^5$

Work as in part (a) of this example to get the graph shown as a dashed line in Figure 40. Both the graph of $f(x) = x^3$ and the graph of $f(x) = x^5$ are symmetric with respect to the origin. (Recall the discussion of odd and even functions from Chapter 3.)

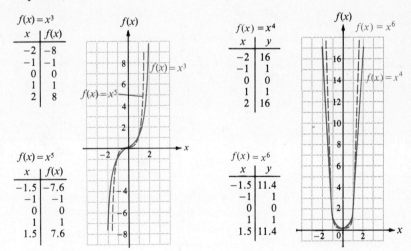

$f(x) = x^3$

x	$f(x)$
-2	-8
-1	-1
0	0
1	1
2	8

$f(x) = x^5$

x	$f(x)$
-1.5	-7.6
-1	-1
0	0
1	1
1.5	7.6

$f(x) = x^4$

x	y
-2	16
-1	1
0	0
1	1
2	16

$f(x) = x^6$

x	y
-1.5	11.4
-1	1
0	0
1	1
1.5	11.4

Figure 40   **Figure 41**

(c) The graphs of $f(x) = x^4$ and $f(x) = x^6$ are shown in Figure 41. These graphs have symmetry about the y-axis, as does the graph of $f(x) = ax^2$ for a nonzero real number a.   ∎

As shown earlier with the graph of $f(x) = ax^2$, the value of a in $f(x) = ax^n$ affects the width of the graph. When $|a| > 1$, the graph is narrower than the graph of $f(x) = x^n$; when $0 < |a| < 1$, the graph is broader. When $a < 0$, the graph of $f(x) = ax^n$ is reflected about the x-axis as compared to the graph of $f(x) = x^n$.

Example 2   Graph the functions defined as follows.

(a) $f(x) = \dfrac{1}{2}x^3$

The graph will be broader than that of $f(x) = x^3$, but will have the same general shape. It includes the points $(-2, -4)$, $(-1, -1/2)$, $(0, 0)$, $(1, 1/2)$, and $(2, 4)$. See Figure 42.

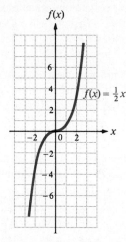

Figure 42

Figure 43

(b) $f(x) = -\dfrac{3}{2}x^4$

The following table gives some ordered pairs.

x	-2	-1	0	1	2
$f(x)$	-24	$-\dfrac{3}{2}$	0	$-\dfrac{3}{2}$	-24

The graph is shown in Figure 43. This graph is narrower than that of $f(x) = x^4$ and opens downward because of the minus sign.   ■

Example 3   Graph the functions defined as follows.

(a) $f(x) = x^5 - 2$

The graph will be the same as that of $f(x) = x^5$, but translated down 2 units. See Figure 44.

(b) $f(x) = -x^3 + 3$

The negative sign causes the graph to be reflected about the x-axis compared to the graph of $f(x) = x^3$. As shown in Figure 45, the graph is also translated up 3 units.

(c) $f(x) = (x + 1)^6$

This function has a graph like that of $f(x) = x^6$, but translated 1 unit to the left, as shown in Figure 46.   ■

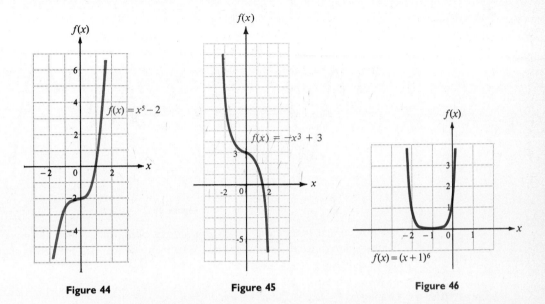

Figure 44 **Figure 45** **Figure 46**

Generalizing from the graphs in Examples 1–3, the domain of a polynomial function is the set of all real numbers. The range of a polynomial function of odd degree is also the set of all real numbers. Some typical graphs of polynomial functions of odd degree are shown in Figure 47. These graphs suggest that for every polynomial function f of odd degree there is at least one real value of x that makes $f(x) = 0$. Such values of x, called the **real zeros** of f, are also the x-intercepts of the graph.

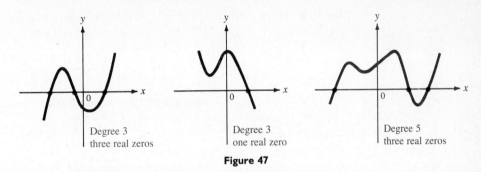

| Degree 3 | Degree 3 | Degree 5 |
| three real zeros | one real zero | three real zeros |

Figure 47

A polynomial function of even degree will have a range that takes the form $(-\infty, k]$ or else $[k, +\infty)$ for some real number k. Figure 48 on the following page shows two typical graphs of polynomial functions of even degree.

The graphs in Figures 47 above and 48 on the next page show that polynomial functions often have **turning points** where the function changes from increasing to decreasing or from decreasing to increasing. A polynomial function of degree n has at most $n - 1$ turning points. The graphs shown above illustrate this.

It is difficult to graph most polynomial functions without the use of calculus. A large number of points must be plotted to get a reasonably accurate graph. However,

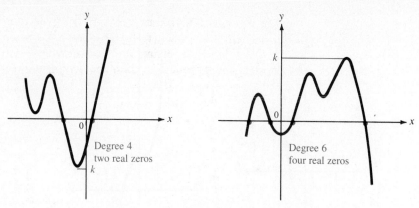

Figure 48

if a polynomial function is defined by an expression that can be factored, its graph can be approximated without plotting very many points; this method is shown in the next examples.

Example 4   Graph $f(x) = (2x + 3)(x - 1)(x + 2)$.

-3 -4 -1

Multiplying out the expression on the right would show that f is a third degree polynomial, or a **cubic** polynomial. To sketch the graph of f, first find its zeros by setting each of the three factors equal to 0 and solving the resulting equations.

$$2x + 3 = 0 \quad\text{or}\quad x - 1 = 0 \quad\text{or}\quad x + 2 = 0$$

$$x = -\frac{3}{2} \qquad\qquad x = 1 \qquad\qquad x = -2$$

The three zeros, $-3/2$, 1, and -2, divide the x-axis into four regions (shown in Figure 49):

$$x < -2, \quad -2 < x < -\frac{3}{2}, \quad -\frac{3}{2} < x < 1, \quad\text{and}\quad 1 < x.$$

Figure 49

In any of these regions, the values of $f(x)$ are either always positive or always negative. To find the sign of $f(x)$ in each region, select an x-value in each region and substitute it into the equation for $f(x)$ to determine if the values of the function are positive or negative in that region. A typical selection of test points and the results of the tests are shown below.

Region	Test point	Value of $f(x)$	Sign of $f(x)$
$x < -2$	-3	-12	negative
$-2 < x < -3/2$	$-7/4$	$11/32$	positive
$-3/2 < x < 1$	0	-6	negative
$1 < x$	2	28	positive

When the values of $f(x)$ are negative, the graph is below the x-axis, and when $f(x)$ takes on positive values, the graph is above the x-axis. All these results suggest that the graph looks something like the sketch in Figure 50. The sketch could be improved by plotting additional points in each region. ■

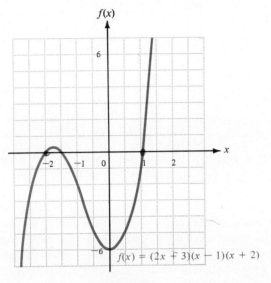

$f(x) = (2x + 3)(x - 1)(x + 2)$

Figure 50

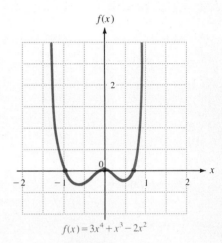

$f(x) = 3x^4 + x^3 - 2x^2$

Figure 51

Example 5   Sketch the graph of $f(x) = 3x^4 + x^3 - 2x^2$.
The polynomial can be factored as follows.

$$3x^4 + x^3 - 2x^2 = x^2(3x^2 + x - 2)$$
$$= x^2(3x - 2)(x + 1)$$

The zeros, 0, 2/3, and -1, divide the x-axis into four regions:

$$x < -1, \qquad -1 < x < 0, \qquad 0 < x < 2/3, \qquad \text{and} \qquad 2/3 < x.$$

Determine the sign of $f(x)$ in each region by substituting a value for x from each region to get the following information.

Region	Test point	Value of $f(x)$	Sign of $f(x)$	Location relative to axis
$x < -1$	-2	32	positive	above
$-1 < x < 0$	$-1/2$	$-.4375$	negative	below
$0 < x < 2/3$	$1/2$	$-.1875$	negative	below
$2/3 < x$	1	2	positive	above

With the values of x used for the test points and the corresponding values of $f(x)$, sketch the graph as shown in Figure 51. ■

4.6 Exercises   help presension   every other odd

Each of the polynomial functions defined as follows is symmetric about a line or a point. For each function,   (a) sketch the graph and   (b) give the line or point of symmetry.

1. $f(x) = \frac{1}{4}x^6$

2. $f(x) = -\frac{2}{3}x^5$

3. $f(x) = \frac{-5}{4}x^5$

4. $f(x) = 2x^4$

5. $f(x) = \frac{1}{2}x^3 + 1$

6. $f(x) = -x^4 + 2$

7. $f(x) = -(x + 1)^3$

8. $f(x) = \frac{1}{3}(x + 3)^4$

9. $f(x) = (x - 1)^4 + 2$ **10.** $f(x) = (x + 2)^3 - 1$

Graph each of the polynomial functions defined as follows.

11. $f(x) = 2x(x - 3)(x + 2)$

12. $f(x) = x^2(x + 1)(x - 1)$

13. $f(x) = x^2(x - 2)(x + 3)^2$

14. $f(x) = x^2(x - 5)(x + 3)(x - 1)$

15. $f(x) = x^3 - x^2 - 2x$

16. $f(x) = -x^3 - 4x^2 - 3x$

17. $f(x) = 3x^4 + 5x^3 - 2x^2$

18. $f(x) = 4x^3 + 2x^2 - 12x$

19. $f(x) = 2x^3(x^2 - 4)(x - 1)$

20. $f(x) = 5x^2(x^3 - 1)(x + 2)$

21. $f(x) = x^4 - 4x^2$

22. $f(x) = -x^4 + 16x^2$

23. $f(x) = x^2(x - 3)^3(x + 1)$

24. $f(x) = x(x - 4)^2(x + 2)^2$

25. The polynomial function A defined by

$$A(x) = -.015x^3 + 1.058x$$

gives the approximate alcohol concentration (in tenths of a percent) in an average person's bloodstream x hr after drinking about 8 oz of 100 proof whiskey. The function is approximately valid for x in the interval $[0, 8]$.
(a) Graph $y = A(x)$.
(b) Using the graph you drew for part (a), estimate the time of maximum alcohol concentration.
(c) In one state, a person is legally drunk if the blood alcohol concentration exceeds .15%. Use the graph of part (a) to estimate the period in which this average person is legally drunk.

26. A technique for measuring cardiac output depends on the concentration of a dye in the bloodstream after a known amount is injected into a vein near the heart. For a normal heart, the concentration of the dye in the bloodstream at time x (in sec) is given by

$$g(x) = -.006x^4 + .140x^3 - .053x^2 + 1.79x.$$

Graph $g(x)$.

27. The pressure of the oil in a reservoir tends to drop with time. By taking sample pressure readings for a particular oil reservoir, petroleum engineers have found that the change in pressure is given by

$$P(t) = t^3 - 25t^2 + 200t,$$

where t is time in years from the date of the first reading.
(a) Graph $P(t)$.
(b) For what time period is the change in pressure (drop) increasing? decreasing?

28. During the early part of the twentieth century, the deer population of the Kaibab Plateau in Arizona experienced a rapid increase, because hunters had reduced the number of natural predators. The increase in population depleted the food resources and eventually caused the population to decline. For the period from 1905 to 1930, the deer population was approximated by

$$D(x) = -.125x^5 + 3.125x^4 + 4000,$$

where x is time in years from 1905.

(a) Use a calculator to find enough points to graph $D(x)$.

(b) From the graph, over what period of time (from 1905 to 1930) was the population increasing? relatively stable? decreasing?

Use the definition of odd and even functions given in Section 3.4 to decide if the polynomial functions defined as follows are odd, even, or neither.

29. $f(x) = 2x^3$

30. $f(x) = -4x^5$

31. $f(x) = .2x^4$

32. $f(x) = -x^6$

33. $f(x) = -x^5$

34. $f(x) = (x - 1)^3$

35. $f(x) = 2x^3 + 3$

36. $f(x) = 4x^3 - x$

37. $f(x) = x^4 + 3x^2 + 5$

38. $f(x) = 11x + 2$

39. $f(x) = (x + 5)^2$

40. $f(x) = (3x - 5)^2$

We can find approximate maximum or minimum values of polynomial functions for given intervals by first evaluating the function at the left endpoint of the given interval. Then add .1 to the value of x and reevaluate the polynomial. Keep doing this until the right endpoint of the interval is reached. Then identify the approximate maximum and minimum value for the polynomial on the interval.

41. $f(x) = x^3 + 4x^2 - 8x - 8$, $[-3.8, -3]$

42. $f(x) = x^3 + 4x^2 - 8x - 8$, $[.3, 1]$

43. $f(x) = 2x^3 - 5x^2 - x + 1$, $[-1, 0]$

44. $f(x) = 2x^3 - 5x^2 - x + 1$, $[1.4, 2]$

45. $f(x) = x^4 - 7x^3 + 13x^2 + 6x - 28$, $[-2, -1]$

46. $f(x) = x^4 - 7x^3 + 13x^2 + 6x - 28$, $[2, 3]$

47. The graphs of $f(x) = x$, $f(x) = x^3$, and $f(x) = x^5$ all pass through the points $(-1, -1)$, $(0, 0)$, and $(1, 1)$. Draw a graph showing these three functions for the domain $[-1, 1]$. Use your result to decide how the graph of $f(x) = x^{11}$ would look for the same interval.

48. The graphs of $f(x) = x^2$, $f(x) = x^4$, and $f(x) = x^6$ all pass through the points $(-1, 1)$, $(0, 0)$, and $(1, 1)$. Draw a graph showing these three functions for the domain $[-1, 1]$. Use your result to decide how the graph of $f(x) = x^{12}$ would look for the same interval.

4.7 **Rational Functions**

A function defined by a rule of the form

$$f(x) = \frac{p(x)}{q(x)},$$

where $p(x)$ and $q(x)$ define polynomial functions with $q(x) \neq 0$, is called a **rational function.** Since any values of x such that $q(x) = 0$ are excluded from the domain, a rational function usually has a graph that has one or more breaks in it.

The simplest rational function with a variable denominator is defined by

$$f(x) = \frac{1}{x}.$$

The domain of this function is the set of all real numbers except 0. To graph the function, first replace x with $-x$, getting $f(-x) = 1/(-x) = -1/x = -f(x)$, showing that f is an odd function. As mentioned earlier, this means that the graph of f is symmetric with respect to the origin.

The number 0 cannot be used as a value of x, but it is helpful to find the values of $f(x)$ for several values of x close to 0. The following table shows what happens to $f(x)$ as x gets closer and closer to 0 from either side.

x	-1	$-.1$	$-.01$	$-.001$	.001	.01	.1	1
$f(x)$	-1	-10	-100	-1000	1000	100	10	1

x approaches 0

$|f(x)|$ gets larger and larger

The table suggests that $|f(x)|$ gets larger and larger as x gets closer and closer to 0, written in symbols as

$$|f(x)| \rightarrow \infty \text{ as } x \rightarrow 0.$$

(The symbol $x \rightarrow 0$ means that x approaches as close as desired to 0, without necessarily ever being equal to 0.) Since x cannot equal 0, the graph of $f(x) = 1/x$ will never intersect the vertical line $x = 0$. For this graph, the line $x = 0$ is called a *vertical asymptote*.

On the other hand, as $|x|$ gets larger and larger, the values of $f(x) = 1/x$ get closer and closer to 0. (See the table.)

x	$-10,000$	-1000	-100	-10	10	100	1000	10,000
$f(x)$	$-.0001$	$-.001$	$-.01$	$-.1$	.1	.01	.001	.0001

Letting $|x|$ get larger and larger without bound (written $|x| \rightarrow \infty$) causes the graph of $y = 1/x$ to move closer and closer to the horizontal line $y = 0$. This line is called a *horizontal asymptote*.

As we said, $f(x) = 1/x$ is an odd function, symmetric with respect to the origin. Choosing some positive values of x and finding the corresponding values of $f(x)$ gives the first quadrant part of the graph shown in Figure 52. The other part of the graph (in the third quadrant) can be found by symmetry.

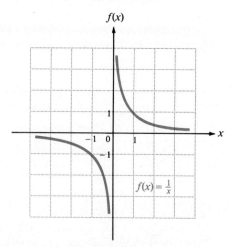

Figure 52

Example 1   Graph $f(x) = \dfrac{-2}{x}$.

Rewrite $f(x)$ as

$$f(x) = -2 \cdot \frac{1}{x}.$$

Compared to $f(x) = 1/x$, the graph will be reflected about the x-axis (because of the negative sign), and each point will be twice as far from the x-axis. See the graph in Figure 53.   ■

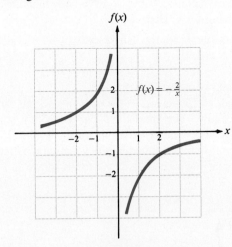

Figure 53   **Figure 54**

Example 2   Graph $f(x) = \dfrac{2}{1 + x}$.

The domain of this function is the set of all real numbers except -1. As shown in Figure 54, the graph is that of $f(x) = 1/x$, translated 1 unit to the left, with each y-value doubled.   ■

The examples above suggest the following definition of vertical and horizontal asymptotes. This definition assumes that the rational function defined by $f(x) = p(x)/q(x)$ is written in lowest terms.

Asymptotes

> For the rational function defined by $y = f(x)$, and for a real number a,
>
> if $|f(x)| \to \infty$ as $x \to a$, then the line $x = a$ is a **vertical asymptote;**
>
> if $f(x) \to a$ as $|x| \to \infty$, then the line $y = a$ is a **horizontal asymptote.**

Example 3   Graph $f(x) = \dfrac{1}{(x - 1)(x + 4)}$.

This graph is not symmetric with respect to the x-axis, the y-axis, or the origin. There are vertical asymptotes at $x = 1$ and $x = -4$, as suggested by the following charts.

x	-4.1	-4.01	-4.001	-3.999	-3.99	-3.9
$f(x)$	1.96	19.96	199.96	-200.04	-20.04	-2.04

x	.9	.99	.999	1.001	1.01	1.1
$f(x)$	-2.04	-20.04	-200.04	199.96	19.96	1.96

Since the given rational function is defined by an expression written in lowest terms, the vertical asymptotes could also have been found by setting the denominator equal to 0 and solving for x. Here,

$$(x - 1)(x + 4) = 0$$
$$x = 1 \quad \text{or} \quad x = -4,$$

the equations of the two vertical asymptotes.

As $|x|$ gets larger and larger, $|(x - 1)(x + 4)|$ also gets larger and larger, bringing y closer and closer to 0. This means that the x-axis is a horizontal asymptote. The vertical asymptotes divide the x-axis into three regions. It is often convenient to consider each region separately. Finding the y-intercept by letting $x = 0$ and plotting a few additional points as needed gives the result shown in Figure 55. ■

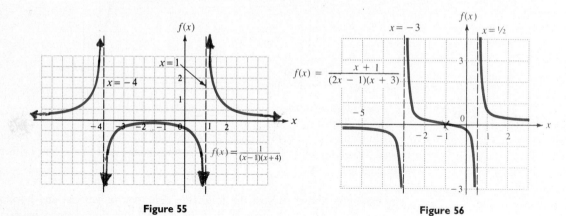

Figure 55   **Figure 56**

Example 4   Graph $f(x) = \dfrac{x + 1}{(2x - 1)(x + 3)}$.

The graph has as vertical asymptotes the lines $x = 1/2$ and $x = -3$. Replacing x with 0 gives $-1/3$ as the y-intercept. For the x-intercept, the value of y must be 0. The only way that y can equal 0 is if the numerator, $x + 1$, is 0. This happens when $x = -1$, making -1 the x-intercept.

To find any horizontal asymptote, multiply the factors in the denominator to get

$$f(x) = \frac{x + 1}{(2x - 1)(x + 3)} = \frac{x + 1}{2x^2 + 5x - 3}.$$

Now divide each term in the numerator and denominator by x^2, since 2 is the highest exponent on x. This gives

$$f(x) = \frac{\dfrac{x}{x^2} + \dfrac{1}{x^2}}{\dfrac{2x^2}{x^2} + \dfrac{5x}{x^2} - \dfrac{3}{x^2}} = \frac{\dfrac{1}{x} + \dfrac{1}{x^2}}{2 + \dfrac{5}{x} - \dfrac{3}{x^2}}.$$

As $|x|$ gets larger and larger, the quotients $1/x$, $1/x^2$, $5/x$, and $3/x^2$ all approach 0, with the value of $f(x)$ approaching

$$\frac{0 + 0}{2 + 0 - 0} = \frac{0}{2} = 0.$$

The line $y = 0$ is therefore a horizontal asymptote. Using these asymptotes and intercepts (the graph has no symmetry with respect to either axis or the origin) and plotting a few points gives the graph in Figure 56. ■

Example 5   Graph $f(x) = \dfrac{2x - 5}{x - 3}$.

Work as in the previous example to find any horizontal asymptote. Since the highest exponent on x is 1, divide each term by x.

$$f(x) = \frac{\overset{\text{horizontal}}{2x - 5}}{\underset{\text{vert}}{x - 3}} = \frac{\dfrac{2x}{x} - \dfrac{5}{x}}{\dfrac{x}{x} - \dfrac{3}{x}} = \frac{2 - \dfrac{5}{x}}{1 - \dfrac{3}{x}}$$

As $|x|$ gets larger and larger, both $3/x$ and $5/x$ approach 0:

$$\frac{2 - 0}{1 - 0} = \frac{2}{1} = 2,$$

showing that the line $y = 2$ is the horizontal asymptote. The vertical asymptote is $x = 3$. See Figure 57.

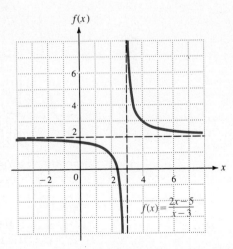

$$f(x) = \frac{2x - 5}{x - 3}$$

Figure 57

Letting $x = 0$ gives the y-intercept 5/3. The x-intercept is found when the numerator $2x - 5 = 0$, so the x-intercept is 5/2. This information was used to get the graph of Figure 57. ∎

There is an alternate way to sketch the graph in Example 5. Use long division to divide $2x - 5$ by $x - 3$, getting

$$\frac{2x - 5}{x - 3} = 2 + \frac{1}{x - 3}.$$

This result shows that the graph of $f(x) = (2x - 5)/(x - 3)$ is the same as that of $1/(x - 3)$, shifted 2 units upward.

Example 6 Graph $f(x) = \dfrac{3(x + 1)(x - 2)}{(x + 4)^2}$.

The only vertical asymptote is the line $x = -4$. To find any horizontal asymptotes, multiply the factors in the numerator and denominator.

$$f(x) = \frac{3x^2 - 3x - 6}{x^2 + 8x + 16}$$

Now divide by x^2.

$$f(x) = \frac{\dfrac{3x^2}{x^2} - \dfrac{3x}{x^2} - \dfrac{6}{x^2}}{\dfrac{x^2}{x^2} + \dfrac{8x}{x^2} + \dfrac{16}{x^2}} = \frac{3 - \dfrac{3}{x} - \dfrac{6}{x^2}}{1 + \dfrac{8}{x} + \dfrac{16}{x^2}}$$

Letting $|x|$ get larger and larger shows that $y = 3/1$, or $y = 3$, is the horizontal asymptote. The y-intercept is $-3/8$, and the x-intercepts are -1 and 2. See the final graph in Figure 58. ∎

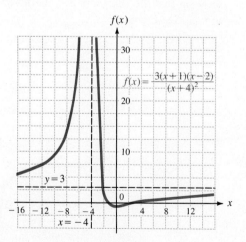

Figure 58

In Figure 58, the graph of the function actually crosses the horizontal asymptote. While it is possible for the graph of a rational function to cross a horizontal asymptote, such a graph can never cross a vertical asymptote. The reason that a graph can cross a horizontal asymptote comes from the definition—only values of x where $|x|$ is relatively large are used in finding horizontal asymptotes. Find the value of x where the graph crosses the horizontal asymptote of Figure 58 by letting $f(x) = 3$, since the horizontal asymptote is the line $y = 3$. Solving $f(x) = 3$ gives $x = -2$, so the graph crosses the horizontal asymptote at $(-2, 3)$.

The next example shows a rational function defined by an expression having a numerator with higher degree than the denominator. Polynomial division must be used for these functions.

Example 7 Graph $f(x) = \dfrac{x^2 + 1}{x - 2}$.

Dividing by x^2 as in the examples above would show that the graph has no horizontal asymptote. The vertical asymptote is the line $x = 2$. Since the numerator has higher degree than the denominator, divide to rewrite the equation for the function in another form.

$$
\begin{array}{r}
x + 2 \\
x - 2 \overline{)\, x^2 + 1} \\
\underline{x^2 - 2x } \\
2x + 1 \\
\underline{2x - 4} \\
5
\end{array}
$$

By this result,

$$f(x) = \frac{x^2 + 1}{x - 2} = x + 2 + \frac{5}{x - 2}.$$

For very large values of $|x|$, $5/(x - 2)$ is close to 0, and the graph approaches the line $y = x + 2$. This line is an **oblique asymptote** (neither vertical nor horizontal) for the graph. The y-intercept is $-1/2$. There is no x-intercept. Using the asymptotes, the y-intercept, and additional points as needed leads to the graph in Figure 59. ∎

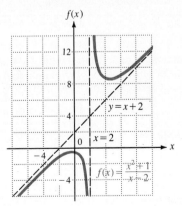

Figure 59

In general, if the degree of the numerator is exactly one more than the degree of the denominator, the rational function may have an oblique asymptote. The equation of this asymptote is found by dividing the numerator by the denominator and dropping the remainder.

To summarize: when graphing a rational function, use the following procedure. (We assume that the rational function is defined by an expression written in lowest terms.)

Graphing Rational Functions

1. Find any vertical asymptotes by setting the denominator equal to 0 and solving for x. If a is a zero of the denominator, then the line $x = a$ is a vertical asymptote.

2. Determine any other asymptotes. There are three possibilities:
   (a) If the numerator has lower degree than the denominator, there is a horizontal asymptote, $y = 0$.
   (b) If the numerator and denominator have the same degree, and the function is defined by an expression of the form

   $$f(x) = \frac{a_n x^n + \ldots + a_0}{b_n x^n + \ldots + b_0}, \qquad b_n \neq 0,$$

   dividing by x^n produces the horizontal asymptote

   $$y = \frac{a_n}{b_n}.$$

   (c) If the numerator is of degree exactly one more than the denominator, there may be an oblique asymptote. To find it, divide the numerator by the denominator and drop any remainder. The rest of the quotient gives the equation of the asymptote.

3. Find any intercepts.
4. Check for symmetry.
5. Plot a few selected points—at least one in each region of the domain determined by the vertical asymptotes.
6. Complete the sketch.

It is fairly common in calculus to draw the graph of a rational function whose numerator and denominator have a common factor. (This is done with functions whose graphs are not "continuous" because of "holes" in the graphs.) Work as in the next example.

Example 8   Graph $f(x) = \dfrac{(x - 1)(x^2 - 3)}{x - 1}$.

First,

$$\frac{(x - 1)(x^2 - 3)}{x - 1} = x^2 - 3, \quad \text{if } x \neq 1.$$

This means that the graph of f is the same as that of the parabola $y = x^2 - 3$, except at $x = 1$. At $x = 1$, there is no value of $f(x)$, so that the graph has no point for $x = 1$. See Figure 60. (Note that $x = 1$ is not a vertical asymptote since replacing x with 1 makes both the numerator and denominator equal to 0.)   ∎

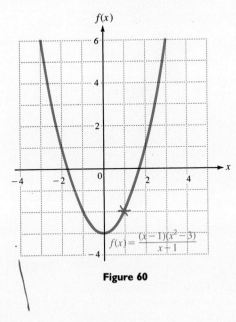

$$f(x) = \frac{(x-1)(x^2-3)}{x+1}$$

Figure 60

4.7 Exercises

Find any vertical, horizontal, or oblique asymptotes for the following.

1. $f(x) = \dfrac{2}{x - 5}$

2. $f(x) = \dfrac{-1}{x + 2}$

3. $f(x) = \dfrac{-8}{3x - 7}$

4. $f(x) = \dfrac{5}{4x - 9}$

5. $f(x) = \dfrac{2 - x}{x + 2}$

6. $f(x) = \dfrac{x - 4}{5 - x}$

7. $f(x) = \dfrac{3x - 5}{2x + 9}$

8. $f(x) = \dfrac{4x + 3}{3x - 7}$

9. $f(x) = \dfrac{2}{x^2 - 4x + 3}$

10. $f(x) = \dfrac{-5}{x^2 - 3x - 10}$

11. $f(x) = \dfrac{x^2 - 1}{x + 3}$

12. $f(x) = \dfrac{x^2 + 4}{x - 1}$

13. $f(x) = \dfrac{(x - 3)(x + 1)}{(x + 2)(2x - 5)}$

14. $f(x) = \dfrac{3(x + 2)(x - 4)}{(5x - 1)(x - 5)}$

15. $f(x) = \dfrac{2x^2 - 32}{x^2 + 1}$

16. $f(x) = \dfrac{-5x^2 + 20}{x^2 + 9}$

17. Sketch the following graphs and compare them with the graph of $f(x) = 1/x^2$.

(a) $f(x) = \dfrac{1}{(x - 3)^2}$

(b) $f(x) = \dfrac{-2}{x^2}$

(c) $f(x) = \dfrac{-2}{(x - 3)^2}$

Graph each of the following.

18. $f(x) = \dfrac{4}{5 + 3x}$

19. $f(x) = \dfrac{1}{(x - 2)(x + 4)}$

20. $f(x) = \dfrac{3}{(x + 4)^2}$

21. $f(x) = \dfrac{3x}{(x + 1)(x - 2)}$

22. $f(x) = \dfrac{2x + 1}{(x + 2)(x + 4)}$

23. $f(x) = \dfrac{5x}{x^2 - 1}$

24. $f(x) = \dfrac{-x}{x^2 - 4}$

25. $f(x) = \dfrac{3x}{x - 1}$

26. $f(x) = \dfrac{4x}{1 - 3x}$

27. $f(x) = \dfrac{x + 1}{x - 4}$

28. $f(x) = \dfrac{x - 3}{x + 5}$

29. $f(x) = \dfrac{x - 5}{x + 3}$

30. $f(x) = \dfrac{3x}{x^2 - 16}$

31. $f(x) = \dfrac{x}{x^2 - 9}$

32. $f(x) = \dfrac{x^2 - 5}{x + 2}$

33. $f(x) = \dfrac{x^2 - 3x + 2}{x - 3}$

34. $f(x) = \dfrac{x^2 + 1}{x + 3}$

35. $f(x) = \dfrac{2x^2 + 3}{x - 4}$

36. $f(x) = \dfrac{x^2 - x}{x + 2}$

37. $f(x) = \dfrac{x^2 + 2x}{2x - 1}$

38. $f(x) = \dfrac{x(x - 2)}{(x + 3)^2}$

39. $f(x) = \dfrac{(x - 3)(x + 1)}{(x - 1)^2}$

40. $f(x) = \dfrac{(x + 4)(x - 1)}{x^2 + 1}$

41. $f(x) = \dfrac{(x - 5)(x - 2)}{x^2 + 9}$

42. $f(x) = \dfrac{1}{x^2 + 1}$

43. $f(x) = \dfrac{-9}{x^2 + 9}$

44. $f(x) = \dfrac{(x - 1)(x + 2)}{x + 2}$

45. $f(x) = \dfrac{(2x - 3)(x - 4)}{x - 4}$

46. $f(x) = \dfrac{(x + 2)(x^2 + 1)}{x + 2}$

47. $f(x) = \dfrac{(x + 3)(x^2 - 4)}{x + 3}$

48. Suppose the average cost per unit, $C(x)$, to produce x units of margarine is given by

$$C(x) = \frac{500}{x + 30}.$$

(a) Find $C(10)$, $C(20)$, $C(50)$, $C(75)$, and $C(100)$.
(b) Would a more reasonable domain for C be $(0, +\infty)$ or $[0, +\infty)$? Why?

49. In a recent year, the cost per ton, y, to build an oil tanker of x thousand deadweight tons was approximated by

$$y = \frac{110,000}{x + 225}.$$

(a) Find y for $x = 25$, $x = 50$, $x = 100$, $x = 200$, $x = 300$, and $x = 400$.
(b) Graph y.

50. Antique-car fans often enter their cars in a *concours d'elegance* in which a maximum of 100 points can be awarded to a particular car. Points are awarded for the general attractiveness of the car. The function defined by

$$C(x) = \frac{10x}{49(101 - x)}$$

expresses the cost, in thousands of dollars, of restoring a car so that it will win x points. Graph the function.

51. In situations involving environmental pollution, a cost-benefit model expresses cost as a function of the percentage of pollutant removed from the environment. Suppose a cost-benefit model is expressed as

$$y = \frac{6.7x}{100 - x},$$

where y is the cost in thousands of dollars of removing x percent of a certain pollutant.
(a) Graph the function.
(b) Is it possible, according to this function, to remove all of the pollutant?

In recent years the economist Arthur Laffer has been a center of controversy because of his
Laffer curve, an idealized version of which is shown here.

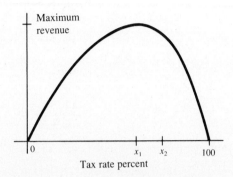

According to this curve, increasing a tax rate, say from x_1 percent to x_2 percent on the graph,
can actually lead to a decrease in government revenue. All economists agree on the
endpoints, 0 revenue at tax rates of both 0% and 100%, but there is much disagreement on
the location of the rate x_1 that produces maximum revenue.

52. Suppose an economist studying the Laffer curve produces the rational function

$$y = \frac{60x - 6000}{x - 120},$$

where y is government revenue in millions of dollars from a tax rate of x percent, with
the function valid for $50 \leq x \leq 100$. Find the revenue from the following tax rates:
(a) 50% **(b)** 60% **(c)** 80% **(d)** 100%. **(e)** Graph the function.

53. Suppose our economist studies a different tax, this time producing

$$y = \frac{80x - 8000}{x - 110},$$

with y giving government revenue in tens of millions of dollars for a tax rate of x
percent, with the function valid for $55 \leq x \leq 100$. Find the revenue for the following tax
rates: **(a)** 55% **(b)** 60% **(c)** 70% **(d)** 90% **(e)** 100%. **(f)** Graph the function.

54. Let $f(x) = p(x)/q(x)$ define a rational function. If the degree of $p(x)$ is m, and the degree
of $q(x)$ is n, and if $m > n$, tell how to find the horizontal or vertical asymptotes of f.

Chapter 4 Summary

Key Words

polynomial function	parabola	hyperbola
linear function	axis	asymptotes of a hyperbola
quadratic function	vertex	fundamental rectangle
slope	ellipse	conic sections
point-slope form	foci	rational function
intercept	vertices of an ellipse	asymptote
slope-intercept form	major axis	

Review Exercises

Find the slope for each of the following lines that has a slope.

1. Through $(8, 7)$ and $(1/2, -2)$
2. Through $(2, -2)$ and $(3, -4)$
3. Through $(5, 6)$ and $(5, -2)$
4. Through $(0, -7)$ and $(3, -7)$
5. $9x - 4y = 2$
6. $11x + 2y = 3$
7. $x - 5y = 0$
8. $x - 2 = 0$
9. $y + 6 = 0$
10. $y = x$

Graph each of the following.

11. $3x + 7y = 14$
12. $2x - 5y = 5$
13. $3y = x$
14. $y = 3$
15. $x = -5$
16. $y = x$

For each of the following lines, write an equation in the form $ax + by = c$.

17. Through $(-2, 4)$ and $(1, 3)$
18. Through $(-2/3, -1)$ and $(0, 4)$
19. Through $(3, -5)$ with slope -2
20. Through $(-4, 4)$ with slope $3/2$
21. Through $(1/5, 1/3)$ with slope $-1/2$
22. x-intercept -3, y-intercept 5
23. No x-intercept, y-intercept $3/4$
24. Through $(2, -1)$, parallel to $3x - y = 1$
25. Through $(0, 5)$, perpendicular to $8x + 5y = 3$
26. Through $(2, -10)$, perpendicular to a line with undefined slope.
27. Through $(3, -5)$ parallel to $y = 4$
28. Through $(-7, 4)$, perpendicular to $y = 8$

Graph each of the following lines.

29. Through $(2, -4)$, $m = 3/4$
30. Through $(0, 5)$, $m = -2/3$
31. Through $(-4, 1)$, $m = 3$
32. Through $(-3, -2)$, $m = -1$

Graph each of the functions defined as follows.

33. $f(x) = \begin{cases} 3x + 1 & \text{if } x < 2 \\ -x + 4 & \text{if } x \geq 2 \end{cases}$
34. $f(x) = \begin{cases} -4x + 2 & \text{if } x \leq 1 \\ 3x - 5 & \text{if } x > 1 \end{cases}$
35. $f(x) = x^2 - 4$
36. $f(x) = 6 - x^2$
37. $f(x) = 3(x + 1)^2 - 5$
38. $f(x) = -\frac{1}{4}(x - 2)^2 + 3$
39. $f(x) = x^2 - 4x + 2$
40. $f(x) = -3x^2 - 12x - 1$

Give the vertex and axis of the graphs of the functions defined as follows.

41. $f(x) = -(x + 3)^2 - 9$
42. $f(x) = (x - 7)^2 + 3$
43. $f(x) = x^2 - 7x + 2$
44. $f(x) = -x^2 - 4x + 1$
45. $f(x) = -3x^2 - 6x + 1$
46. $f(x) = 4x^2 - 4x + 3$

Use parabolas to work each of the following problems.

47. Find two numbers whose sum is 11 and whose product is a maximum.

48. Find two numbers having a sum of 40 such that the sum of the square of one and twice the square of the other is minimum.

49. Find the rectangular region of maximum area that can be enclosed with 180 meters of fencing.

50. Find the rectangular region of maximum area that can be enclosed with 180 meters of fencing if no fencing is needed along one side of the region.

Graph each of the following. Identify each graph.

51. $\dfrac{x^2}{25} + \dfrac{y^2}{4} = 1$

52. $\dfrac{x^2}{3} + \dfrac{y^2}{16} = 1$

53. $\dfrac{x^2}{4} - \dfrac{y^2}{9} = 1$

54. $\dfrac{y^2}{100} - \dfrac{x^2}{25} = 1$

55. $x^2 = 16 + y^2$

56. $4x^2 + 9y^2 = 36$

57. $\dfrac{25x^2}{9} + \dfrac{4y^2}{25} = 1$

58. $\dfrac{100x^2}{49} + \dfrac{9y^2}{16} = 1$

59. $\dfrac{(x-2)^2}{9} + \dfrac{(y+3)^2}{4} = 1$

60. $\dfrac{(x-3)^2}{4} + (y+1)^2 = 1$

61. $\dfrac{(y+2)^2}{4} - \dfrac{(x+3)^2}{9} = 1$

62. $\dfrac{(x+1)^2}{16} - \dfrac{(y-2)^2}{4} = 1$

63. $\dfrac{x}{3} = -\sqrt{1 - \dfrac{y^2}{16}}$

64. $x = -\sqrt{1 - \dfrac{y^2}{36}}$

65. $y = -\sqrt{1 - \dfrac{x^2}{25}}$.

66. $y = -\sqrt{1 + x^2}$

Graph each of the functions defined as follows.

67. $f(x) = x^3 + 5$

68. $f(x) = 1 - x^4$

69. $f(x) = x^2(2x + 1)(x - 2)$

70. $f(x) = (4x - 3)(3x + 2)(x - 1)$

71. $f(x) = 2x^3 + 13x^2 + 15x$

72. $f(x) = x(x - 1)(x + 2)(x - 3)$

73. $f(x) = \dfrac{8}{x}$

74. $f(x) = \dfrac{2}{3x - 1}$

75. $f(x) = \dfrac{4x - 2}{3x + 1}$

76. $f(x) = \dfrac{6x}{(x - 1)(x + 2)}$

77. $f(x) = \dfrac{2x}{x^2 - 1}$

78. $f(x) = \dfrac{x^2 + 4}{x + 2}$

79. $f(x) = \dfrac{x^2 - 1}{x}$

80. $f(x) = \dfrac{x^2 + 6x + 5}{x - 3}$

81. Find an equation of the line with slope 2/3 that goes through the center of the circle $(x - 4)^2 + (y + 2)^2 = 9$.

82. Use slopes to show that the quadrilateral with vertices at $(-2, 2)$, $(4, 2)$, $(4, -1)$, and $(-2, -1)$ is a rectangle.

83. Find a value of k so that $5x - 3y = k$ goes through the point $(1, 4)$.

84. Find k so that $3x + ky = 2$ goes through $(3, -2)$.

85. Find all squares that have $(0, 7)$ and $(12, 12)$ as two of the vertices. (*Hint:* It is sufficient to identify a square by listing its vertices.)

86. Suppose $a > 0$ and $a + h > 0$. Show that the straight line through $(a, \sqrt{a})$ and $(a + h, \sqrt{a + h})$ has slope $1/(\sqrt{a + h} + \sqrt{a})$. (*Hint:* $(\sqrt{B} - \sqrt{A})(\sqrt{B} + \sqrt{A}) = B - A$ if $A > 0$ and $B > 0$.)

87. Suppose that $x^2 + y^2 + Ax + By + C = 0$ and $x^2 + y^2 + ax + by + c = 0$ are different circles that meet at two distinct points. Show that the line through those two points of intersection has the equation

$$(A - a)x + (B - b)y + (C - c) = 0.$$

88. Show that the line with equation

$$Ax + By = r^2$$

is tangent to (touches in only one point) the circle $x^2 + y^2 = r^2$ at (A, B).

5

Exponential and Logarithmic Functions

In this chapter we will study two kinds of functions that are quite different from those studied before, functions that do not involve just the basic operations of addition, subtraction, multiplication, division, and taking roots. The functions discussed earlier are examples of *algebraic functions*. The functions in this chapter are not algebraic, but are *transcendental functions*. Many applications of mathematics, particularly to growth and decay of populations, involve the closely interrelated exponential and logarithmic functions introduced in this chapter. As shown later, these two types of functions are inverses of each other.

(No graphs on TEST)

5.1 Exponential Functions

As shown in Chapter 1, if $a > 0$, the symbol a^m can be defined for any rational value of m. In this section, the definition of a^m is extended to include all *real,* and not just rational, values of the exponent m. For example, what is meant by $2^{\sqrt{3}}$? The exponent, $\sqrt{3}$, can be approximated more and more closely by the numbers 1.7, 1.73, 1.732, and so on, making it reasonable that $2^{\sqrt{3}}$ should be approximated more and more closely by the numbers $2^{1.7}$, $2^{1.73}$, $2^{1.732}$, and so on. (Recall, for example, that $2^{1.7} = 2^{17/10}$, which means $\sqrt[10]{2^{17}}$.) In fact, this is exactly how the number $2^{\sqrt{3}}$ is defined in a more advanced course. To show that this assumption is reasonable, Figure 1 gives the graphs of the function $f(x) = 2^x$ with three different domains.

We shall assume that the meaning given to real exponents is such that all previous rules and theorems for exponents are valid for real number exponents as well as rational ones. In addition to the rules for exponents presented earlier, the following additional properties will prove useful. First, any given real value of x leads to exactly one value of 2^x. For example,

$$2^2 = 4, \quad 2^3 = 8, \quad \text{and} \quad 2^{1/2} = \sqrt{2} \approx 1.4142.$$

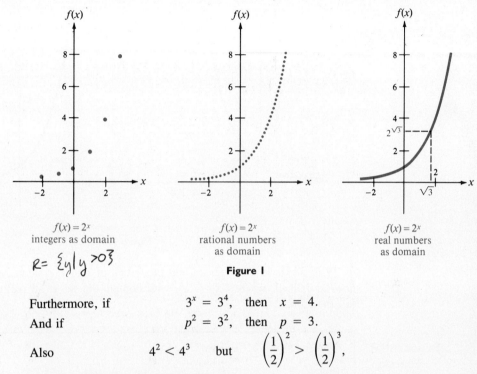

$f(x) = 2^x$
integers as domain

$R = \{y \mid y > 0\}$

$f(x) = 2^x$
rational numbers
as domain

Figure I

$f(x) = 2^x$
real numbers
as domain

Furthermore, if $\qquad\qquad 3^x = 3^4$, then $\quad x = 4.$
And if $\qquad\qquad\qquad p^2 = 3^2$, then $\quad p = 3.$

Also $\qquad\qquad 4^2 < 4^3 \qquad$ but $\qquad \left(\dfrac{1}{2}\right)^2 > \left(\dfrac{1}{2}\right)^3,$

so that when $a > 1$, increasing the exponent on a leads to a *larger* number, but if $0 < a < 1$, increasing the exponent on a leads to a *smaller* number.

These properties are generalized in the next theorem. No proof of these properties is given, since the proof requires more advanced mathematics than that of this course.

Theorem

For any real number $a > 0$, $a \neq 1$, and any real number x

(a) a^x is a unique real number.

(b) $a^b = a^c$ if and only if $b = c$.

(c) If $a > 1$ and $m < n$, then $a^m < a^n$.

(d) If $0 < a < 1$ and $m < n$, then $a^m > a^n$.

Part (a) of the theorem requires $a > 0$ so that a^x is always defined. For example, $(-6)^x$ is not a real number if $x = 1/2$. If $a > 0$, then a^x will always be positive, since a is positive. For part (b) to hold, a must not equal 1 since $1^4 = 1^5$, even though $4 \neq 5$. Figure 2(a) on the next page illustrates part (c): the base, 2, of the exponential 2^x is greater than 1, and as the x-values increase, so do the values of 2^x. Part (d) of the theorem is shown in Figure 2(b) where the base, 1/2, is between 0 and 1, and as the x-values increase, $(1/2)^x$ decreases.

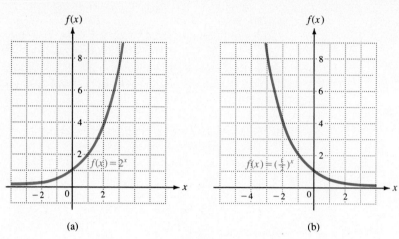

(a)

(b)

Figure 2

With all these assumptions and properties of real exponents, a function, f, can now be defined, where $f(x) = a^x$, with domain the set of all real numbers (and not just the rationals).

Exponential Function

The function f defined by

$$f(x) = a^x, \qquad a > 0 \text{ and } a \neq 1,$$

is the **exponential function with base a.**

(If $a = 1$, the function is the constant function given by $f(x) = 1$.)

Example 1 If $f(x) = 2^x$, find each of the following.

(a) $f(-1)$

Replace x with -1.

$$f(-1) = 2^{-1} = \frac{1}{2}$$

(b) $f(3) = 2^3 = 8$

(c) $f(5/2) = 2^{5/2} = (2^5)^{1/2} = 32^{1/2} = \sqrt{32} = 4\sqrt{2}$ ∎

Example 2 Graph the exponential functions defined as follows.

(a) $f(x) = 2^x$.

The base of this exponential function is 2. Some ordered pairs that satisfy the equation are $(-2, 1/4)$, $(-1, 1/2)$, $(0, 1)$, $(1, 2)$, $(2, 4)$, and $(3, 8)$. Plotting these points and then drawing a smooth curve through them gives the graph in Figure 2(a). As the graph suggests, the domain of the function is the set of all real numbers, and the range is the set of all positive numbers. The function is increasing on its entire domain, making it a one-to-one function. The x-axis is a horizontal asymptote.

(b) $f(x) = (1/2)^x$

Again, plot some ordered pairs and draw a smooth curve through them. For example, $(-3, 8)$, $(-2, 4)$, $(-1, 2)$, $(0, 1)$, and $(1, 1/2)$ are on the graph shown in Figure 2(b). Like the function in part (a), this function also has the set of real numbers as domain and the set of positive real numbers as range. This graph is decreasing on the entire domain. ∎

Starting with $f(x) = 2^x$ and replacing x with $-x$, gives $f(-x) = 2^{-x} = (2^{-1})^x = (1/2)^x$. For this reason, the graphs of $f(x) = 2^x$ and $f(x) = (1/2)^x$ are symmetric with respect to the y-axis. This is also suggested by the graphs in Figures 2(a) and (b).

The graph of $f(x) = 2^x$ is typical of graphs of $f(x) = a^x$ where $a > 1$. For larger values of a, the graphs rise more steeply, but the general shape is similar to the graph in Figure 2(a). When $0 < a < 1$ the graph decreases like the graph of $f(x) = (1/2)^x$ in Figure 2(b). In Figure 3, the graphs of several typical exponential functions illustrate these facts.

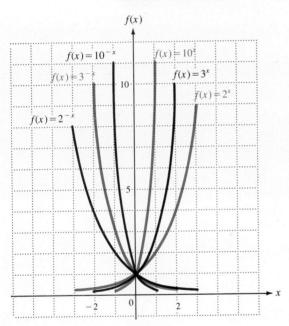

Figure 3

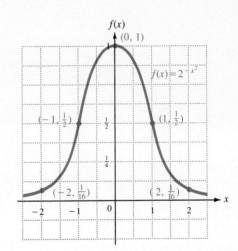

Figure 4

Example 3   Graph $f(x) = 2^{-x^2}$.

Since $f(x) = 2^{-x^2} = 1/(2^{x^2})$, we have $0 < y \le 1$ for all values of x. Plotting some typical points, such as $(-2, 1/16)$, $(-1, 1/2)$, $(0, 1)$, $(1, 1/2)$, and $(2, 1/16)$, and drawing a smooth curve through them gives the graph of Figure 4. This graph is symmetric with respect to the y-axis and has the x-axis as a horizontal asymptote. ∎

The results of the last theorem can be used to solve equations with variable exponents, as shown in the next example.

Example 4   Solve $\left(\dfrac{1}{3}\right)^x = 81$.

First, write $1/3$ as 3^{-1}, so that $(1/3)^x = (3^{-1})^x = 3^{-x}$. Since $81 = 3^4$,

$$\left(\dfrac{1}{3}\right)^x = 81$$

becomes $3^{-x} = 3^4.$

By the second property above,

$$-x = 4, \quad \text{or} \quad x = -4.$$

The solution set of the given equation is $\{-4\}$. (Section 5.4 shows a method for solving equations of this type where both sides cannot be written as powers of the same number.)   ■

Example 5   Find b if $81 = b^{4/3}$.

Since $(b^{4/3})^{3/4} = b^1 = b$, raise both sides of the equation to the $3/4$ power.

$$81 = b^{4/3}$$
$$81^{3/4} = (b^{4/3})^{3/4}$$
$$(\sqrt[4]{81})^3 = b$$
$$3^3 = b$$
$$27 = b$$

Remember that the process of raising both sides of an equation to the same power may result in false "solutions." For this reason, it is necessary to check all proposed solutions. Replacing b with 27 gives

$$27^{4/3} = (\sqrt[3]{27})^4 = 3^4 = 81,$$

which checks; the solution set is $\{27\}$.   ■

The formula for compound interest (interest paid on both principal and interest) defines an exponential function.

Compound Interest

> If P dollars is deposited in an account paying an annual rate of interest i compounded (paid) m times per year, the account will contain
>
> $$A = P\left(1 + \dfrac{i}{m}\right)^{nm}$$
>
> dollars after n years.

Example 6 Suppose $1000 is deposited in an account paying 8% per year compounded quarterly (four times a year). Find the total amount in the account after 10 years if no withdrawals are made. Find the amount of interest earned.

Use the compound interest formula from above with $P = 1000$, $i = .08$, $m = 4$, and $n = 10$.

$$A = P\left(1 + \frac{i}{m}\right)^{nm}$$

$$A = 1000\left(1 + \frac{.08}{4}\right)^{10(4)}$$

$$= 1000(1 + .02)^{40} = 1000(1.02)^{40}$$

The number $(1.02)^{40}$ can be found in financial tables or by using a calculator with a y^x key. To five decimal places, $(1.02)^{40} = 2.20804$. The amount on deposit after 10 years is

$$A = 1000(1.02)^{40} = 1000(2.20804) = 2208.04,$$

or $2208.04. The amount of interest earned is

$$\$2208.04 - \$1000.00 = \$1208.04. \quad \blacksquare$$

The Number e Perhaps the single most useful base for an exponential function is the number e, an irrational number that occurs often in practical applications. To see one way the number e is used, begin with the formula for compound interest. Suppose now that a lucky investment produces annual interest of 100%, so that $i = 1.00$, or $i = 1$. Suppose also that only $1 can be deposited at this rate, and for only one year. Then $P = 1$ and $n = 1$. Substituting into the formula for compound interest gives

$$P\left(1 + \frac{i}{m}\right)^{nm} = 1\left(1 + \frac{1}{m}\right)^{1(m)}$$

$$= \left(1 + \frac{1}{m}\right)^{m}.$$

As interest is compounded more and more often, the value of this expression will increase. If interest is compounded annually, making $m = 1$, the total amount on deposit is

$$\left(1 + \frac{1}{m}\right)^{m} = \left(1 + \frac{1}{1}\right)^{1}$$

$$= 2^1 = 2,$$

so an investment of $1 becomes $2 in one year.

A calculator with a y^x key gives the results in the table on the following page. These results have been rounded to five decimal places.

m	$\left(1 + \dfrac{1}{m}\right)^m$
1	2
2	2.25
5	2.48832
10	2.59374
25	2.66584
50	2.69159
100	2.70481
500	2.71557
1000	2.71692
10,000	2.71815
1,000,000	2.71828

The table suggests that as m increases, the value of $(1 + 1/m)^m$ gets closer and closer to some fixed number. It turns out that this is indeed the case. This fixed number is called e.

Value of e

> To nine decimal places,
>
> $$e \approx 2.718281828.$$

Table 1 in this book gives various powers of e. Also, some calculators will give values of e^x. In Figure 5, the functions defined by $f(x) = 2^x$, $f(x) = e^x$, and $f(x) = 3^x$ are graphed for comparison.

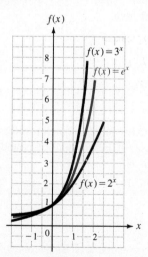

Figure 5

It can be shown that in many situations involving growth or decay of a population, the amount or number present at time t can be closely approximated by an exponential function with base e. The next example illustrates exponential growth.

Example 7   Suppose the population of a midwestern city is approximated by

$$P(t) = 10,000e^{.04t},$$

where t represents time measured in years. Find the population of the city at time
(a) $t = 0$ (b) $t = 5$.

(a) The population at time $t = 0$ is

$$\begin{aligned}
P(0) &= 10,000e^{(.04)0} \\
&= 10,000e^0 \\
&= 10,000(1) \\
&= 10,000,
\end{aligned}$$

written $P_0 = 10,000$.

(b) The population of the city at year $t = 5$ is

$$\begin{aligned}
P(5) &= 10,000e^{(.04)5} \\
&= 10,000e^{.2}.
\end{aligned}$$

The number $e^{.2}$ can be found in Table 1 or by using a suitable calculator. By either of these methods, $e^{.2} = 1.22140$ (to five decimal places), so that

$$P(5) = 10,000(1.22140) = 12,214.$$

In five years the population of the city will be about 12,200.   ∎

5.1 Exercises

Which of the following define exponential functions?

1. $f(x) = 5^x - 1$

2. $f(x) = 2x^5$

3. $f(x) = 4x^3 - 1$

4. $f(x) = 2 \cdot 3^{5x-1}$

Let $f(x) = (2/3)^x$. Find each of the following values.

5. $f(2)$

6. $f(-1)$

7. $f\left(\dfrac{1}{2}\right)$

8. $f(0)$

9. Graph each of the functions defined as follows. Compare the graphs to that of $f(x) = 2^x$.
   (a) $f(x) = 2^x + 1$ (b) $f(x) = 2^x - 4$ (c) $f(x) = 2^{x+1}$ (d) $f(x) = 2^{x-4}$

10. Graph each of the functions defined as follows. Compare the graphs to that of $f(x) = 3^{-x}$.
   (a) $f(x) = 3^{-x} - 2$ (b) $f(x) = 3^{-x} + 4$ (c) $f(x) = 3^{-x-2}$ (d) $f(x) = 3^{-x+4}$

Graph each of the functions defined as follows.

11. $f(x) = 3^x$

12. $f(x) = 4^x$

13. $f(x) = (3/2)^x$

14. $f(x) = 3^{-x}$

15. $f(x) = 10^{-x}$

16. $f(x) = 10^x$

17. $f(x) = 2^{x+1}$

18. $f(x) = 2^{1-x}$

19. $f(x) = 2^{|x|}$

20. $f(x) = 2^{-|x|}$

21. $f(x) = 2^x + 2^{-x}$

22. $f(x) = (1/2)^x + (1/2)^{-x}$

Use a calculator to help graph each of the functions defined as follows.

23. $f(x) = x \cdot 2^x$

24. $f(x) = x^2 \cdot 2^x$

25. $f(x) = \dfrac{e^x - e^{-x}}{2}$

26. $f(x) = \dfrac{e^x + e^{-x}}{2}$

27. $f(x) = (1 - x)e^x$

28. $f(x) = x \cdot e^{-x}$

Solve each of the following equations.

29. $4^x = 2$

30. $125^r = 5$

31. $\left(\dfrac{1}{2}\right)^k = 4$

32. $\left(\dfrac{2}{3}\right)^x = \dfrac{9}{4}$

33. $2^{3-y} = 8$

34. $5^{2p+1} = 25$

35. $\dfrac{1}{27} = b^{-3}$

36. $\dfrac{1}{81} = k^{-4}$

37. $4 = r^{2/3}$

38. $z^{5/2} = 32$

39. $27^{4z} = 9^{z+1}$

40. $32^t = 16^{1-t}$

41. $125^{-x} = 25^{3x}$

42. $216^{3-a} = 36^a$

43. $\left(\dfrac{1}{8}\right)^{-2p} = 2^{p+3}$

44. $3^{-h} = \left(\dfrac{1}{27}\right)^{1-2h}$

45. $\left(\dfrac{1}{2}\right)^{-x} = \left(\dfrac{1}{4}\right)^{x+1}$

46. $\left(\dfrac{2}{3}\right)^{k-1} = \left(\dfrac{81}{16}\right)^{k+1}$

47. $4^{|x|} = 64$

48. $3^{-|x|} = \dfrac{1}{27}$

49. $\left(\dfrac{1}{5}\right)^{|x-2|} = \dfrac{1}{125}$

50. $\left(\dfrac{2}{5}\right)^{|3x-2|} = \dfrac{4}{25}$

51. $e^{-5x} = (e^2)^x$

52. $e^{3(1+x)} = e^{-8x}$

53. For $a > 1$, how does the graph of $y = a^x$ change as a increases? What if $0 < a < 1$?

Use the formula for compound interest,

$$A = P\left(1 + \dfrac{i}{m}\right)^{nm},$$

to find each of the following amounts.

54. $4292 at 6% compounded annually for 10 years

55. $965.43 at 9% compounded annually for 15 years

56. $10,765 at 11% compounded semiannually for 7 years

57. $1593.24 at 10½% compounded quarterly for 14 years

58. $68,922 at 10% compounded daily (365 days) for 4 years

59. $2964.58 at 11¼% compounded daily for 9 years (Ignore leap years.)

60. Suppose $10,000 is left at interest for 3 years at 12%. Find the final amount on deposit if the interest is compounded   **(a)** annually   **(b)** quarterly   **(c)** daily (365 days).

61. Find the final amount on deposit if $5800 is left at interest for 6 years at 13% and interest is compounded   **(a)** annually   **(b)** semiannually   **(c)** daily (365 days).

62. Suppose the population of a city is given by $P(t)$, where

$$P(t) = 1,000,000e^{.02t},$$

where t represents time measured in years from some initial year. Find each of the following: **(a)** $P(0)$ **(b)** $P(2)$ **(c)** $P(4)$ **(d)** $P(10)$.   **(e)** Graph $y = P(t)$.

63. Suppose the quantity in grams of a radioactive substance present at time t is

$$Q(t) = 500e^{-.05t}.$$

Let t be time measured in days from some initial day. Find the quantity present at each

of the following times: (a) $t = 0$ (b) $t = 4$ (c) $t = 8$ (d) $t = 20$.
(e) Graph $y = Q(t)$.

64. Experiments have shown that the sales of a product, under relatively stable market conditions, but in the absence of promotional activities such as advertising, tend to decline at a constant yearly rate. This rate of sales decline varies considerably from product to product, but seems to remain the same for any particular product. The sales decline can be expressed by a function of the form

$$S(t) = S_0 e^{-at},$$

where $S(t)$ is the rate of sales at time t measured in years, S_0 is the rate of sales at time $t = 0$, and a is the sales decay constant. (a) Suppose the sales decay constant for a particular product is $a = 0.10$. Let $S_0 = 50,000$ and find $S(1)$ and $S(3)$. (b) Find $S(2)$ and $S(10)$ if $S_0 = 80,000$ and $a = 0.05$.

65. *Escherichia coli* is a strain of bacteria that occurs naturally in many different organisms. Under certain conditions, the number of these bacteria present in a colony is

$$E(t) = E_0 \cdot 2^{t/30},$$

where $E(t)$ is the number of bacteria present t minutes after the beginning of an experiment, and E_0 is the number present when $t = 0$. Let $E_0 = 2,400,000$ and find the number of bacteria at the following times: (a) $t = 5$ (b) $t = 10$ (c) $t = 60$ (d) $t = 120$.

66. The higher a student's grade-point average, the fewer applications that the student need send to medical schools (other things being equal). Using information given in a guidebook for prospective medical students, we constructed the function defined by $y = 540e^{-1.3x}$ for the number of applications a student should send out. Here y is the number of applications for a student whose grade-point average is x. The domain of x is the interval $[2.0, 4.0]$. Find the number of applications that should be sent out by students having a grade-point average of (a) 2.0 (b) 3.0 (c) 3.5 (d) 3.9.

The pressure of the atmosphere, $p(h)$, in pounds per square inch, is given by

$$p(h) = p_0 e^{-kh},$$

where h is the height above sea level and p_0 and k are constants. The pressure at sea level is 15 pounds per square inch and the pressure is 9 pounds per square inch at a height of 12,000 feet.

67. Find the pressure at an altitude of 6000 feet.

68. What would be the pressure encountered by a spaceship at an altitude of 150,000 feet?

69. In our definition of exponential function, we ruled out negative values of a. However, in a textbook on mathematical economics, the author obtained a "graph" of $y = (-2)^x$ by plotting the following points.

x	-4	-3	-2	-1	0	1	2
y	1/16	$-1/8$	1/4	$-1/2$	1	-2	4

The graph, which occupies a half page in the book, oscillates very neatly from positive to negative values of y. Comment on this approach. (This example shows the dangers of relying solely on point plotting when drawing graphs.)

70. When defining an exponential function, why did we require $a > 0$?

Any points where the graphs of functions f and g cross give solutions of the equation $f(x) = g(x)$. Use this idea to estimate the *number* of solutions of the following equations.

71. $x = 2^x$ **72.** $2^{-x} = -x$ **73.** $3^{-x} = 1 - 2x$ **74.** $3x + 2 = 4^x$

Let $f(x) = a^x$ define an exponential function of base a.

75. Is f odd, even, or neither?

76. Prove that $f(m + n) = f(m) \cdot f(n)$ for any real numbers m and n.

Find examples of a function f satisfying the following conditions.

77. $f(2x) = [f(x)]^2$ **78.** $f(x + 1) = 2 \cdot f(x)$

In calculus, it is shown that

$$e^x = 1 + x + \frac{x^2}{2 \cdot 1} + \frac{x^3}{3 \cdot 2 \cdot 1} + \frac{x^4}{4 \cdot 3 \cdot 2 \cdot 1} + \frac{x^5}{5 \cdot 4 \cdot 3 \cdot 2 \cdot 1} + \cdots$$

79. Use the terms shown here and replace x with 1 to approximate $e^1 = e$ to three decimal places. Then check your results in Table 1 or with a calculator.

80. Use the terms shown here and replace x with $-.05$ to approximate $e^{-.05}$ to four decimal places. Check your results in Table 1 or with a calculator.

81. Let $f(x) = e^x$. Show that $\dfrac{f(x + h) - f(x)}{h} = \dfrac{e^x(e^h - 1)}{h}$.

82. Let $f(x) = 1 + e^x$. Show that $\dfrac{1}{f(x)} + \dfrac{1}{f(-x)} = 1$.

Simplify each expression.

83. $\dfrac{(e^{-x} - e^x)(-e^{-x} + e^x) - (e^{-x} + e^x)(-e^{-x} - e^x)}{(e^{-x} - e^x)^2}$

84. $\dfrac{(e^x + e^{2x})(2e^{2x} + 3e^{3x}) - (e^{2x} + e^{3x})(e^x + 2e^{2x})}{(e^x + e^{2x})^2}$

5.2   Logarithmic Functions

Exponential functions defined by $f(x) = a^x$ for all positive values of a, where $a \neq 1$, were discussed in the previous section. As mentioned there, exponential functions are one-to-one, and so have inverse functions. In this section we discuss the inverses of exponential functions. The equation of the inverse comes from exchanging x and y. Doing this with $y = a^x$ gives

$$x = a^y$$

as the inverse of the exponential function defined by $f(x) = a^x$. To solve $x = a^y$ for *y*, use the following definition.

Definition of Logarithm

> For all real numbers *y*, and all positive numbers *a* and *x*, where $a \neq 1$,
>
> $$y = \log_a x \quad \text{if and only if} \quad x = a^y.$$

Log is an abbreviation for *logarithm*. Read $\log_a x$ as "the logarithm of *x* to the base *a*."

This key definition should be memorized. It is important to remember the location of the base and exponent in each part of the definition.

$$\text{logarithmic form:} \quad \overset{\text{exponent}}{\underset{\text{base}}{y = \log_a x}}$$

$$\text{exponential form:} \quad \overset{\text{exponent}}{\underset{\text{base}}{a^y = x}}$$

By the definition, a logarithm is an exponent: the exponent on the base *a* that will yield the number *x*.

Example 1   The chart below shows several pairs of equivalent statements. The same statement is written in both exponential and logarithmic forms.

Exponential form	Logarithmic form
$2^3 = 8$	$\log_2 8 = 3$
$(1/2)^{-4} = 16$	$\log_{1/2} 16 = -4$
$10^5 = 100{,}000$	$\log_{10} 100{,}000 = 5$
$3^{-4} = 1/81$	$\log_3 (1/81) = -4$
$5^1 = 5$	$\log_5 5 = 1$
$(3/4)^0 = 1$	$\log_{3/4} 1 = 0$ ∎

The definition of logarithm can be used to define the logarithmic function with base *a*.

Logarithmic Function

> If $a > 0$, $a \neq 1$, and $x > 0$, then the function *f* defined by
>
> $$f(x) = \log_a x$$
>
> is the **logarithmic function with base *a*.**

Exponential and logarithmic functions are inverses of each other. Since the domain of an exponential function is the set of all real numbers, the range of a logarithmic function will also be the set of all real numbers. In the same way, both the range of an exponential function and the domain of a logarithmic function are the set of all positive real numbers, so logarithms can be found for positive numbers only.

Example 2 Graph the logarithmic functions defined as follows.

(a) $f(x) = \log_2 x$

One way to graph a logarithmic function is to begin with its inverse function. Here, the inverse has equation $y = 2^x$. The graph of the equation $y = 2^x$ is shown with a dashed curve in Figure 6(a). To get the graph of $y = \log_2 x$, reflect the graph of $y = 2^x$ about the 45° line $y = x$. The graph of the equation $y = \log_2 x$ is shown as a solid curve. As the graph shows, the function defined by $f(x) = \log_2 x$ is increasing for all its domain, is one-to-one, and has the y-axis as a vertical asymptote.

(b) $f(x) = \log_{1/2} x$

The graph of the equation of the inverse, $y = (1/2)^x$, is shown with a dashed curve in Figure 6(b). The graph of $y = \log_{1/2} x$, shown as a solid curve, is found by reflecting the graph of $y = (1/2)^x$ about the line $y = x$. The function defined by $f(x) = \log_{1/2} x$ is decreasing for all its domain, is one-to-one, and has the y-axis for a vertical asymptote. ■

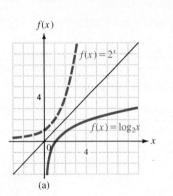

(a)

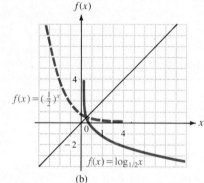

(b)

Figure 6

Example 3 Graph $f(x) = \log_2 (x - 1)$.

The graph of $f(x) = \log_2 (x - 1)$ will be the graph of $f(x) = \log_2 x$, translated one unit to the right. The asymptote is $x = 1$. The domain of the function defined by $f(x) = \log_2 (x - 1)$ is $(1, +\infty)$, since logarithms can be found only for positive numbers. See Figure 7. ■

Example 4 Graph $f(x) = \log_3 |x|$.

Use the definition of logarithm to write the equation $y = \log_3 |x|$ as $|x| = 3^y$. Then choose values for y and find the corresponding x-values. Some ordered pairs for this equation are shown below; the graph is given in Figure 8.

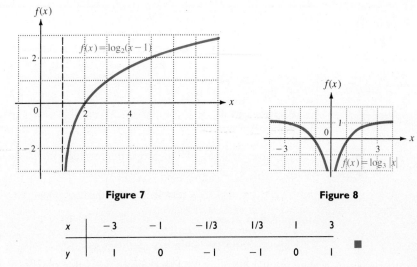

Figure 7   **Figure 8**

x	-3	-1	$-1/3$	$1/3$	1	3
y	1	0	-1	-1	0	1

Equations with logarithms often can be solved by rewriting them in exponential form as shown in the next example.

Example 5   Solve each of the following equations.

(a) $\log_x \dfrac{8}{27} = 3$

First, write the expression in exponential form; then solve.

$$x^3 = \frac{8}{27}$$

$$x^3 = \left(\frac{2}{3}\right)^3$$

$$x = \frac{2}{3}$$

The solution set is $\{2/3\}$.

(b) $\log_4 x = 5/2$

In exponential form, the given statement becomes

$$4^{5/2} = x$$
$$(4^{1/2})^5 = x$$
$$2^5 = x$$
$$32 = x.$$

The solution set is $\{32\}$.   ■

Logarithms were originally important as an aid for numerical calculations, but the availability of inexpensive calculators has greatly reduced the need for this application of logarithms. Yet the principles behind the use of logarithms for calculation are important in other applications; these principles are based on the properties of logarithms discussed in the next theorem.

Properties of Logarithms	If x and y are any positive real numbers, r is any real number, and a is any positive real number, $a \neq 1$, then **(a)** $\log_a xy = \log_a x + \log_a y$ **(b)** $\log_a \dfrac{x}{y} = \log_a x - \log_a y$ **(c)** $\log_a x^r = r \cdot \log_a x$ **(d)** $\log_a a = 1$ **(e)** $\log_a 1 = 0.$

To prove part (a) of the properties of logarithms, let $m = \log_a x$ and $n = \log_a y$. Then, by the definition of logarithm,

$$a^m = x \quad \text{and} \quad a^n = y.$$

Multiplication gives $\qquad a^m \cdot a^n = xy.$

By a property of exponents, $\qquad a^{m+n} = xy.$

Now use the definition of logarithm to write this statement as

$$\log_a xy = m + n.$$

Since $m = \log_a x$ and $n = \log_a y$,

$$\log_a xy = \log_a x + \log_a y.$$

To prove part (b) of the properties, use m and n as defined above. Then

$$\frac{a^m}{a^n} = \frac{x}{y}.$$

Since $\qquad\qquad\qquad \dfrac{a^m}{a^n} = a^{m-n},$

then $\qquad\qquad\qquad a^{m-n} = \dfrac{x}{y}.$

By the definition of logarithm, this statement can be written

$$\log_a \frac{x}{y} = m - n,$$

or $\qquad\qquad\qquad \log_a \dfrac{x}{y} = \log_a x - \log_a y.$

For part (c),

$$(a^m)^r = x^r \quad \text{or} \quad a^{mr} = x^r.$$

Again using the definition of logarithm

$$\log_a x^r = mr,$$

or $\qquad\qquad\qquad \log_a x^r = r \cdot \log_a x.$

Finally, (d) and (e) follow directly from the definition of logarithm since $a^1 = a$ and $a^0 = 1$.

The properties of logarithms are useful for rewriting expressions with logarithms in different forms, as shown in the next examples.

Example 6 Assuming all variables represent positive real numbers, use the properties of logarithms to write each of the following in a different form.

(a) $\log_6 7 \cdot 9 = \log_6 7 + \log_6 9$

(b) $\log_9 \dfrac{15}{7} = \log_9 15 - \log_9 7$

(c) $\log_5 \sqrt{8} = \log_5 8^{1/2} = \dfrac{1}{2} \log_5 8$

(d) $\log_a \dfrac{mnq}{p^2} = \log_a m + \log_a n + \log_a q - 2 \log_a p$

(e) $\log_a \sqrt[3]{m^2} = \dfrac{2}{3} \log_a m$

(f) $\log_b \sqrt[n]{\dfrac{x^3 y^5}{z^m}} = \dfrac{1}{n} \log_b \dfrac{x^3 y^5}{z^m}$

$$= \dfrac{1}{n} (\log_b x^3 + \log_b y^5 - \log_b z^m)$$

$$= \dfrac{3}{n} \log_b x + \dfrac{5}{n} \log_b y - \dfrac{m}{n} \log_b z \quad \blacksquare$$

Example 7 Use the properties of logarithms to write each of the following as a single logarithm with a coefficient of 1. Assume all variables represent positive real numbers.

(a) $\log_3 (x + 2) + \log_3 x - \log_3 2 = \log_3 \dfrac{(x + 2)x}{2}$

(b) $2 \log_a m - 3 \log_a n = \log_a m^2 - \log_a n^3 = \log_a \dfrac{m^2}{n^3}$

(c) $\dfrac{1}{2} \log_b m + \dfrac{3}{2} \log_b 2n - \log_b m^2 n$

$$= \log_b m^{1/2} + \log_b (2n)^{3/2} - \log_b m^2 n$$

$$= \log_b \dfrac{m^{1/2}(2n)^{3/2}}{m^2 n}$$

$$= \log_b \dfrac{2^{3/2} n^{1/2}}{m^{3/2}} \quad \blacksquare$$

Example 8 Assume $\log_{10} 2 = .3010$. Find the base 10 logarithms of 4 and 5.
By the properties of logarithms,

$$\log_{10} 4 = \log_{10} 2^2 = 2 \log_{10} 2 = 2(.3010) = .6020$$

$$\log_{10} 5 = \log_{10} \dfrac{10}{2} = \log_{10} 10 - \log_{10} 2 = 1 - .3010 = .6990. \quad \blacksquare$$

Compositions of the exponential and logarithmic functions can be used to get two more useful properties. If $f(x) = a^x$ and $g(x) = \log_a x$, then

$$f[g(x)] = a^{\log_a x}$$

and

$$g[f(x)] = \log_a a^x.$$

In Section 3.6 it was shown that if functions f and g are inverses of each other, $f[g(x)] = g[f(x)] = x$. Since exponential and logarithmic functions of the same base are inverses of each other, this result gives the following theorem.

Theorem

If $a > 0$, $a \neq 1$, and $x > 0$, then

$$a^{\log_a x} = x \quad \text{and} \quad \log_a a^x = x.$$

Example 9   Simplify each of the following.

(a) $\log_5 5^3 = 3$ **(b)** $7^{\log_7 10} = 10$ **(c)** $\log_r r^{k+1} = k + 1$ ∎

5.2 Exercises

For each of the following statements, write an equivalent statement in logarithmic form.

1. $3^4 = 81$ **2.** $2^5 = 32$ **3.** $10^4 = 10{,}000$ **4.** $8^2 = 64$

5. $(1/2)^{-4} = 16$ **6.** $(2/3)^{-3} = 27/8$ **7.** $10^{-4} = .0001$ **8.** $(1/100)^{-2} = 10{,}000$

For each of the following statements, write an equivalent statement in exponential form.

9. $\log_6 36 = 2$ **10.** $\log_5 5 = 1$ **11.** $\log_{\sqrt{3}} 81 = 8$ **12.** $\log_4 (1/64) = -3$

13. $\log_{10} .0001 = -4$ **14.** $\log_3 \sqrt[3]{9} = 2/3$ **15.** $\log_m k = n$ **16.** $\log_2 r = y$

Find the value of each of the following. Assume all variables represent positive real numbers.

17. $\log_5 25$ **18.** $\log_3 81$ **19.** $\log_8 8$ **20.** $\log_7 1$

21. $\log_{10} 0.001$ **22.** $\log_6 \dfrac{1}{216}$ **23.** $\log_{25} 5$ **24.** $\log_{16} 2$

25. $\log_4 \dfrac{\sqrt[3]{4}}{2}$ **26.** $\log_9 \dfrac{\sqrt[4]{27}}{3}$ **27.** $\log_{1/3} \dfrac{9^{-4}}{3}$ **28.** $\log_{1/4} \dfrac{16^2}{2^{-3}}$

29. $\log_6 36^4$ **30.** $\log_5 125^2$ **31.** $\log_e e^4$ **32.** $\log_e \dfrac{1}{e}$

33. $\log_e \sqrt{e}$ **34.** $\log_e e^x$ **35.** $e^{\log_e 2}$ **36.** $e^{\log_e 5}$

37. $e^{\log_e x}$ **38.** $e^{\log_e (x+2)}$ **39.** $2^{\log_2 9}$ **40.** $8^{\log_8 11}$

Solve each of the following equations.

41. $\log_x 25 = -2$ **42.** $\log_x \dfrac{1}{16} = -2$ **43.** $\log_9 27 = m$ **44.** $\log_8 4 = z$

45. $\log_y 8 = \dfrac{3}{4}$ **46.** $\log_r 7 = 1/2$ **47.** $\log_e x = 0$ **48.** $\log_e x = 1$

Write each of the following as a sum, difference, or product of logarithms. Simplify the result if possible. Assume all variables represent positive real numbers.

49. $\log_3 (2/5)$

50. $\log_4 (6/7)$

51. $\log_2 \dfrac{6x}{y}$

52. $\log_3 \dfrac{4p}{q}$

53. $\log_5 \dfrac{5\sqrt{7}}{3}$

54. $\log_2 \dfrac{2\sqrt{3}}{5}$

55. $\log_4 (2x + 5y)$

56. $\log_6 (7m + 3q)$

57. $\log_k \dfrac{pq^2}{m}$

58. $\log_z \dfrac{x^5 y^3}{3}$

59. $\log_m \sqrt{\dfrac{5r^3}{z^5}}$

60. $\log_p \sqrt[3]{\dfrac{m^5 n^4}{t^2}}$

Write each of the following expressions as a single logarithm with a coefficient of 1. Assume that all variables represent positive real numbers.

61. $\log_a x + \log_a y - \log_a m$

62. $(\log_b k - \log_b m) - \log_b a$

63. $2 \log_m a - 3 \log_m b^2$

64. $\dfrac{1}{2} \log_y p^3 q^4 - \dfrac{2}{3} \log_y p^4 q^3$

65. $-\dfrac{3}{4} \log_x a^6 b^8 + \dfrac{2}{3} \log_x a^9 b^3$

66. $\log_a (pq^2) + 2 \log_a (p/q)$

67. $\log_b (x + 2) + \log_b 7x - \log_b 8$

68. $\log_h (4m + 1) + \log_h 2m - \log_h 3$

69. $2 \log_a (z + 1) + \log_a (3z + 2)$

70. $\log_b (2y + 5) - \dfrac{1}{2} \log_b (y + 3)$

71. $-\dfrac{2}{3} \log_5 5m^2 + \dfrac{1}{2} \log_5 25m^2$

72. $-\dfrac{3}{4} \log_3 16p^4 - \dfrac{2}{3} \log_3 8p^3$

73. Graph $f(x) = \log_3 x$ and $f(x) = 3^x$ on the same axes.

74. Graph $f(x) = \log_4 x$ and $f(x) = 4^x$ on the same axes.

75. Graph each of the following equations. Compare the graphs to that of $y = \log_2 x$.
   (a) $y = (\log_2 x) + 3$ **(b)** $y = \log_2 (x + 3)$ **(c)** $y = |\log_2 (x + 3)|$

76. Graph each of the following equations. Compare the graphs to that of $y = \log_{1/2} x$.
   (a) $y = (\log_{1/2} x) - 2$ **(b)** $y = \log_{1/2} (x - 2)$ **(c)** $y = |\log_{1/2} (x - 2)|$

Graph each of the functions defined as follows.

77. $f(x) = \log_5 x$

78. $f(x) = \log_{10} x$

79. $f(x) = \log_{1/2} (1 - x)$

80. $f(x) = \log_{1/3} (3 - x)$

81. $f(x) = \log_2 x^2$

82. $f(x) = \log_3 (x - 1)$

83. $f(x) = x \cdot \log_{10} x$

84. $f(x) = x^2 \cdot \log_{10} x$

Given $\log_{10} 2 = .3010$ and $\log_{10} 3 = .4771$, find each of the following without using calculators or tables.

85. $\log_{10} 6$

86. $\log_{10} 12$

87. $\log_{10} 9$

88. $\log_{10} 20$

89. $\log_{10} 30$

90. $\log_{10} 36$

91. The population of an animal species that is introduced into a certain area may grow rapidly at first but then grow more slowly as time goes on. A logarithmic function can provide an excellent description of such growth. Suppose that the population of foxes, $F(t)$, in an area t months after the foxes were introduced there is

$$F(t) = 500 \log_{10} (2t + 3).$$

Use a calculator with a log key to find the population of foxes at the following times:
(a) when they are first released into the area (that is, when $t = 0$)   **(b)** after 3
months   **(c)** after 15 months.   **(d)** Graph $y = F(t)$.

The loudness of sounds is measured in a unit called a *decibel*. To measure with this unit, we
first assign an intensity of I_0 to a very faint sound, called the *threshold sound*. If a particular
sound has intensity I, then the decibel rating of this louder sound is

$$d = 10 \cdot \log_{10} \frac{I}{I_0}.$$

92. Find the decibel ratings of sounds having the following intensities: **(a)** $100I_0$
   (b) $1000I_0$ **(c)** $100,000I_0$ **(d)** $1,000,000I_0$.

93. Find the decibel ratings of the following sounds, having intensities as given. (You will
need a calculator with a log key.) Round answers to the nearest whole number.
   (a) Whisper, $115I_0$
   (b) Busy street, $9,500,000I_0$
   (c) Heavy truck, 20 m away, $1,200,000,000I_0$
   (d) Rock music, $895,000,000,000I_0$
   (e) Jetliner at takeoff, $109,000,000,000,000I_0$

94. The intensity of an earthquake, measured on the *Richter Scale*, is given by

$$\log_{10} \frac{I}{I_0},$$

where I_0 is the intensity of an earthquake of a certain (small) size. Find the Richter
Scale ratings of earthquakes having intensity   **(a)** $1000I_0$ **(b)** $1,000,000I_0$
   (c) $100,000,000I_0$.

95. The San Francisco earthquake of 1906 had a Richter Scale rating of 8.6. Use a
calculator with a y^x key to express the intensity of this earthquake as a multiple of I_0
(see Exercise 94).

96. How much more powerful is an earthquake with a Richter Scale rating of 8.6 than one
with a rating of 8.2?

97. Using a calculator*, evaluate $(3^{.003})^{1001}$ and $3^{(.003 \times 1001)}$ by computing the expression
within parentheses first. Did you get the same results? If not, can you explain the
difference? What does this tell you about the laws of exponents as applied to calculator
arithmetic?

98. Using a calculator, evaluate $\log_{10} (2^{.0001})$ and $.0001 \times (\log_{10} 2)$ by computing the
expression within parentheses first. Did you get the same results? If not, can you
explain the difference? What does this tell you about the properties of logarithms as
applied to calculator arithmetic?

99. Let $f(x) = e^x$. Graph $y = f^{-1}(x)$.

100. Prove that $\log_e (x - \sqrt{x^2 - 1}) = -\log_e (x + \sqrt{x^2 - 1})$.

*Exercises 97 and 98 from *Calculus and Analytic Geometry* by Abe Mizrahi and Michael Sullivan. Copyright © 1982 by Wadsworth, Inc., Belmont, CA 94002. Reprinted by permission.

5.3 Natural Logarithms

Since our number system uses base 10, logarithms to base 10 are most convenient for numerical calculation, historically the main application of logarithms. Base 10 logarithms are called **common logarithms.** The common logarithm of the number x, or $\log_{10} x$, is often abbreviated as just log x. Common logarithms are discussed in Section 5.6.

In most other practical applications of logarithms, however, the number $e \approx$ 2.718281828 is used as base. Logarithms to base e are called **natural logarithms,** since they occur in many natural-world applications, such as those involving growth and decay. The abbreviation $\ln x$ is used for the natural logarithm of x, so that $\log_e x = \ln x$. The graph of $f(x) = e^x$ was given in Figure 5. Since the functions defined by $f(x) = e^x$ and $f(x) = \ln x$ are inverses, the graph of $f(x) = \ln x$, shown in Figure 9, can be found by reflecting the graph of $f(x) = e^x$ about the line $y = x$.

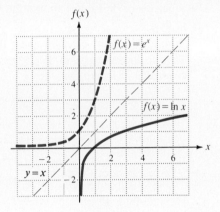

Figure 9

The following results for the natural logarithm function are direct applications of the properties of logarithms.

Properties of Natural Logarithms	$e^{\ln x} = x$ if $x > 0$ $\qquad$ $\ln 1 = 0$ $\ln e^x = x$ $\qquad\qquad\qquad$ $\ln e = 1$

Example I Evaluate each expression.

$\ln e^x = x$

(a) $e^{\ln 5} = 5$

(b) $e^{\ln \sqrt{3}} = \sqrt{3}$

(c) $\ln e^2 = 2$

(d) $\ln e^{-1.5} = -1.5$

(e) $8 \ln 1 - 16 \ln e + 2 \ln e^4$

Since $\ln 1 = 0$, $\ln e = 1$, and $\ln e^4 = 4$,

$$8 \ln 1 - 16 \ln e + 2 \ln e^4 = 8(0) - 16(1) + 2(4) = -8. \quad \blacksquare$$

Values of natural logarithms can be found with a calculator which has a ln key or with a table of natural logarithms. A table of natural logarithms is given in Table 1. Reading directly from this table,

$$\ln 55 = 4.0073,$$
$$\ln 1.9 = .6419,$$

and
$$\ln .4 = -.9163.$$

Example 2   Use a calculator or Table 1 to find the following logarithms.

(a) $\ln 85$

With a calculator, enter 85, press the ln key, and read the result, 4.4427.
Table 1 does not give $\ln 85$. However, the value of $\ln 85$ can be found using the properties of logarithms.

$$\begin{aligned}
\ln 85 &= \ln (8.5 \times 10) \\
&= \ln 8.5 + \ln 10 \\
&\approx 2.1401 + 2.3026 \\
&= 4.4427
\end{aligned}$$

A result found in this way is sometimes slightly different from the answer found using a calculator, due to rounding error.

(b) $\ln 36$

A calculator gives $\ln 36 = 3.5835$. To use the table, first use properties of logarithms, since 36 is not listed in Table 1.

$$\begin{aligned}
\ln 36 &= \ln 6^2 \\
&= 2 \ln 6 \\
&\approx 2(1.7918) \\
&= 3.5836
\end{aligned}$$

Alternatively, $\ln 36$ can be found as follows.

$$\ln 36 = \ln 9 \cdot 4 = \ln 9 + \ln 4 = 2.1972 + 1.3863 = 3.5835 \quad \blacksquare$$

Logarithms to other bases   A calculator or a table will give the values of either natural logarithms (base e) or common logarithms (base 10). However, sometimes it is convenient to use logarithms to other bases. The following theorem can be used to convert logarithms from one base to another.

**Change of Base
Theorem**

> If x is any positive number and if a and b are positive real numbers, $a \neq 1$, $b \neq 1$, then
>
> $$\log_a x = \frac{\log_b x}{\log_b a}.$$

To prove this result, use the definition of logarithm to write $y = \log_a x$ as $x = a^y$ or $x = a^{\log_a x}$ (for positive x and positive a, $a \neq 1$). Now take base b logarithms of both sides of this last equation.

$$\log_b x = \log_b a^{\log_a x}$$

or

$$\log_b x = (\log_a x)(\log_b a),$$

from which

$$\log_a x = \frac{\log_b x}{\log_b a}.$$

Example 3 Use natural logarithms to find each of the following. Round to the nearest hundredth.

(a) $\log_5 27$

Let $x = 27$, $a = 5$, and $b = e$. Substituting into the change of base theorem gives

$$\log_5 27 = \frac{\log_e 27}{\log_e 5}$$

$$= \frac{\ln 27}{\ln 5}.$$

Now use a calculator or Table 1.

$$\log_5 27 \approx \frac{3.2958}{1.6094}$$

$$\approx 2.05$$

To check, use a calculator with a y^x key, along with the definition of logarithm, to verify that $5^{2.05} \approx 27$.

(b) $\log_2 .1$

$$\log_2 .1 = \frac{\ln .1}{\ln 2} \approx \frac{-2.3026}{.6931} \approx -3.32 \quad \blacksquare$$

Example 4 One measure of the diversity of the species in an ecological community is given by

$$H = -[P_1 \log_2 P_1 + P_2 \log_2 P_2 + \cdots + P_n \log_2 P_n],$$

where $P_1, P_2, \ldots, P_n$ are the proportions of a sample belonging to each of n species found in the sample. For example, in a community with two species, where there are 90 of one species and 10 of the other, $P_1 = 90/100 = .9$ and $P_2 = 10/100 = .1$, with

$$H = -[.9 \log_2 .9 + .1 \log_2 .1].$$

The value of $\log_2 .1$ was found in Example 3(b) above. Now find $\log_2 .9$.

$$\log_2 .9 = \frac{\ln .9}{\ln 2} \approx \frac{-.1054}{.6931} \approx -.152$$

Therefore, $H \approx -[(.9)(-.152) + (.1)(-3.32)] \approx .469.$ ■

5.3 Exercises

Evaluate each expression.

1. $\ln e$

2. $\ln 1$

3. $\ln e^4$

4. $\ln e^2$

5. $\ln e^{-3}$

6. $\ln e^{-.017}$

7. $e^{\ln 1.2}$

8. $e^{\ln .003}$

Find each of the following logarithms to four decimal places. Use a calculator or Table 1.

9. $\ln 4$

10. $\ln 6$

11. $\ln 17$

12. $\ln 29$

13. $\ln 350$

14. $\ln 900$

15. $\ln 49$

16. $\ln 64$

17. $\ln 42$

18. $\ln 72$

19. $\ln 81,000$

20. $\ln 121,000$

Find each of the following logarithms to the nearest hundredth.

21. $\log_5 10$

22. $\log_9 12$

23. $\log_{15} 5$

24. $\log_6 8$

25. $\log_{1/2} 3$

26. $\log_{12} 62$

27. $\log_{100} 83$

28. $\log_{200} 175$

29. $\log_{2.9} 7.5$

30. $\log_{5.8} 12.7$

31. $\log_{.6} 13.2$

32. $\log_{.9} 5.77$

Graph each of the functions defined as follows.

33. $f(x) = \ln x$

34. $f(x) = |\ln x|$

35. $f(x) = x \ln x$

36. $f(x) = x^2 \ln x$

Work the following problems. Refer to Example 4 for Exercises 37 and 38.

37. Suppose a sample of a small community shows two species with 50 individuals each. Find the index of diversity H.

38. A virgin forest in northwestern Pennsylvania has 4 species of large trees with the following proportions of each: hemlock, .521; beech, .324; birch, .081; maple, .074. Find the index of diversity H.

39. The number of species in a sample is given by

$$S(n) = a \ln \left(1 + \frac{n}{a} \right).$$

Here n is the number of individuals in the sample and a is a constant that indicates the diversity of species in the community. If $a = .36$, find $S(n)$ for the following values of n. **(a)** 100 **(b)** 200 **(c)** 150 **(d)** 10

40. In Exercise 39, find n if $S(n) = 9$ and $a = .36$.

41. Suppose the number of rabbits in a colony is

$$y = y_0 e^{.4t},$$

where t represents time in months and y_0 is the rabbit population when $t = 0$.
(a) If $y_0 = 100$, find the number of rabbits present at time $t = 4$.
(b) Find y_0 if there are 30 rabbits after 4 months.

42. A Midwestern city finds its residents moving to the suburbs. Its population is declining according to the relationship

$$P = P_0 e^{-.04t},$$

where t is time measured in years and P_0 is the population at time $t = 0$. Assume $P_0 = 1,000,000$.

(a) Find the population at time $t = 1$.

(b) If the decline continues at the same rate, what will the population be after 10 years?

43. If an object is fired vertically upward and is subject only to the force of gravity, g, and to air resistance, then the maximum height, H, attained by the object is

$$H = \frac{1}{K}\left(V_0 - \frac{g}{K} \ln \frac{g + V_0 K}{g}\right),$$

where V_0 is the initial velocity of the object and K is a constant. Find H if $K = 2.5$, $V_0 = 1000$ feet per second, and $g = 32$ feet per second per second.

44. The pull, P, of a tracked vehicle on dry sand under certain conditions is approximated by

$$P = W\left[.2 + .16 \ln \frac{G(bl)^{3/2}}{W}\right]$$

where G is an index of sand strength, W is the load on the vehicle, b is the width of the track, and l is the length of the track.* Find P if $W = 10$, $G = 5$, $b = 30.5$ cm, and $l = 61.0$ cm.

45. The following formula† can be used to estimate the population of the United States, where t is time in years measured from 1914 (times before 1914 are negative):

$$N = \frac{197,273,000}{1 + e^{-.03134t}}$$

(a) Complete the following chart.

Year	Observed population	Predicted population
1790	3,929,000	3,929,000
1810	7,240,000	
1860	31,443,000	30,412,000
1900	75,995,000	
1970	204,000,000	
1980	224,000,000	

(b) Estimate the number of years after 1914 that it would take for the population to increase to 197,273,000.

---

*Gerald W. Turnage, *Prediction of Track Pull Performance in Desert Sand*, unpublished MS thesis, The Florida State University, 1971. †The formula is given in *Elements of Mathematical Biology*, by Alfred J. Lotka, 1957, p. 67. Reprinted by permission of Dover Publications, Inc.

To find the maximum permitted levels of certain pollutants in fresh water, the EPA has established the functions defined in Exercises 46–47, where $M(h)$ is the maximum permitted level of pollutant for a water hardness of h milligrams per liter. Find $M(h)$ in each case. (These results give the maximum permitted average concentration in micrograms per liter for a 24-hour period.)

46. Pollutant: copper; $M(h) = e^r$, where $r = .65 \cdot \ln h - 1.94$ and $h = 9.7$

47. Pollutant: lead; $M(h) = e^r$, where $r = 1.51 \cdot \ln h - 3.37$ and $h = 8.4$

In the central Sierra Nevada Mountains of California, the percent of moisture that falls as snow rather than rain is approximated reasonably well by $p = 86.3 \ln h - 680$, where p is the percent of snow at an altitude h (in feet). (Assume $h \geq 3000$.)

48. Find the percent of moisture that falls as snow at the following altitudes: **(a)** 3000 ft **(b)** 4000 ft **(c)** 7000 ft. **(d)** Graph p.

In Exercises 49 and 50, assume a and b represent positive numbers other than 1, and x is any real number.

49. Show that $\dfrac{1}{\log_a b} = \log_b a$.

50. Show that $a^x = e^{x \ln a}$.

5.4 Exponential and Logarithmic Equations

In Section 5.1 we solved exponential equations such as $(1/3)^x = 81$ by writing each side of the equation as a power of 3. However, that method cannot be used to solve an equation such as $7^x = 12$, since 12 cannot easily be written as a power of 7. However, the equation $7^x = 12$ can be solved by taking the logarithm of each side, a process that depends on the fact that a logarithmic function is one-to-one.

If $x > 0$, $y > 0$, $b > 0$, $b \neq 1$, then

$$x = y \quad \text{if and only if} \quad \log_b x = \log_b y.$$

Example 1 Solve the equation $7^x = 12$.

While the result above is valid for any appropriate base b, the best practical base to use is often base e. Taking base e (natural) logarithms of both sides gives

$$\ln 7^x = \ln 12$$
$$x \cdot \ln 7 = \ln 12$$
$$x = \frac{\ln 12}{\ln 7}.$$

To get a decimal approximation for x, use Table 1 or a calculator.

$$x = \frac{\ln 12}{\ln 7} \approx \frac{2.4849}{1.9459}$$

Using a calculator to divide 2.4849 by 1.9459 gives

$$x \approx 1.277.$$

A calculator with a y^x key can be used to check this answer. Evaluate $7^{1.277}$; the result should be approximately 12. This step verifies that, to the nearest thousandth, the solution set is {1.277}. ∎

See. 5.1

Example 2 Solve $3^{2x-1} = 4^{x+2}$.

Taking natural logarithms on both sides gives

$$\ln 3^{2x-1} = \ln 4^{x+2}$$

Now use properties of logarithms.

$$(2x - 1) \ln 3 = (x + 2) \ln 4$$

$$2x \ln 3 - \ln 3 = x \ln 4 + 2 \ln 4$$

$$2x \ln 3 - x \ln 4 = 2 \ln 4 + \ln 3.$$

Factor out x on the left to get

$$x(2 \ln 3 - \ln 4) = 2 \ln 4 + \ln 3$$

or

$$x = \frac{2 \ln 4 + \ln 3}{2 \ln 3 - \ln 4}.$$

Using properties of logarithms,

$$x = \frac{\ln 16 + \ln 3}{\ln 9 - \ln 4}$$

or, finally,

$$x = \frac{\ln 48}{\ln \frac{9}{4}}.$$

This quotient could be approximated by a decimal if desired.

$$x = \frac{\ln 48}{\ln 2.25} \approx \frac{3.8712}{.8109} \approx 4.774$$

To the nearest thousandth, the solution set is {4.774}. To find $\ln 2.25$ with Table 1, write $\ln 2.25$ as $\ln 1.5^2 = 2 \ln 1.5$. ∎

Example 3 Solve $e^{x^2} = 200$.

Take natural logarithms on both sides; then use properties of logarithms.

$$e^{x^2} = 200$$

$$\ln e^{x^2} = \ln 200$$

$$x^2 \ln e = \ln 200$$

Since $\ln e = 1$,

$$x^2 = \ln 200$$

$$x = \pm \sqrt{\ln 200}$$

$$x \approx \pm 2.302.$$

The solution set is {±2.302}, rounding to the nearest thousandth. ∎

Example 4 Solve $3 = 5(1 - e^x)$.

First solve for e^x.

$$3 = 5(1 - e^x)$$

$$\frac{3}{5} = 1 - e^x$$

$$e^x = 1 - \frac{3}{5} = \frac{2}{5}$$

Now take the natural logarithm on each side.

$$\ln e^x = \ln \frac{2}{5}$$

$$x \ln e = \ln \frac{2}{5}$$

Replace $\ln e$ with 1 to get

$$x = \ln \frac{2}{5} = \ln .4 \approx -.916.$$

To the nearest thousandth, the solution set is $\{-.916\}$. ■

Logarithmic Equations The properties of logarithms given in Section 5.2 are useful in solving logarithmic equations, as shown in the next examples.

Example 5 Solve $\log_a (x + 4) - \log_a (x + 2) = \log_a x$.

Using a property of logarithms, rewrite the equation as

$$\log_a \frac{x + 4}{x + 2} = \log_a x.$$

Then

$$\frac{x + 4}{x + 2} = x$$

$$x + 4 = x(x + 2)$$

$$x + 4 = x^2 + 2x$$

$$x^2 + x - 4 = 0.$$

By the quadratic formula,

$$x = \frac{-1 \pm \sqrt{1 + 16}}{2}$$

so that

$$x = \frac{-1 + \sqrt{17}}{2} \quad \text{or} \quad x = \frac{-1 - \sqrt{17}}{2}.$$

$\log_a x$ cannot be evaluated for $x = (-1 - \sqrt{17})/2$, since this number is negative and not in the domain of $\log_a x$. By substitution verify that $x = (-1 + \sqrt{17})/2$ is a solution, giving the solution set $\{(-1 + \sqrt{17})/2\}$. ■

Example 6   Solve log $(3x + 2)$ + log $(x - 1) = 1$.

   Since log x is an abbreviation for $\log_{10} x$, and $1 = \log_{10} 10$, the properties of logarithms give

$$\log (3x + 2)(x - 1) = \log 10$$
$$(3x + 2)(x - 1) = 10$$
$$3x^2 - x - 2 = 10$$
$$3x^2 - x - 12 = 0.$$

Now use the quadratic formula to arrive at

$$x = \frac{1 \pm \sqrt{1 + 144}}{6}.$$

If $x = (1 - \sqrt{145})/6$, then $x - 1 < 0$ and log $(x - 1)$ does not exist. For this reason, $(1 - \sqrt{145})/6$ must be discarded as a solution. A calculator can help to show that $(1 + \sqrt{145})/6$ is a solution, so the solution set is $\{(1 + \sqrt{145})/6\}$.   ∎

Example 7   Solve $\log_3 (3m^2)^{1/4} - 1 = 2$

   Solve for the logarithm first.

$$\log_3 (3m^2)^{1/4} - 1 = 2$$
$$\log_3 (3m^2)^{1/4} = 3$$

Now write the expression in exponential form as

$$(3m^2)^{1/4} = 3^3$$
$$3^{1/4}m^{1/2} = 3^3$$
$$m^{1/2} = 3^{11/4}$$

Square both sides to get

$$m = (3^{11/4})^2 = 3^{11/2}$$

or

$$m \approx 420.888.$$

To the nearest thousandth the solution set is $\{420.888\}$.   ∎

Example 8   Suppose

$$P(t) = 10{,}000e^{.4t}$$

gives the population of a city at time t (in years). In how many years will the population double?

   Replace t with 0 to show that the population of the city is 10,000 when $t = 0$. We want to find the value of t for which $P(t)$ doubles to 20,000. Do this by substituting 20,000 for $P(t)$ and solving for $e^{.4t}$.

$$20{,}000 = 10{,}000e^{.4t}$$
$$2 = e^{.4t}.$$

Now take natural logarithms on both sides.

$$\ln 2 = \ln e^{.4t}$$

Since $\ln e^{.4t} = (.4t)\ln e = (.4t)(1) = .4t$,

$$\ln 2 = .4t,$$

with

$$t = \frac{\ln 2}{.4}.$$

Using a calculator or Table 1 to find $\ln 2$ and then using a calculator to find the quotient gives $t = 1.733$ to the nearest thousandth. The population of the city will double in about 1.733 years. ■

Example 9   Solve the equation $y = \dfrac{1 - e^x}{1 - e^{-x}}$ for x.

Begin by multiplying both sides by $1 - e^{-x}$.

$$y(1 - e^{-x}) = 1 - e^x$$

or

$$y - ye^{-x} = 1 - e^x.$$

Get 0 alone on one side of the equation.

$$e^x + y - 1 - ye^{-x} = 0.$$

Multiply both sides by e^x.

$$e^{2x} + (y - 1)e^x - y = 0.$$

Rewrite this equation as

$$(e^x)^2 + (y - 1)e^x - y = 0,$$

a quadratic equation in e^x. Solve this equation by using the quadratic formula with $a = 1$, $b = y - 1$, and $c = -y$.

$$e^x = \frac{-(y - 1) \pm \sqrt{(y - 1)^2 - 4(1)(-y)}}{2(1)}$$

$$= \frac{-(y - 1) \pm \sqrt{y^2 - 2y + 1 + 4y}}{2}$$

$$= \frac{-(y - 1) \pm \sqrt{y^2 + 2y + 1}}{2}$$

$$= \frac{-(y - 1) \pm \sqrt{(y + 1)^2}}{2}$$

$$e^x = \frac{-y + 1 \pm (y + 1)}{2}.$$

First, use the $+$ sign:

$$e^x = \frac{-y + 1 + y + 1}{2} = \frac{2}{2} = 1.$$

If $e^x = 1$, $x = 0$ and $e^{-x} = 1$ also. This leads to a zero denominator in the original equation, so the $+$ sign leads to no solution.

Try the $-$ sign.

$$e^x = \frac{-y + 1 - y - 1}{2} = \frac{-2y}{2} = -y$$

If $e^x = -y$, take natural logarithms on both sides to get

$$\ln e^x = \ln(-y)$$
$$x \cdot \ln e = \ln(-y)$$
$$x \cdot 1 = \ln(-y)$$
$$x = \ln(-y).$$

This result will be satisfied by all values of y less than 0.   ■

**Solving
Exponential or
Logarithmic
Equations**

In summary, to solve an exponential or logarithmic equation, first use the properties of algebra to change the given equation into one of the following forms, where a and b are real numbers.

1. $a^{f(x)} = b$

   To solve, take logarithms to base a on both sides.

2. $\log_a f(x) = b$

   Solve by changing to the exponential form $a^b = f(x)$.

3. $\log_a f(x) = \log_a g(x)$

   From the given equation, obtain the equation $f(x) = g(x)$, then solve algebraically.

4. In a more complicated equation, such as the one in Example 9, it is necessary to first solve for $e^{f(x)}$ or $\log_a f(x)$ and then solve the resulting equation using one of the methods given above.

5.4 Exercises

Solve the following equations. Give answers as decimals rounded to the nearest thousandth.

1. $3^x = 6$

2. $4^x = 12$

3. $7^x = 8$

4. $13^p = 55$

5. $3^{a+2} = 5$

6. $5^{2-x} = 12$

7. $6^{1-2k} = 8$

8. $2^{k-3} = 11$

9. $4^{3m-1} = 12^{m+2}$

10. $3^{2m-5} = 13^{m-1}$

11. $e^{k-1} = 4$

12. $e^{2-y} = 12$

13. $2e^{5a+2} = 8$

14. $10e^{3z-7} = 5$

15. $2^x = -3$

16. $(1/4)^p = -4$

17. $\left(1 + \dfrac{r}{2}\right)^5 = 9$

18. $\left(1 + \dfrac{n}{4}\right)^3 = 12$

19. $100(1 + .02)^{3+n} = 150$

20. $500(1 + .05)^{p/4} = 200$

21. $2^{x^2-1} = 12$

22. $3^{2-x^2} = 4$

23. $2(e^x + 1) = 10$

24. $5(e^{2x} - 2) = 15$

25. $\log(t - 1) = 1$

26. $\log q^2 = 1$

27. $\log(x - 3) = 1 - \log x$

28. $\log(z - 6) = 2 - \log(z + 15)$

29. $\ln(y + 2) = \ln(y - 7) + \ln 4$

30. $\ln p - \ln(p + 1) = \ln 5$

31. $\ln(3x - 1) - \ln(2 + x) = \ln 2$

32. $\ln(8k - 7) - \ln(3 + 4k) = \ln(9/11)$

33. $\ln(5 + 4y) - \ln(3 + y) = \ln 3$

34. $\ln m + \ln(2m + 5) = \ln 7$

35. $\ln x + 1 = \ln(x - 4)$

36. $\ln(4x - 2) = \ln 4 - \ln(x - 2)$

37. $2\ln(x - 3) = \ln(x + 5) + \ln 4$

38. $\ln(k+5) + \ln(k+2) = \ln 14k$

39. $\log_5(r+2) + \log_5(r-2) = 1$

40. $\log_4(z+3) + \log_4(z-3) = 1$

41. $\log_3(a-3) = 1 + \log_3(a+1)$

42. $\log w + \log(3w-13) = 1$

43. $\log_2 \sqrt{2y^2-1} = 1/2$

44. $\log_2(\log_2 x) = 1$

45. $\log z = \sqrt{\log z}$

46. $\log x^2 = (\log x)^2$

47. $\log_x 5.87 = 2$

48. $\log_x 11.9 = 3$

49. $1.8^{p+4} = 9.31$

50. $3.7^{5z-1} = 5.88$

51. The amount of a radioactive specimen present at time t (measured in seconds) is $A(t) = 5000(10)^{-.02t}$, where $A(t)$ is measured in grams. Find the half-life of the specimen, that is, the time it will take for exactly half the specimen to remain.

A large cloud of radioactive debris from a nuclear explosion has floated over the Pacific Northwest, contaminating much of the hay supply. Consequently, farmers in the area are concerned that the cows who eat this hay will give contaminated milk. (The tolerance level for radioactive iodine in milk is 0.) The percent of the initial amount of radioactive iodine still present in the hay after t days is approximated by $P(t) = 100\, e^{-.1t}$, where t is time measured in days.

52. Some scientists feel that the hay is safe after the percent of radioactive iodine has declined to 10% of the original amount. Find the number of days before the hay could be used.

53. Other scientists believe that the hay is not safe until the level of radioactive iodine has declined to only 1% of the level. Find the number of days this would take.

Solve the following equations for x. (*Hint:* In Exercises 56–57 multiply by e^x.)

54. $2^{2x} - 3 \cdot 2^x + 2 = 0$

55. $5^{2x} + 2 \cdot 5^x - 3 = 0$

56. $e^x - 5 + 6e^{-x} = 0$

57. $e^x - 6 + 5e^{-x} = 0$

58. $y = \dfrac{1 + e^{-x}}{1 + e^x}$

59. $y = \dfrac{e^x}{1 - e^x}$

60. $y = \dfrac{e^x + e^{-x}}{2}$

61. $y = \dfrac{e^x - e^{-x}}{2}$

Solve each of the following equations for the indicated variables. Use logarithms to the appropriate bases.

62. $P = P_0 e^{kt/1000}$,   for t

63. $I = \dfrac{E}{R}(1 - e^{-Rt/2})$,   for t

64. $T = T_0 + (T_1 - T_0)\, 10^{-kt}$,   for t

65. $A = \dfrac{Pi}{1 - (1+i)^{-n}}$,   for n

66. $(\log_{10} x) - y = \log_{10}(3x - 1)$,   for x

67. $\log_{10}(x - y) = \log_{10}(3x - 1)$,   for x

Solve each inequality for x.

68. $\log_2 x < 4$

69. $\log_3 x > -1$

70. $\log_x 16 < 4$

71. $\log_x .1 < -1$

72. $\log_2 x < \log_3 x$

73. $\log_5 x < \log_{10} x$

74. $\log_6 |5 - 2x| < 0$

75. $\log_2 |x^6 - 1| \le 1$

76. Recall (from the exercises for Section 5.2) the formula for the decibel rating of a sound:

$$d = 10 \log \frac{I}{I_0}.$$

Solve this formula for I.

77. A few years ago, there was a controversy about a proposed government limit on factory noise—one group wanted a maximum of 89 decibels, while another group wanted 86. This difference seemed very small to many people. Find the percent by which the 89 decibel intensity exceeds that for 86 decibels.

The formula for compound interest is

$$A = P\left(1 + \frac{i}{m}\right)^{nm}.$$

Use natural logarithms and solve for the following.

78. i **79.** n

80. The turnover of legislators is a problem of interest to political scientists. One model of legislative turnover in the U.S. House of Representatives is given by

$$M = 434e^{-.08t},$$

where M is the number of continuously serving members at time t.* This model is based on the 1965 membership of the House. Find the number of continuously serving members in each of the following years: **(a)** 1969 **(b)** 1973 **(c)** 1979.

81. Solve the formula in Exercise 80 for t.

82. The growth of bacteria in food products makes it necessary to time-date some products (such as milk) so that they will be sold and consumed before the bacteria count is too high. Suppose for a certain product that the number of bacteria present is given by

$$f(t) = 500e^{.1t},$$

under certain storage conditions, where t is time in days after packing of the product and the value of $f(t)$ is in millions. Find the number of bacteria present at each of the following times: **(a)** 2 days **(b)** 1 week **(c)** 2 weeks.
(d) Suppose the product cannot be safely eaten after the bacteria count reaches 3,000,000,000. How long will this take?
(e) If $t = 0$ corresponds to January 1, what date should be placed on the product?

5.5 Exponential Growth and Decay (Optional) no

In many situations that occur in biology, economics, and the social sciences, a quantity changes at a rate proportional to the amount present. In such cases the amount present at time t is a function of t called the **exponential growth function.**

Exponential Growth Function

> Let y_0 be the amount or number present at time $t = 0$. Then, under certain conditions, the amount present at any time t is given by
>
> $$y = y_0 e^{kt},$$
>
> where k is a constant.

*From "Exponential Models of Legislative Turnover" by Thomas W. Casstevens, *UMAP*, Unit 296. Reprinted by permission of COMAP, Inc.

This section shows how to determine the equation from given data.

Radioactive decay is an important application; it has been shown that radioactive substances decay exponentially—that is, according to

$$y = y_0 e^{kt},$$

where k is a negative number.

Example 1   If 600 grams of a radioactive substance are present initially and 3 years later only 300 grams remain, how much of the substance will be present after 6 years?

From the statement of the problem, $y = 600$ when $x = 0$ (that is, initially), so

$$600 = y_0 e^{k(0)}$$

$$600 = y_0,$$

giving the exponential decay equation

$$y = 600\, e^{kt}.$$

Since there are 300 grams after 3 years, use the fact that $y = 300$ when $x = 3$ to find k.

$$300 = 600\, e^{3k}$$

$$\frac{1}{2} = e^{3k}$$

Take natural logarithms on both sides, then solve for k.

$$\ln \frac{1}{2} = \ln e^{3k}$$

$$\ln .5 = 3k \qquad \text{(Since } \ln e^x = x\text{)}$$

$$\frac{\ln .5}{3} = k$$

$$k \approx -.231,$$

giving $y = 600 e^{-.231t}$

as the exponential decay equation. To find the amount present after 6 years let $t = 6$.

$$y = 600\, e^{-.231(6)}$$

$$\approx 600\, e^{-1.386}$$

$$\approx 150.$$

After 6 years, 150 grams of the substance remain.   ∎

As mentioned in Exercise 51 of Section 5.4, the *half-life* of a radioactive substance is the time it takes for half of a given amount of the substance to decay.

Example 2   The amount in grams of a certain radioactive substance at time t is given by

$$y = y_0\, e^{-.1t},$$

where t is in days. Find the half-life of the substance.

Find the time t when y will equal $y_0/2$. That is, solve the equation

$$\frac{y_0}{2} = y_0 e^{-.1t}.$$

Divide both sides by y_0 to get

$$\frac{1}{2} = e^{-.1t}$$

Now take natural logarithms on both sides and solve for t.

$$\ln \frac{1}{2} = \ln e^{-.1t}$$

$$\ln \frac{1}{2} = -.1t \qquad \text{(Using } \ln e^x = x\text{)}$$

$$t = \frac{-\ln 1/2}{.1}$$

From Table 1 or a calculator, $\ln 1/2 = \ln .5 = -.6931$, so that

$$t \approx 6.9 \text{ days.} \quad \blacksquare$$

Example 3 Carbon 14 is a radioactive isotope of carbon which has a half-life of about 5600 years. The atmosphere contains much carbon, mostly in the form of carbon dioxide, with small traces of carbon 14. Most of this atmospheric carbon is in the form of the nonradioactive isotope carbon 12. The ratio of carbon 14 to carbon 12 is virtually constant in the atmosphere. However, as a plant absorbs carbon dioxide from the air in the process of photosynthesis, the carbon 12 stays in the plant while the carbon 14 decays by conversion to nitrogen. Thus, the ratio of carbon 14 to carbon 12 is smaller in the plant than it is in the atmosphere. Even when the plant is eaten by an animal, this ratio will continue to decrease. Based on these facts, a method of dating objects called carbon 14 dating has been developed.

Let R be the (nearly constant) ratio of carbon 14 to carbon 12 found in the atmosphere, and let r be the ratio found in a fossil. It can be shown that the relationship between R and r is given by

$$\frac{R}{r} = e^{(t \ln 2)/5600},$$

where t is the age of the fossil in years.

(a) Verify the formula for $t = 0$.

If $t = 0$,

$$\frac{R}{r} = e^0 = 1.$$

The quotient R/r can equal 1 only when $R = r$, so that the ratio in the fossil is the same as the ratio in the atmosphere. This is true only when $t = 0$.

(b) Verify the formula for $t = 5600$.

Substitute 5600 for t. Then

$$\frac{R}{r} = e^{(5600 \ln 2)/5600}$$

$$\frac{R}{r} = e^{\ln 2} = 2 \qquad \text{(Recall that } e^{\ln x} = x\text{)}$$

$$r = \frac{1}{2}R.$$

From this last result, the ratio in the fossil is half the ratio in the atmosphere. Since the half-life of carbon 14 is 5600 years, only half of it would remain at the end of that time. The formula gives the correct result for $t = 5600$.

(c) If the ratio in a fossil is 40% of R, how old is the fossil?

Let $r = .40R$, so

$$\frac{R}{r} = \frac{R}{.4R} = \frac{1}{.4} = 2.5.$$

Now use the equation $R/r = e^{t \ln 2/5600}$ and solve for t.

$$\frac{R}{r} = e^{t \ln 2/5600}$$

$$2.5 = e^{t \ln 2/5600}$$

$$\ln 2.5 = \ln e^{t \ln 2/5600}$$

$$\ln 2.5 = \frac{t \ln 2}{5600} \qquad \text{(Since } \ln e^x = x\text{)}$$

$$\frac{5600}{\ln 2} (\ln 2.5) = t$$

$$t \approx 7400$$

The fossil is about 7400 years old. ∎

The compound interest formula

$$A = P\left(1 + \frac{i}{m}\right)^{nm}$$

was discussed in Section 5.1. The table presented there shows that increasing the frequency of compounding makes smaller and smaller differences in the amount of interest earned. In fact, it can be shown that even if interest is compounded at intervals of time as small as one chooses (such as each hour, each minute, or each second), the total amount of interest earned will be only slightly more than for daily compounding. This is true even for a process called **continuous compounding,** which can be described loosely as compounding every instant. It turns out (although we shall not prove it) that the formula for continuous compounding involves the number e.

Continuous Compounding

> If P dollars is deposited at a rate of interest i compounded continuously for n years, the final amount on deposit is
>
> $$A = Pe^{ni}$$
>
> dollars.

Example 4   Suppose $5000 is deposited in an account paying 8% compounded continuously for five years. Find the total amount on deposit at the end of five years.

Let $P = 5000$, $n = 5$, and $i = .08$. Then

$$A = 5000e^{5(.08)} = 5000e^{.4}.$$

From Table 1 or a calculator, $e^{.4} \approx 1.49182$, and

$$A = 5000(1.49182) = 7459.10,$$

or $7459.10. Check that daily compounding would have produced a compound amount about 30¢ less.   ■

Example 5   How long will it take for the money in an account that is compounded continuously at 8% interest to double?

Use the formula for continuous compounding, $A = Pe^{ni}$, to find the time n that makes $A = 2P$. Substitute $2P$ for A and .08 for i; then solve for n.

$$A = Pe^{ni}$$
$$2P = Pe^{.08n}$$
$$2 = e^{.08n}$$

Taking natural logarithms on both sides gives

$$\ln 2 = \ln e^{.08n}.$$

Use the property $\ln e^x = x$ to get $\ln e^{.08n} = .08n$, and

$$\ln 2 = .08n$$
$$\frac{\ln 2}{.08} = n$$
$$8.664 = n.$$

It will take about 8 2/3 years for the amount to double.   ■

Example 6   Suppose inflation is about 10% per year. Assuming this rate continues, in how many years would prices double?

Use the equation for continuous compounding, with $A = 2P$ and $i = .10$.

$$A = Pe^{ni}$$
$$2P = Pe^{.1n}$$
$$2 = e^{.1n}$$

Solve for n by taking natural logarithms on both sides.

$$\ln 2 = \ln e^{.1n}$$

$$\ln 2 = .1n$$

$$n = \frac{\ln 2}{.1} \approx 6.9$$

Prices will double in about 7 years. ∎

5.5 Exercises

1. A population of lice is growing exponentially. After 2 months the population has increased from 100 to 125. In how many months will the population reach 500 lice?

2. A population of bacteria in a culture is increasing exponentially. The original culture of 25,000 bacteria contains 40,000 bacteria after 10 hours. How long will it be until there are 60,000 bacteria in the culture?

3. A radioactive substance is decaying exponentially. The substance is reduced from 800 grams to 400 grams after 4 days. How much remains after 6 days?

4. When a bactericide is introduced into a culture of bacteria, the number of bacteria decreases exponentially. After 9 hours there are only 20,000 bacteria. In how many hours will the original population of 50,000 bacteria be reduced to half?

5. The amount of a certain chemical that will dissolve in a solution increases exponentially as the temperature is increased. At 0°C 1 gram dissolved and at a temperature of 10°C 11 grams dissolved. At what temperature will 15 grams dissolve?

6. The amount of a certain radioactive specimen present at time t (in days) decreases exponentially. If 5000 grams decreased to 4000 grams in 5 days, find the half-life of the specimen.

Exercises 7–12 refer to the carbon dating process of Example 3 in the text.

7. Suppose an Egyptian mummy is discovered in which the ratio of carbon 14 to carbon 12 is only about half the ratio found in the atmosphere. About how long ago did the Egyptian die?

8. If the ratio of carbon 14 to carbon 12 in an object is 1/4 the atmospheric ratio, how old is the object? How old if the ratio is 1/8?

9. Verify the formula of Example 3 for $t = 11,200$.

10. Solve the formula for t.

11. In the Lascaux caves of France, the ratio of carbon 14 to carbon 12 is only about 15% of the ratio found in the atmosphere. Estimate the age of the caves.

12. Suppose a specimen is found in which $r = (2/3)R$. Estimate the age of the specimen.

Find each of the amounts in Exercises 13–18, assuming continuous compounding.

13. $2,000 at 8% for 1 year

14. $2,000 at 8% for 5 years

15. $12,700 at 10% for 3 years

16. $175.25 at 11% for 8 years

17. $5,800 at 13% for 6 years

18. $10,000 at 12% for 3 years

19. Assuming an inflation rate of 5% compounded continuously, how long will it take for prices to double?

20. In Exercise 19, how long will it take if the inflation rate is 9%?

Solve $A = Pe^{ni}$ for the following.

21. i **22.** n

Newton's law of cooling says that the rate at which a body cools is proportional to the difference in temperature between the body and the environment into which it is introduced. The temperature $f(t)$ of the body at time t in hours after being introduced into an environment having constant temperature T_0 is

$$f(t) = T_0 + Ce^{-kt},$$

where C and k are constants. Use this result in Exercises 23–26.

23. A piece of metal is heated to 300°C and then placed in a cooling liquid at 50°C. After 4 minutes, the metal has cooled to 175°C. Find its temperature after 12 minutes.

24. Boiling water, at 100°C, is placed in a freezer at 0°C. The temperature of the water is 50°C after 24 minutes. Find the temperature of the water after 96 minutes.

25. A volcano discharges lava at 800°C. The surrounding air has a temperature of 20°C. The lava cools to 410°C in five hours. Find its temperature after 15 hours.

26. Paisley refuses to drink coffee cooler than 95°F. She makes coffee with a temperature of 170°F in a room with a temperature of 70°F. The coffee cools to 120°F in 10 minutes. What is the longest time she can let the coffee sit before she drinks the coffee?

Many environmental situations place effective limits on the growth of the number of an organism in an area. Many such limited growth situations are described by the *logistic function*, defined by

$$G(t) = \frac{m \cdot G_0}{G_0 + (m - G_0)e^{-kmt}},$$

where G_0 is the initial number present, m is the maximum possible size of the population, and k is a positive constant. Assume $G_0 = 1000$, $m = 2500$, $k = .0004$, and t is time in decades (10-year periods).

27. Find $G(1)$.   **28.** Find $G(2)$.

29. When will the population double?   **30.** When will the population reach 5000?

5.6   Common Logarithms (Optional)

As we said earlier, base 10 logarithms are called **common logarithms.** It is customary to abbreviate $\log_{10} x$ as simply $\log x$. This convention started when base 10 logarithms were used extensively for calculation. (But be careful: some advanced books use $\log x$ as an abbreviation for $\log_e x$.) Examples of common logarithms include

$$\log 1000 = \log 10^3 = 3$$
$$\log 100 = \log 10^2 = 2$$

$$\log 10 = \log 10^1 = 1$$
$$\log 1 = \log 10^0 = 0$$
$$\log .1 = \log 10^{-1} = -1$$
$$\log .001 = \log 10^{-3} = -3.$$

Though it can be shown that there is no rational number x such that $10^x = 6$ (and thus no rational number x such that $x = \log 6$), a table of common logarithms or a calculator can be used to find a decimal approximation for $\log 6$. From Table 2 in the Appendix, a decimal approximation of $\log 6$ is

$$\log 6 \approx .7782,$$

or, equivalently, $10^{.7782} \approx 6$. Since most logarithms are approximations anyway, it is common to replace $\approx$ with $=$ and write

$$\log 6 = .7782.$$

Table 2 gives the logarithms of numbers between 1 and 10. Since every positive number can be written in scientific notation as the product of a number between 1 and 10 and a power of 10, the logarithm of any positive number can be found by using the table and the properties of logarithms.

Example 1   Find log 6.24.

Locate the first two digits, 6.2, in the left column of the table. Then find the third digit, 4, across the top of the table. You should find

$$\log 6.24 = .7952. \quad \blacksquare$$

Example 2   Find log 6240.

Write 6240 using scientific notation, as

$$6240 = 6.24 \times 10^3.$$

Then use the properties of logarithms.

$$\log 6240 = \log (6.24 \times 10^3)$$
$$= \log 6.24 + \log 10^3$$
$$= \log 6.24 + 3 \log 10$$
$$= \log 6.24 + 3$$

From Example 1, $\log 6.24 = .7952$, so

$$\log 6240 = .7952 + 3 = 3.7952. \quad \blacksquare$$

The decimal part of the logarithm, .7952 in Example 2, is called the **mantissa,** and the integer part, 3 here, is the **characteristic.** When using a table of logarithms, always make sure the mantissa is positive. The characteristic can be any integer, positive, negative, or zero.

Example 3   Find log .00587.

Use scientific notation and the properties of logarithms to get

$$\begin{aligned}
\log .00587 &= \log (5.87 \times 10^{-3}) \\
&= \log 5.87 + \log 10^{-3} \\
&= .7686 + (-3) \\
&= .7686 - 3.
\end{aligned}$$

The logarithm is usually left in this form. A calculator would give the answer as -2.2314, the algebraic sum of .7686 and -3. The decimal portion in the calculator answer is a negative number. This is not the best form for the logarithm when using tables, since it is not clear which number is the mantissa.

It is possible to write the characteristic in other forms. For example, log .00587 could be written as

$$\log .00587 = 7.7686 - 10.$$

The best choice depends on the anticipated use of the logarithm. ■

Example 4   Find each of the following.

(a) $\log (2.73)^4$

Use a property of logarithms to get

$$\begin{aligned}
\log (2.73)^4 &= 4 \log 2.73 \\
&= 4(.4362) \\
&= 1.7448.
\end{aligned}$$

(b) $\log \sqrt[3]{.0762}$

Here

$$\log \sqrt[3]{.0762} = \log (.0762)^{1/3}$$

$$= \frac{1}{3} \log .0762 = \frac{1}{3}(.8820 - 2).$$

To preserve the characteristic as an integer, change the characteristic to a multiple of 3 before multiplying by 1/3. One way to do this is to add and subtract 1 (which adds to 0) as follows.

$$\begin{aligned}
.8820 - 2 &= 1 - 1 + .8820 - 2 \\
&= 1.8820 - 3
\end{aligned}$$

Now complete the work above.

$$\log \sqrt[3]{.0762} = \frac{1}{3}(1.8820 - 3)$$

$$= .6273 - 1 \quad ■$$

Sometimes the logarithm of a number is known and the number itself must be found. The number is called the **antilogarithm,** sometimes abbreviated **antilog.** For example, .756 is the antilogarithm of .8785 − 1, since

$$\log .756 = .8785 - 1.$$

To find the antilogarithm, look for .8785 in the body of the logarithm table. You should find 7.5 at the left and 6 at the top. Since the characteristic of .8785 − 1 is −1, the antilogarithm is

$$7.56 \times 10^{-1} = .756.$$

In exponential notation, since

$$\log .756 = .8785 - 1 = -.1215,$$

then $\qquad\qquad\qquad .756 = 10^{-.1215}.$

Example 5 Find each of the following antilogarithms.

(a) $\log x = 2.5340$

Find .5340 in the body of the table; 3.4 is at the left and 2 is at the top.

$$\begin{aligned} \log x &= 2 + .5340 \\ &= \log 10^2 + \log 3.42 \qquad \text{(From Table 2)} \\ &= \log (3.42 \times 10^2) = \log 342 \end{aligned}$$

Since $\qquad\qquad \log x = \log 342,$
$$x = 342,$$

and 342 is the antilogarithm of 2.5340.

(b) $\log x = .7536 - 3$

Table 2 shows that the antilogarithm of .7536 − 3 is

$$5.67 \times 10^{-3} = .00567.$$

(c) $\log x = -4.0670 = -4 + (-.0670)$
$$= -4 + (-1) + [1 + (-.0670)] = -5 + .9330$$
$$\log x = \log .0000857$$
$$x = .0000857 \quad \blacksquare$$

The next example uses Table 2 and the properties of logarithms for a numerical calculation.

Example 6 Find $(\sqrt[3]{42})(76.9)(.00283)$.

Use the properties of logarithms and Table 2.

$$\log (\sqrt[3]{42})(76.9)(.00283) = \frac{1}{3} \log 42 + \log 76.9 + \log .00283$$

$$= \frac{1}{3}(1.6232) + 1.8859 + (.4518 - 3)$$

$$= .5411 + 1.8859 + (.4518 - 3)$$

$$= 2.8788 - 3 = .8788 - 1.$$

From the logarithm table, .756 is the antilogarithm of .8788 − 1, and so

$$(\sqrt[3]{42})(76.9)(.00283) \approx .756. \quad \blacksquare$$

Example 7 The cost in dollars, $C(x)$, of manufacturing x picture frames, where x is measured in thousands, is

$$C(x) = 5000 + 2000 \log (x + 1).$$

Find the cost of manufacturing 19,000 frames.

To find the cost of producing 19,000 frames, let $x = 19$. This gives

$$
\begin{aligned}
C(19) &= 5000 + 2000 \log (19 + 1) \\
&= 5000 + 2000 \log 20 \\
&= 5000 + 2000(1.3010) \\
&= 7602.
\end{aligned}
$$

Thus, 19,000 frames cost a total of about $7600 to produce. $\quad \blacksquare$

Example 8 In chemistry, the pH of a solution is defined as

$$pH = -\log [H_3O^+],$$

where $[H_3O^+]$ is the hydronium ion concentration in moles per liter. The number pH is a measure of the acidity or alkalinity of solutions. Pure water has a pH of 7.0 with values greater than that indicating alkalinity and values less than 7.0 indicating acidity. Find the following.

(a) The pH of a solution with $[H_3O^+] = 2.5 \times 10^{-4}$

$$
\begin{aligned}
pH &= -\log [H_3O^+] \\
pH &= -\log (2.5 \times 10^{-4}) \\
&= -(\log 2.5 + \log 10^{-4}) \\
&= -(.3979 - 4) \\
&= -.3979 + 4 \\
&\approx 3.6
\end{aligned}
$$

It is customary to round pH values to the nearest tenth.

(b) The hydronium ion concentration of a solution with $pH = 7.1$

$$
\begin{aligned}
pH &= -\log [H_3O^+] \\
7.1 &= -\log [H_3O^+] \\
-7.1 &= \log [H_3O^+]
\end{aligned}
$$

To find the antilogarithm of the number -7.1, first write -7.1 as $-7.1 + 8 - 8 = .9 - 8$. Look up the mantissa .9 in Table 2. Write the antilogarithm in scientific notation, rounding to the nearest tenth.

$$H_3O^+ = 7.9 \times 10^{-8} \quad \blacksquare$$

5.6 Exercises

Find the characteristic of the logarithms of each of the following.

1. 875   **2.** 9462   **3.** 2,400,000   **4.** 875,000

5. .00023   **6.** .098   **7.** .000042   **8.** .000000257

Find the common logarithms of the following numbers. A calculator is needed for Exercises 15–18.

9. .000893   **10.** .00376   **11.** 68,200   **12.** 103,000

13. 7.63   **14.** 9.37   **15.** 42,967   **16.** 1.51172

17. .008094   **18.** .00004397

Find the common antilogarithms of each of the following logarithms to three significant digits.

19. 1.5366 : 34.4/ **20.** 2.9253   **21.** .8733 − 2   **22.** .2504 − 1

23. 3.4947   **24.** 4.6863   **25.** .8039 − 3   **26.** .8078 − 2

Use the change-of-base formula from Section 5.3 with common logarithms and Table 2 to find each of the following logarithms. Round to the nearest hundredth.

27. ln 125   **28.** ln 63.1   **29.** ln .98   **30.** ln 1.53

31. ln 275   **32.** ln 39.8   **33.** $\log_2 10$ **34.** $\log_5 8$

35. $\log_3 180$ **36.** $\log_2 9.65$ **37.** $\log_4 12$ **38.** $\log_7 4$

Use common logarithms to solve the following equations. Round to the nearest hundredth.

39. $10^x = 2.5$ **40.** $10^{x-1} = 143$ **41.** $100^{2x+1} = 17$ **42.** $1000^{5x} = 2010$

43. $50^{x+2} = 8$ **44.** $200^{x+3} = 12$ **45.** $.01^{3x} = .005$ **46.** $.001^{2x} = .0004$

Use common logarithms to find approximations to three significant digits for each of the following.

47. $(8.16)^{1/3}$ **48.** $\sqrt{276}$ **49.** $\dfrac{(7.06)^3}{(31.7)(\sqrt{1.09})}$ **50.** $\dfrac{(2.51)^2}{(\sqrt{1.52})(3.94)}$

51. $(115)^{1/2} + (35.2)^{2/3}$ **52.** $(778)^{1/3} + (159)^{3/4}$

Use logarithms to solve each of the following applications.

53. The number of years, n, since two independently evolving languages split off from a common ancestral language is approximated by

$$n \approx -7600 \log r,$$

where r is the proportion of words from the ancestral language common to both languages. Find n if **(a)** $r = .9$; **(b)** $r = .3$; **(c)** How many years have elapsed since the split if half of the words of the ancestral language are common to both languages?

54. Midwest Creations finds that its total sales, $T(x)$, from the distribution of x catalogs, measured in thousands, is approximated by $T(x) = 5000 \log (x + 1)$. Find the total sales resulting from the distribution of **(a)** 0 catalogs **(b)** 5000 catalogs **(c)** 24,000 catalogs **(d)** 49,000 catalogs.

Find the pH of each of the following substances, using the given hydronium ion concentration.

55. Grapefruit, 6.3×10^{-4}

56. Crackers, 3.9×10^{-9}

57. Limes, 1.6×10^{-2}

58. Sodium hydroxide (lye), 3.2×10^{-14}

Find $[H_3O^+]$ for each of the following substances, using the given pH.

59. Soda pop, 2.7   **60.** Wine, 3.4   **61.** Beer, 4.8   **62.** Drinking water, 6.5

The area of a triangle having sides of length a, b, and c is given by

$$A = \sqrt{s(s - a)(s - b)(s - c)},$$

where $s = (a + b + c)/2$. Use logarithms to find the areas of the following triangles.

63. $a = 114, b = 196, c = 153$

64. $a = .0941, b = .0873, c = .0896$

65. A common problem in archaeology is to determine estimates of populations at a particular site. Several methods have been proposed to do this. One method relates the total surface area of a site to the number of occupants. If P represents the population of a site which covers an area of a square units, then

$$\log P = k \log a,$$

where k is an appropriate constant which varies for hilly, coastal, or desert environments, or for sites with single family dwellings or multiple family dwellings.* Find the population of sites with the following areas (use .8 for k):   **(a)** 230 m²   **(b)** 95 m²   **(c)** 20,000 m².

66. In Exercise 65, find the population of a site with an area 100,000 m² for the following values of k: **(a)** 1.2   **(b)** .5   **(c)** .7.

Appendix: Interpolation

The table of common logarithms included in this text contains decimal approximations of common logarithms to four places of accuracy. If greater accuracy is necessary, a calculator or more accurate tables can be used. However, if desired, more accuracy can be obtained from the table included in this text by the process of **linear interpolation.** As an example, use linear interpolation to approximate log 75.37. First,

$$\log 75.3 < \log 75.37 < \log 75.4$$

Figure 10 on the next page shows the portion of the curve $y = \log x$ between $x = 75.3$ and $x = 75.4$. We shall use the line segment PR to approximate the logarithm curve (this approximation is usually adequate for values of x relatively close to one another). From the figure, log 75.37 is given by the length of segment MQ, which

*From "The Quantitative Investigation of Indian Mounds" by S. F. Cook, Berkeley: *University of California Publications in American Archaeology and Ethnology,* 40: 231–33, 1950. Reprinted by permission of the University of California Press.

cannot be found directly from Table 2. We can, however, find MN, which we shall use as our approximation to log 75.37. By properties of similar triangles,

$$\frac{PS}{PT} = \frac{SN}{TR}.$$

In this case, $PS = 75.37 - 75.3 = .07$, $PT = 75.4 - 75.3 = .1$, and $RT = \log 75.4 - \log 75.3 = 1.8774 - 1.8768 = .0006$. Hence,

$$\frac{.07}{.1} = \frac{SN}{.0006}$$

or $$SN = .7(.0006) \approx .0004.$$

(This is 7/10 of the difference of the two logarithms.) Since log 75.37 = MN = $MS + SN$, and since $MS = \log 75.3 = 1.8768$,

$$\log 75.37 = 1.8768 + .0004 = 1.8772.$$

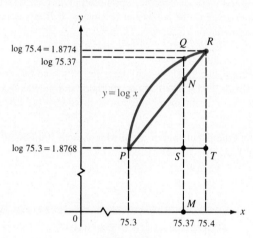

Figure 10

Example 1 Find log 8726.

All the work of the example above can be condensed as follows.

$$10\left\{6\left\{\begin{matrix}\log 8720 = 3.9405 \\ \log 8726 = \end{matrix}\right\}x\\ \log 8730 = 3.9410\end{matrix}\right\}.0005$$

From this display,

$$\frac{6}{10} = \frac{x}{.0005}$$

$$x = \frac{6(.0005)}{10}$$

$$x = .0003,$$

and log 8726 = 3.9405 + .0003 = 3.9408. ∎

Example 2 Find log .0005958.

Work as shown above.

$$10\left\{8\begin{cases} \log .0005950 = -4 + .7745 \\ \log .0005958 = \\ \log .0005960 = -4 + .7752 \end{cases}x\right\}.0007$$

Then

$$\frac{8}{10} = \frac{x}{.0007},$$

or

$$x = .0006,$$

and

$$\log .0005958 = (-4 + .7745) + .0006$$
$$= -4 + .7751 \quad \blacksquare$$

Interpolation also can be used when finding an antilogarithm, as shown in the next example.

Example 3 Find x such that log x = .3275.

From the logarithm table,

$$\log 2.12 = .3263 < .3275 < .3284 = \log 2.13.$$

Set up the work as follows.

$$.01\left\{y\begin{cases} \log 2.12 = .3263 \\ \log x \quad = .3275 \end{cases}.0012\right\}.0021$$
$$\log 2.13 = .3284$$

$$\frac{y}{.01} = \frac{.0012}{.0021}$$

$$y = \frac{(.01)(.0012)}{.0021}$$

$$y = .006$$

Thus, $x = 2.12 + .006 = 2.126$, and

$$\log 2.126 = .3275. \quad \blacksquare$$

Appendix Exercises

Interpolate to find common logarithms of the following numbers.

1. 2345   **2.** 1.732   **3.** 48.26   **4.** 351.9

5. .06273   **6.** .003471   **7.** 27.05   **8.** 342.6

Use interpolation to find the antilogarithms of the following common logarithms.

9. 1.7942   **10.** 3.9225   **11.** 7.6565 − 10   **12.** 8.7296 − 10

13. 5.6930   **14.** 12.6268   **15.** −3.7778   **16.** −4.1323

Use common logarithms (and interpolation) to find approximations to four-digit accuracy for each of the following computations.

17. $\dfrac{(26.13)(5.427)}{101.6}$ **18.** $\dfrac{(32.68)(142.8)}{973.4}$ **19.** $\sqrt{\dfrac{6.532}{2.718}}$ **20.** $\left(\dfrac{27.46}{58.29}\right)^3$

21. $(.2374)^{.05}$ **22.** $(1.792)^{.23}$ **23.** $(49.83)^{1/2} + (2.917)^2$ **24.** $(38.42)^{1/3} + (86.13)^{1/4}$

25. The maximum load L that a cylindrical column can hold is given by

$$L = \frac{d^4}{h^2},$$

where d is the diameter of the cylindrical cross section and h is the height of the column. Find L if $d = 2.143$ feet and $h = 12.25$ feet.

26. Two electrons repel each other with a force $F = kd^{-2}$, where d is the distance between the electrons. Find F if $d = .0005241$ cm and $k = 1$.

Chapter 5 Summary

Key Words

exponential function	common logarithm	characteristic
logarithm	natural logarithm	antilogarithm
logarithmic function	mantissa	interpolation

Review Exercises

Graph each of the functions defined as follows.

1. $f(x) = 2^x$ **2.** $f(x) = 2^{-x}$ **3.** $f(x) = (1/2)^{x+1}$ **4.** $f(x) = 4^x + 4^{-x}$

5. $f(x) = (x-1)e^{-x}$ **6.** $f(x) = \log_2(x-1)$ **7.** $f(x) = \log_3 2x$ **8.** $f(x) = \ln x^{-2}$

Solve each of the following equations.

9. $8^p = 32$ **10.** $9^{2y-1} = 27^y$ **11.** $\dfrac{8}{27} = b^{-3}$ **12.** $\dfrac{1}{2} = \left(\dfrac{b}{4}\right)^{1/4}$

The amount of a certain radioactive material, in grams, present after t days, is given by

$$A(t) = 800e^{-.04t}.$$

Find $A(t)$ if

13. $t = 0$. **14.** $t = 5$.

How much would $1200 amount to at 10% compounded continuously for the following number of years?

15. 4 years **16.** 10 years

17. Historically, the consumption of electricity has increased at a continuous rate of 6% per year. If it continued to increase at this rate, find the number of years before exactly twice as much electricity would be needed.

18. Suppose a conservation campaign together with higher rates caused demand for electricity to increase at only 2% per year. (See Exercise 17.) Find the number of years before twice as much electricity would be needed as is needed today.

Write each of the following expressions in logarithmic form.

19. $2^5 = 32$
20. $100^{1/2} = 10$
21. $(1/16)^{1/4} = 1/2$
22. $(3/4)^{-1} = 4/3$

23. $10^{.4771} = 3$
24. $e^{2.4849} = 12$
25. $e^{.1} = 1.1052$
26. $2^{2.322} = 5$

Write each of the following logarithms in exponential form.

27. $\log_{10} .001 = -3$
28. $\log_2 \sqrt{32} = 5/2$
29. $\log 3.45 = .537819$
30. $\ln 45 = 3.806662$

Use properties of logarithms to write each of the following as a sum, difference, or product of logarithms. Assume all variables represent positive real numbers.

31. $\log_3 \dfrac{mn}{p}$
32. $\log_2 \dfrac{\sqrt{5}}{3}$
33. $\log_5 x^2 y^4 \sqrt[5]{m^3 p}$
34. $\log_7 (7k + 5r^2)$

Find the common logarithm of each of the following numbers.

35. 8.47
36. .00421
37. 1050
38. 69,800

Find the antilogarithm of each of the following numbers to three significant digits.

39. 3.4983
40. 9.7243
41. .6493 − 2
42. .4232 − 3

Use common logarithms to approximate each of the following.

43. $\dfrac{6^{2.1}}{\sqrt{52}}$
44. $2.43^{3.2}$
45. $\sqrt[5]{\dfrac{27.1}{4.33}}$
46. $(2^{5.71})(5.43^3)$

The height, in meters, of the members of a certain tribe is approximated by

$$h(t) = .5 + \log t,$$

where t is the tribe member's age in years, and $1 \le t \le 20$. Find the height of a tribe member of each of the following ages.

47. 2 years
48. 5 years
49. 10 years
50. 20 years

51. A person learning certain skills involving repetition tends to learn quickly at first. Then learning tapers off and approaches some upper limit. Suppose the number of symbols per minute a keypunch operator can produce is given by

$$p(t) = 250 - 120(2.8)^{-.5t}$$

where t is the number of months the operator has been in training. Find each of the following: **(a)** $p(2)$ **(b)** $p(4)$ **(c)** $p(10)$. **(d)** Graph $p(t)$.

52. The concentration of pollutants, in grams per liter, in the east fork of the Big Weasel River is approximated by

$$P(x) = .04 \, e^{-4x},$$

where x is the number of miles downstream from a paper mill that the measurement is taken. Find **(a)** $P(.5)$ **(b)** $P(1)$ **(c)** the concentration of pollutants 2 miles downstream.

Solve each of the following equations. Round to the nearest thousandth.

53. $5^r = 11$
54. $10^{2r-3} = 17$
55. $e^{p+1} = 10$
56. $(1/2)^{3k+1} = 3$
57. $6^{2-m} = 2^{3m+1}$
58. $4(1 + e^x) = 8$

59. $\log_{64} y = 1/3$

60. $\log_2 (y + 3) = 5$

61. $\ln 6x - \ln (x + 1) = \ln 4$

62. $\log_{16} \sqrt{x + 1} = 1/4$

63. $\log (3p - 1) = 1 - \log p$

64. $\log_2 (2b - 1)^2 = 4$

Solve for the indicated variable.

65. $y = \dfrac{5^x - 5^{-x}}{2}$, for x

66. $y = \dfrac{1}{2(5^x - 5^{-x})}$, for x

67. $2 \log_a (x - 2) = 1 + \log_a (x + 1)$, for x

68. $r = r_0 e^{nt}$, for t

69. $N = a + b \ln \dfrac{c}{d}$, for c

70. $P = \dfrac{k}{1 + e^{-rt}}$, for t

Find each of the following logarithms.

71. $\ln e^{-5.3}$

72. $\ln e^{.04}$

73. $\ln 89$

74. $\ln .000050$

75. $\ln 8$

76. $\log_{3.4} 15.8$

77. $\log_{1/2} 9.45$

78. $\log_3 769$

Use interpolation to find each of the following logarithms.

79. $\log 18.99$

80. $\log 4.763$

81. $\log .009814$

The formula

$$A = P\left(1 + \frac{i}{m}\right)^{mn}$$

in Section 5.1 gives the amount of money in an account after n years if P dollars are deposited at a rate of interest i compounded m times per year. If A is known, then P is called the *present value* of A. (Think of present value as the value today of a sum of money to be received at some time in the future.) Find the present value of the following sums.

82. $4500 at 6% compounded annually for 9 years

83. $11,500 at 4% compounded annually for 12 years

84. $2000 at 4% compounded semiannually for 11 years

85. $2000 at 6% compounded quarterly for 8 years

86. How long would it take for $1000 at 5% compounded quarterly to double?

87. In Exercise 86, how long would it take at 10%?

88. If the inflation rate were 10%, use the formula for continuous compounding to find the number of years for a $1 item to cost $2.

89. In Exercise 88, find the number of years if the rate of inflation were 13%.

If R dollars is deposited at the end of each year in an account paying a rate of interest of i per year compounded annually, then after n years the account will contain a total of

$$R\left[\frac{(1 + i)^n - 1}{i}\right]$$

dollars. Find the final amount on deposit for each of the following. Use logarithms or a calculator. (Such a sequence of payments is called an *annuity*.)

90. $800, 12%, 10 years

91. $1500, 14%, 7 years

92. $375, 10%, 12 years

93. Manual deposits $10,000 at the end of each year for 12 years in an account paying 12% compounded annually. He then puts this total amount on deposit in another account paying 10% compounded semiannually for another 9 years. Find the total amount on deposit after the entire 21-year period.

94. Scott Hardy deposits $12,000 at the end of each year for 8 years in an account paying 14% compounded annually. He then leaves the money alone with no further deposits for an additional 6 years. Find the total amount on deposit after the entire 14-year period.

95. A radioactive substance is decaying exponentially. A sample of the substance is reduced from 500 grams to 400 grams after 4 days.
(a) How much is left after 10 days?
(b) How long will it take for the substance to decay to 100 grams?

96. The population of a boomtown is increasing exponentially. There were 10,000 people in town when the boom began. Two years later the population had reached 12,000. Assume this growth rate continues.
(a) What will be the population after 5 years?
(b) How long will it take for the population to double?

97. Natural logarithms can be calculated on a hand calculator even if the calculator does not have an ln key. If $1/2 \leq x \leq 3/2$ and if $A = (x - 1)/(x + 1)$, then a good approximation to $\ln x$ is given by

$$\ln x \approx ((3A^2/5 + 1)A^2/3 + 1)2A.$$

(a) Use this formula to calculate ln .8 and ln 1.2.
(b) Using facts about logarithms, use the formula to calculate (approximately)

$$\ln 2 = \ln ((3/2)(4/3)).$$

(c) Using (b), calculate ln 3 and ln 8.

98. The exponential e^x can be estimated for x in $[-1/2, 1/2]$ by the formula

$$e^x \approx \left(\left(\left(\left(\frac{x}{5} + 1\right)\frac{x}{4} + 1\right)\frac{x}{3} + 1\right)\frac{x}{2} + 1\right)x + 1.$$

(a) Calculate an approximate value for $e^{.13}$.
(b) Calculate an approximate value for $e^{-.37}$.
(c) Calculate an approximate value for $e^{4.13}$.
(d) Calculate an approximate value for $e^{-2.63}$.
   (*Hint:* Use part (b).)

99. The quantity $n! = n(n - 1)(n - 2) \cdots 3 \cdot 2 \cdot 1$ grows very rapidly as n increases. According to *Stirling's Formula*, when n is large,

$$n! \approx \sqrt{2\pi n} \left(\frac{n}{e}\right)^n.$$

Use Stirling's Formula to estimate 100! and 200! (*Hint:* Use common logarithms.)

100. Watch carefully. Suppose $0 < A < B$. Because the logarithm is an increasing function, we have
(a) $\log A < \log B$.   Then
(b) $10A \cdot \log A < 10B \cdot \log B$,
(c) $\log A^{10A} < \log B^{10B}$,
(d) $A^{10A} < B^{10B}$.
On the other hand, we run into trouble with particular choices of A and B. For instance, choose $A = 1/10$ and $B = 1/2$. Clearly $0 < A < B$, but $A^{10A} = (1/10)^1 = 1/10$ is greater than $B^{10B} = (1/2)^5 = 1/32$. Where was the first false step made?

6

Systems of Equations and Inequalities

Many applications of mathematics require the simultaneous solution of a large number of equations having many variables. Such a set of equations is called a **system of equations.** The definition of a linear equation given earlier can be extended to more variables: Any equation of the form

$$a_1x_1 + a_2x_2 + \cdots + a_nx_n = b$$

for real numbers $a_1, a_2, \ldots, a_n$ (not all of which are 0), and b, is a **linear equation.** If all the equations in a system are linear, the system is a **system of linear equations,** or a **linear system.** In this chapter, methods of solving systems of equations or inequalities are discussed. The matrix techniques discussed in the second section have gained particular importance with the increasing availability of computers.

6.1 Linear Systems

Any solution of the system of equations

$$5x + 3y = 95$$
$$2x - 7y = -3$$

will be an ordered pair of numbers (x, y) that satisfies both equations. One way to find the solution of such a system involves substitution; this substitution method is illustrated in the following example.

Example 1 Solve the system

$$5x + 3y = 95 \qquad \textbf{(1)}$$
$$2x - 7y = -3. \qquad \textbf{(2)}$$

First solve either equation for one variable. Let us solve equation (2) for x.

$$2x - 7y = -3$$
$$2x = 7y - 3$$
$$x = \frac{7y - 3}{2} \tag{3}$$

Now substitute the result for x in equation (1).

$$5x + 3y = 95$$
$$5\left(\frac{7y - 3}{2}\right) + 3y = 95$$

To solve for y, first multiply both sides of the equation by 2 to eliminate the denominator.

$$5(7y - 3) + 6y = 190$$
$$35y - 15 + 6y = 190$$
$$41y = 205$$
$$y = 5$$

Find x by substituting 5 for y in equation (3).

$$x = \frac{7y - 3}{2} = \frac{7(5) - 3}{2} = 16.$$

Check by substitution in the original system that the solution is the ordered pair (16, 5), so that the solution set is $\{(16, 5)\}$.   ■

Another way to solve a system of two equations, called the **elimination method,** involves using addition to eliminate a variable from both equations. This elimination can be done if the coefficients of that variable in the two equations are additive inverses of each other. To achieve this, properties of algebra are used to change the system to obtain an **equivalent system,** one with the same solution set as the given system. There are three transformations that may be applied to a system to get an equivalent system.

Transformation of a Linear System

1. Any two equations of the system may be exchanged.
2. Both sides of any equation of the system may be multiplied by any nonzero real number.
3. Any equation of the system may be replaced by the sum of that equation and a multiple of another equation in the system.

The next example illustrates the use of these transformations in the elimination method to solve a linear system.

Example 2   Solve the system $3x - 4y = 1$ **(1)**

$2x + 3y = 12.$ **(2)**

The goal is to use the transformations to change one or both equations so the coefficients of one variable in the two equations are additive inverses of each other. Then addition of the two equations will eliminate that variable. One way to eliminate a variable in this example is to use the second transformation and multiply both sides of equation (2) by -3, giving the equivalent system

$$3x - 4y = 1$$
$$-6x - 9y = -36.$$ **(3)**

Now multiply both sides of equation (1) by 2, and use the third transformation to add the result to equation (3), eliminating x. The addition of

$$6x - 8y = 2 \quad \text{to} \quad -6x - 9y = -36 \quad \text{to get} \quad -17y = -34$$

is done mentally or to the side. The result is the system

$$3x - 4y = 1$$
$$-17y = -34.$$ **(4)**

Dividing both sides of equation (4) by -17 gives the equivalent system

$$3x - 4y = 1$$
$$y = 2.$$

Now substitute 2 for y in equation (1) to get

$$3x - 4(2) = 1$$
$$3x - 8 = 1$$
$$3x = 9$$
$$x = 3.$$

The solution of the original system is (3, 2), which gives the solution set $\{(3, 2)\}$. The graphs of both equations of the system are sketched in Figure 1. As the graph suggests, (3, 2) satisfies both equations of the system.   ■

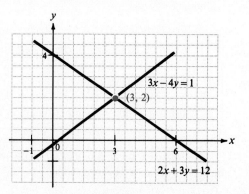

Figure 1

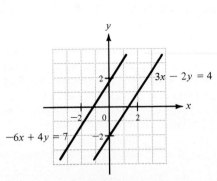

Figure 2   **Figure 3**

Example 3 Solve the system $3x - 2y = 4$ **(1)**

   $-6x + 4y = 7.$ **(2)**

Multiply both sides of equation (1) by 2, and add to equation (2), giving the equivalent system

$$3x - 2y = 4$$
$$0 = 15.$$

Since $0 = 15$ is never true, the system has no solution. As suggested by Figure 2, this means that the graphs of the equations of the system never intersect (the lines are parallel). The solution set for this system is $\emptyset$. ∎

A system of equations with no solution is **inconsistent,** and the graphs of an inconsistent linear system are parallel lines. If the equations of a system are all equivalent, the equations are said to be **dependent equations,** and their graphs will coincide as in Figure 3. In this case the system has an infinite number of solutions.

Example 4 Solve the system $8x - 2y = -4$ **(1)**

   $-4x + y = 2.$ **(2)**

Divide both sides of equation (1) by 2, and add the result to equation (2), to get the equivalent system

$$8x - 2y = -4$$
$$0 = 0.$$

The second equation, $0 = 0$, is always true, which indicates that the equations of the original system are equivalent. (With this system, the second transformation can be used to change either equation into the other.) Any ordered pair (x, y) that satisfies either equation will satisfy the system. From equation (2),

$$-4x + y = 2,$$

or

$$y = 2 + 4x.$$

By this result, the solution of the given system is the (infinite) set of ordered pairs of the form $(x, 2 + 4x)$, where x is a real number. The solution set can be written as $\{(x, y)|y = 2 + 4x\}$ or $\{(x, 2 + 4x)\}$. Typical ordered pairs in the solution set are $(0, 2 + 4 \cdot 0) = (0, 2)$, $(-4, 2 + 4(-4)) = (-4, -14)$, $(3, 14)$, and $(7, 30)$. As shown in Figure 3, both equations of the original system lead to the same straight line graph. ∎

Transformations can also be used to solve a system of three linear equations in three unknowns. It is a good idea to proceed systematically, as follows.

Solving a System of Three Equations

Use transformations to change a system having variables x, y, and z as follows.

1. If necessary, change the coefficient of x in the first equation to 1.
2. In the second equation, eliminate the x term by using the equation from Step 1 and, if necessary, change the coefficient of y to 1.
3. In the third equation, eliminate the x and y terms by using the equations from Steps 1 and 2 and, if necessary, change the coefficient of z to 1.

As a preliminary step, rearrange the order of the three given equations if that will eliminate the need for some of the steps. The next example shows how the transformations are used.

Example 5 Solve the system

$$3x + 9y + 6z = 3 \quad (1) \tag{1}$$
$$2x + y - z = 2 \quad (2) \tag{2}$$
$$x + y + z = 2. \quad (3) \tag{3}$$

Divide both sides of equation (1) by 3 to change the coefficient of x to 1. (Equations (1) and (3) could have been exchanged instead.)

$$x + 3y + 2z = 1 \quad (4) \tag{4}$$
$$2x + y - z = 2 \quad (2)$$
$$x + y + z = 2. \quad (3)$$

To eliminate x in equation (2), multiply both sides of equation (4) by -2 and add the results to equation (2). This gives the system

$$x + 3y + 2z = 1 \quad (4)$$
$$-5y - 5z = 0 \quad (5) \tag{5}$$
$$x + y + z = 2. \quad (3)$$

Now divide both sides of equation (5) by -5 to get

$$x + 3y + 2z = 1 \quad (4)$$
$$y + z = 0 \quad (5) \tag{6}$$
$$x + y + z = 2. \quad (3)$$

Eliminate x in equation (3) by multiplying both sides of equation (4) by -1 and adding the results to equation (3).

$$x + 3y + 2z = 1 \quad (4)$$
$$y + z = 0 \quad (5)$$
$$-2y - z = 1. \quad (6 \; 7) \tag{7}$$

To eliminate y in equation (7), multiply both sides of equation (6) by 2 and add the result to equation (7). The new system is

$$x + 3y + 2z = 1$$
$$y + z = 0$$
$$z = 1. \tag{8}$$

This last system is in **triangular form.** From the last equation, $z = 1$. Substitute $z = 1$ into equation (6) to get $y = -1$ and then substitute these values into equation (4) to find that $x = 2$. The solution of the system is the **ordered triple** $(2, -1, 1)$, which gives the solution set $\{(2, -1, 1)\}$. The process of substituting the known values of the variables back into the equations to find the remaining unknown variables is called **back-substitution.** ∎

A system of equations will sometimes have more variables than equations. In this case, there usually is no single solution, and in fact there might be no solution at all. The next example shows how to handle these systems.

Example 6   Solve the system $x + 2y + z = 4$ (1)
$3x - y - 4z = -9.$ (2)

Eliminate x from equation (2) by multiplying both sides of equation (1) by -3 and adding the result to equation (2).

$$x + 2y + z = 4$$
$$-7y - 7z = -21 \tag{3}$$

Divide both sides of equation (3) by -7, giving the simplified system

$$x + 2y + z = 4$$
$$y + z = 3. \tag{4}$$

It is not possible to eliminate more variables here—any attempt to do so would reintroduce x into the second equation. Instead, solve equation (4) for y. (We could have used z.)

$$y = 3 - z$$

Solve equation (1) for x after first substituting $3 - z$ for y.

$$x + 2(3 - z) + z = 4$$
$$x + 6 - 2z + z = 4$$
$$x - z = -2$$
$$x = -2 + z$$

For any value of z that might be chosen, y is found from the fact that $y = 3 - z$, and x from the fact that $x = -2 + z$. The solution set can be written as the set of ordered triples $\{(-2 + z, 3 - z, z)\}$, where z is a real number. Typical values can be found by choosing values of z. For example, if z is 4, then $y = 3 - z = 3 - 4 = -1$, and $x = -2 + z = -2 + 4 = 2$, giving as one solution $(2, -1, 4)$. The solution set is an infinite set of ordered triples. Compare this with Example 4.

Had we solved equation (4) for z instead of y, the solution would have had the form $(1 - y, y, 3 - y)$, where y is a real number. Verify that this alternate solution leads to the same set of ordered triples. The solution set, $\{(-2 + z, 3 - z, z)\}$, is said to have z **arbitrary.** ■

Example 7   Solve the system

$$x + 2y + z = 0 \tag{1}$$
$$4x - y + z = 0 \tag{2}$$
$$-x - 2y - z = 0. \tag{3}$$

Equation (3) is a multiple of equation (1), so the system can be reduced to

$$x + 2y + z = 0$$
$$4x - y + z = 0.$$

This system of two equations in three variables can be solved by the method illustrated in Example 6. Verify that the solution set with z arbitrary is $\{(-z/3, -z/3, z)\}$, where z is a real number. ■

In the system of equations given in Example 7, each equation has a constant term of 0. By inspection, the ordered triple $(0, 0, 0)$ is a solution of the system. However, in this case, there are infinitely many other solutions. A system of three equations in three unknowns with all constant terms equal to zero is called a **homogeneous system** of three equations in three unknowns. The ordered triple $(0, 0, 0)$ is always a solution of a homogeneous system; this solution is the **trivial solution.**

In Section 3.2 the general form of the equation of a circle was given as

$$x^2 + y^2 + Cx + Dy + E = 0.$$

This general form is used in the next example.

Example 8   Find an equation for all the circles going through $(2, 1)$ and $(-4, -1)$.

A circle is completely determined if any three distinct points on the circle are given. Here only two points are known. There are infinitely many circles through these two points. The set of all the circles going through these two points is called a **family of circles.** To find the equation of this family of circles, start with the general form of the equation of a circle,

$$x^2 + y^2 + Cx + Dy + E = 0.$$

Substitute the values from the two given points, $(2, 1)$, and $(-4, -1)$ into this equation, producing the following system of two equations in three unknowns.

$$4 + 1 + 2C + D + E = 0$$
$$16 + 1 - 4C - D + E = 0$$

These equations may be rewritten as

$$E + D + 2C = -5 \tag{1}$$
$$E - D - 4C = -17. \tag{2}$$

Multiplying both sides of equation (1) by -1 and adding the results to equation (2), gives

$$E + D + 2C = -5$$
$$-2D - 6C = -12. \qquad (3)$$

Divide both sides of equation (3) by -2 to get

$$E + D + 2C = -5$$
$$D + 3C = 6. \qquad (4)$$

Solve equation (4) for D: $D = -3C + 6$. Substitute this value for D into equation (1) and solve for E.

$$E + (-3C + 6) + 2C = -5$$
$$E - C = -11$$
$$E = C - 11$$

The solution is $(C, -3C + 6, C - 11)$, so that the equation of the family of circles is

$$x^2 + y^2 + Cx + (-3C + 6)y + (C - 11) = 0.$$

To find the equation for a particular circle that goes through the two given points, choose a value for C. For example, choosing $C = 3$ gives $D = -3(3) + 6 = -3$ and $E = 3 - 11 = -8$. Substituting these values into the general equation for the family of circles gives the particular equation

$$x^2 + y^2 + 3x - 3y - 8 = 0.$$

Verify that this circle goes through the two given points by showing that both ordered pairs, $(2, 1)$ and $(-4, -1)$ satisfy the equation. Figure 4 shows this circle and two others that also go through the points $(2, 1)$ and $(-4, -1)$. ∎

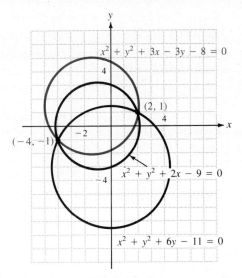

Figure 4

In the last two examples, the systems discussed had one more variable than equation. If there are two more variables than equations, there will be two arbitrary variables, and so on.

Applications of mathematics often require the solution of a system of equations, as the next example shows.

Example 9   An animal feed is made from three ingredients: corn, soybeans, and cottonseed. One unit of each ingredient provides units of protein, fat, and fiber as shown in the table below. How many units of each ingredient should be used to make a feed which contains 22 units of protein, 28 units of fat, and 18 units of fiber?

	Protein	Fat	Fiber
Corn	.25	.4	.3
Soybeans	.4	.2	.2
Cottonseed	.2	.3	.1

Let x represent the number of units of corn; y, the number of units of soybeans; and z, the number of units of cottonseed that are required. Since the total amount of protein is to be 22 units,

$$.25x + .4y + .2z = 22.$$

Also, for the 28 units of fat,

$$.4x + .2y + .3z = 28,$$

and, for the 18 units of fiber,

$$.3x + .2y + .1z = 18.$$

Multiply the first equation on both sides by 100, and the second and third equations by 10 to get the system

$$25x + 40y + 20z = 2200$$
$$4x + 2y + 3z = 280$$
$$3x + 2y + z = 180.$$

Verify that (40, 15, 30) is the solution of this system, so that the feed should contain 40 units of corn, 15 units of soybeans, and 30 units of cottonseed to meet the given requirements. ■

6.1 Exercises

Solve each of the following systems by the substitution method.

1. $x - 5y = 8$
  $x = 4y$

2. $6x - y = 5$
  $y = 11x$

3. $4x - 5y = -11$
  $2x + y = 5$

4. $-3x + 2y = 11$
  $2x + 7y = 11$

5. $7x - y = -10$
  $3y - x = 10$

6. $4x + 5y = 7$
  $9y = 31 + 2x$

7. $5x = y - 8$
  $3y = 4x + 46$

8. $-2x = 6y + 18$
  $-29 = 5y - 3x$

9. $x = 2y - z$
  $3x + y = 3z + 19$
  $2x - y + 4z = -1$

10.
$$y = 3x + z - 13$$
$$x - y = z + 3$$
$$x + 2y - z = -3$$

11. $x - 2y + z = 3$
$3x + y - 2z = -1$
$x - 4y + z = 3$

12. $-2x + 5y + 3z = -6$
$4x - y - 2z = -6$
$3x + 4y - 2z = -14$

Solve each system by elimination.

13. $3x + 2y = -6$
$5x - 2y = -10$

14. $3x - y = -4$
$x + 3y = 12$

15. $2x - 3y = -7$
$5x + 4y = 17$

16. $4x + 3y = -1$
$2x + 5y = 3$

17. $5x + 7y = 6$
$10x - 3y = 46$

18. $12x - 5y = 9$
$3x - 8y = -18$

19. $6x + 7y = -2$
$7x - 6y = 26$

20. $2x + 9y = 3$
$5x + 7y = -8$

21. $\dfrac{x}{2} + \dfrac{y}{3} = 8$
$\dfrac{2x}{3} + \dfrac{3y}{2} = 17$

22. $\dfrac{x}{5} + 3y = 31$
$2x - \dfrac{y}{5} = 8$

23. $\dfrac{3x}{2} - \dfrac{y}{3} = 5$
$\dfrac{5x}{2} + \dfrac{2y}{3} = 12$

24. $\dfrac{4x}{5} + \dfrac{y}{4} = -2$
$\dfrac{x}{5} + \dfrac{y}{8} = 0$

25. $0.05x - 0.02y = -0.18$
$0.04x + 0.06y = -0.22$

26. $0.08x + 0.03y = -0.24$
$0.04x - 0.01y = -0.32$

27. $0.6x + 0.3y = 0.087$
$0.5x - 0.4y = 0.378$

28. $0.7x - 0.5y = -0.884$
$0.1x + 0.3y = 0.13$

29. $1.9x - 4.8y = 11.2$
$-4.37x + 11.04y = -7.6$

30. $3.7x + 1.82y = -9.7$
$2.96x + 1.456y = 4.8$

31. $\dfrac{2x - 1}{3} + \dfrac{y + 2}{4} = 4$
$\dfrac{x + 3}{2} - \dfrac{x - y}{3} = 3$

32. $\dfrac{x + 6}{5} + \dfrac{2y - x}{10} = 1$
$\dfrac{x + 2}{4} - \dfrac{4y + 2}{5} = -3$

33. $x + y + z = 2$
$2x + y - z = 5$
$x - y + z = -2$

34. $2x + y + z = 9$
$-x - y + z = 1$
$3x - y + z = 9$

35. $x + 3y + 4z = 14$
$2x - 3y + 2z = 10$
$3x - y + z = 9$

36. $4x - y + 3z = -2$
$3x + 5y - z = 15$
$-2x + y + 4z = 14$

37. $x + 2y + 3z = 8$
$3x - y + 2z = 5$
$-2x - 4y - 6z = 5$

38. $3x - 2y - 8z = 1$
$9x - 6y - 24z = -2$
$x - y + z = 1$

39. $\bigcirc + 3y + 2z = 6$
$x - y + \bigcirc = 0$
$4x + \bigcirc + z = 8$

40. $x + 3z = 0$
$2y - z = 9$
$4x + y = -7$

41. $5.3x - 4.7y + 5.9z = 1.14$
$-2.5x + 3.2y - 1.4z = 7.22$
$2.25x - 2.88y + 1.26z = 4.88$

42. $7.77x - 8.61y + 4.2z = 15.96$
$11.6x - 4.9y + 0.8z = 12.4$
$3.7x - 4.1y + 2z = 10.6$

Solve each system by elimination. (*Hint:* In Exercises 43–46 let $1/x = t$ and $1/y = u$.)

43. $\dfrac{2}{x} + \dfrac{1}{y} = \dfrac{3}{2}$
$\dfrac{3}{x} - \dfrac{1}{y} = 1$

44. $\dfrac{1}{x} + \dfrac{3}{y} = \dfrac{16}{5}$
$\dfrac{5}{x} + \dfrac{4}{y} = 5$

45. $\dfrac{2}{x} + \dfrac{1}{y} = 11$
$\dfrac{3}{x} - \dfrac{5}{y} = 10$

46. $\dfrac{2}{x} + \dfrac{3}{y} = 18$
$\dfrac{4}{x} - \dfrac{5}{y} = -8$

47. $\dfrac{4}{x + 2} - \dfrac{3}{y - 1} = 1$
$\dfrac{1}{x + 2} + \dfrac{2}{y - 1} = 1$

48. $\dfrac{1}{x - 3} + \dfrac{5}{y + 2} = -6$
$\dfrac{-3}{x - 3} + \dfrac{2}{y + 2} = 1$

49. $\dfrac{2}{x} + \dfrac{3}{y} - \dfrac{2}{z} = -1$
$\dfrac{8}{x} - \dfrac{12}{y} + \dfrac{5}{z} = 5$
$\dfrac{6}{x} + \dfrac{3}{y} - \dfrac{1}{z} = 1$

50. $-\dfrac{5}{x} + \dfrac{4}{y} + \dfrac{3}{z} = 2$
$\dfrac{10}{x} + \dfrac{3}{y} - \dfrac{6}{z} = 7$
$\dfrac{5}{x} + \dfrac{2}{y} - \dfrac{9}{z} = 6$

Solve each of the following systems in terms of the arbitrary variable x.

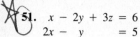

51. $\quad x - 2y + 3z = 6$
$\quad 2x - y \quad\quad = 5$

52. $3x + 4y - z = 8$
$\quad x \quad\quad + 2z = 4$

53. $\quad 5x - 4y + z = 0$
$\quad\quad x + y \quad\quad = 0$
$\quad -10x + 8y - 2z = 0$

54. $2x + y - 3z = 0$
$\quad 4x + 2y - 6z = 0$
$\quad x - y + z = 0$

55. $4x + y + z = 6$
$\quad\quad y \quad = 2x$

56. $3x - 5y - 4z = 6$
$\quad 3x \quad\quad = z$

Write a system of linear equations for each of the following, and then use the system to solve the problem.

57. At the Sharp Ranch, 6 goats and 5 sheep sell for $305 while 2 goats and 9 sheep cost $285. Find the cost of a goat and the cost of a sheep.

58. Linda Ramirez is a building contractor. If she hires 7 day laborers and 2 concrete finishers, her payroll for the day is $608, while 1 day laborer and 5 concrete finishers cost $464. Find the daily-wage charge of each type of worker.

59. During summer vacation Hector and Ann earned a total of $1088. Hector worked 8 days less than Ann and earned $2 per day less. Find the number of days he worked and the daily wage he made if the total number of days worked by both was 72.

60. The perimeter of a rectangle is 42 centimeters. The longer side has a length of 7 centimeters more than the shorter side. Find the length of the longer side.

61. Thirty liters of a 50% alcohol solution are to be made by mixing 70% solution and 20% solution. How many liters of each solution should be used?

62. A merchant wishes to make one hundred pounds of a coffee blend that can be sold for $4 per pound. This blend is to be made by mixing coffee worth $6 a pound with coffee worth $3 a pound. How many pounds of each will be needed?

63. Ms. Kelley inherits $25,000. She deposits part at 8% annual interest and part at 10%. Her total annual income from these investments is $2300. How much is invested at each rate?

64. Chuck Sullivan earned $100,000 in a lottery. He invested part of the money at 10% and part at 12%. His total annual income from the two investments is $11,000. How much did he have invested at each rate?

65. Mr. Caminiti has some money invested at 8% and three times as much invested at 12%. His total annual income from the two investments is $2200. How much is invested at each rate?

66. A cash drawer contains only fives and twenties. There are eight more fives than twenties. The total value of the money is $215. How many of each type of bill are there?

67. How many gallons of milk (4.5% butterfat) must be mixed with 250 gallons of skim milk (0% butterfat) to get lowfat milk (2% butterfat)?

68. By boat Tom can go 72 miles upstream to a fishing hole in 4 hours. Returning, he needs only 3 hours. What is the speed of the current in the stream?

69. Two cars start together and travel in opposite directions. At the end of four hours, the cars are 656 kilometers apart. If one car travels 20 kilometers per hour faster than the other, find the speed of each car.

70. Two trains leave towns 192 kilometers apart, traveling toward one another. One train travels 40 kilometers per hour faster than the other. They pass one another two hours later. What is the speed of each train?

71. The perimeter of a triangle is 33 centimeters. The longest side is 3 centimeters longer than the medium side. The medium side is twice the shortest side. Find the length of each side of the triangle.

72. Carrie O'Sullivan invests in three ways the $30,000 she won in a lottery. With part of the money, she buys a mutual fund, paying 9% per year. The second part, $2000 more than the first, is used to buy utility bonds paying 10% per year. The rest is invested in a tax-free 5% bond. The first year her investments bring a return of $2500. How much is invested at each rate?

73. Find three numbers whose sum is 50, if the first is 2 more than the second, and the third is two-thirds of the second.

74. The sum of three numbers is 15. The second number is the negative of one-sixth of the first number, while the third number is 3 more than the second. Find the three numbers.

Solve each system for x and y in terms of a and b. Assume no denominators are zero.

75. $\dfrac{x-1}{a} + \dfrac{y-2}{b} = 1$

$\dfrac{x-1}{a} - \dfrac{y-2}{b} = 1$

76. $a^2x + b^2y = 1$

$b^2x + a^2y = 1$

The position of a particle moving in a straight line is given by $s = at^2 + bt + c$, where t is time in seconds and a, b, and c are real numbers.

77. If $s(0) = 5$, $s(1) = 23$, and $s(2) = 37$, find $s(8)$.

78. If $s(0) = -10$, $s(1) = 6$, and $s(2) = 30$, find $s(10)$.

6.2 Matrix Solution of Linear Systems

Read

Since systems of linear equations occur in so many practical situations, methods of efficiently solving linear systems by computer have been developed. Computer solutions of linear systems depend on the idea of a **matrix**: a rectangular array of numbers enclosed in brackets or parentheses. For example

$$\begin{bmatrix} 2 & 3 & 7 \\ 5 & -1 & 10 \end{bmatrix}$$

is a matrix. Each number is called an **element** of the matrix. Matrices in general are discussed in more detail later in this chapter. In this section, a method is developed for solving linear systems using matrices. As an example, start with the system

$$x + 3y + 2z = 1$$
$$2x + y - z = 2$$
$$x + y + z = 2,$$

and write the coefficients of the variables and the constants as a matrix, called the **augmented matrix** of the system.

$$\begin{bmatrix} 1 & 3 & 2 & | & 1 \\ 2 & 1 & -1 & | & 2 \\ 1 & 1 & 1 & | & 2 \end{bmatrix}$$

The vertical line, which is optional, is used only to separate the coefficients from the constants. This matrix has 3 rows (horizontal) and 4 columns (vertical). To refer to a number in the matrix, use its row and column numbers. For example, the number 3 is in the first row, second column position.

The rows of this matrix can be treated just like the equations of a system of linear equations. Since the augmented matrix is nothing more than a short form of the system, any transformation of the matrix which results in an equivalent system of equations can be performed. Operations that produce such transformations are given below.

Matrix Row Transformations

For any augmented matrix of a system of linear equations, the following row transformations will result in the matrix of an equivalent system.

1. Any two rows may be interchanged.
2. The elements of any row may be multiplied by a nonzero real number.
3. Any row may be changed by adding to its elements a multiple of the elements of another row.

These transformations are just a restatement in matrix form of the transformations of systems discussed in the previous section. From now on, when referring to the third transformation, "a multiple of the elements of a row" will be abbreviated as "a multiple of a row."

Before using matrices to solve a linear system, the system must be arranged in the proper form, with variable terms on the left side of the equation and constant terms on the right. The variable terms must be in the same order in each of the equations. The following example illustrates the matrix method and compares it with the elimination method of the previous section.

Example 1   Solve the system $3x - 4y = 1$
                                    $5x + 2y = 19.$

The procedure for matrix solution is parallel to the elimination method used in the previous section, except for the last step. First, write the augmented matrix for the system. Here the system is already in the proper form.

Elimination method

$3x - 4y = 1$ **(1)**
$5x + 2y = 19$ **(2)**

Matrix method

$$\begin{bmatrix} 3 & -4 & | & 1 \\ 5 & 2 & | & 19 \end{bmatrix}$$

Divide both sides of equation (1) by 3 so that x has a coefficient of 1.

$$x - \frac{4}{3}y = \frac{1}{3} \qquad \textbf{(3)}$$

$$5x + 2y = 19$$

Eliminate x from equation (2) by adding -5 times equation (3) to equation (2).

$$x - \frac{4}{3}y = \frac{1}{3}$$

$$\frac{26}{3}y = \frac{52}{3} \qquad \textbf{(4)}$$

Multiply both sides of equation (4) by 3/26 to get $y = 2$.

$$x - \frac{4}{3}y = \frac{1}{3}$$

$$y = 2$$

Using row transformation (2) multiply each element of row 1 by 1/3.

$$\begin{bmatrix} 1 & -\dfrac{4}{3} & \Big| & \dfrac{1}{3} \\ 5 & 2 & \Big| & 19 \end{bmatrix}$$

Using row transformation (3), add -5 times the elements of row 1 to the elements of row 2.

$$\begin{bmatrix} 1 & -\dfrac{4}{3} & \Big| & \dfrac{1}{3} \\ 0 & \dfrac{26}{3} & \Big| & \dfrac{52}{3} \end{bmatrix}$$

Multiply the elements of row 2 by 3/26, using row transformation (2).

$$\begin{bmatrix} 1 & -\dfrac{4}{3} & \Big| & \dfrac{1}{3} \\ 0 & 1 & \Big| & 2 \end{bmatrix}$$

Write the corresponding equations:

$$x - \frac{4}{3}y = \frac{1}{3}$$

$$y = 2$$

Finish the solution in either method by substituting 2 for y in the first equation to get $x = 3$. The solution of the system is (3, 2) with solution set $\{(3, 2)\}$. ∎

Example 2 Use row transformations to solve the linear system

$$2x + 6y = 28$$

$$4x - 3y = -19.$$

Begin with the augmented matrix

$$\begin{bmatrix} 2 & 6 & \Big| & 28 \\ 4 & -3 & \Big| & -19 \end{bmatrix}.$$

It is best to work vertically, in columns, beginning in each column with the element which is to become 1. This is the same order used in the previous section to arrange a system of equations in triangular form. The augmented matrix has a 2 in the first row, first column position. To get 1 in this position, use the second transformation and multiply each entry in the first row by 1/2. This is indicated below with the notation $\frac{1}{2}R_1$ next to the new first row.

$$\begin{bmatrix} 1 & 3 & \Big| & 14 \\ 4 & -3 & \Big| & -19 \end{bmatrix} \qquad \frac{1}{2}R_1$$

To get 0 in the second row, first column, add -4 times the first row to the second row.

$$\begin{bmatrix} 1 & 3 & | & 14 \\ 0 & -15 & | & -75 \end{bmatrix} \qquad -4R_1 + R_2$$

To get 1 in the second row, second column, multiply each element of the second row by $-1/15$, which gives

$$\begin{bmatrix} 1 & 3 & | & 14 \\ 0 & 1 & | & 5 \end{bmatrix}. \qquad -\frac{1}{15}R_2$$

This matrix corresponds to the system

$$x + 3y = 14 \qquad\qquad\qquad (1)$$
$$y = 5. \qquad\qquad\qquad (2)$$

Substitute 5 for y in equation (1) to get $x = -1$. The solution set of the system is thus $\{(-1, 5)\}$. ∎

Example 3 Use matrix methods to solve the system

$$x - y + 5z = -6$$
$$3x + 3y - z = 10$$
$$x + 3y + 2z = 5.$$

Begin by writing the augmented matrix of the linear system.

$$\begin{bmatrix} 1 & -1 & 5 & | & -6 \\ 3 & 3 & -1 & | & 10 \\ 1 & 3 & 2 & | & 5 \end{bmatrix}$$

There is already a 1 in row 1, column 1. The next thing to do is get 0's in the rest of column 1. First, add to row 2 the results of multiplying row 1 by -3.

$$\begin{bmatrix} 1 & -1 & 5 & | & -6 \\ 0 & 6 & -16 & | & 28 \\ 1 & 3 & 2 & | & 5 \end{bmatrix} \qquad -3R_1 + R_2$$

Now add to row 3 the results of multiplying row 1 by -1.

$$\begin{bmatrix} 1 & -1 & 5 & | & -6 \\ 0 & 6 & -16 & | & 28 \\ 0 & 4 & -3 & | & 11 \end{bmatrix} \qquad -1R_1 + R_3$$

To get 1 in row 2, column 2, multiply row 2 by 1/6.

$$\begin{bmatrix} 1 & -1 & 5 & | & -6 \\ 0 & 1 & -\frac{8}{3} & | & \frac{14}{3} \\ 0 & 4 & -3 & | & 11 \end{bmatrix} \qquad \frac{1}{6}R_2$$

Next, to get 0 in row 3, column 2, add to row 3 the results of multiplying row 2 by -4.

$$\begin{bmatrix} 1 & -1 & 5 & \bigm| & -6 \\ 0 & 1 & -\dfrac{8}{3} & \bigm| & \dfrac{14}{3} \\ 0 & 0 & \dfrac{23}{3} & \bigm| & -\dfrac{23}{3} \end{bmatrix} \quad -4R_2 + R_3$$

Finally, multiply the last row by $\dfrac{3}{23}$ to get 1 in row 3, column 3.

$$\begin{bmatrix} 1 & -1 & 5 & \bigm| & -6 \\ 0 & 1 & -\dfrac{8}{3} & \bigm| & \dfrac{14}{3} \\ 0 & 0 & 1 & \bigm| & -1 \end{bmatrix} \quad \dfrac{3}{23}R_3$$

The **main diagonal** of this matrix is the diagonal (to the left of the vertical bar) where all elements are 1. Since each element below the main diagonal is 0, the matrix is written in **triangular form.**

The final matrix above produces the system of equations

$$x - y + 5z = -6 \qquad\qquad (1)$$

$$y - \frac{8}{3}z = \frac{14}{3} \qquad\qquad (2)$$

$$z = -1. \qquad\qquad (3)$$

Use back-substitution into equations (1) and (2) to find the solution set of the system, $\{(1, 2, -1)\}$. ■

Example 4   Solve the system $x +\ \ y = 2$
$2x + 2y = 5.$

Start with the augmented matrix.

$$\begin{bmatrix} 1 & 1 & \bigm| & 2 \\ 2 & 2 & \bigm| & 5 \end{bmatrix}$$

Next, add to row 2 the results of multiplying row 1 by -2.

$$\begin{bmatrix} 1 & 1 & \bigm| & 2 \\ 0 & 0 & \bigm| & 1 \end{bmatrix} \quad -2R_1 + R_2$$

This matrix gives the system of equations

$$x + y = 2$$

$$0 = 1,$$

a system with no solution. Whenever a row of the augmented matrix is of the form $0\ 0\ 0\ .\ .\ .\ .\,|a$, (where $a \neq 0$), there will be no solution since this row corresponds to the equation $0 = a$. The solution set of the given system is $\emptyset$. ■

Example 5 Solve the system $x +\ \ y + 3z = 5$
$\qquad\qquad\qquad\qquad\quad 3x - 4y +\ \ z = 6.$

The augmented matrix is

$$\begin{bmatrix} 1 & 1 & 3 & 5 \\ 3 & -4 & 1 & 6 \end{bmatrix}.$$

Get 0 in row 2, column 1.

$$\begin{bmatrix} 1 & 1 & 3 & 5 \\ 0 & -7 & -8 & -9 \end{bmatrix} \qquad -3R_1 + R_2$$

Now get 1 in row 2, column 2.

$$\begin{bmatrix} 1 & 1 & 3 & 5 \\ 0 & 1 & \dfrac{8}{7} & \dfrac{9}{7} \end{bmatrix} \qquad -\dfrac{1}{7}R_2$$

It is not possible to go further since there are just two rows, so write the system of equations

$$x + y + 3z = 5 \tag{1}$$

$$y + \frac{8}{7}z = \frac{9}{7}. \tag{2}$$

From equation (2), solve for y to get

$$y = -\frac{8}{7}z + \frac{9}{7}.$$

Substitute into equation (1) to get

$$x + \left(-\frac{8}{7}z + \frac{9}{7} \right) + 3z = 5$$

$$x = -\frac{13}{7}z + \frac{26}{7}.$$

The arbitrary variable is z; the solution set can be written

$$\left\{ \left(\frac{-13z + 26}{7}, \frac{-8z + 9}{7}, z \right) \right\},$$

where z is any real number. ■

The cases that might occur when matrix methods are used to solve a system of linear equations are summarized on the next page.

When the matrix is written in triangular form:

1. If the number of rows having nonzero elements to the left of the vertical line is equal to the number of unknowns in the system, then the system has a single solution.
2. If the number of rows having nonzero elements to the left of the vertical line is less than the number of rows in the augmented matrix with nonzero elements, then the system has no solution.

   For example, the system that produced the matrix

$$\begin{bmatrix} 1 & -2 & 4 & | & 2 \\ 0 & 1 & 3 & | & -1 \\ 0 & 0 & 0 & | & 4 \end{bmatrix}$$

   has no solution. Here 2 rows have nonzero elements to the left of the vertical line but 3 rows in the entire matrix have nonzero elements. See Example 4.
3. If the number of rows of the matrix containing nonzero elements is less than the number of unknowns, then there is an infinite number of solutions for the system. This infinite number of solutions should be given in terms of an arbitrary variable. See Example 5.

The method of using matrices to solve a system of linear equations that was developed in this section is called the **Gaussian reduction method** after the mathematician K. F. Gauss (1777–1855).

6.2 Exercises

Write the augmented matrix for each of the following systems. Do not solve.

1. $2x + 3y = 11$
$\quad\ x + 2y = 8$

2. $3x + 5y = -13$
$\quad\ 2x + 3y = -9$

3. $x + 5y = 6$
$\qquad\quad y = 1$

4. $2x + 7y = 1$
$\quad\ 5x \qquad = -15$

5. $2x + y + z = 3$
$\quad\ 3x - 4y + 2z = -7$
$\quad\ x + y + z = 2$

6. $4x - 2y + 3z = 4$
$\quad\ 3x + 5y + z = 7$
$\quad\ 5x - y + 4z = 7$

7. $x + y \qquad = 2$
$\qquad 2y + z = -4$
$\qquad\qquad z = 2$

8. $x \qquad\qquad = 6$
$\qquad y + 2z = 2$
$\quad x \qquad - 3z = 6$

Write the system of equations associated with each of the following augmented matrices. Do not solve.

9. $\begin{bmatrix} 2 & 1 & | & 1 \\ 3 & -2 & | & -9 \end{bmatrix}$

10. $\begin{bmatrix} 1 & -5 & | & -18 \\ 6 & 2 & | & 20 \end{bmatrix}$

11. $\begin{bmatrix} 1 & 0 & 0 & | & 2 \\ 0 & 1 & 0 & | & 3 \\ 0 & 0 & 1 & | & -2 \end{bmatrix}$

12. $\begin{bmatrix} 1 & 0 & 1 & | & 4 \\ 0 & 1 & 0 & | & 2 \\ 0 & 0 & 1 & | & 3 \end{bmatrix}$

13. $\begin{bmatrix} 3 & 2 & 1 & | & 1 \\ 0 & 2 & 4 & | & 22 \\ -1 & -2 & 3 & | & 15 \end{bmatrix}$

14. $\begin{bmatrix} 2 & 1 & 3 & | & 12 \\ 4 & -3 & 0 & | & 10 \\ 5 & 0 & -4 & | & -11 \end{bmatrix}$

Use the Gaussian reduction method to solve each of the following systems of equations. For systems with dependent equations in 3 variables, give the solution with z arbitrary; for any such equations with 4 variables, let w be the arbitrary variable.

15. $x + y = 5$
$x - y = -1$

16. $x + 2y = 5$
$2x + y = -2$

17. $x + y = -3$
$2x - 5y = -6$

18. $3x - 2y = 4$
$3x + y = -2$

19. $2x - 3y = 10$
$2x + 2y = 5$

20. $6x + y = 5$
$5x + y = 3$

21. $2x - 5y = 10$
$3x + y = 15$

22. $4x - y = 3$
$-2x + 3y = 1$

23. $x + y = -1$
$y + z = 4$
$x + z = 1$

24. $x - z = -3$
$y + z = 9$
$x + z = 7$

25. $x + y - z = 6$
$2x - y + z = -9$
$x - 2y + 3z = 1$

26. $x + 3y - 6z = 7$
$2x - y + z = 1$
$x + 2y + 2z = -1$

27. $-x + y = -1$
$y - z = 6$
$x + z = -1$

28. $x + y = 1$
$2x - z = 0$
$y + 2z = -2$

29. $2x - y + 3z = 0$
$x + 2y - z = 5$
$2y + z = 1$

30. $4x + 2y - 3z = 6$
$x - 4y + z = -4$
$-x + 2z = 2$

31. $x - 2y + z = 5$
$2x + y - z = 2$
$-2x + 4y - 2z = 2$

32. $3x + 5y - z = 0$
$4x - y + 2z = 1$
$-6x - 10y + 2z = 0$

33. $x + y + z = 6$
$2x - y - z = 3$

34. $5x - 3y + z = 1$
$2x + y - z = 4$

35. $x - 8y + z = 4$
$3x - y + 2z = -1$

36. $-3x + y - z = 8$
$2x + y + 4z = 0$

37. $x + 3y - z = 0$
$2y - z = 4$

38. $2x - y + 4z = 1$
$y + z = 3$

39. $3x + 2y - w = 0$
$2x + z + 2w = 5$
$x + 2y - z = -2$
$2x - y + z + w = 2$

40. $x + 3y - 2z - w = 9$
$4x + y + z + 2w = 2$
$-3x - y + z - w = -5$
$x - y - 3z - 2w = 2$

41. $x - y + 2z + w = 4$
$y + z = 3$
$z - w = 2$

42. $3x + y = 1$
$y - 4z + w = 0$
$z - 3w = 1$

43. $3.5x + 2.9y = 12.91$
$1.7x - 3.8y = -6.23$

44. $-4.3x + 1.1y = 10.1$
$4.8x - 2.7y = -17.31$

45. $9.1x + 2.3y - z = -3.06$
$4.7x + y - 3.8z = -21.15$
$5.1y - 4.7z = -31.77$

46. $7.2x + 4.8y + 3.1z = -14.96$
$3x - y + 11.5z = -15.6$
$1.1x + 2.4y = -1.24$

Solve each of the following word problems.

47. A working couple earned a total of $4352. The wife earned $64 per day; the husband earned $8 per day less. Find the number of days each worked if the total number of days worked by both was 72.

48. Midtown Manufacturing Company makes two products, plastic plates and plastic cups. Both require time on two machines; plates: 1 hour on machine A and 2 hours on machine B, cups: 3 hours on machine A and 1 hour on machine B. Both machines operate 15 hours a day. How many of each product can be produced in a day under these conditions?

49. A company produces two models of bicycles, model 201 and model 301. Model 201 requires 2 hours of assembly time, and model 301 requires 3 hours of assembly time. The parts for model 201 cost $25 per bike; those for model 301 cost $30 per bike. If the company has a total of 34 hours of assembly time and $365 available per day for these two models, how many of each can be made in a day?

50. Juanita Wilson invests $10,000 in three ways. With one part, she buys mutual funds which offer a return of 8% per year. The second part, which amounts to twice the first, is used to buy government bonds at 9% per year. She puts the rest in the bank at 5% annual interest. The first year her investments bring a return of $830. How much did she invest each way?

51. To get the necessary funds for a planned expansion, a small company took out three loans totaling $25,000. The company was able to borrow some of the money at 8%. They borrowed $2000 more than one-half the amount of the 8% loan at 10%, and the rest at 9%. The total annual interest was $2220. How much did they borrow at each rate?

At rush hours, substantial traffic congestion is encountered at the traffic intersections shown in the figure. (All streets are one-way.)

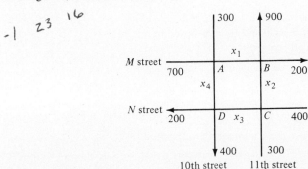

The city wishes to improve the signals at these corners so as to speed the flow of traffic. The traffic engineers first gather data. As the figure shows, 700 cars per hour come down M Street to intersection A; 300 cars per hour come to intersection A on 10th Street. A total of x_1 of these cars leave A on M Street, while x_4 cars leave A on 10th Street. The number of cars entering A must equal the number leaving, so that

$$x_1 + x_4 = 700 + 300$$

or

$$x_1 + x_4 = 1000.$$

For intersection B, x_1 cars enter B on M Street, and x_2 cars enter B on 11th Street. The figure shows that 900 cars leave B on 11th while 200 leave on M. We have

$$x_1 + x_2 = 900 + 200$$

$$x_1 + x_2 = 1100.$$

At intersection C, 400 cars enter on N Street, 300 on 11th Street, while x_2 leave on 11th Street and x_3 leave on N Street. This gives

$$x_2 + x_3 = 400 + 300$$

$$x_2 + x_3 = 700.$$

Finally, intersection D has x_3 cars entering at N and x_4 entering on 10th. There are 400 cars leaving D on 10th and 200 leaving on N.

52. Set up an equation for intersection D.

53. Use the four equations to set up an augmented matrix, and then use the Gaussian method to solve it.

54. Since you got a row of all zeros, the system of equations does not have a unique solution. Write three equations, corresponding to the three nonzero rows of the matrix. Solve each of the equations for x_4.

55. One of your equations should have been $x_4 = 1000 - x_1$. What is the largest possible value of x_1 so that x_4 is not negative?

56. Your second equation should have been $x_4 = x_2 - 100$. Find the smallest possible value of x_2 so that x_4 is not negative.

57. Find the largest possible values of x_3 and x_4 so that neither variable is negative.

58. Use the results of Exercises 52–57 to give a solution for the problem in which all the equations are satisfied and all variables are nonnegative. Is the solution unique?

Solve each of the following problems.

59. Three small boxes and 8 large boxes cost $13.30. If 6 small and 16 large boxes are purchased, the cost is $27.20. Find the cost of a small box and the cost of a large box.

60. The sum of three numbers is 20. One number is one more than another. Find the three numbers.

61. If three numbers are added, the result is -10. The second number, added to twice the third, gives 5. Find the three numbers.

62. If half of one number is added to another, the sum is twice a third number. The third number is half the second number. Find the three numbers.

For each of the following equations, determine the constants A, B, and C, that make the equation an identity.

63. $\dfrac{x}{(x+1)(x-1)} = \dfrac{A}{x+1} + \dfrac{B}{x-1}$

64. $\dfrac{2y+1}{y(y+1)} = \dfrac{A}{y} + \dfrac{B}{y+1}$

65. $\dfrac{2m^2 - 3m + 3}{m(m-1)^2} = \dfrac{A}{m} + \dfrac{B}{m-1} + \dfrac{C}{(m-1)^2}$

66. $\dfrac{2k^2 + 13k + 18}{(k+3)^3} = \dfrac{A}{k+3} + \dfrac{B}{(k+3)^2} + \dfrac{C}{(k+3)^3}$

67. $\dfrac{3x^2 + 3x + 4}{x(x^2 + 2)} = \dfrac{A}{x} + \dfrac{Bx + C}{x^2 + 2}$

68. $\dfrac{2x^2 + 7x + 3}{(x-1)(x^2 + x + 1)} = \dfrac{A}{x-1} + \dfrac{Bx + C}{x^2 + x + 1}$

6.3 Properties of Matrices

The use of matrix notation in solving a system of linear equations was shown in the previous section. In this section, the algebraic properties of matrices are discussed.

It is customary to use capital letters to name matrices. Also, subscript notation is often used to name the elements of a matrix, as in the following matrix A.

$$A = \begin{bmatrix} a_{11} & a_{12} & a_{13} & \cdots & a_{1n} \\ a_{21} & a_{22} & a_{23} & \cdots & a_{2n} \\ a_{31} & a_{32} & a_{33} & \cdots & a_{3n} \\ \cdot & \cdot & \cdot & & \cdot \\ \cdot & \cdot & \cdot & & \cdot \\ \cdot & \cdot & \cdot & & \cdot \\ a_{m1} & a_{m2} & a_{m3} & & a_{mn} \end{bmatrix}$$

With this notation, the first row, first column element is a_{11}; the second row, third column element is a_{23}; and in general, the ith row, jth column element is a_{ij}. As a shorthand notation, matrix A above is sometimes written as just $A = (a_{ij})$.

Matrices are classified by their dimension or order, that is, by the number of rows and columns that they contain. For example, the matrix

$$\begin{bmatrix} 2 & 7 & -5 \\ 3 & -6 & 0 \end{bmatrix}$$

has two rows (horizontal) and three columns (vertical), and has *dimension* 2×3 or *order* 2×3; a matrix with m rows and n columns has **dimension $m \times n$** or **order $m \times n$**. The number of rows is always given first.

Certain matrices have special names: an $n \times n$ matrix is a **square matrix** of order n. Also, a matrix with just one row is a **row matrix,** and a matrix with just one column is a **column matrix.**

Two matrices are **equal** if they are of the same order and if each corresponding element, position by position, is equal. Using this definition, the matrices

$$\begin{bmatrix} 2 & 1 \\ 3 & -5 \end{bmatrix} \quad \text{and} \quad \begin{bmatrix} 1 & 2 \\ -5 & 3 \end{bmatrix}$$

are *not* equal (even though they contain the same elements and are of the same order), since the corresponding elements differ.

Example 1   **(a)** From the definition of equality given above, the only way that the statement

$$\begin{bmatrix} 2 & 1 \\ p & q \end{bmatrix} = \begin{bmatrix} x & y \\ -1 & 0 \end{bmatrix}$$

can be true is if $2 = x$, $1 = y$, $p = -1$, and $q = 0$.

(b) The statement

$$\begin{bmatrix} x \\ y \end{bmatrix} = \begin{bmatrix} 1 \\ 4 \\ 0 \end{bmatrix}$$

can never be true, since the two matrices are of different order. (One is 2×1 and the other is 3×1.)   ∎

Addition of matrices is defined as follows.

Addition of Matrices

To **add** two matrices of the same order, add corresponding elements. Only matrices of the same order can be added.

It can be shown that matrix addition (for matrices of the same order) satisfies the commutative, associative, closure, identity, and inverse properties. (See Exercises 41–45.)

Example 2 Find each of the following sums.

(a) $\begin{bmatrix} 5 & -6 \\ 8 & 9 \end{bmatrix} + \begin{bmatrix} -4 & 6 \\ 8 & -3 \end{bmatrix} = \begin{bmatrix} 5 + (-4) & -6 + 6 \\ 8 + 8 & 9 + (-3) \end{bmatrix} = \begin{bmatrix} 1 & 0 \\ 16 & 6 \end{bmatrix}$

(b) $\begin{bmatrix} 2 \\ 5 \\ 8 \end{bmatrix} + \begin{bmatrix} -6 \\ 3 \\ 12 \end{bmatrix} = \begin{bmatrix} -4 \\ 8 \\ 20 \end{bmatrix}$

(c) The matrices $A = \begin{bmatrix} 5 & 8 \\ 6 & 2 \end{bmatrix}$ and $B = \begin{bmatrix} 3 & 9 & 1 \\ 4 & 2 & 5 \end{bmatrix}$

are of different orders, so the sum $A + B$ does not exist. ■

A matrix containing only zero elements is called a **zero matrix.** For example, $[0 \quad 0 \quad 0]$ is the 1×3 zero matrix, while

$$\begin{bmatrix} 0 & 0 & 0 \\ 0 & 0 & 0 \end{bmatrix}$$

is the 2×3 zero matrix.

By the additive inverse property in Chapter 1, each real number has an additive inverse: if a is a real number, there is a real number $-a$ such that

$$a + (-a) = 0 \quad \text{and} \quad -a + a = 0.$$

What about matrices? Given the matrix

$$A = \begin{bmatrix} -5 & 2 & -1 \\ 3 & 4 & -6 \end{bmatrix},$$

is there a matrix $-A$ such that

$$A + (-A) = 0 \quad \text{and} \quad -A + A = 0,$$

where 0 is the 2×3 zero matrix? The answer is yes: the matrix $-A$ has as elements the additive inverses of the elements of A. (Remember, each element of A is a real number and, therefore, has an additive inverse.)

$$-A = \begin{bmatrix} 5 & -2 & 1 \\ -3 & -4 & 6 \end{bmatrix}$$

To check, first test that $A + (-A)$ equals the zero matrix, 0.

$$A + (-A) = \begin{bmatrix} -5 & 2 & -1 \\ 3 & 4 & -6 \end{bmatrix} + \begin{bmatrix} 5 & -2 & 1 \\ -3 & -4 & 6 \end{bmatrix} = \begin{bmatrix} 0 & 0 & 0 \\ 0 & 0 & 0 \end{bmatrix}$$

Then test that $-A + A$ is also 0. Matrix $-A$ is called the **additive inverse,** or **negative,** of matrix A. Every matrix, no matter the order, has an additive inverse.

The real number b is subtracted from the real number a, written $a - b$, by adding a and the additive inverse of b. That is,

$$a - b = a + (-b).$$

The same definition works for **subtraction** of matrices.

Subtraction of Matrices

> If A and B are two matrices of the same order, then
>
> $$A - B = A + (-B).$$

Example 3   Find each of the following differences.

(a) $\begin{bmatrix} -5 & 6 \\ 2 & 4 \end{bmatrix} - \begin{bmatrix} -3 & 2 \\ 5 & -8 \end{bmatrix} = \begin{bmatrix} -5 & 6 \\ 2 & 4 \end{bmatrix} + \begin{bmatrix} 3 & -2 \\ -5 & 8 \end{bmatrix}$

(Here we found the additive inverse of each element in the second matrix.)

$$= \begin{bmatrix} -2 & 4 \\ -3 & 12 \end{bmatrix}$$

(b) $[8 \quad 6 \quad -4] - [3 \quad 5 \quad -8] = [5 \quad 1 \quad 4]$

(c) The matrices

$$\begin{bmatrix} -2 & 5 \\ 0 & 1 \end{bmatrix} \quad \text{and} \quad \begin{bmatrix} 3 \\ 5 \end{bmatrix}$$

are of different orders and cannot be subtracted.   ■

In work with matrices, a real number is called a **scalar** to distinguish it from a matrix. The product of a scalar k and a matrix X is the matrix kX, each of whose elements is k times the corresponding element of X.

Example 4
(a) $5 \begin{bmatrix} 2 & -3 \\ 0 & 4 \end{bmatrix} = \begin{bmatrix} 10 & -15 \\ 0 & 20 \end{bmatrix}$

(b) $\dfrac{3}{4} \begin{bmatrix} 20 & 36 \\ 12 & -16 \end{bmatrix} = \begin{bmatrix} 15 & 27 \\ 9 & -12 \end{bmatrix}$ ■

The proofs of the following properties of scalar multiplication are left for Exercises 49–52 below.

Properties of Scalar Multiplication

> If A and B are matrices of the same order and c and d are real numbers,
>
> $$(c + d)A = cA + dA$$
> $$c(A + B) = cA + cB$$
> $$(cd)A = c(dA).$$

We have seen how to multiply a real number (scalar) and a matrix. Now we define the product of two matrices. The procedure developed below for finding the product of two matrices may seem artificial, but it is useful in applications. The method will be illustrated before a formal rule is given. To find the product of

$$A = \begin{bmatrix} -3 & 4 & 2 \\ 5 & 0 & 4 \end{bmatrix} \quad \text{and} \quad B = \begin{bmatrix} -6 & 4 \\ 2 & 3 \\ 3 & -2 \end{bmatrix},$$

first, locate *row* 1 of *A* and *column* 1 of *B,* shown shaded below.

$$A = \begin{bmatrix} -3 & 4 & 2 \\ 5 & 0 & 4 \end{bmatrix} \quad B = \begin{bmatrix} -6 & 4 \\ 2 & 3 \\ 3 & -2 \end{bmatrix}$$

Multiply corresponding elements, and find the sum of the products.

$$(-3)(-6) + (4)(2) + (2)(3) = 32$$

This result is the element for row 1, column 1 of the product matrix.

Now use *row* 1 of *A* and *column* 2 of *B* (shown shaded below) to determine the element in row 1 and column 2 of the product matrix.

$$\begin{bmatrix} -3 & 4 & 2 \\ 5 & 0 & 4 \end{bmatrix} \quad \begin{bmatrix} -6 & 4 \\ 2 & 3 \\ 3 & -2 \end{bmatrix}$$

Multiply corresponding elements and add the products:

$$(-3)(4) + (4)(3) + (2)(-2) = -4,$$

which is the row 1, column 2 element of the product matrix.

Next, use *row* 2 of *A* and *column* 1 of *B;* this will give the row 2, column 1 entry of the product matrix.

$$(5)(-6) + (0)(2) + (4)(3) = -18$$

Finally, use *row* 2 of *A* and *column* 2 of *B* to find the entry for row 2, column 2 of the product matrix.

$$(5)(4) + (0)(3) + (4)(-2) = 12$$

The product matrix can now be written.

$$\begin{bmatrix} -3 & 4 & 2 \\ 5 & 0 & 4 \end{bmatrix} \begin{bmatrix} -6 & 4 \\ 2 & 3 \\ 3 & -2 \end{bmatrix} = \begin{bmatrix} +32 & -4 \\ -18 & +12 \end{bmatrix}$$

As in this example, the product of a 2 × 3 matrix and a 3 × 2 matrix is a 2 × 2 matrix.

By definition, the **product** *AB* of an *m* × *n* matrix *A* and an *n* × *p* matrix *B* is found as follows. Multiply each element of the first row of *A* by the corresponding element of the first column of *B*. The sum of these *n* products is the first row, first column element of *AB*.

Also, the sum of the products found by multiplying the elements of the first row of *A* times the corresponding elements of the second column of *B* gives the first row, second column element of *AB*, and so on.

To find the *i*th row, *j*th column element of *AB*, multiply each element in the *i*th row of *A* by the corresponding element in the *j*th column of *B* (note colored areas in matrices on the following page). The sum of these products will give the element of row *i*, column *j* of *AB*.

$$A = \begin{bmatrix} a_{11} & a_{12} & a_{13} & \cdots & a_{1n} \\ a_{21} & a_{22} & a_{23} & \cdots & a_{2n} \\ & & \vdots & & \\ a_{i1} & a_{i2} & a_{i3} & \cdots & a_{in} \\ & & \vdots & & \\ a_{m1} & a_{m2} & a_{m3} & \cdots & a_{mn} \end{bmatrix} \qquad B = \begin{bmatrix} b_{11} & b_{12} & \cdots & b_{1j} & \cdots & b_{1p} \\ b_{21} & b_{22} & \cdots & b_{2j} & \cdots & b_{2p} \\ \vdots & & & & & \\ b_{n1} & b_{n2} & \cdots & b_{nj} & \cdots & b_{np} \end{bmatrix}$$

Matrix Multiplication

If the number of columns of matrix A is the same as the number of rows of matrix B, then entry c_{ij} of the product matrix $C = AB$ is found as follows:

$$c_{ij} = a_{i1}b_{1j} + a_{i2}b_{2j} + \cdots + a_{in}b_{nj}.$$

The final product will have as many rows as A and as many columns as B.

Example 5 Suppose matrix A is 3×2, while matrix B is 2×4. Can the product AB be calculated? What is the order of the product? Can the product BA be calculated? What is the order of BA?

The following diagram helps answer the questions about the product AB.

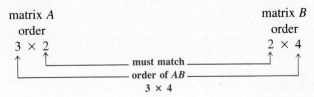

The product AB exists, since the number of columns of A equals the number of rows of B (both are 2). The product has order 3×4. Make a similar diagram for BA.

$$\begin{array}{ccc} \text{matrix } B & & \text{matrix } A \\ \text{order} & & \text{order} \\ 2 \times 4 & & 3 \times 2 \end{array}$$
$$\underbrace{\qquad\qquad\qquad \text{different} \qquad\qquad\qquad}$$

The product BA does not exist. ∎

As Example 5 shows, matrix multiplication is not commutative.

Example 6 Find AB and BA, if possible, where

$$A = \begin{bmatrix} 1 & -3 \\ 7 & 2 \end{bmatrix} \quad \text{and} \quad B = \begin{bmatrix} 1 & 0 & -1 & 2 \\ 3 & 1 & 4 & -1 \end{bmatrix}.$$

Use the definition of matrix multiplication.

$$AB = \begin{bmatrix} 1 & -3 \\ 7 & 2 \end{bmatrix} \begin{bmatrix} 1 & 0 & -1 & 2 \\ 3 & 1 & 4 & -1 \end{bmatrix}$$

$$= \begin{bmatrix} 1(1) + (-3)(3) & 1(0) + (-3)1 & 1(-1) + (-3)4 & 1(2) + (-3)(-1) \\ 7(1) + 2(3) & 7(0) + 2(1) & 7(-1) + 2(4) & 7(2) + 2(-1) \end{bmatrix}$$

$$= \begin{bmatrix} -8 & -3 & -13 & 5 \\ 13 & 2 & 1 & 12 \end{bmatrix}$$

Since B is a 2×4 matrix, and A is a 2×2 matrix, the product BA cannot be found. ∎

Example 7 If $A = \begin{bmatrix} 1 & 3 \\ -2 & 5 \end{bmatrix}$ and $B = \begin{bmatrix} -2 & 7 \\ 0 & 2 \end{bmatrix}$, then the definition of matrix multiplication

can be used to show that

$$AB = \begin{bmatrix} -2 & 13 \\ 4 & -4 \end{bmatrix} \quad \text{and} \quad BA = \begin{bmatrix} -16 & 29 \\ -4 & 10 \end{bmatrix}. \quad ∎$$

Example 7 shows that even when both AB and BA can be found, they may not be equal. In general, for matrices A and B, $AB \neq BA$, so that, as mentioned above, matrix multiplication is not commutative.

Matrix multiplication does, however, satisfy the associative and distributive properties.

Properties of Matrix Multiplication

If A, B, and C are matrices such that all the following products and sums exist, then

$$(AB)C = A(BC)$$
$$A(B + C) = AB + AC$$
$$(B + C)A = BA + CA.$$

For proofs of these results for the special cases when A, B, and C are square matrices, see Exercises 46 and 47 below. The identity and inverse properties for matrix multiplication are discussed in the next section of this chapter.

6.3 Exercises

Find the values of the variables in each of the following.

1. $\begin{bmatrix} 2 & 1 \\ 4 & 8 \end{bmatrix} = \begin{bmatrix} x & 1 \\ y & z \end{bmatrix}$

2. $\begin{bmatrix} 2 & 5 & 6 \\ 1 & m & n \end{bmatrix} = \begin{bmatrix} z & y & w \\ 1 & 8 & -2 \end{bmatrix}$

3. $\begin{bmatrix} x + 6 & y + 2 \\ 8 & 3 \end{bmatrix} = \begin{bmatrix} -9 & 7 \\ 8 & k \end{bmatrix}$

4. $\begin{bmatrix} 9 & 7 \\ r & 0 \end{bmatrix} = \begin{bmatrix} m - 3 & n + 5 \\ 8 & 0 \end{bmatrix}$

5. $\begin{bmatrix} -7 + z & 4r & 8s \\ 6p & 2 & 5 \end{bmatrix} + \begin{bmatrix} -9 & 8r & 3 \\ 2 & 5 & 4 \end{bmatrix} = \begin{bmatrix} 2 & 36 & 27 \\ 20 & 7 & 12a \end{bmatrix}$

6. $\begin{bmatrix} a + 2 & 3z + 1 & 5m \\ 4k & 0 & 3 \end{bmatrix} + \begin{bmatrix} 3a & 2z & 5m \\ 2k & 5 & 6 \end{bmatrix} = \begin{bmatrix} 10 & -14 & 80 \\ 10 & 5 & 9 \end{bmatrix}$

Perform each of the following operations, whenever possible.

7. $\begin{bmatrix} 6 & -9 & 2 \\ 4 & 1 & 3 \end{bmatrix} - \begin{bmatrix} -8 & 2 & 5 \\ 6 & -3 & 4 \end{bmatrix}$

8. $\begin{bmatrix} 9 & 4 \\ -8 & 2 \end{bmatrix} + \begin{bmatrix} -3 & 2 \\ -4 & 7 \end{bmatrix}$

9. $\begin{bmatrix} -6 & 8 \\ 0 & 0 \end{bmatrix} - \begin{bmatrix} 0 & 0 \\ -4 & -2 \end{bmatrix}$

10. $\begin{bmatrix} 1 & -4 \\ 2 & -3 \\ -8 & 4 \end{bmatrix} - \begin{bmatrix} -6 & 9 \\ -2 & 5 \\ -7 & -12 \end{bmatrix}$

11. $\begin{bmatrix} -8 & 4 & 0 \\ 2 & 5 & 0 \end{bmatrix} + \begin{bmatrix} 6 & 3 \\ 8 & 9 \end{bmatrix}$

12. $\begin{bmatrix} 2 \\ 3 \end{bmatrix} - \begin{bmatrix} 8 & 1 \\ 9 & 4 \end{bmatrix}$

13. $\begin{bmatrix} -4x + 2y & -3x + y \\ 6x - 3y & 2x - 5y \end{bmatrix} + \begin{bmatrix} -8x + 6y & 2x \\ 3y - 5x & 6x + 4y \end{bmatrix}$

14. $\begin{bmatrix} 4k - 8y \\ 6z - 3x \\ 2k + 5a \\ -4m + 2n \end{bmatrix} - \begin{bmatrix} 5k + 6y \\ 2z + 5x \\ 4k + 6a \\ 4m - 2n \end{bmatrix}$

15. When John inventoried his screw collection, he found that he had 7 flat-head long screws, 9 flat-head medium, 8 flat-head short, 2 round-head long, no round-head medium, and 6 round-head short. Write this information first as a 3 × 2 matrix and then as a 2 × 3 matrix.

16. At the grocery store, Miguel bought 4 quarts of milk, 2 loaves of bread, 4 chickens, and an apple. Mary bought 2 quarts of milk, a loaf of bread, 5 chickens, and 4 apples. Write this information first as a 2 × 4 matrix and then as a 4 × 2 matrix.

Let $A = \begin{bmatrix} -2 & 4 \\ 0 & 3 \end{bmatrix}$ and $B = \begin{bmatrix} -6 & 2 \\ 4 & 0 \end{bmatrix}$. Find each of the following.

17. $2A$

18. $-3B$

19. $2A - B$

20. $-2A + 4B$

21. $-A + \frac{1}{2}B$

22. $\frac{3}{4}A - B$

Find each of the following matrix products, whenever possible.

23. $\begin{bmatrix} 1 & 2 \\ 3 & 4 \end{bmatrix}\begin{bmatrix} -1 \\ 7 \end{bmatrix}$

24. $\begin{bmatrix} 3 & -4 & 1 \\ 5 & 0 & 2 \end{bmatrix}\begin{bmatrix} -1 \\ 4 \\ 2 \end{bmatrix}$

25. $\begin{bmatrix} -1 & 2 & 0 \\ 0 & 3 & 2 \\ 0 & 1 & 4 \end{bmatrix}\begin{bmatrix} 2 & -1 & 2 \\ 0 & 2 & 1 \\ 3 & 0 & -1 \end{bmatrix}$

26. $\begin{bmatrix} -2 & -3 & -4 \\ 2 & -1 & 0 \\ 4 & -2 & 3 \end{bmatrix}\begin{bmatrix} 0 & 1 & 4 \\ 1 & 2 & -1 \\ 3 & 2 & -2 \end{bmatrix}$

27. $\begin{bmatrix} -2 & 1 & 4 \\ 0 & 1 & 2 \end{bmatrix}\begin{bmatrix} -2 & 1 & 0 \\ 0 & -2 & 0 \\ 4 & 1 & 2 \end{bmatrix}$

28. $\begin{bmatrix} -1 & 0 & 0 \\ 2 & 1 & 4 \end{bmatrix}\begin{bmatrix} 4 & -2 & 5 \\ 0 & 1 & 4 \\ 2 & -9 & 0 \end{bmatrix}$

29. $\begin{bmatrix} -3 & 0 & 2 & 1 \\ 4 & 0 & 2 & 6 \end{bmatrix}\begin{bmatrix} -4 & 2 \\ 0 & 1 \end{bmatrix}$

30. $\begin{bmatrix} -1 & 2 & 4 & 1 \\ 0 & 2 & -3 & 5 \end{bmatrix}\begin{bmatrix} 1 & 2 & 4 \\ -2 & 5 & 1 \end{bmatrix}$

31. $[-2 \ \ 4 \ \ 6] \begin{bmatrix} 3 \\ -2 \\ 1 \end{bmatrix}$ 32. $[4 \ \ 0 \ \ 2] \begin{bmatrix} -5 \\ 1 \\ 6 \end{bmatrix}$ 33. $\begin{bmatrix} 3 \\ -2 \\ 1 \end{bmatrix} [-2 \ \ 4 \ \ 6]$ 34. $\begin{bmatrix} -5 \\ 1 \\ 6 \end{bmatrix} [4 \ \ 0 \ \ 2]$

35. The Bread Box, a small neighborhood bakery, sells four main items: sweet rolls, bread, cake, and pie. The amount of certain major ingredients (measured in cups except for eggs) required to make these items is given in matrix A.

$$A = \begin{array}{c} \\ \\ \\ \\ \\ \\ \end{array} \begin{array}{ccccc} \text{eggs} & \text{flour} & \text{sugar} & \overset{\text{short-}}{\text{ening}} & \text{milk} \\ \end{array}$$

$$A = \begin{bmatrix} 1 & 4 & \frac{1}{4} & \frac{1}{4} & 1 \\ 0 & 3 & 0 & \frac{1}{4} & 0 \\ 4 & 3 & 2 & 1 & 1 \\ 0 & 1 & 0 & \frac{1}{3} & 0 \end{bmatrix} \begin{array}{l} \text{rolls} \\ \text{(dozen)} \\ \text{bread} \\ \text{(loaves)} \\ \text{cake (1)} \\ \text{pie (1)} \end{array}$$

The cost (in cents) for each ingredient when purchased in large lots and in small lots is given by matrix B.

$$B = \begin{array}{c} \text{cost} \\ \begin{array}{cc} \text{large lot} & \text{small lot} \end{array} \\ \begin{bmatrix} 5 & 5 \\ 8 & 10 \\ 10 & 12 \\ 12 & 15 \\ 5 & 6 \end{bmatrix} \begin{array}{l} \text{eggs} \\ \text{flour} \\ \text{sugar} \\ \text{shortening} \\ \text{milk} \end{array} \end{array}$$

(a) Use matrix multiplication to find a matrix representing the comparative costs per item under the two purchase options.

Suppose a day's orders consist of 20 dozen sweet rolls, 200 loaves of bread, 50 cakes, and 60 pies.

(b) Represent these orders as a 1×4 matrix and use matrix multiplication to write as a matrix the amount of each ingredient required to fill the day's orders.

(c) Use matrix multiplication to find a matrix representing the costs under the two purchase options to fill the day's orders.

Find each of the following products.

36. $\begin{bmatrix} .8 & .4 \\ -.7 & -.22 \end{bmatrix} \begin{bmatrix} -.72 & -.8 \\ .4 & -.2 \end{bmatrix}$ 37. $\begin{bmatrix} -.6 & .93 \\ .8 & .47 \end{bmatrix} \begin{bmatrix} .9 & .4 \\ .6 & -.8 \end{bmatrix}$ 38. $\begin{bmatrix} -.42 & .6 \\ .9 & .3 \end{bmatrix} \begin{bmatrix} .1 & .8 \\ -.4 & .11 \end{bmatrix}$

39. $\begin{bmatrix} .8 & .72 & .44 \\ -.41 & .83 & .29 \\ -.77 & .61 & -.42 \end{bmatrix} \begin{bmatrix} -3 & .8 & -.4 \\ -7 & -.1 & -.2 \\ -1 & .3 & .9 \end{bmatrix}$ 40. $\begin{bmatrix} .71 & -.22 & .88 \\ .62 & -.79 & .33 \\ .43 & -.11 & .19 \end{bmatrix} \begin{bmatrix} .08 & -.73 & -.07 \\ .04 & -.64 & .90 \\ .91 & 1.33 & -.22 \end{bmatrix}$

For Exercises 51–54, let

$$A = \begin{bmatrix} a_{11} & a_{12} \\ a_{21} & a_{22} \end{bmatrix}, \quad B = \begin{bmatrix} b_{11} & b_{12} \\ b_{21} & b_{22} \end{bmatrix}, \quad \text{and} \quad C = \begin{bmatrix} c_{11} & c_{12} \\ c_{21} & c_{22} \end{bmatrix},$$

where all the elements are real numbers. Decide which of the following statements are true for these three matrices. If a statement is true, prove that it is true. If it is false, give a numerical example to show it false.

41. $A + B = B + A$ (commutative property)   42. $A + (B + C) = (A + B) + C$ (associative property)

43. $A + B$ is a 2×2 matrix (closure property)

44. There exists a matrix 0 such that $A + 0 = A$ and $0 + A = A$. (identity property)

45. There exists a matrix $-A$ such that $A + (-A) = 0$ and $-A + A = 0$. (inverse property)

46. $(AB)C = A(BC)$ (associative property)

47. $A(B + C) = AB + AC$ (distributive property)

48. AB is a 2×2 matrix. (closure property)

49. $c(A + B) = cA + cB$ for any real number c

50. $(c + d)A = cA + dA$ for any real numbers c and d

51. $c(A)d = cd(A)$

52. $(cd)A = c(dA)$

53. $(A + B)(A - B) = A^2 - B^2$ (where $A^2 = AA$)

54. $(A + B)^2 = A^2 + 2AB + B^2$

The *transpose*, A^T, of a matrix A is found by exchanging the rows and columns of A. That is, if

$$A = \begin{bmatrix} a & b \\ c & d \end{bmatrix}, \quad \text{then} \quad A^T = \begin{bmatrix} a & c \\ b & d \end{bmatrix}.$$

Show that each of the following equations are true for matrices A and B, where

$$B = \begin{bmatrix} m & n \\ p & q \end{bmatrix}.$$

55. $(A^T)^T = A$

56. $(A + B)^T = A^T + B^T$

57. $(AB)^T = B^T A^T$

6.4 Matrix Inverses

As shown in the exercises for the previous section, the commutative, associative, closure, identity, and inverse properties hold for *addition* of matrices of the same order. Also, the associative and closure properties hold for *multiplication* of square matrices of the same order. There is no commutative property for multiplication, but the distributive property is valid for matrices of the proper order.

In this section two additional properties are discussed, the identity and inverse properties for multiplication. For the identity property to hold, there must be a matrix I such that

$$AI = A \quad \text{and} \quad IA = A$$

for any square matrix A. (Compare these products to the statement of the identity property for real numbers: $a \cdot 1 = a$ and $1 \cdot a = a$ for any real number a.) It can be shown that for 2×2 matrices,

2 × 2 Identity

if I_2 represents the 2×2 identity, then

$$I_2 = \begin{bmatrix} 1 & 0 \\ 0 & 1 \end{bmatrix}.$$

To verify that I_2 is the 2×2 identity matrix show that $AI = A$ and $IA = A$ for any 2×2 matrix. Let

$$A = \begin{bmatrix} x & y \\ z & w \end{bmatrix}.$$

Then
$$AI = \begin{bmatrix} x & y \\ z & w \end{bmatrix}\begin{bmatrix} 1 & 0 \\ 0 & 1 \end{bmatrix} = \begin{bmatrix} x \cdot 1 + y \cdot 0 & x \cdot 0 + y \cdot 1 \\ z \cdot 1 + w \cdot 0 & z \cdot 0 + w \cdot 1 \end{bmatrix}$$

$$= \begin{bmatrix} x & y \\ z & w \end{bmatrix} = A;$$

and
$$IA = \begin{bmatrix} 1 & 0 \\ 0 & 1 \end{bmatrix}\begin{bmatrix} x & y \\ z & w \end{bmatrix} = \begin{bmatrix} 1 \cdot x + 0 \cdot z & 1 \cdot y + 0 \cdot w \\ 0 \cdot x + 1 \cdot z & 0 \cdot y + 1 \cdot w \end{bmatrix}$$

$$= \begin{bmatrix} x & y \\ z & w \end{bmatrix} = A.$$

Generalizing from this example, there is an $n \times n$ identity matrix having 1's on the main diagonal and 0's elsewhere.

$n \times n$ Identity Matrix

The $n \times n$ identity matrix is given by I_n where

$$I_n = \begin{bmatrix} 1 & 0 & \cdots & 0 \\ 0 & 1 & \cdots & 0 \\ & & & \\ & & a_{ij} & \\ & & & \\ 0 & 0 & \cdots & 1 \end{bmatrix}.$$

The element $a_{ij} = 1$ when $i = j$ (the diagonal elements) and $a_{ij} = 0$ otherwise.

For every nonzero real number a, there is a multiplication inverse $1/a$ such that

$$a \cdot \frac{1}{a} = 1 \quad \text{and} \quad \frac{1}{a} \cdot a = 1.$$

(Recall: $1/a$ is also written a^{-1}.) In a similar way, if A is an $n \times n$ matrix, then its **multiplication inverse,** written A^{-1}, must satisfy both

$$AA^{-1} = I_n \quad \text{and} \quad A^{-1}A = I_n.$$

This means that only a square matrix can have a multiplication inverse. The matrix A^{-1} can be found by using the row operations introduced in Section 2 of this chapter.

As an example, let us find the inverse of

$$A = \begin{bmatrix} 2 & 4 \\ 1 & -1 \end{bmatrix}.$$

Let the unknown inverse matrix be

$$A^{-1} = \begin{bmatrix} x & y \\ z & w \end{bmatrix}.$$

By the definition of matrix inverse, $AA^{-1} = I_2$, or

$$AA^{-1} = \begin{bmatrix} 2 & 4 \\ 1 & -1 \end{bmatrix}\begin{bmatrix} x & y \\ z & w \end{bmatrix} = \begin{bmatrix} 1 & 0 \\ 0 & 1 \end{bmatrix}.$$

By matrix multiplication,

$$\begin{bmatrix} 2x + 4z & 2y + 4w \\ x - z & y - w \end{bmatrix} = \begin{bmatrix} 1 & 0 \\ 0 & 1 \end{bmatrix}.$$

Setting corresponding elements equal gives the system of equations

$$2x + 4z = 1 \tag{1}$$
$$2y + 4w = 0 \tag{2}$$
$$x - z = 0 \tag{3}$$
$$y - w = 1. \tag{4}$$

Since equations (1) and (3) involve only x and z, while equations (2) and (4) involve only y and w, these four equations lead to two systems of equations,

$$\begin{array}{cc} \begin{array}{c} 2x + 4z = 1 \\ x - z = 0 \end{array} & \text{and} & \begin{array}{c} 2y + 4w = 0 \\ y - w = 1. \end{array} \end{array}$$

Writing the two systems as augmented matrices gives

$$\begin{bmatrix} 2 & 4 & | & 1 \\ 1 & -1 & | & 0 \end{bmatrix} \quad \text{and} \quad \begin{bmatrix} 2 & 4 & | & 0 \\ 1 & -1 & | & 1 \end{bmatrix}.$$

Each of these systems can be solved by the Gaussian method. However, since the elements to the left of the vertical bar are identical, the two systems can be combined into one matrix

$$\begin{bmatrix} 2 & 4 & | & 1 & 0 \\ 1 & -1 & | & 0 & 1 \end{bmatrix}$$

and solved simultaneously using matrix row transformations. Start by exchanging the two rows to get a 1 in the upper left corner.

$$\begin{bmatrix} 1 & -1 & | & 0 & 1 \\ 2 & 4 & | & 1 & 0 \end{bmatrix}$$

Multiply row one by -2 and add the results to row two to get

$$\begin{bmatrix} 1 & -1 & | & 0 & 1 \\ 0 & 6 & | & 1 & -2 \end{bmatrix} \qquad -2R_1 + R_2$$

Now, to get a 1 in the second row, second column position, multiply row two by 1/6.

$$\begin{bmatrix} 1 & -1 & | & 0 & 1 \\ 0 & 1 & | & \frac{1}{6} & -\frac{1}{3} \end{bmatrix} \qquad \frac{1}{6}R_2$$

Finally, add row two to row one to get a 0 in the first row, second column.

$$\begin{bmatrix} 1 & 0 & | & \frac{1}{6} & \frac{2}{3} \\ 0 & 1 & | & \frac{1}{6} & -\frac{1}{3} \end{bmatrix} \qquad R_2 + R_1$$

The numbers in the first column to the right of the vertical bar give the values of x and z. The second column gives the values of y and w. That is,

$$\begin{bmatrix} 1 & 0 & | & x & y \\ 0 & 1 & | & z & w \end{bmatrix} = \begin{bmatrix} 1 & 0 & | & \frac{1}{6} & \frac{2}{3} \\ 0 & 1 & | & \frac{1}{6} & -\frac{1}{3} \end{bmatrix}$$

so that

$$A^{-1} = \begin{bmatrix} x & y \\ z & w \end{bmatrix} = \begin{bmatrix} \frac{1}{6} & \frac{2}{3} \\ \frac{1}{6} & -\frac{1}{3} \end{bmatrix}.$$

To check, multiply A by A^{-1}. The result should be I_2.

$$AA^{-1} = \begin{bmatrix} 2 & 4 \\ 1 & -1 \end{bmatrix}\begin{bmatrix} \frac{1}{6} & \frac{2}{3} \\ \frac{1}{6} & -\frac{1}{3} \end{bmatrix} = \begin{bmatrix} \frac{1}{3}+\frac{2}{3} & \frac{4}{3}-\frac{4}{3} \\ \frac{1}{6}-\frac{1}{6} & \frac{2}{3}+\frac{1}{3} \end{bmatrix} = \begin{bmatrix} 1 & 0 \\ 0 & 1 \end{bmatrix} = I_2.$$

Verify that $A^{-1}A = I_2$, also. Finally,

$$A^{-1} = \begin{bmatrix} \frac{1}{6} & \frac{2}{3} \\ \frac{1}{6} & -\frac{1}{3} \end{bmatrix}.$$

In summary, to use this method of finding inverses, first form the augmented matrix $[A|I_n]$, where A is any $n \times n$ matrix and I_n is the $n \times n$ multiplication identity matrix. Find A^{-1} by performing row transformations on $[A|I_n]$ until a matrix of the form $[I_n|B]$ is found. Matrix B is the desired matrix A^{-1}. If it is not possible to get a matrix of the form $[I_n|B]$, then A^{-1} does not exist.

Example 1 Find A^{-1} if $A = \begin{bmatrix} 1 & 0 & 1 \\ 2 & -2 & -1 \\ 3 & 0 & 0 \end{bmatrix}$.

Use row transformations, going through as many steps as needed.

Step 1 Write the augmented matrix $[A|I_3]$:

$$\begin{bmatrix} 1 & 0 & 1 & | & 1 & 0 & 0 \\ 2 & -2 & -1 & | & 0 & 1 & 0 \\ 3 & 0 & 0 & | & 0 & 0 & 1 \end{bmatrix}.$$

Step 2 Since 1 is already in the upper left-hand corner as necessary, begin by selecting the row operation which will result in a 0 for the first element in the second row. Add to each element in the second row the result of multiplying the first row by -2.

$$\left[\begin{array}{ccc|ccc} 1 & 0 & 1 & 1 & 0 & 0 \\ 0 & -2 & -3 & -2 & 1 & 0 \\ 3 & 0 & 0 & 0 & 0 & 1 \end{array}\right] \qquad -2R_1 + R_2$$

Step 3   To get 0 for the first element in the third row, add to the third row the results of multiplying each element of the first row by -3.

$$\left[\begin{array}{ccc|ccc} 1 & 0 & 1 & 1 & 0 & 0 \\ 0 & -2 & -3 & -2 & 1 & 0 \\ 0 & 0 & -3 & -3 & 0 & 1 \end{array}\right] \qquad -3R_1 + R_3$$

Step 4   To get 1 for the second element in the second row, multiply the second row by $-1/2$.

$$\left[\begin{array}{ccc|ccc} 1 & 0 & 1 & 1 & 0 & 0 \\ 0 & 1 & \frac{3}{2} & 1 & -\frac{1}{2} & 0 \\ 0 & 0 & -3 & -3 & 0 & 1 \end{array}\right] \qquad -\frac{1}{2}R_2$$

Step 5   To get 1 for the third element in the third row, multiply the third row by $-1/3$.

$$\left[\begin{array}{ccc|ccc} 1 & 0 & 1 & 1 & 0 & 0 \\ 0 & 1 & \frac{3}{2} & 1 & -\frac{1}{2} & 0 \\ 0 & 0 & 1 & 1 & 0 & -\frac{1}{3} \end{array}\right] \qquad -\frac{1}{3}R_3$$

Step 6   To get 0 for the third element in the first row, add to the first row the results of multiplying each element in row three by -1.

$$\left[\begin{array}{ccc|ccc} 1 & 0 & 0 & 0 & 0 & \frac{1}{3} \\ 0 & 1 & \frac{3}{2} & 1 & -\frac{1}{2} & 0 \\ 0 & 0 & 1 & 1 & 0 & -\frac{1}{3} \end{array}\right] \qquad -1R_3 + R_1$$

Step 7   To get 0 for the third element in the second row, add to the second row the results of multiplying each element of row three by $-3/2$.

$$\left[\begin{array}{ccc|ccc} 1 & 0 & 0 & 0 & 0 & \frac{1}{3} \\ 0 & 1 & 0 & -\frac{1}{2} & -\frac{1}{2} & \frac{1}{2} \\ 0 & 0 & 1 & 1 & 0 & -\frac{1}{3} \end{array}\right] \qquad -\frac{3}{2}R_3 + R_2$$

From the last transformation, we get the desired inverse.

$$A^{-1} = \left[\begin{array}{ccc} 0 & 0 & \frac{1}{3} \\ -\frac{1}{2} & -\frac{1}{2} & \frac{1}{2} \\ 1 & 0 & -\frac{1}{3} \end{array}\right]$$

Confirm this by forming the products $A^{-1}A$, and AA^{-1}, each of which should be equal to I_3.   ∎

Example 2 Find A^{-1} given $A = \begin{bmatrix} 2 & -4 \\ 1 & -2 \end{bmatrix}$.

Using row operations to transform the first column of the augmented matrix

$$\left[\begin{array}{rr|rr} 2 & -4 & 1 & 0 \\ 1 & -2 & 0 & 1 \end{array}\right],$$

results in the following matrices:

$$\left[\begin{array}{rr|rr} 1 & -2 & \frac{1}{2} & 0 \\ 1 & -2 & 0 & 1 \end{array}\right], \qquad \frac{1}{2}R_1$$

$$\left[\begin{array}{rr|rr} 1 & -2 & \frac{1}{2} & 0 \\ 0 & 0 & -\frac{1}{2} & 1 \end{array}\right]. \qquad -1\ R_1 + R_2$$

At this point, the matrix should be changed so that the second row, second column element will be 1. Since that element is now 0, there is no way to complete the desired transformation, so matrix A^{-1} does not exist. ∎

If the inverse of a matrix exists, it is unique. That is, any given square matrix has no more than one inverse. The proof of this is left to Exercise 48 of this section.

Solving Systems by Inverses Matrix inverses can be used to solve square linear systems of equations. (A square system has the same number of equations as variables.) For example, given the linear system

$$a_{11}x + a_{12}y + a_{13}z = d_1$$
$$a_{21}x + a_{22}y + a_{23}z = d_2$$
$$a_{31}x + a_{32}y + a_{33}z = d_3,$$

the definition of matrix multiplication can be used to rewrite the system as

$$\begin{bmatrix} a_{11} & a_{12} & a_{13} \\ a_{21} & a_{22} & a_{23} \\ a_{31} & a_{32} & a_{33} \end{bmatrix} \cdot \begin{bmatrix} x \\ y \\ z \end{bmatrix} = \begin{bmatrix} d_1 \\ d_2 \\ d_3 \end{bmatrix}. \qquad (1)$$

(To see this, multiply the matrices on the left.)

$$\text{If } A = \begin{bmatrix} a_{11} & a_{12} & a_{13} \\ a_{21} & a_{22} & a_{23} \\ a_{31} & a_{32} & a_{33} \end{bmatrix}, \qquad X = \begin{bmatrix} x \\ y \\ z \end{bmatrix}, \qquad \text{and} \qquad B = \begin{bmatrix} d_1 \\ d_2 \\ d_3 \end{bmatrix},$$

the system given in (1) becomes

$$AX = B.$$

If A^{-1} exists, then both sides of $AX = B$ can be multiplied on the left to get

$$A^{-1}(AX) = A^{-1}B$$
$$(A^{-1}A)X = A^{-1}B$$
$$I_3X = A^{-1}B$$
$$X = A^{-1}B.$$

Matrix $A^{-1}B$ gives the solution of the system.

Example 3   Use the method of matrix inverses to solve the system

$$2x - 3y = 4$$
$$x + 5y = 2.$$

To represent the system as a matrix equation, use one matrix for the coefficients, one for the variables, and one for the constants, as follows.

$$A = \begin{bmatrix} 2 & -3 \\ 1 & 5 \end{bmatrix}, \qquad X = \begin{bmatrix} x \\ y \end{bmatrix}, \qquad \text{and} \qquad B = \begin{bmatrix} 4 \\ 2 \end{bmatrix}.$$

The system can then be written in matrix form as the equation $AX = B$, since

$$AX = \begin{bmatrix} 2 & -3 \\ 1 & 5 \end{bmatrix} \begin{bmatrix} x \\ y \end{bmatrix} = \begin{bmatrix} 2x - 3y \\ x + 5y \end{bmatrix} = \begin{bmatrix} 4 \\ 2 \end{bmatrix} = B.$$

To solve the system, first find A^{-1}. Verify that

$$A^{-1} = \begin{bmatrix} 5/13 & 3/13 \\ -1/13 & 2/13 \end{bmatrix}.$$

Next, find the product $A^{-1}B$.

$$A^{-1}B = \begin{bmatrix} 5/13 & 3/13 \\ -1/13 & 2/13 \end{bmatrix} \begin{bmatrix} 4 \\ 2 \end{bmatrix} = \begin{bmatrix} 2 \\ 0 \end{bmatrix}.$$

Since $X = A^{-1}B$,

$$X = \begin{bmatrix} x \\ y \end{bmatrix} = \begin{bmatrix} 2 \\ 0 \end{bmatrix}.$$

The final matrix shows that the solution set of the system is $\{(2, 0)\}$.   ■

This method of using matrix inverses to solve systems of equations is useful when the inverse is already known or when many systems of the form $AX = B$ must be solved and only B changes.

6.4 Exercises

Decide whether or not the given matrices are inverses of each other. (Check to see if their product is the identity matrix I_n.)

1. $\begin{bmatrix} 2 & 3 \\ 1 & 1 \end{bmatrix}$ and $\begin{bmatrix} -1 & 3 \\ 1 & -2 \end{bmatrix}$

2. $\begin{bmatrix} 5 & 7 \\ 2 & 3 \end{bmatrix}$ and $\begin{bmatrix} 3 & -7 \\ -2 & 5 \end{bmatrix}$

3. $\begin{bmatrix} 2 & 1 \\ 3 & 2 \end{bmatrix}$ and $\begin{bmatrix} 2 & 1 \\ -3 & 2 \end{bmatrix}$

4. $\begin{bmatrix} -1 & 2 \\ 3 & -5 \end{bmatrix}$ and $\begin{bmatrix} -5 & -2 \\ -3 & -1 \end{bmatrix}$

5. $\begin{bmatrix} 1 & 2 & 0 \\ 0 & 1 & 0 \\ 0 & 1 & 0 \end{bmatrix}$ and $\begin{bmatrix} 1 & -2 & 0 \\ 0 & 1 & 0 \\ 0 & -1 & 1 \end{bmatrix}$

6. $\begin{bmatrix} 0 & 1 & 0 \\ 0 & 0 & -2 \\ 1 & -1 & 0 \end{bmatrix}$ and $\begin{bmatrix} 1 & 0 & 1 \\ 1 & 0 & 0 \\ 0 & -1 & 0 \end{bmatrix}$

7. $\begin{bmatrix} 1 & 3 & 3 \\ 1 & 4 & 3 \\ 1 & 3 & 4 \end{bmatrix}$ and $\begin{bmatrix} 7 & -3 & -3 \\ -1 & 1 & 0 \\ -1 & 0 & 1 \end{bmatrix}$

8. $\begin{bmatrix} -1 & 0 & 2 \\ 3 & 1 & 0 \\ 0 & 2 & -3 \end{bmatrix}$ and $\begin{bmatrix} -\frac{1}{5} & \frac{4}{15} & -\frac{2}{15} \\ \frac{3}{5} & \frac{1}{5} & \frac{2}{5} \\ \frac{2}{5} & \frac{2}{15} & -\frac{1}{15} \end{bmatrix}$

Find the inverse, if it exists, for each matrix.

9. $\begin{bmatrix} 1 & -1 \\ 2 & 0 \end{bmatrix}$ 　　　**10.** $\begin{bmatrix} -1 & 2 \\ -2 & -1 \end{bmatrix}$ 　　　**11.** $\begin{bmatrix} 3 & -1 \\ -5 & 2 \end{bmatrix}$ 　　　**12.** $\begin{bmatrix} -1 & -2 \\ 3 & 4 \end{bmatrix}$

13. $\begin{bmatrix} -6 & 4 \\ -3 & 2 \end{bmatrix}$ 　　　**14.** $\begin{bmatrix} 5 & 10 \\ -3 & -6 \end{bmatrix}$ 　　　**15.** $\begin{bmatrix} 1 & 0 & 0 \\ 0 & -1 & 0 \\ 1 & 0 & 1 \end{bmatrix}$ 　　　**16.** $\begin{bmatrix} 1 & 0 & 1 \\ 0 & -1 & 0 \\ 2 & 1 & 1 \end{bmatrix}$

17. $\begin{bmatrix} -1 & -1 & -1 \\ 4 & 5 & 0 \\ 0 & 1 & -3 \end{bmatrix}$ 　　　**18.** $\begin{bmatrix} 2 & 0 & 4 \\ 3 & 1 & 5 \\ -1 & 1 & -2 \end{bmatrix}$ 　　　**19.** $\begin{bmatrix} 1 & 2 & 3 \\ -3 & -2 & -1 \\ -1 & 0 & 1 \end{bmatrix}$

20. $\begin{bmatrix} 2 & 0 & 4 \\ 1 & 0 & -1 \\ 3 & 0 & -2 \end{bmatrix}$ 　　　**21.** $\begin{bmatrix} 2 & 4 & 6 \\ -1 & -4 & -3 \\ 0 & 1 & -1 \end{bmatrix}$ 　　　**22.** $\begin{bmatrix} 2 & 2 & -4 \\ 2 & 6 & 0 \\ -3 & -3 & 5 \end{bmatrix}$

23. $\begin{bmatrix} 1 & -2 & 3 & 0 \\ 0 & 1 & -1 & 1 \\ -2 & 2 & -2 & 4 \\ 0 & 2 & -3 & 1 \end{bmatrix}$ 　　　**24.** $\begin{bmatrix} 1 & 1 & 0 & 2 \\ 2 & -1 & 1 & -1 \\ 3 & 3 & 2 & -2 \\ 1 & 2 & 1 & 0 \end{bmatrix}$

Solve each system of equations by using the inverse of the coefficient matrix.

25. $2x + 3y = 10$
$x - y = -5$

26. $-x + 2y = 15$
$-2x - y = 20$

27. $2x + y = 5$
$5x + 3y = 13$

28. $-x - 2y = 8$
$3x + 4y = 24$

29. $-x + y = 1$
$2x - y = 1$

30. $3x - 6y = 1$
$-5x + 9y = -1$

31. $-x - 8y = 12$
$3x + 24y = -36$

32. $x + 3y = -14$
$2x - y = 7$

Solve each system of equations by using the inverse of the coefficient matrix. The inverses for the first four problems are found in Exercises 17, 18, 21, and 22 above.

33. $-x - y - z = 1$
$4x + 5y = -2$
$y - 3z = 3$

34. $2x + 4z = -8$
$3x + y + 5z = 2$
$-x + y - 2z = 4$

35. $2x + 4y + 6z = 4$
$-x - 4y - 3z = 8$
$y - z = -4$

36. $2x + 2y - 4z = 12$
$2x + 6y = 16$
$-3x - 3y + 5z = -20$

37. $x + 2y + 3z = 5$
$2x + 3y + 2z = 2$
$-x - 2y - 4z = -1$

38. $x + y - 3z = 4$
$2x + 4y - 4z = 8$
$-x + y + 4z = -3$

39. $2x - 2y = 5$
$ 4y + 8z = 7$
$x + 2z = 1$

40. $x + z = 3$
$ y + 2z = 8$
$-x + y = 4$

Solve each system of equations by using the inverse of the coefficient matrix. The inverses were found in Exercises 23 and 24.

41. $x - 2y + 3z = 4$
$ y - z + w = -8$
$-2x + 2y - 2z + 4w = 12$
$ 2y - 3z + w = -4$

42. $x + y + 2w = 3$
$2x - y + z - w = 3$
$3x + 3y + 2z - 2w = 5$
$x + 2y + z = 3$

Let $A = \begin{bmatrix} a & b \\ c & d \end{bmatrix}$ and let O be the 2×2 matrix of all zeros. Show that statements 43–46 are true.

43. $A \cdot O = O$ **44.** $AA^{-1} = I_2$ **45.** $A^{-1}A = I_2$.

46. For square matrices A and B of the same order, if $AB = O$ and if A^{-1} exists, then $B = O$.

47. The Bread Box Bakery sells three types of cakes, each requiring the amounts of the basic ingredients shown in the following matrix.

$$
\begin{array}{r}
\text{Flour (cups)} \\
\text{Sugar (cups)} \\
\text{Eggs}
\end{array}
\overset{\begin{array}{ccc} \text{I} & \text{II} & \text{III} \end{array}}{\begin{bmatrix} 2 & 4 & 2 \\ 2 & 1 & 2 \\ 2 & 1 & 3 \end{bmatrix}}
$$

Types of cake

To fill its daily orders for these three kinds of cake, the bakery uses 72 cups of flour, 48 cups of sugar, and 60 eggs.
(a) Write a 3×1 matrix for the amounts used daily.
(b) Let the number of daily orders for cakes be a 3×1 matrix X with entries x_1, x_2, and x_3. Write a matrix equation which you can solve for X, using the given matrix and the matrix from part (a).
(c) Solve the equation you wrote in part (b) to find the number of daily orders for each type of cake.

48. Prove that, if it exists, the inverse of a matrix is unique. *Hint:* Assume there are two inverses B and C for some matrix A, so that $AB = BA = I$ and $AC = CA = I$. Multiply the first equation by C and the second by B.

49. Give an example of two matrices A and B where $(AB)^{-1} \neq A^{-1}B^{-1}$.

50. Suppose A and B are matrices where A^{-1}, B^{-1}, and AB all exist. Show that $(AB)^{-1} = B^{-1}A^{-1}$.

51. Let $A = \begin{bmatrix} a & 0 \\ 0 & b \end{bmatrix}$. Under what conditions on a and b does A^{-1} exist?

52. Derive a formula for the inverse of $\begin{bmatrix} a & 0 \\ 0 & d \end{bmatrix}$, where $ad \neq 0$.

53. Let $A = \begin{bmatrix} a & 0 & 0 \\ 0 & b & 0 \\ 0 & 0 & c \end{bmatrix}$, where a, b and c are nonzero real numbers. Find A^{-1}.

6.5 Determinants

Given any square matrix A, there is a unique real number associated with A—the **determinant** of A. Determinants are defined in this section, and in the next section determinants are used to solve a system of linear equations.

The determinant of a matrix A is written $|A|$. The determinant of a 2×2 matrix is defined as on the following page.

**Determinant of a
2 × 2 Matrix**

$$\text{If } A = \begin{bmatrix} a_{11} & a_{12} \\ a_{21} & a_{22} \end{bmatrix}, \text{ then } |A| = \begin{vmatrix} a_{11} & a_{12} \\ a_{21} & a_{22} \end{vmatrix} = a_{11}a_{22} - a_{21}a_{12}.$$

Example 1   Let $A = \begin{bmatrix} -3 & 4 \\ 6 & 8 \end{bmatrix}$. Find $|A|$.

Use the definition above:

$$|A| = \begin{vmatrix} -3 & 4 \\ 6 & 8 \end{vmatrix} = -3(8) - 6(4) = -48. \quad \blacksquare$$

The determinant of a 3 × 3 matrix A is defined as follows.

**Determinant of a
3 × 3 Matrix**

If $A = \begin{bmatrix} a_{11} & a_{12} & a_{13} \\ a_{21} & a_{22} & a_{23} \\ a_{31} & a_{32} & a_{33} \end{bmatrix}$,

then

$$|A| = \begin{vmatrix} a_{11} & a_{12} & a_{13} \\ a_{21} & a_{22} & a_{23} \\ a_{31} & a_{32} & a_{33} \end{vmatrix} = (a_{11}a_{22}a_{33} + a_{12}a_{23}a_{31} + a_{13}a_{21}a_{32}) \\ - (a_{31}a_{22}a_{13} + a_{32}a_{23}a_{11} + a_{33}a_{21}a_{12}).$$

The terms on the right side of the equation in the definition of $|A|$ can be rearranged to get

$$\begin{vmatrix} a_{11} & a_{12} & a_{13} \\ a_{21} & a_{22} & a_{23} \\ a_{31} & a_{32} & a_{33} \end{vmatrix} = a_{11}(a_{22}a_{33} - a_{32}a_{23}) - a_{21}(a_{12}a_{33} - a_{32}a_{13}) \\ + a_{31}(a_{12}a_{23} - a_{22}a_{13}).$$

Each of the quantities in parentheses above represents a determinant of a 2 × 2 matrix which is the part of the 3 × 3 matrix left when the row and column of the multiplier are eliminated as shown below.

$$a_{11}(a_{22}a_{33} - a_{32}a_{23}) \qquad \begin{bmatrix} a_{11} & a_{12} & a_{13} \\ a_{21} & a_{22} & a_{23} \\ a_{31} & a_{32} & a_{33} \end{bmatrix}$$

$$a_{21}(a_{12}a_{33} - a_{32}a_{13}) \qquad \begin{bmatrix} a_{11} & a_{12} & a_{13} \\ a_{21} & a_{22} & a_{23} \\ a_{31} & a_{32} & a_{33} \end{bmatrix}$$

$$a_{31}(a_{12}a_{23} - a_{22}a_{13}) \qquad \begin{bmatrix} a_{11} & a_{12} & a_{13} \\ a_{21} & a_{22} & a_{23} \\ a_{31} & a_{32} & a_{33} \end{bmatrix}$$

These determinants of 2 × 2 matrices are called **minors** of an element in the 3 × 3 matrix. The symbol M_{ij} represents the determinant of the matrix that results when row i and column j are eliminated. The following list gives some of the minors from the matrix above.

Element	Minor	Element	Minor
a_{11}	$M_{11} = \begin{vmatrix} a_{22} & a_{23} \\ a_{32} & a_{33} \end{vmatrix}$	a_{22}	$M_{22} = \begin{vmatrix} a_{11} & a_{13} \\ a_{31} & a_{33} \end{vmatrix}$
a_{21}	$M_{21} = \begin{vmatrix} a_{12} & a_{13} \\ a_{32} & a_{33} \end{vmatrix}$	a_{23}	$M_{23} = \begin{vmatrix} a_{11} & a_{12} \\ a_{31} & a_{32} \end{vmatrix}$
a_{31}	$M_{31} = \begin{vmatrix} a_{12} & a_{13} \\ a_{22} & a_{23} \end{vmatrix}$	a_{33}	$M_{33} = \begin{vmatrix} a_{11} & a_{12} \\ a_{21} & a_{22} \end{vmatrix}$

In a 4×4 matrix, the minors are determinants of 3×3 matrices, and an $n \times n$ matrix has minors that are determinants of $(n - 1) \times (n - 1)$ matrices.

To find the determinant of a 3×3 or larger matrix, first choose any row or column. Then the minor of each element in that row or column must be multiplied by $+1$ or -1, depending on whether the sum of the row numbers and column numbers is even or odd. The product of a minor and the number $+1$ or -1 is called a **cofactor.**

Cofactor

> Let M_{ij} be the minor for element a_{ij} in an $n \times n$ matrix. The *cofactor* of a_{ij}, written A_{ij}, is
> $$A_{ij} = (-1)^{i+j} \cdot M_{ij}.$$

Finally, the determinant of a 3×3 or larger matrix is found as follows.

Finding the Determinant of a Matrix

> Multiply each element in any row or column of the matrix by its cofactor. The sum of these products gives the value of the determinant.

The process of forming this sum of products is called **expansion by a given row or column.** (See Exercise 55–56.)

Example 2   Evaluate $\begin{vmatrix} 2 & -3 & -2 \\ -1 & -4 & -3 \\ -1 & 0 & 2 \end{vmatrix}$. Expand by the second column.

To find this determinant, first get the minors of each element in the second column.

$$M_{12} = \begin{vmatrix} -1 & -3 \\ -1 & 2 \end{vmatrix} = -1(2) - (-1)(-3) = -5$$

$$M_{22} = \begin{vmatrix} 2 & -2 \\ -1 & 2 \end{vmatrix} = 2(2) - (-1)(-2) = 2$$

$$M_{32} = \begin{vmatrix} 2 & -2 \\ -1 & -3 \end{vmatrix} = 2(-3) - (-1)(-2) = -8$$

Now find the cofactor of each of these minors.

$$A_{12} = (-1)^{1+2} \cdot M_{12} = (-1)^3 \cdot (-5) = (-1)(-5) = 5$$
$$A_{22} = (-1)^{2+2} \cdot M_{22} = (-1)^4 \cdot (2) = 1 \cdot 2 = 2$$
$$A_{32} = (-1)^{3+2} \cdot M_{32} = (-1)^5 \cdot (-8) = (-1)(-8) = 8$$

The determinant is found by multiplying each cofactor by its corresponding element in the matrix and finding the sum of these products.

$$\begin{vmatrix} 2 & -3 & -2 \\ -1 & -4 & -3 \\ -1 & 0 & 2 \end{vmatrix} = a_{12} \cdot A_{12} + a_{22} \cdot A_{22} + a_{32} \cdot A_{32}$$
$$= -3(5) + (-4)(2) + (0)(8)$$
$$= -15 + (-8) + 0 = -23 \quad \blacksquare$$

Exactly the same answer would be found using any row or column of the determinant. One reason that column 2 was used here is that it contains a 0 element, so that it was not really necessary to calculate M_{32} and A_{32} above. One learns quickly that 0's are friends in work with determinants.

Instead of calculating $(-1)^{i+j}$ for a given element, the following sign checkerboards can be used.

Array of Signs

<table>
<tr><td colspan="3">For 3 × 3 matrices</td><td colspan="4">For 4 × 4 matrices</td></tr>
<tr><td>+</td><td>−</td><td>+</td><td>+</td><td>−</td><td>+</td><td>−</td></tr>
<tr><td>−</td><td>+</td><td>−</td><td>−</td><td>+</td><td>−</td><td>+</td></tr>
<tr><td>+</td><td>−</td><td>+</td><td>+</td><td>−</td><td>+</td><td>−</td></tr>
<tr><td></td><td></td><td></td><td>−</td><td>+</td><td>−</td><td>+</td></tr>
</table>

The signs alternate for each row and column, beginning with + in the first row, first column position. Thus, these arrays of signs can be reproduced as needed. If we expand a 3 × 3 matrix about row 3, for example, the first minor would have a + sign associated with it, the second minor a − sign, and the third minor a + sign. These arrays of signs can be extended in this way for determinants of 5 × 5, 6 × 6, and larger matrices.

Example 3 Evaluate $\begin{vmatrix} -1 & -2 & 3 & 2 \\ 0 & 1 & 4 & -2 \\ 3 & -1 & 4 & 0 \\ 2 & 1 & 0 & 3 \end{vmatrix}$.

Expand about the fourth row, and do the arithmetic that has been left out.

$$-2 \begin{vmatrix} -2 & 3 & 2 \\ 1 & 4 & -2 \\ -1 & 4 & 0 \end{vmatrix} + 1 \begin{vmatrix} -1 & 3 & 2 \\ 0 & 4 & -2 \\ 3 & 4 & 0 \end{vmatrix} - 0 \begin{vmatrix} -1 & -2 & 2 \\ 0 & 1 & -2 \\ 3 & -1 & 0 \end{vmatrix} + 3 \begin{vmatrix} -1 & -2 & 3 \\ 0 & 1 & 4 \\ 3 & -1 & 4 \end{vmatrix}$$
$$= -2(6) + 1(-50) - 0 + 3(-41) = -185. \quad \blacksquare$$

There are several theorems which make it easier to calculate determinants. The theorems are true for square matrices of any order, but they are proved here only for determinants of 3×3 matrices.

Theorem

> If every element in a row (or column) of matrix A is 0, then $|A| = 0$.

Example 4

$$\begin{vmatrix} -3 & 7 & 0 \\ 4 & 9 & 0 \\ -6 & 8 & 0 \end{vmatrix} = 0 \quad \blacksquare$$

To prove the theorem, expand the given determinant by the row (or column) of zeros. Each term of this expansion will have a zero factor, making the final determinant 0. For example,

$$\begin{vmatrix} a_{11} & a_{12} & a_{13} \\ 0 & 0 & 0 \\ a_{31} & a_{32} & a_{33} \end{vmatrix} = -0 \cdot \begin{vmatrix} a_{12} & a_{13} \\ a_{32} & a_{33} \end{vmatrix} + 0 \cdot \begin{vmatrix} a_{11} & a_{13} \\ a_{31} & a_{33} \end{vmatrix} - 0 \cdot \begin{vmatrix} a_{11} & a_{12} \\ a_{31} & a_{32} \end{vmatrix} = 0.$$

Theorem

> If the rows of matrix A are the corresponding columns of matrix B, then $|B| = |A|$.

Example 5 (a) Let $A = \begin{bmatrix} 2 & 1 \\ 3 & 4 \end{bmatrix}$. Interchange the rows and columns of A, to get matrix B:

$$B = \begin{bmatrix} 2 & 3 \\ 1 & 4 \end{bmatrix}.$$

Check that $|A| = 5$ and $|B| = 5$, so that $|B| = |A|$.

(b) $\begin{vmatrix} 2 & 1 & 6 \\ 3 & 0 & 5 \\ -4 & 6 & 9 \end{vmatrix} = \begin{vmatrix} 2 & 3 & -4 \\ 1 & 0 & 6 \\ 6 & 5 & 9 \end{vmatrix} \quad \blacksquare$

To prove the theorem, let

$$A = \begin{bmatrix} a_{11} & a_{12} & a_{13} \\ a_{21} & a_{22} & a_{23} \\ a_{31} & a_{32} & a_{33} \end{bmatrix} \quad \text{and} \quad B = \begin{bmatrix} a_{11} & a_{21} & a_{31} \\ a_{12} & a_{22} & a_{32} \\ a_{13} & a_{23} & a_{33} \end{bmatrix},$$

where B was obtained by interchanging the corresponding rows and columns of A. Find $|A|$ by expansion about row 1. Then find $|B|$ by expansion about column 1. You should find that $|B| = |A|$.

Theorem

> If any two rows (or columns) of matrix A are interchanged to form matrix B, then $|B| = -|A|$.

Example 6 **(a)** Let $A = \begin{bmatrix} 2 & 5 \\ 3 & 4 \end{bmatrix}$. Exchange the two columns of A to get the matrix

$B = \begin{bmatrix} 5 & 2 \\ 4 & 3 \end{bmatrix}$. Check that $|A| = -7$ and $|B| = 7$, so that $|B| = -|A|$.

(b) $\begin{vmatrix} 2 & 1 & 6 \\ 3 & 0 & 5 \\ -4 & 6 & 9 \end{vmatrix} = - \begin{vmatrix} -4 & 6 & 9 \\ 3 & 0 & 5 \\ 2 & 1 & 6 \end{vmatrix}$ ∎

This theorem is proved by steps very similar to those used to prove the previous theorem. (See Exercise 57.)

> **Theorem**
>
> Suppose matrix B is formed by multiplying every element of a row (or column) of matrix A by the real number k. Then $|B| = k \cdot |A|$.

Example 7 Let $A = \begin{bmatrix} 2 & -3 \\ 4 & 1 \end{bmatrix}$. Form the new matrix B by multiplying each element of the

second row of A by -5:

$$B = \begin{bmatrix} 2 & -3 \\ 4(-5) & 1(-5) \end{bmatrix} = \begin{bmatrix} 2 & -3 \\ -20 & -5 \end{bmatrix}.$$

Check that $|B| = -5 \cdot |A|$. ∎

The proof of this theorem is left for Exercise 58.

> **Theorem**
>
> If two rows (or columns) of a matrix A are identical, then $|A| = 0$.

Example 8 Since two rows are identical, $\begin{vmatrix} -4 & 2 & 3 \\ 0 & 1 & 6 \\ -4 & 2 & 3 \end{vmatrix} = 0.$ ∎

To prove this theorem, note that if two rows or columns of a matrix A are interchanged to form matrix B, then $|A| = -|B|$; while if two rows of matrix A are identical and are interchanged, we still have matrix A. But then $|A| = -|A|$, which can only happen if $|A| = 0$.

The last theorem of this section is perhaps the most useful of all.

> **Theorem**
>
> If matrix B is obtained from matrix A by adding to the elements of a row (or column) of A the results of multiplying the elements of any other row (or column) of A by the same number, then $|B| = |A|$.

This theorem is proved in much the same way as the others in this section. (See Exercises 59–60 below.) It is the most important theorem of the section and provides a powerful method for simplifying the work of finding a 3×3 or larger determinant, as shown in the next example.

Example 9 Let $A = \begin{bmatrix} -2 & 4 & 1 \\ 2 & 1 & 5 \\ 4 & 0 & 2 \end{bmatrix}$. Find $|A|$.

First obtain a new matrix B by adding row 1 to row 2 and then adding 2 times row 1 to row 3.

$$B = \begin{bmatrix} -2 & 4 & 1 \\ 0 & 5 & 6 \\ 0 & 8 & 4 \end{bmatrix} \qquad \begin{matrix} R_1 + R_2 \\ 2R_1 + R_3 \end{matrix}$$

Now find $|B|$ by expanding about the first column.

$$|B| = -2 \begin{vmatrix} 5 & 6 \\ 8 & 4 \end{vmatrix} = -2(20 - 48) = 56$$

By the theorem above, $|B| = |A|$ so $|A| = 56$. ∎

The following examples show how the properties of determinants are used to simplify the calculation of determinants.

Example 10 Without expanding, show that the value of the following determinant is 0.

$$\begin{vmatrix} 2 & 5 & -1 \\ 1 & -15 & 3 \\ -2 & 10 & -2 \end{vmatrix}$$

Examining the columns of the array shows that each element in the second column is -5 times the corresponding element in the third column. By the last theorem above, add to the elements of the second column the results of multiplying the elements of the third column by 5 (abbreviated below as $5C_3 + C_2$) to get the equivalent determinant

$$\begin{vmatrix} 2 & 0 & -1 \\ 1 & 0 & 3 \\ -2 & 0 & -2 \end{vmatrix}. \qquad 5C_3 + C_2$$

The value of this determinant is 0 by the first theorem of this section. ∎

Example 11 Find $|A|$ if

$$|A| = \begin{vmatrix} 4 & 2 & 1 & 0 \\ -2 & 4 & -1 & 7 \\ -5 & 2 & 3 & 1 \\ 6 & 4 & -3 & 2 \end{vmatrix}.$$

The goal is to change the first row (any row or column could be selected) to a row in which every element but one is 0. To begin, multiply the elements of the second column by -2 and add the results to the elements of the first column.

$$\begin{vmatrix} 0 & 2 & 1 & 0 \\ -10 & 4 & -1 & 7 \\ -9 & 2 & 3 & 1 \\ -2 & 4 & -3 & 2 \end{vmatrix} \qquad -2C_2 + C_1$$

Add to the elements of the second column the results of multiplying the elements of column 3 by -2.

$$\begin{vmatrix} 0 & 0 & 1 & 0 \\ -10 & 6 & -1 & 7 \\ -9 & -4 & 3 & 1 \\ -2 & 10 & -3 & 2 \end{vmatrix} \qquad -2C_3 + C_2$$

The first row has only one nonzero number, so expand about the first row.

$$|A| = +1 \begin{vmatrix} -10 & 6 & 7 \\ -9 & -4 & 1 \\ -2 & 10 & 2 \end{vmatrix}$$

Now change the third column to a column with two zeros.

$$\begin{vmatrix} 53 & 34 & 0 \\ -9 & -4 & 1 \\ -2 & 10 & 2 \end{vmatrix} \qquad -7R_2 + R_1$$

$$\begin{vmatrix} 53 & 34 & 0 \\ -9 & -4 & 1 \\ 16 & 18 & 0 \end{vmatrix} \qquad -2R_2 + R_3$$

Finally, expand about the third column to find the value of $|A|$.

$$|A| = -1 \begin{vmatrix} 53 & 34 \\ 16 & 18 \end{vmatrix} = -1(954 - 544) = -410 \qquad \blacksquare$$

In this example, working with *rows* led to a *column* with only one nonzero number and working with *columns* led to a *row* with one nonzero number.

6.5 Exercises

Find the value of each determinant. All variables represent real numbers.

1. $\begin{vmatrix} 3 & 4 \\ 5 & -2 \end{vmatrix}$

2. $\begin{vmatrix} -9 & 7 \\ 2 & 6 \end{vmatrix}$

3. $\begin{vmatrix} 0 & 4 \\ 4 & 0 \end{vmatrix}$

4. $\begin{vmatrix} 1 & 0 \\ 0 & 2 \end{vmatrix}$

5. $\begin{vmatrix} y & 2 \\ 8 & y \end{vmatrix}$

6. $\begin{vmatrix} 3 & 8 \\ m & n \end{vmatrix}$

7. $\begin{vmatrix} x & y \\ y & x \end{vmatrix}$

8. $\begin{vmatrix} 2m & 8n \\ 8n & 2m \end{vmatrix}$

Find the cofactor of each element in the second row for the following determinants.

9. $\begin{vmatrix} -2 & 0 & 1 \\ 3 & 2 & -1 \\ 1 & 0 & 2 \end{vmatrix}$
10. $\begin{vmatrix} 0 & -1 & 2 \\ 1 & 0 & 2 \\ 0 & -3 & 1 \end{vmatrix}$
11. $\begin{vmatrix} 1 & 2 & -1 \\ 2 & 3 & -2 \\ -1 & 4 & 1 \end{vmatrix}$
12. $\begin{vmatrix} 2 & -1 & 4 \\ 3 & 0 & 1 \\ -2 & 1 & 4 \end{vmatrix}$

Find the value of each determinant. All variables represent real numbers.

13. $\begin{vmatrix} 1 & 0 & 0 \\ 0 & -1 & 0 \\ 1 & 0 & 1 \end{vmatrix}$
14. $\begin{vmatrix} -2 & 0 & 1 \\ 0 & 1 & 0 \\ 0 & 0 & -1 \end{vmatrix}$
15. $\begin{vmatrix} 1 & -2 & 3 \\ 0 & 0 & 0 \\ 1 & 10 & -12 \end{vmatrix}$
16. $\begin{vmatrix} 2 & 3 & 0 \\ 1 & 9 & 0 \\ -1 & -2 & 0 \end{vmatrix}$

17. $\begin{vmatrix} 3 & 3 & -1 \\ 2 & 6 & 0 \\ -6 & -6 & 2 \end{vmatrix}$
18. $\begin{vmatrix} 5 & -3 & 2 \\ -5 & 3 & -2 \\ 1 & 0 & 1 \end{vmatrix}$
19. $\begin{vmatrix} 3 & 2 & 0 \\ 0 & 1 & x \\ 2 & 0 & 0 \end{vmatrix}$
20. $\begin{vmatrix} 0 & 3 & y \\ 0 & 4 & 2 \\ 1 & 0 & 1 \end{vmatrix}$

21. $\begin{vmatrix} i & j & k \\ -1 & 2 & 4 \\ 3 & 0 & 5 \end{vmatrix}$
22. $\begin{vmatrix} i & j & k \\ 0 & -4 & 2 \\ -1 & 3 & 1 \end{vmatrix}$
23. $\begin{vmatrix} 4 & 0 & 0 & 2 \\ -1 & 0 & 3 & 0 \\ 2 & 4 & 0 & 1 \\ 0 & 0 & 1 & 2 \end{vmatrix}$
24. $\begin{vmatrix} -2 & 0 & 4 & 2 \\ 3 & 6 & 0 & 4 \\ 0 & 0 & 0 & 3 \\ 9 & 0 & 2 & -1 \end{vmatrix}$

25. $\begin{vmatrix} 0.4 & -0.8 & 0.6 \\ 0.3 & 0.9 & 0.7 \\ 3.1 & 4.1 & -2.8 \end{vmatrix}$
26. $\begin{vmatrix} -0.3 & -0.1 & 0.9 \\ 2.5 & 4.9 & -3.2 \\ -0.1 & 0.4 & 0.8 \end{vmatrix}$

Tell why each determinant has a value of 0. All variables represent real numbers.

27. $\begin{vmatrix} 2 & 3 \\ 2 & 3 \end{vmatrix}$
28. $\begin{vmatrix} -5 & -5 \\ 6 & 6 \end{vmatrix}$
29. $\begin{vmatrix} 2 & 0 \\ 3 & 0 \end{vmatrix}$
30. $\begin{vmatrix} -8 & 0 \\ -6 & 0 \end{vmatrix}$

31. $\begin{vmatrix} -1 & 2 & 4 \\ 4 & -8 & -16 \\ 3 & 0 & 5 \end{vmatrix}$
32. $\begin{vmatrix} 1 & 0 & 0 \\ 1 & 0 & 1 \\ 3 & 0 & 0 \end{vmatrix}$
33. $\begin{vmatrix} m & 2 & 2m \\ 3n & 1 & 6n \\ 5p & 6 & 10p \end{vmatrix}$
34. $\begin{vmatrix} 7z & 8x & 2y \\ z & x & y \\ 7z & 7x & 7y \end{vmatrix}$

Use the appropriate theorems from this section to tell why each statement is true. Do not evaluate the determinants. All variables represent real numbers.

35. $\begin{vmatrix} 2 & 1 & 6 \\ 3 & 0 & 2 \\ 4 & 1 & 8 \end{vmatrix} = \begin{vmatrix} 2 & 3 & 4 \\ 1 & 0 & 1 \\ 6 & 2 & 8 \end{vmatrix}$
36. $\begin{vmatrix} 4 & -2 \\ 3 & 8 \end{vmatrix} = \begin{vmatrix} 4 & 3 \\ -2 & 8 \end{vmatrix}$

37. $\begin{vmatrix} 2 & 6 \\ 3 & 5 \end{vmatrix} = - \begin{vmatrix} 3 & 5 \\ 2 & 6 \end{vmatrix}$
38. $\begin{vmatrix} -1 & 8 & 9 \\ 0 & 2 & 1 \\ 3 & 2 & 0 \end{vmatrix} = - \begin{vmatrix} 8 & -1 & 9 \\ 2 & 0 & 1 \\ 2 & 3 & 0 \end{vmatrix}$

39. $3\begin{vmatrix} 6 & 0 & 2 \\ 4 & 1 & 3 \\ 2 & 8 & 6 \end{vmatrix} = \begin{vmatrix} 6 & 0 & 2 \\ 4 & 3 & 3 \\ 2 & 24 & 6 \end{vmatrix}$
40. $-\dfrac{1}{2}\begin{vmatrix} 5 & -8 & 2 \\ 3 & -6 & 9 \\ 2 & 4 & 4 \end{vmatrix} = \begin{vmatrix} 5 & 4 & 2 \\ 3 & 3 & 9 \\ 2 & -2 & 4 \end{vmatrix}$

41. $\begin{vmatrix} 3 & -4 \\ 2 & 5 \end{vmatrix} = \begin{vmatrix} 3 & -4 \\ 5 & 1 \end{vmatrix}$
42. $\begin{vmatrix} -1 & 6 \\ 3 & -5 \end{vmatrix} = \begin{vmatrix} -1 & 6 \\ 2 & 1 \end{vmatrix}$

43. $2\begin{vmatrix} 4 & 2 & -1 \\ m & 2n & 3p \\ 5 & 1 & 0 \end{vmatrix} = \begin{vmatrix} 4 & 2 & -1 \\ 2m & 4n & 6p \\ 5 & 1 & 0 \end{vmatrix}$
44. $\begin{vmatrix} -4 & 2 & 1 \\ 3 & 0 & 5 \\ -1 & 4 & -2 \end{vmatrix} = \begin{vmatrix} -4 & 2 & 1 + (-4)k \\ 3 & 0 & 5 + 3k \\ -1 & 4 & -2 + (-1)k \end{vmatrix}$

Use the method of Examples 10 and 11 to find the value of each determinant.

45. $\begin{vmatrix} 2 & 4 \\ 3 & 6 \end{vmatrix}$

46. $\begin{vmatrix} -5 & 10 \\ 6 & -12 \end{vmatrix}$

47. $\begin{vmatrix} 4 & 8 & 0 \\ -1 & -2 & 1 \\ 2 & 4 & 3 \end{vmatrix}$

48. $\begin{vmatrix} 6 & 8 & -12 \\ -1 & 0 & 2 \\ 4 & 0 & -8 \end{vmatrix}$

49. $\begin{vmatrix} 3 & 1 & 2 \\ 2 & 0 & 1 \\ 1 & 0 & -2 \end{vmatrix}$

50. $\begin{vmatrix} -2 & 2 & 3 \\ 0 & 2 & 1 \\ -1 & 4 & 0 \end{vmatrix}$

51. $\begin{vmatrix} -4 & 2 & 3 \\ 2 & 0 & 1 \\ 0 & 4 & 2 \end{vmatrix}$

52. $\begin{vmatrix} 6 & 3 & 2 \\ 1 & 0 & 2 \\ -1 & 4 & 1 \end{vmatrix}$

53. $\begin{vmatrix} 1 & 0 & 2 & 2 \\ 2 & 4 & 1 & -1 \\ 1 & -3 & 1 & 0 \\ 1 & 1 & 0 & 1 \end{vmatrix}$

54. $\begin{vmatrix} 2 & -1 & 1 & 0 \\ 1 & 1 & 0 & 1 \\ 0 & -1 & 1 & 1 \\ 1 & 2 & 1 & 2 \end{vmatrix}$

Let $A = \begin{bmatrix} a_{11} & a_{12} & a_{13} \\ a_{21} & a_{22} & a_{23} \\ a_{31} & a_{32} & a_{33} \end{bmatrix}$.

55. Find $|A|$ by expansion about row 3. Show that your result is really equal to $|A|$ as given in the second box of this section.

56. Repeat Exercise 55 for column 3.

57. Obtain matrix B by exchanging columns 1 and 3 of matrix A. Show that $|A| = -|B|$.

58. Obtain matrix B by multiplying each element of row 3 of A by the real number k. Show that $|B| = k \cdot |A|$.

59. Obtain matrix B by adding to column 1 of matrix A the result of multiplying each element of column 2 of A by the real number k. Show that $|A| = |B|$.

60. Obtain matrix B by adding to row 1 of matrix A the result of multiplying each element of row 3 of A by the real number k. Show that $|A| = |B|$.

61. Let A and B be any 2×2 matrices. Show that $|AB| = |A| \cdot |B|$, where $|AB|$ is the determinant of matrix AB.

62. Let A be an $n \times n$ matrix. Suppose matrix B is found by multiplying every element of A by the real number k. Express $|B|$ in terms of $|A|$.

63. Show that $\begin{vmatrix} a_{11} & a_{12} & a_{13} & a_{14} \\ 0 & a_{22} & a_{23} & a_{24} \\ 0 & 0 & a_{33} & a_{34} \\ 0 & 0 & 0 & a_{44} \end{vmatrix} = a_{11}a_{22}a_{33}a_{44}.$

64. Show that $\begin{vmatrix} a_{11} & a_{12} & a_{13} & a_{14} \\ a_{21} & a_{22} & a_{23} & a_{24} \\ 0 & 0 & a_{33} & a_{34} \\ 0 & 0 & a_{43} & a_{44} \end{vmatrix} = \begin{vmatrix} a_{11} & a_{12} \\ a_{21} & a_{22} \end{vmatrix} \cdot \begin{vmatrix} a_{33} & a_{34} \\ a_{43} & a_{44} \end{vmatrix}.$

Determinants can be used to find the area of a triangle, given the coordinates of its vertices. Given a triangle PQR with vertices (x_1, y_1), (x_2, y_2), and (x_3, y_3), as shown in the following figure, we can introduce line segments PM, RN, and QS perpendicular to the x-axis, forming trapezoids $PMNR$, $NSQR$, and $PMSQ$. Recall that the area of a trapezoid is given by the product of half the sum of the parallel bases and the altitude. For example, the area of trapezoid $PMSQ$ is $(1/2)(y_1 + y_2)(x_2 - x_1)$. The area of triangle PQR can be found by subtracting the area of *PMSQ* from the sum of the areas of *PMNR* and *RNSQ*. Thus, for the area A of triangle *PQR*,

$$A = \frac{1}{2}(x_3y_1 - x_1y_3 + x_2y_3 - x_3y_2 + x_1y_2 - x_2y_1).$$

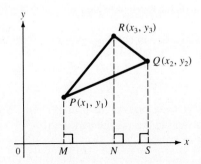

By evaluating the determinant $\begin{vmatrix} x_1 & y_1 & 1 \\ x_2 & y_2 & 1 \\ x_3 & y_3 & 1 \end{vmatrix}$ it can be shown that the area of the triangle is

given by A, where

$$A = \frac{1}{2}\begin{vmatrix} x_1 & y_1 & 1 \\ x_2 & y_2 & 1 \\ x_3 & y_3 & 1 \end{vmatrix}.$$

The points (x_1, y_1), (x_2, y_2), (x_3, y_3) must be taken in counterclockwise order; if this is not done then A may have the wrong sign. Alternatively, we could define A as the absolute value of 1/2 the determinant shown above.

Use the formula given to find the area of the following triangles.

65. $P(0, 1)$, $Q(2, 0)$, $R(1, 3)$

66. $P(2, 5)$, $Q(-1, 3)$, $R(4, 0)$

67. $P(2, -2)$, $Q(0, 0)$, $R(-3, -4)$

68. $P(4, 7)$, $Q(5, -2)$, $R(1, 1)$

69. $P(3,8)$, $Q(-1, 4)$, $R(0, 1)$

70. $P(-3, -1)$, $Q(4, 2)$, $R(3, -3)$

71. Prove that the straight line through the distinct points (x_1, y_1) and (x_2, y_2) has equation

$$\begin{vmatrix} x & y & 1 \\ x_1 & y_1 & 1 \\ x_2 & y_2 & 1 \end{vmatrix} = 0.$$

72. Use the result of Exercise 71 to show that three distinct points (x_1, y_1), (x_2, y_2) and (x_3, y_3) lie on a straight line if

$$\begin{vmatrix} x_1 & y_1 & 1 \\ x_2 & y_2 & 1 \\ x_3 & y_3 & 1 \end{vmatrix} = 0.$$

73. Show that the lines $a_1x + b_1y = c_1$ and $a_2x + b_2y = c_2$, when $c_1 \neq c_2$, are parallel if

$$\begin{vmatrix} a_1 & b_1 \\ a_2 & b_2 \end{vmatrix} = 0.$$

74. Prove that

$$\begin{vmatrix} 1 & 1 & 1 \\ a & b & c \\ a^2 & b^2 & c^2 \end{vmatrix} = (a - b)(b - c)(c - a).$$

Evaluate each determinant for real numbers a, b, c, and d.

75. $\begin{vmatrix} a & b & c \\ a & a+b & a+b+c \\ a & 2a+b & 3a+2b+c \end{vmatrix}$

76. $\begin{vmatrix} a & b & c & d \\ a & -b & -c & -d \\ a & b & -c & -d \\ a & b & c & -d \end{vmatrix}$

6.6 Cramer's Rule

We have now seen how to solve a system of n linear equations with n variables using the following methods: elimination, substitution, row transformations of matrices, and matrix inverses. Most of these systems can also be solved with determinants, as shown in this section.

To see how determinants arise in solving a system, write the linear system

$$a_{11}x + a_{12}y = b_1$$
$$a_{21}x + a_{22}y = b_2,$$

where each equation has at least one nonzero coefficient. Solve this system using matrix methods; begin by writing the augmented matrix

$$\begin{bmatrix} a_{11} & a_{12} & b_1 \\ a_{21} & a_{22} & b_2 \end{bmatrix}.$$

Multiply each element of row 1 by $1/a_{11}$. (Here we assume $a_{11} \neq 0$—see Exercise 42.) This gives the matrix of an equivalent system:

$$\begin{bmatrix} 1 & \dfrac{a_{12}}{a_{11}} & \dfrac{b_1}{a_{11}} \\ a_{21} & a_{22} & b_2 \end{bmatrix}. \qquad \dfrac{1}{a_{11}}\,\text{R}_1$$

Multiply each element of row 1 by $-a_{21}$, and add the result to the corresponding element of row 2.

$$\begin{bmatrix} 1 & \dfrac{a_{12}}{a_{11}} & \dfrac{b_1}{a_{11}} \\ 0 & a_{22} - \dfrac{a_{21}a_{12}}{a_{11}} & b_2 - \dfrac{a_{21}b_1}{a_{11}} \end{bmatrix} \qquad -a_{21}\,\text{R}_1 + \text{R}_2$$

Multiply each element of row 2 by a_{11}.

$$\begin{bmatrix} 1 & \dfrac{a_{12}}{a_{11}} & \dfrac{b_1}{a_{11}} \\ 0 & a_{11}a_{22} - a_{21}a_{12} & a_{11}b_2 - a_{21}b_1 \end{bmatrix} \qquad a_{11}\,\text{R}_2$$

This matrix leads to the system of equations

$$x + \dfrac{a_{12}}{a_{11}}y = \dfrac{b_1}{a_{11}}$$
$$(a_{11}a_{22} - a_{21}a_{12})y = a_{11}b_2 - a_{21}b_1.$$

From the second equation of this system,

$$y = \dfrac{a_{11}b_2 - a_{21}b_1}{a_{11}a_{22} - a_{21}a_{12}}.$$

Both the numerator and denominator here may be written as determinants:

$$y = \frac{\begin{vmatrix} a_{11} & b_1 \\ a_{21} & b_2 \end{vmatrix}}{\begin{vmatrix} a_{11} & a_{12} \\ a_{21} & a_{22} \end{vmatrix}}. \tag{1}$$

Inserting this value of y into the first equation above shows that x can also be written with determinants as

$$x = \frac{\begin{vmatrix} b_1 & a_{12} \\ b_2 & a_{22} \end{vmatrix}}{\begin{vmatrix} a_{11} & a_{12} \\ a_{21} & a_{22} \end{vmatrix}}. \tag{2}$$

The denominator for finding both x and y is just the determinant of the matrix of coefficients of the original system. This determinant is often denoted D, so that

$$D = \begin{vmatrix} a_{11} & a_{12} \\ a_{21} & a_{22} \end{vmatrix}.$$

In equation (1), the numerator is the determinant of a matrix obtained by replacing the coefficients of y in D with the respective constants: D_y is defined as

$$D_y = \begin{vmatrix} a_{11} & b_1 \\ a_{21} & b_2 \end{vmatrix}.$$

In the same way, from equation (2), D_x is defined as

$$D_x = \begin{vmatrix} b_1 & a_{12} \\ b_2 & a_{22} \end{vmatrix}.$$

With this notation, the solution of the given system is

$$x = \frac{D_x}{D} \quad \text{and} \quad y = \frac{D_y}{D}.$$

The system has a single solution as long as $D \neq 0$. We have now proved much of the next theorem, called **Cramer's rule.**

Cramer's Rule for 2 Equations in 2 Variables

Let

$$a_{11}x + a_{12}y = b_1$$
$$a_{21}x + a_{22}y = b_2$$

be a system of equations, where each equation has at least one nonzero coefficient. The system has a unique solution if $D \neq 0$; this solution is

$$x = \frac{D_x}{D} \quad \text{and} \quad y = \frac{D_y}{D}.$$

If $D = 0$, the system is dependent if $D_x = 0$ and $D_y = 0$; otherwise it is inconsistent.

(For proof of this last statement, see Exercises 41–42 below.)

Example 1   Use Cramer's rule to solve the system $5x + 7y = -1$
$$6x + 8y = 1.$$

To use Cramer's rule, first evaluate D, D_x, and D_y.

$$D = \begin{vmatrix} 5 & 7 \\ 6 & 8 \end{vmatrix} = 5(8) - 6(7) = -2.$$

$$D_x = \begin{vmatrix} -1 & 7 \\ 1 & 8 \end{vmatrix} = (-1)(8) - (1)(7) = -15.$$

$$D_y = \begin{vmatrix} 5 & -1 \\ 6 & 1 \end{vmatrix} = 5(1) - (6)(-1) = 11.$$

By Cramer's rule, $x = -15/(-2) = 15/2$, and $y = 11/(-2) = -11/2$. The solution set is $\{(15/2, -11/2)\}$, as can be verified by substituting within the given system.   ■

By much the same method as used above, Cramer's rule can be generalized to a system of n linear equations with n variables.

General Form of Cramer's Rule

Let an $n \times n$ system have linear equations of the form

$$a_{11}x_1 + a_{12}x_2 + a_{13}x_3 + \ldots + a_{1n}x_n = b_1.$$

Define D as the determinant of the $n \times n$ matrix of all coefficients of the variables. Define D_{x1} as the determinant obtained from D by replacing the entries in column 1 of D with the constants of the system. Define D_{xi} as the determinant obtained from D by replacing the entries in column i with the constants of the system. If $D \neq 0$, the unique solution of the system is given by

$$x_1 = \frac{D_{x1}}{D}, \ x_2 = \frac{D_{x2}}{D}, \ x_3 = \frac{D_{x3}}{D}, \ldots, \ x_n = \frac{D_{xn}}{D}.$$

Example 2   Use Cramer's rule to solve the system

$$x + y - z + 2 = 0$$
$$2x - y + z + 5 = 0$$
$$x - 2y + 3z - 4 = 0.$$

To use Cramer's rule, the system must be rewritten in the form

$$x + y - z = -2$$
$$2x - y + z = -5$$
$$x - 2y + 3z = 4.$$

The determinant of coefficients, D, is

$$D = \begin{vmatrix} 1 & 1 & -1 \\ 2 & -1 & 1 \\ 1 & -2 & 3 \end{vmatrix}.$$

To find D_x, replace the elements of the first column of D with the constants of the system. Find D_y and D_z in a similar way.

$$D_x = \begin{vmatrix} -2 & 1 & -1 \\ -5 & -1 & 1 \\ 4 & -2 & 3 \end{vmatrix}, \quad D_y = \begin{vmatrix} 1 & -2 & -1 \\ 2 & -5 & 1 \\ 1 & 4 & 3 \end{vmatrix}, \quad D_z = \begin{vmatrix} 1 & 1 & -2 \\ 2 & -1 & -5 \\ 1 & -2 & 4 \end{vmatrix}$$

Verify that $D = -3$, $D_x = 7$, $D_y = -22$, and $D_z = -21$. Then, by Cramer's rule,

$$x = \frac{D_x}{D} = \frac{7}{-3} = \frac{-7}{3}, \quad y = \frac{D_y}{D} = \frac{-22}{-3} = \frac{22}{3}, \quad \text{and } z = \frac{D_z}{D} = \frac{-21}{-3} = 7.$$

The solution set of the system is $\{(-7/3, 22/3, 7)\}$. ■

Example 3   Use Cramer's rule to solve the system

$$2x - 3y + 4z = 10$$
$$6x - 9y + 12z = 24$$
$$x + 2y - 3z = 5.$$

Here $D = 0$. Also, $D_x = 6$. Since $D = 0$ and at least one of the other determinants is not zero, the system is inconsistent. (Had all the determinants been 0, the system would have been dependent.) ■

Several different methods for solving systems of equations have now been shown. In general, if a small system of linear equations must be solved by pencil and paper, substitution is the best method if the various equations can easily be solved in terms of each other. This happens rarely. The next choice, perhaps the best choice of all, is the elimination method. Some people like the Gaussian reduction method, which is really just a systematic way of doing the elimination method. The Gaussian reduction method is probably superior where four or more equations are involved. *Cramer's rule is seldom the method of choice simply because it involves more calculations than any other method.*

6.6 Exercises

Use Cramer's rule to solve each of the following systems of linear equations.

1. $x + y = 4$
   $2x - y = 2$

2. $3x + 2y = -4$
   $2x - y = -5$

3. $4x + 3y = -7$
   $2x + 3y = -11$

4. $3x + 2y = -4$
   $5x - y = 2$

5. $2x - 3y = -5$
   $x + 5y = 17$

6. $x + 9y = -15$
   $3x + 2y = 5$

7. $5x + 2y = 7$
   $6x + y = 8$

8. $7x + 3y = 5$
   $2x + 4y = 3$

9. $10x - 8y = 1$
   $-15x + 12y = 4$

10. $8x - 6y = 10$
   $20x - 15y = 6$

11. $4x - y + 3z = -3$
   $3x + y + z = 0$
   $2x - y + 4z = 0$

12. $5x + 2y + z = 15$
   $2x - y + z = 9$
   $4x + 3y + 2z = 13$

13. $2x - y + 4z = -2$
   $3x + 2y - z = -3$
   $x + 4y + 2z = 17$

14. $x + y + z = 4$
   $2x - y + 3z = 4$
   $4x + 2y - z = -15$

15. $4x - 3y + z = -1$
   $5x + 7y + 2z = -2$
   $3x - 5y - z = 1$

16. $2x - 3y + z = 8$
   $-x - 5y + z = -4$
   $3x - 5y + 2z = 12$

17. $x + 2y + 3z = 4$
   $4x + 3y + 2z = 1$
   $-x - 2y - 3z = 0$

18. $2x - y + 3z = 1$
   $-2x + y - 3z = 2$
   $5x - y + z = 2$

19. $x - 2y + 3z = 4$
   $5x + 7y - z = 2$
   $2x + 2y - 5z = 3$

20. $-3x - 2y - z = 4$
   $4x + y + z = 5$
   $3x - 2y + 2z = 1$

21. $2x + 3y = 13$
   $2y - z = 5$
   $x + 2z = 4$

22. $3x - z = -10$
   $y + 4z = 8$
   $x + 2z = -1$

23. $5x - y = -4$
   $3x + 2z = 4$
   $4y + 3z = 22$

24. $3x + 5y = -7$
   $2x + 7z = 2$
   $4y + 3z = -8$

25. $x + 2y = 10$
   $3x + 4z = 7$
   $-y - z = 1$

26. $5x - 2y = 3$
   $4y + z = 8$
   $x + 2z = 4$

27. $x + z = 0$
   $y + 2z + w = 0$
   $2x - w = 0$
   $x + 2y + 3z = -2$

28. $x - y + z + w = 6$
   $2y - w = -7$
   $x - z = -1$
   $y + w = 1$

29. $.4x - .6y = .4$
   $.3x + .2y = -.22$

30. $-.5x + .4y = .43$
   $.1x + .2y = .11$

31. $.4x - .2y + .3z = -.08$
   $-.5x + .3y - .7z = .19$
   $.8x - .7y + .1z = -.21$

32. $-.6x + .2y - .5z = -.16$
   $.8x - .3y - .2z = -.48$
   $.4x + .6y - .8z = -.8$

Use Cramer's rule to solve each of the following.

33. Paying $96, Mari Hall bought 5 shirts and 2 pairs of pants. Later, paying $66, she bought one more shirt and 3 more pairs of pants. Find the cost of a shirt and the cost of a pair of pants.

34. Barbara Maring received $50,000 from the sale of a business. She invested part of the money at 16% and the rest at 18%, which resulted in her earning $8550 per year in interest. Find the amount invested at each rate.

35. A gold merchant has some 12-carat gold (12/24 pure gold), and some 22-carat gold (22/24 pure). How many grams of each should be mixed to get 25 grams of 15-carat gold?

36. A chemist has some 40% acid solution and some 60% solution. How many liters of each should be used to get 40 liters of a 45% solution?

37. How many pounds of tea worth $4.60 a pound should be mixed with tea worth $6.50 a pound to get 8 pounds of a mixture worth $5.20 a pound?

38. A solution of a drug with a strength of 5% is to be mixed with some of a 15% solution to get 15 ml of an 8% solution. How many ml of each solution should be used?

39. The cashier at an amusement park has a total of $2480, made up of fives, tens, and twenties. The total number of bills is 290, and the value of the tens is $60 more than the value of the twenties. How many of each type of bill does the cashier have?

40. Ms. Levy invests $50,000 three ways—at 8%, 8.5%, and 11%. In total, she receives $4436.25 per year in interest. The interest from the 11% investment is $80 more than the interest on the 8% investment. Find the amount she has invested at each rate.

For the following two exercises, use the system of equations

$$a_1x + b_1y = c_1$$
$$a_2x + b_2y = c_2.$$

41. Assume $D_x = 0$ and $D_y = 0$. Show that if $c_1c_2 \neq 0$, then $D = 0$, and the equations are dependent.

42. Assume $D = 0$, $D_x = 0$, and $b_1b_2 \neq 0$. Show that $D_y = 0$.

Use Cramer's rule to solve each system for (x, y). Assume all variables represent real numbers and all denominators are nonzero.

43. $ax + by = a^2$
$bx + ay = b^2$

44. $ax + by = \dfrac{1}{a}$
$bx + ay = \dfrac{1}{b}$

45. $\dfrac{x}{2} - \dfrac{y}{b} = 1$
$2x + by = 4b$

46. $a^2x + b^2y = a^2$
$bx + ay = a$

6.7 Nonlinear Systems ⁿᵒ   yes!

A system of equations in which at least one equation is *not* linear is called a **nonlinear system**. The substitution method works well for solving many such systems, as the next example shows.

Example 1   Solve the system $x^2 - y = 4$ *solve for y*   **(1)**
$$x + y = -2. \qquad \qquad \textbf{(2)}$$

When one of the equations in a nonlinear system is linear, it is usually best to begin by solving the linear equation for any variable. With this system, begin by solving equation (2) for x, giving

$$x = -y - 2.$$

Now substitute this result for x in equation (1) to get

$$(-y - 2)^2 - y = 4$$
$$y^2 + 4y + 4 - y = 4$$
$$y^2 + 3y = 0$$
$$y(y + 3) = 0$$
$$y = 0 \quad \text{or} \quad y + 3 = 0$$
$$y = 0 \quad \text{or} \quad y = -3.$$

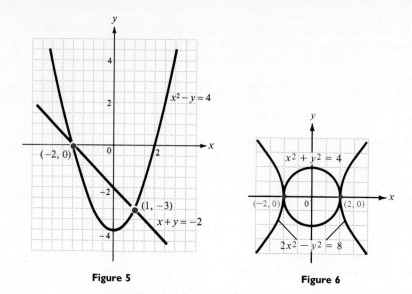

Figure 5   **Figure 6**

Substituting 0 for y in equation (2) gives $x = -2$. Also, if $y = -3$, then $x = 1$, making the solution set of the given system $\{(-2, 0), (1, -3)\}$. A graph of the system is shown in Figure 5.   ∎

Some nonlinear systems are best solved by elimination, as shown in the next example.

Example 2   Solve the system $x^2 + y^2 = 4$ (1)
$\qquad\qquad\qquad\qquad 2x^2 - y^2 = 8.$ (2)

Adding equation (1) to equation (2) (to eliminate y) gives the new system

$$x^2 + y^2 = 4$$
$$3x^2 \qquad\ = 12.$$ (3)

Solve equation (3) for x:

$$x^2 = 4$$
$$x = 2 \quad \text{or} \quad x = -2.$$

Find y by substituting back into equation (1). If $x = 2$, then $y = 0$, and if $x = -2$, then $y = 0$. The solution set of the given system is $\{(2, 0), (-2, 0)\}$. See Figure 6.

The elimination method works with this system since the system can be thought of as a system of linear equations where the variables are x^2 and y^2. In other words, the system is **linear in x^2 and y^2**. To see this, substitute u for x^2 and v for y^2. The resulting system is linear in u and v.   ∎

Some systems require a combination of the elimination method and substitution, as in the next example.

Example 3   Solve the system $x^2 + 3xy + y^2 = 22$ **(1)**
$$x^2 - xy + y^2 = 6.$$ **(2)**

Begin by multiplying both sides of equation (2) by -1, and adding the results to equation (1), giving the system

$$4xy = 16$$ **(3)**

$$x^2 - xy + y^2 = 6.$$

Now solve equation (3) for either x or y and substitute the result into one of the given equations. Solving for y gives

$$y = \frac{4}{x} \quad (x \neq 0).$$ **(4)**

(If $x = 0$ there is no value of y which satisfies the system.) Substituting back for y in equation (2) and simplifying gives

$$x^2 - x\left(\frac{4}{x}\right) + \left(\frac{4}{x}\right)^2 = 6$$

$$x^2 - 4 + \frac{16}{x^2} = 6$$

$$x^4 - 4x^2 + 16 = 6x^2$$

$$x^4 - 10x^2 + 16 = 0$$

$$(x^2 - 2)(x^2 - 8) = 0$$

$$x^2 = 2 \quad \text{or} \quad x^2 = 8$$

$$x = \sqrt{2} \text{ or } x = -\sqrt{2} \text{ or } x = 2\sqrt{2} \text{ or } x = -2\sqrt{2}.$$

Substitute these values of x into equation (4) to find the corresponding values for y.

$$\text{If } x = \sqrt{2}, y = \frac{4}{\sqrt{2}} = 2\sqrt{2}.$$

$$\text{If } x = -\sqrt{2}, y = -\frac{4}{\sqrt{2}} = -2\sqrt{2}.$$

$$\text{If } x = 2\sqrt{2}, y = \frac{4}{2\sqrt{2}} = \sqrt{2}.$$

$$\text{If } x = -2\sqrt{2}, y = -\frac{4}{2\sqrt{2}} = -\sqrt{2}.$$

The solution set of the system is $\{(\sqrt{2}, 2\sqrt{2}), (-\sqrt{2}, -2\sqrt{2}), (2\sqrt{2}, \sqrt{2}), (-2\sqrt{2}, -\sqrt{2})\}$.   ∎

6.7 Exercises

Solve each system by the substitution method.

1. $2x^2 = 3y + 23$
$\quad y = 2x - 5$

2. $x^2 - 3y^2 = 22$
$\quad x + 3y = 2$

3. $x = y - 2$
$\quad y^2 = 2x^2 + 8$

4. $x^2 + y^2 = 45$
   $x + y = -3$

5. $x^2 - y = -1$
   $3x = y - 11$

6. $3y = 2x + 5$
   $2y^2 = 2x + 4$

7. $y = x^2$
   $x + y = 2$

8. $y = -x^2 + 2$
   $x - y = 0$

9. $y = x^2 - 2x + 1$
   $x - 3y = -1$

10. $y = x^2 + 6x + 9$
   $x + 2y = -2$

11. $3x^2 + 2y^2 = 5$
   $x - y = -2$

12. $x^2 + y^2 = 5$
   $-3x + 4y = 2$

Solve the following systems by any method.

13. $x^2 + y^2 = 8$
   $x^2 - y^2 = 0$

14. $x^2 + y^2 = 10$
   $2x^2 - y^2 = 17$

15. $5x^2 - y^2 = 0$
   $3x^2 + 4y^2 = 0$

16. $x^2 + y^2 = 4$
   $2x^2 - 3y^2 = -12$

17. $2x^2 + 3y^2 = 5$
   $3x^2 - 4y^2 = -1$

18. $3x^2 + 5y^2 = 17$
   $2x^2 - 3y^2 = 5$

19. $2x^2 + 2y^2 = 20$
   $3x^2 + 3y^2 = 30$

20. $x^2 + y^2 = 4$
   $5x^2 + 5y^2 = 28$

21. $xy = 6$
   $x + y = 5$

22. $xy = -4$
   $2x + y = -7$

23. $xy = -15$
   $4x + 3y = 3$

24. $xy = 8$
   $3x + 2y = -16$

25. $2xy + 1 = 0$
   $x + 16y = 2$

26. $-5xy + 2 = 0$
   $x - 15y = 5$

27. $x^2 + 4y^2 = 25$
   $xy = 6$

28. $5x^2 - 2y^2 = 6$
   $xy = 2$

29. $x^2 + 2xy - y^2 = 14$
   $x^2 - \qquad y^2 = -16$

30. $3x^2 + xy + 3y^2 = 7$
   $x^2 + \qquad y^2 = 2$

31. $x^2 - xy + y^2 = 5$
   $2x^2 + xy - y^2 = 10$

32. $3x^2 + 2xy - y^2 = 9$
   $x^2 - xy + y^2 = 9$

33. $x^2 + 2xy - y^2 + y = 1$
   $3x + y = 6$

34. $x^2 - 4x + y + xy = -10$
   $2x - y = 10$

35. $y = 3^x$
   $y = 9^{2x}$

36. $y = 5^{3x}$
   $y = 125^{4x}$

37. $y = \log(x - 2)$
   $y = -1 + \log(8x + 4)$

38. $y = -1 + \log(18x + 10)$
   $y = \log(x + 5)$

Use any method to solve the following problems.

39. Find two numbers whose sum is 12 and whose product is 36.

40. Find two numbers whose sum is 17 and whose squares differ by 17.

41. Find two numbers whose ratio is 5:3 and whose product is 135.

42. Find two numbers whose squares have a sum of 100 and a difference of 28.

43. The longest side of a right triangle is 13 meters in length. One of the other sides is 7 meters longer than the shortest side. Find the length of each of the two shorter sides of the triangle.

44. Does the straight line $3x - 2y = 9$ intersect the circle $x^2 + y^2 = 25$? (*Hint:* To find out, try to solve the system made up of these two equations.)

45. Do the parabola $y = x^2 + 2$ and the ellipse $2x^2 + y^2 - 4x - 4y = 0$ have any points in common?

46. For what value of b will the line $x + 2y = b$ touch the circle $x^2 + y^2 = 9$ in only one point?

47. For what values of a do the circle $x^2 + y^2 = 25$ and the ellipse $x^2/a^2 + y^2/25 = 1$ have exactly two points in common?

48. For what values of k do the parabolas $y = x^2 - 4$ and $y = -x^2 + k$ have exactly two points in common?

Solve each system.

49. $y = -\sqrt{1 + x}$
$(x - 3)^2 + y^2 = 16$

50. $x^2 + y^2 = 2$
$\sqrt{x + y} = \sqrt{x - y} + 2$

51. $\dfrac{1}{x^2} - \dfrac{3}{y^2} = 14$
$\dfrac{2}{x^2} + \dfrac{1}{y^2} = 35$

52. $\dfrac{8}{\sqrt{x}} + \dfrac{9}{\sqrt{y}} = 5$
$\dfrac{12}{\sqrt{x}} - \dfrac{18}{\sqrt{y}} = -3$

Solve each system for x and y if a and b represent real numbers.

53. $x^3 - y^3 = a^3$
$x - y = 6$

54. $x^3 + y^3 = a^2$
$x + y = 1$

6.8 Systems of Inequalities

Many mathematical descriptions of real world situations are best expressed as inequalities, rather than equalities. For example, a firm might be able to use a machine *no more* than 12 hours a day, while a production of *at least* 500 cases of a certain product might be required to meet a contract. Perhaps the simplest way to see the solution of an inequality in two variables is to draw its graph.

A line divides a plane into three sets of points—the points of the line itself and the points belonging to the two regions determined by the line. Each of these two regions is called a **half-plane.** In Figure 7 line r divides the plane into three different sets of points, line r, half-plane P and half-plane Q. The points of r belong to neither P nor Q. Line r is the boundary of each half-plane.

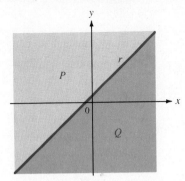

Figure 7

A **linear inequality in two variables** is an inequality of the form

$$Ax + By \le C,$$

where A, B, and C are real numbers, with A and B not both equal to 0. (The symbol $\le$ could be replaced with $\ge$, $<$, or $>$.) The graph of a linear inequality is a half-plane, sometimes including its boundary. For example, to graph the linear inequality $3x - 2y \leq 6$, first graph the boundary, $3x - 2y = 6$, as shown in Figure 8. Since the points of the line $3x - 2y = 6$ satisfy $3x - 2y \leq 6$, this boundary line is part of the solution. To decide which half-plane—the one above the line $3x - 2y = 6$ or the one below the line—is part of the solution, solve the original inequality for y.

$$3x - 2y \leq 6$$

$$-2y \leq -3x + 6$$

$$y \geq \frac{3}{2}x - 3 \qquad \text{Change} \leq \text{to} \geq$$

For a particular value of x, the inequality will be satisfied by all values of y which are *greater than* or equal to $(3/2)x - 3$. This means that the solution contains the half-plane *above* the line, as shown in Figure 9.

Example 1   Graph $x + 4y > 4$.

The boundary here is the straight line $x + 4y = 4$. Since the points on this line do not satisfy $x + 4y > 4$, it is customary to make the line dashed, as in Figure 10. To decide which half-plane represents the solution, solve for y.

$$x + 4y > 4$$

$$4y > -x + 4$$

$$y > -\frac{1}{4}x + 1$$

Since y is *greater than* $(-1/4)x + 1$, the graph of the solution is the half-plane above the boundary, as shown in Figure 10.

As a check, choose a point not on the boundary line and substitute into the inequality. The point $(0, 0)$ is a good choice if it does not lie on the boundary, since the substitution is easily done. Here, substitution of $(0, 0)$ into the original inequality gives

$$x + 4y > 4$$

$$0 + 4(0) > 4$$

$$0 > 4,$$

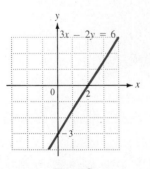

Figure 8

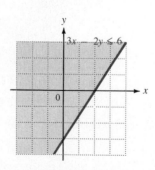

Figure 9

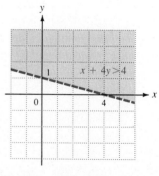

Figure 10

a false statement. Since the point $(0, 0)$ is below the line, the points which satisfy the inequality must be above the line, which agrees with the result above.   ∎

The methods used to graph linear inequalities can be used for other inequalities of the form $y \leq f(x)$ as summarized here. (Similar statements can be made for $<$, $>$, and $\geq$.)

Graphing Inequalities

> For a function f,
> the graph of $y < f(x)$ is made up of all the points which are *below* the graph of $y = f(x)$;
> the graph of $y > f(x)$ is made up of all the points which are *above* the graph of $y = f(x)$.

The solution set of a **system of inequalities,** such as

$$x > 6 - 2y$$
$$x^2 < 2y,$$

is the intersection of the solution set of its members. This intersection is found by graphing the solution sets of both inequalities on the same coordinate axes and identifying, by shading, the region common to all graphs.

Example 2   Graph the solution set of the system above.

Figure 11 shows the graphs of both $x > 6 - 2y$ and $x^2 < 2y$. The methods of the previous section can be used to show that the boundaries cross at the points $(2, 2)$ and $(-3, 9/2)$. The solution set of the system includes all points in the heavily shaded area. The points on the boundaries of $x > 6 - 2y$ and $x^2 < 2y$ do not belong to the graph of the solution. For this reason, the boundaries are dashed lines.   ∎

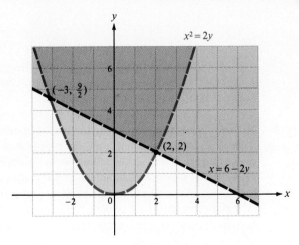

Figure 11

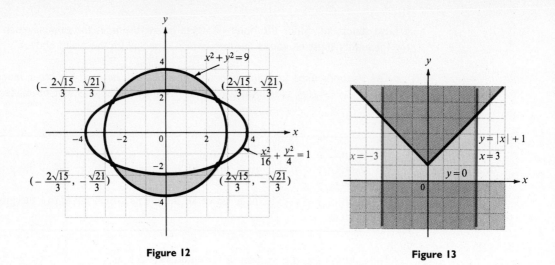

Figure 12   **Figure 13**

Example 3 Graph the solution set of the system

$$x^2 + y^2 \le 9$$
$$\frac{x^2}{16} + \frac{y^2}{4} \ge 1.$$

The graph is found by graphing both inequalities on the same axes and shading the common region as shown in Figure 12. The boundaries meet in four points, $(2\sqrt{15}/3, \sqrt{21}/3)$, $(2\sqrt{15}/3, -\sqrt{21}/3)$, $(-2\sqrt{15}/3, \sqrt{21}/3)$, and $(-2\sqrt{15}/3, -\sqrt{21}/3)$. ■

Example 4 Graph the solution set of the system

$$|x| \le 3$$
$$y \le 0$$
$$y \ge |x| + 1.$$

The system is graphed in Figure 13. The graph shows that the solution sets of $y \le 0$ and $y \ge |x| + 1$ have no points in common; therefore the solution set for the system is $\emptyset$. ■

Example 5 Midtown Manufacturing Company makes plastic plates and cups, both of which require time on two machines. Plates require one hour on machine A and two hours on machine B. Cups require three hours on machine A and one hour on machine B. Suppose that the two machines, A and B, are operated for at most 15 hours per day. How many plates and cups can be made under those conditions?

Let x represent the number of plates, and y the number of cups produced in one day. Each plate requires one hour on machine A, while each cup requires three hours on this same machine. The total number of hours required on machine A, then, is $x + 3y$. Since this machine operates for at most 15 hours per day,

$$x + 3y \leq 15.$$

In the same way, the total number of hours per day required on machine B is $2x + y$. It, too, operates at most 15 hours per day, so that

$$2x + y \leq 15.$$

Both x and y must be nonnegative integers (why?) so that

$$x \geq 0 \quad \text{and} \quad y \geq 0.$$

These four inequalities form the system

$$x + 3y \leq 15$$
$$2x + y \leq 15$$
$$x \geq 0$$
$$y \geq 0.$$

A graph of the solution of this system is shown in Figure 14. Any point in the shaded region satisfies all the conditions of the original problem. ■

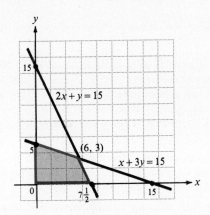

Figure 14

6.8 Exercises

Graph the solution of each of the following systems of inequalities.

1. $x + y \leq 4$
　　$x - 2y \geq 6$

2. $2x + y > 2$
　　$x - 3y < 6$

3. $4x + 3y < 12$
　　$y + 4x > -4$

4. $3x + 5y \leq 15$
　　$x - 3y \geq 9$

5. $x + 2y \leq 4$
　　$y \geq x^2 - 1$

6. $4x - 3y \leq 12$
　　$y \leq x^2$

7. $y \leq -x^2$
　　$y \geq x^2 - 6$

8. $x^2 + y^2 \leq 9$
　　$x \leq -y^2$

9. $x^2 - y^2 < 1$
　　$-1 < y < 1$

10. $x^2 + y^2 \leq 36$
　　$-4 \leq x \leq 4$

11. $2x^2 - y^2 > 4$
　　$2y^2 - x^2 > 4$

12. $y \geq x^2 + 4x + 4$
   $y < -x^2$

13. $\dfrac{x^2}{16} + \dfrac{y^2}{9} \leq 1$
   $\dfrac{x^2}{4} - \dfrac{y^2}{16} \geq 1$

14. $y \geq (x - 2)^2 + 3$
   $16x^2 + y^2 \leq 16$

15. $x + y \leq 4$
   $x - y \leq 5$
   $4x + y \leq -4$

16. $3x - 2y \geq 6$
   $x + y \leq -5$
   $y \leq 4$

17. $-2 < x < 3$
   $-1 \leq y \leq 5$
   $2x + y < 6$

18. $-2 < x < 2$
   $y > 1$
   $x - y > 0$

19. $2y + x \geq -5$
   $y \leq 3 + x$
   $x \leq 0$
   $y \leq 0$

20. $2x + 3y \leq 12$
   $2x + 3y > -6$
   $3x + y < 4$
   $x \geq 0$
   $y \geq 0$

21. $\dfrac{x^2}{4} + \dfrac{y^2}{9} > 1$
   $x^2 - y^2 \geq 1$
   $-4 \leq x \leq 4$

22. $2x - 3y < 6$
   $4x^2 + 9y^2 < 36$
   $x \geq -1$

23. $y \geq 3^x$
   $y \geq 2$

24. $y \leq \left(\dfrac{1}{2}\right)^x$
   $y \geq 4$

25. $|x| \geq 2$
   $|y| \geq 4$
   $y < x^2$

26. $|x| + 2 \geq 4$
   $|y| \leq 1$
   $\dfrac{x^2}{9} + \dfrac{y^2}{16} \leq 1$

27. $y \leq |x + 2|$
   $\dfrac{x^2}{16} - \dfrac{y^2}{9} \leq 1$

28. $y \leq \log x$
   $y \geq |x - 2|$

29. $y \geq |4 - x|$
   $y \geq |x|$

30. $|x + 2| < y$
   $|x| \leq 3$

Write a system of inequalities for each of the following problems and then graph the solution of the system.

31. A pizza company makes two kinds of pizza, basic and plain. Basic contains cheese and beef, while plain contains onions and beef. The company sells at least three units a day of basic, and at least two units of plain. The beef costs $5 per unit for basic, and $4 per unit for plain. They can spend no more than $50 per day on beef. Dough for basic is $2 per unit, while dough for plain is $1 per unit. The company can spend no more than $16 per day on dough.

32. A farmer raises only pigs and geese. She wants to raise no more than 16 animals, including no more than 12 geese. She spends $5 to raise a pig and $2 to raise a goose. She has available $50 for this purpose.

33. George takes vitamin pills. Each day he must have at least 16 units of vitamin A, 5 units of vitamin B_1, and 20 units of vitamin C. He can choose between red pills which contain 8 units of A, 1 of B_1, and 2 of C, and green pills which contain 2 units of A, 1 of B_1 and 7 of C.

34. Sue is on a diet and wishes to restrict herself to no more than 1600 calories per day. Her diet consists of foods chosen from three food groups: I, meats and dairy products; II, fruits and vegetables; III, breads and other starches. These three groups contain calories per serving as shown below.

Group	I	II	III
Calories	170	50	140

Sue wishes to include four servings daily from group I and more servings from group II than from the other groups combined.

35. A manufacturer of refrigerators must ship at least 100 refrigerators to its two West Coast warehouses. Each warehouse holds a maximum of 100 refrigerators. Warehouse A holds

25 refrigerators already, while warehouse B has 20 on hand. It costs $12 to ship a refrigerator to warehouse A and $10 to ship one to warehouse B. The total shipping cost is budgeted at a maximum of $1200.

36. The California Almond Growers have 2400 boxes of almonds to be shipped from their plant in Sacramento to Des Moines and San Antonio. The Des Moines market needs at least 1000 boxes, while the San Antonio market must have at least 800 boxes.

6.9 Linear Programming ~~no~~

An important application of mathematics to business and social science is called linear programming. **Linear programming** is used to find such things as minimum cost and maximum profit. The basic ideas of this technique will be explained with an example.

Example 1 The Smith Company makes two products, tape decks and amplifiers. Each tape deck gives a profit of $3, while each amplifier produces $7. The company must manufacture at least one tape deck per day to satisfy one of its customers, but no more than five because of production problems. Also, the number of amplifiers produced cannot exceed six per day. As a further requirement, the number of tape decks cannot exceed the number of amplifiers. How many of each should the company manufacture in order to obtain the maximum profit?

To begin, translate the statement of the problem into symbols by assuming

$$x = \text{number of tape decks to be produced daily}$$
$$y = \text{number of amplifiers to be produced daily.}$$

According to the statement of the problem given above, the company must produce at least one tape deck (one or more), so that

$$x \geq 1.$$

No more than 5 tape decks may be produced means that

$$x \leq 5.$$

Since not more than 6 amplifiers may be made in one day,

$$y \leq 6.$$

The number of tape decks may not exceed the number of amplifiers translates as

$$x \leq y.$$

The number of tape decks and of amplifiers cannot be negative, so that

$$x \geq 0 \quad \text{and} \quad y \geq 0.$$

Listing all the restrictions, or **constraints,** that are placed on production, gives

$$x \geq 1, \quad x \leq 5, \quad y \leq 6, \quad x \leq y, \quad x \geq 0, \quad y \geq 0.$$

To find the maximum possible profit that the company can make, subject to these constraints, begin by sketching the graph of each constraint. The only feasible values of x and y are those that satisfy all constraints; that is, the values that lie in the intersection of the graphs of the constraints. The intersection is shown in Figure 15. Any point lying inside the shaded region or on the boundary in Figure 15 satisfies the restrictions as to the number of tape decks and amplifiers that may be produced. This region is called the **region of feasible solutions.**

Since each tape deck gives a profit of $3, the daily profit from the production of x tape decks is $3x$ dollars. Also, the profit from the production of y amplifiers will be $7y$ dollars per day. The total daily profit is thus

$$\text{profit} = 3x + 7y.$$

The problem of the Smith Company may now be stated as follows: find values of x and y in the shaded region of Figure 15 that will produce the maximum possible value of $3x + 7y$. To locate the point (x, y) that gives the maximum profit, add to the graph of Figure 15 several lines which represent some possible profit functions:

$$3x + 7y = 0,$$
$$3x + 7y = 10,$$
$$3x + 7y = 30,$$
$$3x + 7y = 60.$$

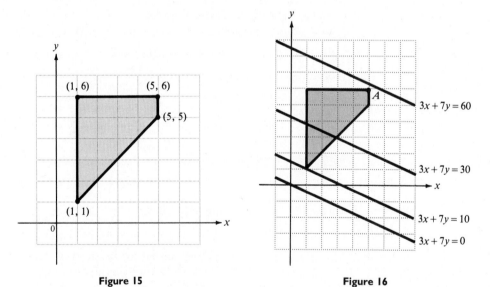

Figure 15

Figure 16

Figure 16 shows the region of feasible solutions together with these lines. From the figure we see that the profit cannot be as large as 60 because the graph for $3x + 7y = 60$ is outside the region of feasible solutions. The point representing maximum possible profit will be found on a line, parallel to the others, between the lines $3x + 7y = 30$ and $3x + 7y = 60$. Profit can be maximized while still satisfying all constraints if this line just touches the feasible region.

This occurs at point A, a **vertex** (or corner) point of the region of feasible solutions. Since the coordinates of this point are (5, 6), the maximum profit is obtained when five tape decks and six amplifiers are produced each day. The maximum profit will be $3(5) + 7(6) = 57$ dollars per day.

As suggested by the parallel lines of Figure 16, the optimum value will always be at a vertex point if the region of feasible solutions is bounded (enclosed).   ■

Example 2   Robin, who is ill, takes vitamin pills. Each day, she must have at least 16 units of Vitamin A, at least 5 units of Vitamin B_1, and at least 20 units of Vitamin C. She can choose between red pills, costing 10¢ each, which contain 8 units of A, 1 of B_1, and 2 of C; and blue pills, costing 20¢ each, which contain 2 units of A, 1 of B_1, and 7 of C. How many of each pill should she buy in order to minimize her cost and yet fulfill her daily requirements?

Let x represent the number of red pills to buy, and let y represent the number of blue pills to buy. Then the cost in pennies per day is given by

$$\text{cost} = 10x + 20y.$$

Since Robin buys x of the 10¢ pills and y of the 20¢ pills, she gets Vitamin A as follows: 8 units from each red pill and 2 units from each blue pill. Altogether, she gets $8x + 2y$ units of A per day. Since she must get at least 16 units,

$$8x + 2y \geq 16.$$

Each red pill and each blue pill supplies 1 unit of Vitamin B_1. Robin needs at least 5 units per day, so that

$$x + y \geq 5.$$

For Vitamin C the inequality is

$$2x + 7y \geq 20.$$

Also, $x \geq 0$ and $y \geq 0$, since Robin cannot buy negative numbers of the pills.

Again, total cost of the pills is minimized by the solution of the system of inequalities formed by the constraints. (See Figure 17.) The solution to this minimizing

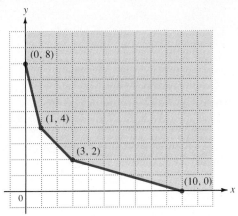

Figure 17

problem will also occur at a vertex point. By substituting the coordinates of the vertex points in the cost function, the lowest cost can be found.

Point	Cost $= 10x + 20y$	
(10, 0)	$10(10) + 20(0) = 100$	
(3, 2)	$10(3) + 20(2) = 70$	← minimum
(1, 4)	$10(1) + 20(4) = 90$	
(0, 8)	$10(0) + 20(8) = 160$	

Robin's best bet is to buy 3 red pills and 2 blue ones, for a total cost of 70¢ per day. She receives just the minimum amounts of Vitamins B_1 and C, but an excess of Vitamin A. Even though she has an excess of A, this is still the best buy. ∎

6.9 Exercises

The graphs in Exercises 1–4 show regions of feasible solutions. Find the maximum and minimum values of the given expressions.

1. $3x + 5y$

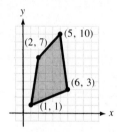

2. $6x + y$

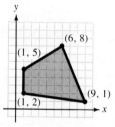

3. $40x + 75y$

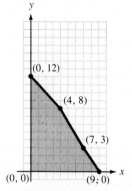

4. $35x + 125y$

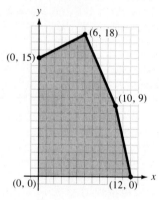

In Exercises 5–10, use graphical methods to solve each problem.

5. Find $x \geq 0$ and $y \geq 0$ such that
$$2x + 3y \leq 6$$
$$4x + y \leq 6$$
and $5x + 2y$ is maximized.

6. Find $x \geq 0$ and $y \geq 0$ such that
$$x + y \leq 10$$
$$5x + 2y \geq 20$$
$$2y \geq x$$
and $x + 3y$ is minimized.

7. Find $x \geq 2$ and $y \geq 5$ such that
$$3x - y \geq 12$$
$$x + y \leq 15$$
and $2x + y$ is minimized.

8. Find $x \geq 10$ and $y \geq 20$ such that
$$2x + 3y \leq 100$$
$$5x + 4y \leq 200$$
and $x + 3y$ is maximized.

9. Find $x \geq 0$ and $y \geq 0$ such that
$$x - y \leq 10$$
$$5x + 3y \leq 75$$
and $4x + 2y$ is maximized.

10. Find $x \geq 0$ and $y \geq 0$ such that
$$10x - 5y \leq 100$$
$$20x + 10y \geq 150$$
and $4x + 5y$ is minimized.

11. Maximize $10x + 12y$ subject separately to each set of constraints.

(a) $x + y \leq 20$
$x + 3y \leq 24$
$x \geq 0$
$y \geq 0$

(b) $3x + y \leq 15$
$x + 2y \leq 18$
$x \geq 0$
$y \geq 0$

(c) $2x + 5y \geq 22$
$4x + 3y \leq 28$
$2x + 2y \leq 17$
$x \geq 0$
$y \geq 0$

12. Minimize $3x + 2y$ subject separately to each set of constraints.

(a) $10x + 7y \leq 42$
$4x + 10y \geq 35$
$x \geq 0$
$y \geq 0$

(b) $6x + 5y \geq 25$
$2x + 6y \geq 15$
$x \geq 0$
$y \geq 0$

(c) $x + 2y \geq 10$
$2x + y \geq 12$
$x - y \leq 8$
$x \geq 0$
$y \geq 0$

Solve each of the following linear programming problems.

13. Farmer Jones raises only pigs and geese. She wants to raise no more than 16 animals with no more than 12 geese. She spends \$5 to raise a pig and \$2 to raise a goose. She has \$50 available for this purpose. Find the maximum profit she can make if she makes a profit of \$8 per goose and \$4 per pig.

14. A wholesaler of party goods wishes to display her products at a convention of social secretaries in such a way that she gets the maximum number of inquiries about her whistles and hats. Her booth at the convention has 12 square meters of floor space to be used for display purposes. A display unit for hats requires 2 square meters, and for whistles, 4 square meters. Experience tells the wholesaler that she should never have more than a total of 5 units of whistles and hats on display at one time. If she receives three inquiries for each unit of hats and two inquiries for each unit of whistles on display, how many of each should she display in order to get the maximum number of inquiries?

15. An office manager wants to buy some filing cabinets. She knows that cabinet #1 costs \$10 each, requires 6 square feet of floor space, and holds 8 cubic feet of files. Cabinet #2 costs \$20 each, requires 8 square feet of floor space, and holds 12 cubic feet. She can spend no more than \$140 due to budget limitations, while her office has room for no more than 72 square feet of cabinets. She wants the maximum storage capacity within the limits imposed by funds and space. How many of each type of cabinet should she buy?

16. The manufacture of "Smokey the Bear" ashtrays and "Superman" cufflinks requires two machines, a drill and a saw. Each ashtray needs one minute on the drill and two minutes on the saw; each cufflink set needs two minutes on the drill and one minute on the saw. The drill is available at most for 12 minutes per day, and the saw is available no more than 15 minutes per day. Each product provides a profit of \$1. How many of each should be manufactured in order to maximize profit?

17. The manufacturing process requires that oil refineries must manufacture at least two barrels of gasoline for every one of fuel oil. To meet the winter demand for fuel oil, at least 3 million barrels a day must be produced. The demand for gasoline is no more than

6.4 million barrels per day. If the price of gasoline is \$1.90 and the price of fuel oil is \$1.50 per gallon, how much of each should be produced to maximize revenue?

18. Theo, who is dieting, requires two food supplements, I and II. He can get these supplements from two different products, *A* and *B,* as shown in the following table.

		Supplement (grams per serving)	
		I	II
Product	*A*	3	2
	B	2	4

Theo's physician has recommended that he include at least 15 grams of each supplement in his daily diet. If product *A* costs 25¢ per serving and product *B* costs 40¢ per serving, how can he satisfy his requirements most economically?

19. A small country can grow only two crops for export, coffee and cocoa. The country has 500,000 hectares* of land available for the crops. Long-term contracts require that at least 100,000 hectares be devoted to coffee and at least 200,000 hectares to cocoa. Cocoa must be processed locally, and production bottlenecks limit cocoa to 270,000 hectares. Coffee requires two workers per hectare; cocoa requires five. No more than 1,750,000 people are available for these crops. Coffee produces a profit of \$220 per hectare, and cocoa a profit of \$310 per hectare. How many hectares should the country devote to each crop in order to maximize profit?

20. In a small town in South Carolina, zoning rules require that the window space (in square feet) in a house be at least one-sixth of the space used up by solid walls. The cost to heat the house is 2¢ for each square foot of solid walls and 8¢ for each square foot of windows. Find the maximum total area (windows plus walls) if \$16 is available to pay for heat.

21. A machine shop manufactures two types of bolts. Each can be made on any of three groups of machines, but the time required on each group differs, as shown in the table below.

		Machine groups		
		I	II	III
Bolts	Type 1	0.4 hour	0.5 hour	0.2 hour
	Type 2	0.3 hour	0.2 hour	0.4 hour

Production schedules are made up for one week at a time. In this period there are 1200 hours of machine time available for each machine group. Type 1 bolts sell for 10¢ and type 2 bolts for 12¢. How many of each type of bolt should be manufactured per week to maximize revenue? What is the maximum revenue?

22. Seall Manufacturing Company makes color television sets. It produces a bargain set that sells for \$100 profit and a deluxe set that sells for \$150 profit. On the assembly line the bargain set requires 3 hours, while the deluxe set takes 5 hours. The cabinet shop spends one hour on the cabinet for the bargain set and 3 hours on the cabinet for the deluxe set. Both sets require 2 hours of time for testing and packing. On a particular production run the Seall Company has available 3900 work hours on the assembly line, 2100 work hours in the cabinet shop, and 2200 work hours in the testing and packing department. How many sets of each type should it produce to make maximum profit? What is the maximum profit?

*A hectare is a unit of land measurement; one hectare is equivalent to approximately 2.47 acres.

Chapter 6 Summary

Key Words

system of linear equations	Gaussian reduction method	column matrix
elimination method	square matrix	identity matrix
substitution method	determinant	multiplication inverse of a
nonlinear system	minor	matrix
matrix	cofactor	Cramer's rule
order	zero matrix	region of feasible solutions
row transformations	row matrix	

Review Exercises

Use the elimination or substitution method to solve each of the following linear systems. Identify any dependent or inconsistent systems.

1. $3x - 5y = -18$
$\quad 2x + 7y = 19$

2. $6x + 5y = 53$
$\quad 4x - 3y = 29$

3. $\dfrac{2}{3}x - \dfrac{3}{4}y = 13$
$\quad \dfrac{1}{2}x + \dfrac{2}{3}y = -5$

4. $3x + 7y = 10$
$\quad 18x + 42y = 50$

5. $\dfrac{1}{x} + \dfrac{1}{y} = \dfrac{7}{10}$
$\quad \dfrac{3}{x} - \dfrac{5}{y} = \dfrac{1}{2}$

6. $.9x - .2y = .8$
$\quad .3x + .7y = 4.1$

7. $2x - 3y + z = -5$
$\quad x + 4y + 2z = 13$
$\quad 5x + 5y + 3z = 14$

8. $x - 3y \quad = 12$
$\quad 2y + 5z = 1$
$\quad 4x \quad + z = 25$

9. A student bought some candy bars, paying 25¢ each for some and 50¢ each for others. The student bought a total of 22 bars, paying a total of $8.50. How many of each kind of bar did he buy?

10. Ink worth $25 a bottle is to be mixed with ink worth $18 per bottle to get 12 bottles of ink worth $20 each. How much of each type of ink should be used?

11. Donna Sharp wins $50,000 in a lottery. She invests part of the money at 6%, twice as much at 7%, with $10,000 more than the amount invested at 6% invested at 9%. Total annual interest is $3800. How much is invested at each rate?

12. The sum of three numbers is 23. The second number is 3 more than the first. The sum of the first and twice the third is 4. Find the three numbers.

Find solutions for the following systems in terms of an arbitrary variable.

13. $3x - 4y + z = 2$
$\quad 2x + y - 4z = 1$

14. $ax + by + cz = 5$
$\quad dx + y \quad = 1$

Solve each of the following systems.

15. $x^2 - 4y^2 = 19$
$\quad x^2 + y^2 = 29$

16. $xy = 4$
$\quad x - 6y = 2$

17. $x^2 + 2xy + y^2 = 4$
$$x = 3y - 2$$

18. $y = 5^{x+5}$
$$y = 25^{3x}$$

19. Do the circle $x^2 + y^2 = 144$ and the line $x + 2y = 8$ have any points in common? If so, what are they?

20. Find a value of b so that the straight line $3x - y = b$ touches the circle $x^2 + y^2 = 25$ in only one point.

Find the values of all variables in the following.

21. $\begin{bmatrix} 5 & x + 2 \\ -6y & z \end{bmatrix} = \begin{bmatrix} a & 3x - 1 \\ 5y & 9 \end{bmatrix}$

22. $\begin{bmatrix} -6 + k & 2 & a + 3 \\ -2 + m & 3p & 2r \end{bmatrix} + \begin{bmatrix} 3 - 2k & 5 & 7 \\ 5 & 8p & 5r \end{bmatrix} = \begin{bmatrix} 5 & y & 6a \\ 2m & 11 & -35 \end{bmatrix}$

Perform each of the following operations whenever possible.

23. $\begin{bmatrix} 3 & -4 & 2 \\ 5 & -1 & 6 \end{bmatrix} + \begin{bmatrix} -3 & 2 & 5 \\ 1 & 0 & 4 \end{bmatrix}$

24. $\begin{bmatrix} 3 \\ 2 \\ 5 \end{bmatrix} - \begin{bmatrix} 8 \\ -4 \\ 6 \end{bmatrix} + \begin{bmatrix} 1 \\ 0 \\ 2 \end{bmatrix}$

25. $\begin{bmatrix} 2 & 5 & 8 \\ 1 & 9 & 2 \end{bmatrix} - \begin{bmatrix} 3 & 4 \\ 7 & 1 \end{bmatrix}$

26. $\begin{bmatrix} -3 & 4 \\ 2 & 8 \end{bmatrix}\begin{bmatrix} -1 & 0 \\ 2 & 5 \end{bmatrix}$

27. $\begin{bmatrix} 3 & 2 & -1 \\ 4 & 0 & 6 \end{bmatrix}\begin{bmatrix} -2 & 0 \\ 0 & 2 \\ 3 & 1 \end{bmatrix}$

28. $\begin{bmatrix} 1 & -2 & 4 & 2 \\ 0 & 1 & -1 & 8 \end{bmatrix}\begin{bmatrix} -1 \\ 2 \\ 0 \\ 1 \end{bmatrix}$

Use the Gaussian reduction method to solve each of the following.

29. $2x + 3y = 10$
$$-3x + y = 18$$

30. $5x + 2y = -10$
$$3x - 5y = -6$$

31. $2x - y + 4z = -1$
$$-3x + 5y - z = 5$$
$$2x + 3y + 2z = 3$$

32. $5x - 8y + z = 1$
$$3x - 2y + 4z = 3$$
$$10x - 16y + 2z = 3$$

Find the inverse of each of the following matrices that have inverses.

33. $\begin{bmatrix} 2 & 1 \\ 5 & 3 \end{bmatrix}$

34. $\begin{bmatrix} -4 & 2 \\ 0 & 3 \end{bmatrix}$

35. $\begin{bmatrix} 2 & -1 & 0 \\ 1 & 0 & 1 \\ 1 & -2 & 0 \end{bmatrix}$

36. $\begin{bmatrix} 2 & 3 & 5 \\ -2 & -3 & -5 \\ 1 & 4 & 2 \end{bmatrix}$

Use the method of matrix inverses to solve each of the following.

37. $2x + y = 5$
$$3x - 2y = 4$$

38. $x + y + z = 1$
$$2x - y = -2$$
$$3y + z = 2$$

39. $x = -3$
$$y + z = 6$$
$$2x - 3z = -9$$

40. $3x - 2y + 4z = 1$
$$4x + y - 5z = 2$$
$$-6x + 4y - 8z = -2$$

Find each of the following determinants.

41. $\begin{vmatrix} -1 & 8 \\ 2 & 9 \end{vmatrix}$

42. $\begin{vmatrix} -2 & 4 \\ 0 & 3 \end{vmatrix}$

43. $\begin{vmatrix} -2 & 4 & 1 \\ 3 & 0 & 2 \\ -1 & 0 & 3 \end{vmatrix}$

44. $\begin{vmatrix} -1 & 2 & 3 \\ 4 & 0 & 3 \\ 5 & -1 & 2 \end{vmatrix}$

45. $\begin{vmatrix} -1 & 0 & 2 & -3 \\ 0 & 4 & 4 & -1 \\ -6 & 0 & 3 & -5 \\ 0 & -2 & 1 & 0 \end{vmatrix}$

Explain why each of the following statements is true.

46. $\begin{vmatrix} 8 & 9 & 2 \\ 0 & 0 & 0 \\ 3 & 1 & 4 \end{vmatrix} = 0$

47. $\begin{vmatrix} 4 & 6 \\ 3 & 5 \end{vmatrix} = \begin{vmatrix} 4 & 3 \\ 6 & 5 \end{vmatrix}$

48. $\begin{vmatrix} 8 & 2 \\ 4 & 3 \end{vmatrix} = 2 \begin{vmatrix} 4 & 1 \\ 4 & 3 \end{vmatrix}$

49. $\begin{vmatrix} 4 & 6 & 2 \\ -3 & 8 & -5 \\ 4 & 6 & 2 \end{vmatrix} = 0$

50. $\begin{vmatrix} 5 & -1 & 2 \\ 3 & -2 & 0 \\ -4 & 1 & 2 \end{vmatrix} = \begin{vmatrix} 5 & -1 & 2 \\ 8 & -3 & 2 \\ -4 & 1 & 2 \end{vmatrix}$

51. $\begin{vmatrix} 8 & 2 & -5 \\ -3 & 1 & 4 \\ 2 & 0 & 5 \end{vmatrix} = - \begin{vmatrix} 8 & -5 & 2 \\ -3 & 4 & 1 \\ 2 & 5 & 0 \end{vmatrix}$

Solve each of the following systems by Cramer's rule. Identify any dependent or inconsistent systems.

52. $3x + y = -1$
$5x + 4y = 10$

53. $3x + 7y = 2$
$5x - y = -22$

54. $3x + 2y + z = 2$
$4x - y + 3z = -16$
$x + 3y - z = 12$

55. $5x - 2y - z = 8$
$-5x + 2y + z = -8$
$x - 4y - 2z = 0$

Graph the solution of each of the following systems of inequalities.

56. $x + y \leq 6$
$2x - y \geq 3$

57. $x - 3y \geq 6$
$y^2 \leq 16 - x^2$

58. $9x^2 + 16y^2 \geq 144$
$x^2 - y^2 \leq 16$

59. A bakery makes both cakes and cookies. Each batch of cakes requires two hours in the oven and three hours in the decorating room. Each batch of cookies needs one and a half hours in the oven and two-thirds of an hour in the decorating room. The oven is available no more than 16 hours a day, while the decorating room can be used no more than 12 hours per day. Set up a system of inequalities expressing this information, and then graph the system.

60. A candy company has 100 kilograms of chocolate-covered nuts and 125 kilograms of chocolate-covered raisins to be sold as two different mixtures. One mix will contain 1/2 nuts and 1/2 raisins, while the other mix will contain 1/3 nuts and 2/3 raisins. Set up a system of inequalities expressing this information, and then graph the system.

In each of the following problems, use graphical methods to find nonnegative values of x and y that meet the constraints.

61. Maximize $5x + 7y$ subject to

$$2x + 4y \geq 40$$
$$3x + 2y \leq 60.$$

62. Maximize $2x + 3y$ subject to

$$x + 2y \leq 24$$
$$3x + 4y \leq 60.$$

63. Minimize $x + 2y$ subject to

$$5x + 4y \geq 20$$
$$2x + 6y \geq 24$$
$$x + y \leq 12.$$

64. Minimize $4x + 3y$ subject to

$$x + 2y \leq 12$$
$$4x + 3y \geq 12$$
$$y \leq 2x.$$

65. In Exercise 59, a batch of cookies produces a profit of $20; the profit on a batch of cakes is $30. Find the number of batches of each item which will maximize profit.

66. In Exercise 60, how much of each mixture should be made to maximize revenue if the first mix sells for $6.00 per kilogram and the second mix sells for $4.80 per kilogram?

67. Given the line $y = mx + b$ tangent to the circle $x^2 + y^2 = r^2$, find an equation for x involving m, b, and r.

68. Find an equation of the circle having as its diameter the common chord of the two circles $x^2 + y^2 + 2x - 2y - 14 = 0$ and $x^2 + y^2 - 4x + 4y - 2 = 0$.

7

Zeros of Polynomials

Recall from Chapter 4 that f is a *polynomial function of degree n* if

$$f(x) = a_n x^n + a_{n-1} x^{n-1} + \cdots + a_1 x + a_0,$$

where n is a nonnegative integer and $a_n \neq 0$. In Chapter 4 the coefficients a_0, a_1, $\cdots$, a_n were all real numbers. In this chapter, we extend the definition to include coefficients that are complex numbers.* The number a_n is the **leading coefficient** of $f(x)$. If all the coefficients of a polynomial expression are 0, the polynomial is 0 and is called the **zero polynomial.** The zero polynomial has no degree. However, a polynomial $f(x) = a_0$ for a nonzero complex number a_0 has degree 0.

The values of x that satisfy the equation $f(x) = 0$ are called the **zeros** of $f(x)$. Methods for finding zeros of first- and second-degree polynomials were given in Chapter 2. In this chapter several theorems are presented that help to find, or at least approximate, the zeros of polynomials of higher degree.

7.1 / Polynomial Division

The quotient of two polynomials can be found with a **division algorithm** very similar to that used for dividing whole numbers. (An *algorithm* is a step-by-step procedure for working a problem.)

---

*Remember, *complex numbers* are imaginary numbers like $5i$ or $-i\sqrt{3}$, real numbers, or numbers such as $4 - 2i$ or $6 + i\sqrt{7}$.

Example 1 Divide $4m^3 - 8m^2 + 4m + 6$ by $2m - 1$.

Work as follows.

$4m^3$ divided by $2m$ is $2m^2$

$- 6m^2$ divided by $2m$ is $-3m$

m divided by $2m$ is $\frac{1}{2}$

$$
\begin{array}{r}
2m^2 - 3m + \frac{1}{2} \\
2m - 1\overline{)4m^3 - 8m^2 + 4m + 6} \\
\underline{4m^3 - 2m^2} \\
- 6m^2 + 4m \\
\underline{- 6m^2 + 3m} \\
m + 6 \\
\underline{m - \frac{1}{2}} \\
\frac{13}{2}
\end{array}
$$

$2m^2(2m - 1) = 4m^3 - 2m^2$

Subtract; bring down the next term

$-3m(2m - 1) = -6m^2 + 3m$

Subtract; bring down the next term

$(1/2)(2m - 1) = m - (1/2)$

Subtract. The remainder is 13/2.

By this work,

$$\frac{4m^3 - 8m^2 + 4m + 6}{2m - 1} = 2m^2 - 3m + \frac{1}{2} + \frac{13/2}{2m - 1},$$

or

$$4m^3 - 8m^2 + 4m + 6 = (2m - 1)\left(2m^2 - 3m + \frac{1}{2}\right) + \frac{13}{2}.$$

In dividing these polynomials, $4m^3 - 2m^2$ is subtracted from $4m^3 - 8m^2 + 4m + 6$. The result, $-6m^2 + 4m + 6$, should be written under the line. However, it is customary to save work and "bring down" just the $4m$, the only term needed for the next step. ■

The polynomial $3x^3 - 2x^2 - 150$ has a missing term, the term in which the power of x is 1. When a polynomial with a missing term is divided, it is useful to allow for that term by leaving space for it as shown in the next example.

Example 2 Divide $3x^3 - 2x^2 - 150$ by $x - 4$.

$$
\begin{array}{r}
3x^2 + 10x + 40 \\
x - 4\overline{)3x^3 - 2x^2 \qquad - 150} \\
\underline{3x^3 - 12x^2} \\
10x^2 \\
\underline{10x^2 - 40x} \\
40x - 150 \\
\underline{40x - 160} \\
10
\end{array}
$$

The result of this division can be written

$$\frac{3x^3 - 2x^2 - 150}{x - 4} = 3x^2 + 10x + 40 + \frac{10}{x - 4}$$

or

$$3x^3 - 2x^2 - 150 = (x - 4)(3x^2 + 10x + 40) + 10. \quad ■$$

The following theorem generalizes the division process illustrated in Examples 1 and 2.

Division Algorithm

Let $f(x)$ and $g(x)$ be polynomials with $g(x)$ of lower degree than $f(x)$ and $g(x) \neq 0$. There exist unique polynomials $q(x)$ and $r(x)$ such that

$$f(x) = g(x) \cdot q(x) + r(x),$$

where $r(x) = 0$ or $r(x)$ is of degree less than the degree of $g(x)$.

The polynomial $f(x)$ in the division algorithm is the **dividend** and $g(x)$ is the **divisor**. The polynomial $q(x)$ is the **quotient polynomial** or the **quotient,** while $r(x)$ is the **remainder polynomial** or the **remainder.**

The division algorithm applies to the polynomials in Examples 1 and 2. In Example 1,

$$\frac{4m^3 - 8m^2 + 4m + 6}{2m - 1} = 2m^2 - 3m + \frac{1}{2} + \frac{13/2}{2m - 1}.$$

from which

$$4m^3 - 8m^2 + 4m + 6 = \underbrace{(2m - 1)}_{} \cdot \underbrace{\left(2m^2 - 3m + \frac{1}{2} \right)}_{} + \underbrace{\frac{13}{2}}_{}.$$

$$\underbrace{}_{f(x)} = \underbrace{}_{g(x)} \cdot \underbrace{}_{q(x)} + \underbrace{}_{r(x)}$$

Identify $q(x)$ and $r(x)$ in Example 2.

Synthetic Division   It is common to divide a polynomial by a first-degree binomial of the form $x - k$, where the coefficient of x is 1. A helpful shortcut for division problems of this type can be developed. To illustrate, Example 2 is reworked on the left below. On the right the division process is simplified by omitting all variables and writing only coefficients, with 0 used to represent the coefficient of any missing terms. Since the coefficient of x in the divisor is always 1 in these divisions it too can be omitted. These omissions simplify the problem as shown on the right below.

$$
\begin{array}{r}
3x^2 + 10x + 40 \\
x - 4 \overline{)3x^3 - 2x^2 - 150} \\
\underline{3x^3 - 12x^2} \\
10x^2 \\
\underline{10x^2 - 40x} \\
40x - 150 \\
\underline{40x - 160} \\
10
\end{array}
\qquad
\begin{array}{r}
3 \quad 10 \quad 40 \\
-4 \overline{)3 - 2 + 0 - 150} \\
\underline{3 - 12} \\
10 \\
\underline{10 - 40} \\
40 - 150 \\
\underline{40 - 160} \\
10
\end{array}
$$

The numbers in color are repetitions of the numbers directly above and can also be omitted.

$$\begin{array}{r}
3 \quad10 \quad40 \\
-4)\overline{3 \;-\; 2 \;+\; 0 \;-\; 150} \\
\underline{-\;12} \\
10 \\
\end{array}$$

$$\begin{array}{r}
-\;40 \\
\underline{} \\
40 \;-\; 150 \\
\underline{-\;160} \\
10 \\
\end{array}$$

The entire problem can now be condensed vertically, and the top row of numbers omitted since it duplicates the bottom row.

$$\begin{array}{r|rrrr}
-4) & 3 & -2 & 0 & -150 \\
& & -12 & -40 & -160 \\
\hline
& 3 & 10 & 40 & 10 \\
\end{array}$$

The bottom row is obtained by subtracting -12, -40, and -160 from the corresponding terms above. For reasons that will become clear in the discussion of the next theorem, the -4 at the left is changed to 4, which also changes the sign of the numbers in the second row. To compensate for this change, subtraction is changed to addition. Doing this gives the following result.

$$\begin{array}{r|rrrr}
4) & 3 & -2 & 0 & -150 \\
& & 12 & 40 & 160 \\
\hline
& 3 & 10 & 40 & 10 \\
\end{array}$$

This shortcut process is called **synthetic division.**

In summary, to use synthetic division to divide a polynomial by a binomial of the form $x - k$, begin by writing the coefficients of the polynomial in decreasing powers of the variable, using 0 as the coefficient of any missing powers. The number k is written to the left in the same row. In the example above, $x - k$ is $x - 4$, so k is 4. Next bring down the leading coefficient of the polynomial, 3 in the example above, as the first number in the last row. Multiply the 3 by 4 to get the first number in the second row, 12. Add 12 to -2; this gives 10, the second number in the third row. Multiply 10 by 4 to get 40, the next number in the second row. Add 40 to 0 to get the third number in the third row, and so on. This process of multiplying each result in the third row by k and adding the product to the number in the next column is repeated until there is a number in the last row for each coefficient in the first row. To avoid incorrect results, it is very important to use a 0 for any missing terms when setting up the division.

Example 3   Use synthetic division to divide $5m^3 - 6m^2 - 28m - 2$ by $m + 2$.

Begin by writing

$$\begin{array}{r|rrrr}
-2) & 5 & -6 & -28 & -2. \\
\end{array}$$

The 2 is changed to -2 since k is found by writing $m + 2$ as $m - (-2)$. Next, bring down the 5.

$$\begin{array}{r|rrrr}
-2) & 5 & -6 & -28 & -2 \\
\hline
& 5 & & & \\
\end{array}$$

Now, multiply -2 by 5 to get -10, and add it to the -6 in the first row. The result is -16.

$$
\begin{array}{r|rrrr}
-2) & 5 & -6 & -28 & -2 \\
    &   & -10 &   &   \\
\hline
    & 5 & -16 &   &
\end{array}
$$

Next, $(-2)(-16) = 32$. Add this to the -28 in the first row.

$$
\begin{array}{r|rrrr}
-2) & 5 & -6 & -28 & -2 \\
    &   & -10 & 32 &   \\
\hline
    & 5 & -16 & 4 &
\end{array}
$$

Finally, $(-2)(4) = -8$, giving

$$
\begin{array}{r|rrrr}
-2) & 5 & -6 & -28 & -2 \\
    &   & -10 & 32 & -8 \\
\hline
    & 5 & -16 & 4 & -10.
\end{array}
$$

The coefficients of the quotient polynomial and the remainder are read directly from the bottom row. Since the degree of the quotient will always be one less than the degree of the polynomial to be divided,

$$\frac{5m^3 - 6m^2 - 28m - 2}{m + 2} = 5m^2 - 16m + 4 + \frac{-10}{m + 2}. \quad \blacksquare$$

The result of the division in Example 3 can be written as

$$5m^3 - 6m^2 - 28m - 2 = (m + 2)(5m^2 - 16m + 4) + (-10)$$

by multiplying both sides by the denominator $m + 2$. The following theorem is a generalization of the division process illustrated above.

> For any polynomial $f(x)$ and any complex number k, there exists a unique polynomial $q(x)$ and number r such that
>
> $$f(x) = (x - k)q(x) + r.$$

For example, in the synthetic division above,

$$5m^3 - 6m^2 - 28m - 2 = (m + 2)(5m^2 - 16m + 4) + (-10).$$

$$\underbrace{}_{f(x)} = \underbrace{(x - k)}_{} \cdot \underbrace{}_{q(x)} + \underbrace{r}_{}$$

This theorem is a special case of the division algorithm given earlier. Here $g(x)$ is the first-degree polynomial $x - k$.

Evaluating $f(k)$ By the division algorithm, $f(x) = (x - k)q(x) + r$. This equality is true for all complex values of x, so it is true for $x = k$. Replacing x with k gives

$$f(k) = (k - k)q(k) + r$$
$$f(k) = r.$$

This proves the following remainder theorem.

<table>
<tr><td>**Remainder Theorem**</td><td>If the polynomial $f(x)$ is divided by $x - k$, the remainder is $f(k)$.</td></tr>
</table>

As an example of this theorem, we have seen that the polynomial $f(m) = 5m^3 - 6m^2 - 28m - 2$ can be written as

$$f(m) = (m + 2)(5m^2 - 16m + 4) + (-10).$$

To find $f(-2)$, substitute -2 for m.

$$f(-2) = (-2 + 2)[5(-2)^2 - 16(-2) + 4] + (-10)$$
$$= 0[5(-2)^2 - 16(-2) + 4] + (-10)$$
$$= -10$$

By the remainder theorem, instead of replacing m by -2 to find $f(-2)$, divide $f(m)$ by $m + 2$ using synthetic division. Then $f(-2)$ is the remainder, -10. Work as follows.

$$
\begin{array}{r|rrrr}
-2) & 5 & -6 & -28 & -2 \\
    &   & -10 & 32 & -8 \\
\hline
    & 5 & -16 & 4 & -10 \quad \leftarrow f(-2)
\end{array}
$$

Example 4   Let $f(x) = -x^4 + 3x^2 - 4x - 5$. Find $f(-2)$.
Use the remainder theorem and synthetic division.

$$
\begin{array}{r|rrrrr}
-2) & -1 & 0 & 3 & -4 & -5 \\
    &   & 2 & -4 & 2 & 4 \\
\hline
    & -1 & 2 & -1 & -2 & -1
\end{array}
$$

The remainder when $f(x)$ is divided by $x - (-2) = x + 2$ is -1, so $f(-2) = -1$.   ■

The remainder theorem gives a quick way to decide if a number k is a zero of a polynomial $f(x)$. Use synthetic division to find $f(k)$; if the remainder is zero, then $f(k) = 0$ and k is a zero of $f(x)$. When $f(k) = 0$, the number k is also called a **root** or **solution** of the equation $f(x) = 0$.

Example 5   Decide whether or not the given number is a zero of the given polynomial.

(a) 2; $f(x) = x^3 - 4x^2 + 9x - 10$
Use synthetic division.

$$
\begin{array}{r|rrrr}
2) & 1 & -4 & 9 & -10 \\
   &   & 2 & -4 & 10 \\
\hline
   & 1 & -2 & 5 & 0
\end{array}
$$

Since the remainder is 0, $f(2) = 0$, and 2 is a zero of the polynomial $f(x) = x^3 - 4x^2 + 9x - 10$.

(b) -4; $f(x) = x^4 + x^2 - 3x + 1$

Remember to use a coefficient of 0 for the missing x^3 term in the synthetic division.

$$
\begin{array}{r|rrrrr}
-4) & 1 & 0 & 1 & -3 & 1 \\
& & -4 & 16 & -68 & 284 \\
\hline
& 1 & -4 & 17 & -71 & 285
\end{array}
$$

The remainder is not 0, so -4 is not a zero of $f(x) = x^4 + x^2 - 3x + 1$. In fact, $f(-4) = 285$.

(c) $1 + 2i$; $f(x) = x^4 - 2x^3 + 4x^2 + 2x - 5$

Use synthetic division and operations with complex numbers.

$$
\begin{array}{r|rrrrr}
1 + 2i) & 1 & -2 & 4 & 2 & -5 \\
& & 1 + 2i & -5 & -1 - 2i & 5 \\
\hline
& 1 & -1 + 2i & -1 & 1 - 2i & 0
\end{array}
$$

The zero remainder indicates that $1 + 2i$ is a zero of $f(x) = x^4 - 2x^3 + 4x^2 + 2x - 5$. ■

7.1 Exercises

Use synthetic division to perform each of the following divisions.

1. $\dfrac{x^3 + 2x^2 - 17x - 10}{x + 5}$

2. $\dfrac{a^4 + 4a^3 + 2a^2 + 9a + 4}{a + 4}$

3. $\dfrac{m^4 - 3m^3 - 4m^2 + 12m}{m - 2}$

4. $\dfrac{p^4 - 3p^3 - 5p^2 + 2p - 16}{p + 2}$

5. $\dfrac{3x^3 - 11x^2 - 20x + 3}{x - 5}$

6. $\dfrac{4p^3 + 8p^2 - 16p - 9}{p + 3}$

7. $\dfrac{4m^3 - 3m - 2}{m + 1}$

8. $\dfrac{3q^3 - 4q + 2}{q - 1}$

9. $\dfrac{x^5 + 3x^4 + 2x^3 + 2x^2 + 3x + 1}{x + 2}$

10. $\dfrac{m^6 - 3m^4 + 2m^3 - 6m^2 - 5m + 3}{m + 2}$

11. $\dfrac{\frac{1}{3}x^3 - \frac{2}{9}x^2 + \frac{1}{27}x + 1}{x - \frac{1}{3}}$

12. $\dfrac{x^3 + x^2 + \frac{1}{2}x + \frac{1}{8}}{x + \frac{1}{2}}$

13. $\dfrac{9z^3 - 8z^2 + 7z - 2}{z - 2}$

14. $\dfrac{-11p^4 + 2p^3 - 8p^2 - 4}{p + 1}$

15. $\dfrac{y^3 - 1}{y - 1}$

16. $\dfrac{r^5 - 1}{r - 1}$

17. $\dfrac{x^4 - 1}{x - 1}$

18. $\dfrac{x^7 + 1}{x + 1}$

Express each polynomial in the form $f(x) = (x - k)q(x) + r$ for the given value of k.

19. $f(x) = x^3 + x^2 + x - 8$; $k = 1$

20. $f(x) = 2x^3 + 3x^2 + 4x - 10$; $k = -1$

21. $f(x) = -x^3 + 2x^2 + 4$; $k = -2$

22. $f(x) = -4x^3 + 2x^2 - 3x - 10$; $k = 2$

23. $f(x) = x^4 - 3x^3 + 2x^2 - x$; $k = 3$

24. $f(x) = 2x^4 + x^3 - 15x^2 + 3x$; $k = -3$

25. $f(x) = 3x^4 + 4x^3 - 10x^2 + 15$; $k = -1$

26. $f(x) = -5x^4 + x^3 + 2x^2 + 3x + 1$; $k = 1$

For each of the following polynomials, use the remainder theorem to find $f(k)$.

27. $k = 5$; $f(x) = -x^2 + 2x + 7$

28. $k = -3$; $f(x) = 3x^2 + 8x + 5$

29. $k = 3$; $f(x) = x^2 - 4x + 5$

30. $k = -2$; $f(x) = x^2 + 5x + 6$

31. $k = -1$; $f(x) = x^3 - 4x^2 + 2x + 1$

32. $k = 2$; $f(x) = 2x^3 - 3x^2 - 5x + 4$

33. $k = 2 + i$; $f(x) = x^2 - 5x + 1$

34. $k = 3 - 2i$; $f(x) = x^2 - x + 3$

35. $k = 1 - i$; $f(x) = x^3 + x^2 - x + 1$

36. $k = 2 - 3i$; $f(x) = x^3 + 2x^2 + x - 5$

Use synthetic division to decide whether or not the given number is a zero of the given polynomial.

37. 2; $f(x) = x^2 + 2x - 8$

38. -1; $f(m) = m^2 + 4m - 5$

39. 2; $f(g) = g^3 - 3g^2 + 4g - 4$

40. -3; $f(m) = m^3 + 2m^2 - m + 6$

41. 4; $f(r) = 2r^3 - 6r^2 - 9r + 4$

42. -6; $f(y) = 2y^3 + 9y^2 - 16y + 12$

43. -5; $f(x) = x^3 + 7x^2 + 10x$

44. 3; $f(y) = 2y^3 - 3y^2 - 5y$

45. $-3/2$; $f(m) = 2m^4 + 3m^3 - 8m^2 - 2m + 15$

46. $-4/3$; $f(a) = 3a^4 + 2a^3 - 5a + 10$

47. $2/5$; $f(z) = 5z^4 + 2z^3 - z + 3$

48. $1/2$; $f(w) = 2w^4 - 3w^2 + 4$

49. $1 - i$; $f(y) = y^2 - 2y + 2$

50. $3 - 2i$; $f(r) = r^2 - 6r + 13$

51. $2 + i$; $f(k) = k^2 + 3k + 4$

52. $1 - 2i$; $f(z) = z^2 - 3z + 5$

53. i; $f(x) = x^3 + 2ix^2 + 2x + i$

54. $-i$; $f(p) = p^3 - ip^2 + 3p + 5i$

55. $1 + i$; $f(p) = p^3 + 3p^2 - p + 1$

56. $2 - i$; $f(r) = 2r^3 - r^2 + 3r - 5$

Show that each of the following is true.

57. $f(x) = 2x^4 + 4x^2 + 1$ can have no factor $x - k$ where k is a real number.

58. $f(x) = -x^4 - 5x^2 - 3$ can have no factor $x - k$ where k is a real number.

59. $x - c$ is a factor of $x^n - c^n$ for all positive integers n.

60. $x + c$ is a factor of $x^n + c^n$ for all odd positive integers n.

61. Let $f(x)$ be a polynomial having a zero of c. Let a be a number "close" to c. Write $f(x)$ as $(x - a) \cdot q(x) + r$. Find the following approximation for c:

$$c \approx a - \frac{f(a)}{q(a)}.$$

Use the result of Exercise 61 along with the given value of a to find an approximation to the nearest hundredth of a zero for the following polynomials.

62. $f(x) = x^3 + 7x^2 - 2x - 14$; $a = -1.4$

63. $f(x) = x^3 + 4x^2 - 5x - 5$; $a = 1.5$

64. $f(x) = 2x^3 + 3x^2 - 22x - 33$; $a = 3.2$

7.2 Complex Zeros of Polynomials

By the remainder theorem, if $f(k) = 0$, then the remainder when $f(x)$ is divided by $x - k$ is zero. This means that $x - k$ is a factor of $f(x)$. Conversely, if $x - k$ is a factor of $f(x)$, then $f(k)$ must equal 0. This is summarized in the following *factor theorem*.

Factor Theorem

> The polynomial $x - k$ is a factor of the polynomial $f(x)$ if and only if $f(k) = 0$.

Example 1 Is $x - 1$ a factor of $f(x) = 2x^4 + 3x^2 - 5x + 7$?

By the factor theorem, $x - 1$ will be a factor of $f(x)$ only if $f(1) = 0$. Use synthetic division and the remainder theorem to decide.

$$
\begin{array}{r|rrrr}
1) & 2 & 0 & 3 & -5 & 7 \\
   &   & 2 & 2 & 5 & 0 \\
\hline
   & 2 & 2 & 5 & 0 & 7
\end{array}
$$

Since the remainder is 7, $f(1) = 7$, and not 0, so $x - 1$ is not a factor of $f(x)$. ■

Example 2 Is $x - i$ a factor of $f(x) = 3x^3 + (-4 - 3i)x^2 + (5 + 4i)x - 5i$?

The only way $x - i$ can be a factor of $f(x)$ is for $f(i)$ to be 0. To see if this is the case, use synthetic division.

$$
\begin{array}{r|rrrr}
i) & 3 & -4 - 3i & 5 + 4i & -5i \\
   &   & 3i & -4i & 5i \\
\hline
   & 3 & -4 & 5 & 0
\end{array}
$$

The remainder is 0, so $f(i) = 0$, and $x - i$ is a factor of $f(x)$. The quotient $3x^2 - 4x + 5$ gives the other factor; the polynomial $f(x)$ can be factored as

$$f(x) = (x - i)(3x^2 - 4x + 5).$$ ■

The next theorem says that every polynomial of degree 1 or more has a zero, which means that every such polynomial can be factored. This theorem was first proved by the mathematician K. F. Gauss in his doctoral thesis in 1799 when he was 22 years old. Although many proofs of this result have been given, all of them involve mathematics beyond the algebra in this book, so no proof is included here.

Fundamental Theorem of Algebra

> Every polynomial of degree 1 or more has at least one complex zero.

From the fundamental theorem, if $f(x)$ is of degree 1 or more then there is some number k such that $f(k) = 0$. By the factor theorem, then

$$f(x) = (x - k) \cdot q(x)$$

for some polynomial $q(x)$. The fundamental theorem and the factor theorem can be used to factor $q(x)$ in the same way. Assuming that $f(x)$ has degree n and repeating this process n times gives

$$f(x) = a(x - k_1)(x - k_2) \ldots (x - k_n),$$

where a is the leading coefficient of $f(x)$. Each of these factors leads to a zero of $f(x)$, so $f(x)$ has the n zeros $k_1, k_2, k_3, \ldots, k_n$.

This result is used to prove the next theorem.

Theorem | A polynomial of degree n has at most n distinct zeros.

To prove the theorem, an *indirect proof* is used. To prove a statement indirectly, assume that the opposite of the statement is true. Then show that this leads to a contradiction. Here, to prove that a polynomial of degree $n \geq 1$ has at most n distinct zeros, begin by assuming the opposite—that the polynomial has more than n distinct zeros. Assume that a polynomial $f(x)$ of degree n has $n + 1$ distinct zeros. Since, by the factor theorem,

$$f(x) = a(x - k_1)(x - k_2) \cdots (x - k_n), \tag{1}$$

n of the zeros are the numbers $k_1, k_2, \cdots, k_n$. By the assumption that $f(x)$ has $n + 1$ distinct zeros, there must be another zero, say k, that is not in the list $k_1, k_2, \cdots, k_n$. Since k is a zero of $f(x)$,

$$f(k) = 0.$$

From equation (1), replacing x with k gives

$$f(k) = a(k - k_1)(k - k_2) \cdots (k - k_n)$$

or $$0 = a(k - k_1)(k - k_2) \cdots (k - k_n). \tag{2}$$

Since $k \neq k_1$, $k \neq k_2$, . . . , $k \neq k_n$, we have $k - k_1 \neq 0$, $k - k_2 \neq 0$, . . . , $k - k_n \neq 0$, so that the right side of equation (2) is a product of factors, all of which are nonzero. Since the product of nonzero factors cannot be zero, this is a contradiction. The contradiction shows that the original assumption of more than n distinct zeros is false. This means that a polynomial of degree n has at most n distinct zeros.

The theorem says that there exist *at most* n distinct zeros. For example, the polynomial $f(x) = x^3 + 3x^2 + 3x + 1 = (x + 1)^3$ is of degree 3 but has only one zero, -1. Actually, the zero -1 occurs three times, since there are three factors of $x + 1$; this zero is called a **zero of multiplicity 3.**

Example 3 Find a polynomial $f(x)$ of degree 3 that satisfies the following conditions.

(a) Zeros of -1, 2, and 4; $f(1) = 3$

These three zeros give $x - (-1) = x + 1$, $x - 2$, and $x - 4$ as factors of $f(x)$. Since $f(x)$ is to be of degree 3, these are the only possible factors by the theorem just above. Therefore, $f(x)$ has the form

$$f(x) = a(x + 1)(x - 2)(x - 4)$$

for some real number a. To find a, use the fact that $f(1) = 3$.

$$f(1) = a(1 + 1)(1 - 2)(1 - 4) = 3$$
$$a(2)(-1)(-3) = 3$$
$$6a = 3$$
$$a = \frac{1}{2}$$

Thus, $$f(x) = \frac{1}{2}(x + 1)(x - 2)(x - 4),$$

or, by multiplication,

$$f(x) = \frac{1}{2}x^3 - \frac{5}{2}x^2 + x + 4.$$

(b) -2 is a zero of multiplicity 3; $f(-1) = 4$
The polynomial $f(x)$ has the form

$$f(x) = a(x + 2)(x + 2)(x + 2)$$
$$= a(x + 2)^3.$$

Since $f(-1) = 4$,

$$f(-1) = a(-1 + 2)^3 = 4,$$
$$a(1)^3 = 4$$

or $$a = 4,$$

and $f(x) = 4(x + 2)^3 = 4x^3 + 24x^2 + 48x + 32.$ ■

The remainder theorem can be used to show that both $2 + i$ and $2 - i$ are zeros of $f(x) = x^3 - x^2 - 7x + 15$. It is not just coincidence that both $2 + i$ and its conjugate $2 - i$ are zeros of this polynomial. It turns out that if $a + bi$ is a zero of a polynomial function with *real* coefficients, then so is $a - bi$. To prove this requires the following properties of complex conjugates. Let $z = a + bi$, and write $\bar{z}$ for the conjugate of z, so that $\bar{z} = a - bi$. For example, if $z = -5 + 2i$, then $\bar{z} = -5 - 2i$. The proof of the following equalities is left for the exercises (see Exercises 63–66).

Properties of Conjugates

For any complex numbers c and d,

$$\overline{c + d} = \bar{c} + \bar{d}$$
$$\overline{c \cdot d} = \bar{c} \cdot \bar{d}$$
$$\overline{c^n} = (\bar{c})^n.$$

Now start with the polynomial

$$f(x) = a_n x^n + a_{n-1}x^{n-1} + \ldots + a_1 x + a_0,$$

where all coefficients are real numbers. If $z = a + bi$ is a zero of $f(x)$, then

$$f(z) = a_n z^n + a_{n-1}z^{n-1} + \ldots + a_1 z + a_0 = 0.$$

Taking the conjugate of both sides of this last equation gives

$$\overline{a_n z^n + a_{n-1}z^{n-1} + \ldots + a_1 z + a_0} = \bar{0}.$$

Using generalizations of the properties $\overline{c + d} = \bar{c} + \bar{d}$ and $\overline{c \cdot d} = \bar{c} \cdot \bar{d}$ gives

$$\overline{a_n z^n} + \overline{a_{n-1}z^{n-1}} + \ldots + \overline{a_1 z} + \overline{a_0} = 0$$
or $$\overline{a_n}\, \overline{z^n} + \overline{a_{n-1}}\, \overline{z^{n-1}} + \ldots + \overline{a_1}\, \overline{z} + \overline{a_0} = 0.$$

Now use the third property from above and the fact that for any real number a, $\bar{a} = a$, to get

$$a_n(\bar{z})^n + a_{n-1}(\bar{z})^{n-1} + \ldots + a_1(\bar{z}) + a_0 = 0.$$

Hence $\bar{z}$ is also a zero of $f(x)$, which completes the proof of the *conjugate zeros theorem*.

Conjugate Zeros Theorem

> If $f(x)$ is a polynomial having <u>only real coefficients</u> and if $a + bi$ is a zero of $f(x)$, where a and b are real numbers, then $a - bi$ is also a zero of $f(x)$.

The requirement that the polynomial have only real coefficients is very important. For example, $f(x) = x - (1 + i)$ has $1 + i$ as a zero, but the conjugate $1 - i$ is not a zero.

Example 4   Find a polynomial of lowest degree having only real coefficients and zeros 3 and $2 + i$.

The complex number $2 - i$ also must be a zero, so the polynomial has at least three zeros, 3, $2 + i$, and $2 - i$. For the polynomial to be of lowest degree these must be the only zeros. By the factor theorem there must be three factors, $x - 3$, $x - (2 + i)$, and $x - (2 - i)$. A polynomial of lowest degree is

$$\begin{aligned}
f(x) &= (x - 3)[x - (2 + i)][x - (2 - i)] \\
&= (x - 3)(x - 2 - i)(x - 2 + i) \\
&= x^3 - 7x^2 + 17x - 15.
\end{aligned}$$

Other polynomials, such as $2(x^3 - 7x^2 + 17x - 15)$ or $\sqrt{5}(x^3 - 7x^2 + 17x - 15)$, for example, also satisfy the given conditions on zeros. The information on zeros given in the problem is not enough to give a specific value for the leading coefficient. ■

The theorem on conjugate zeros is important in helping predict the number of real zeros of polynomials with real coefficients. A polynomial with real coefficients of odd degree n, where $n \geq 1$, must have at least one real zero (since zeros of the form $a + bi$, where $b \neq 0$, occur in conjugate pairs.) On the other hand, a polynomial with real coefficients of even degree n may have no real zeros.

Example 5   Find all zeros of $f(x) = x^4 - 7x^3 + 18x^2 - 22x + 12$, given that $1 - i$ is a zero.

Since $1 - i$ is a zero, by the conjugate zeros theorem $1 + i$ is also a zero. To find the remaining zeros, first divide the original polynomial by $1 - i$:

$$
\begin{array}{r|rrrrr}
1 - i) & 1 & -7 & 18 & -22 & 12 \\
& & 1 - i & -7 + 5i & 16 - 6i & -12 \\
\hline
& 1 & -6 - i & 11 + 5i & -6 - 6i & 0
\end{array}
$$

Then divide the quotient by $1 + i$, as follows.

$$
1 + i \overline{)1 \quad -6 - i \quad 11 + 5i \quad -6 - 6i}
$$
$$
\underline{\quad\quad 1 + i \quad -5 - 5i \quad 6 + 6i}
$$
$$
1 \quad -5 \quad\quad 6 \quad\quad\quad 0
$$

Now find the zeros of the quadratic polynomial $x^2 - 5x + 6$. Factoring the polynomial shows that the zeros are 2 and 3, so the four zeros of $f(x)$ are $1 - i$, $1 + i$, 2, and 3. ∎

Example 6 Factor $f(x)$ into linear factors (factors of the form $ax - b$ for complex numbers a and b), given that k is a zero of $f(x)$.

(a) $f(x) = 6x^3 + 19x^2 + 2x - 3$; $k = -3$
Since $k = -3$ is a zero of $f(x)$, $x - (-3) = x + 3$ is a factor. Use synthetic division to divide $f(x)$ by $x + 3$.

$$
-3 \overline{)6 \quad\quad 19 \quad\quad 2 \quad -3}
$$
$$
\underline{\quad\quad -18 \quad -3 \quad\quad 3}
$$
$$
6 \quad\quad 1 \quad -1 \quad\quad 0
$$

The quotient is $6x^2 + x - 1$, so

$$f(x) = (x + 3)(6x^2 + x - 1).$$

Factor $6x^2 + x - 1$ as $(2x + 1)(3x - 1)$ to get

$$f(x) = (x + 3)(2x + 1)(3x - 1),$$

where all factors are linear.

(b) $f(x) = 3x^3 + (-1 + 3i)x^2 + (-12 + 5i)x + 4 - 2i$; $k = 2 - i$
One factor is $x - (2 - i)$ or $x - 2 + i$. Divide $f(x)$ by $x - (2 - i)$.

$$
2 - i \overline{)3 \quad -1 + 3i \quad -12 + 5i \quad 4 - 2i}
$$
$$
\underline{\quad\quad 6 - 3i \quad\quad 10 - 5i \quad -4 + 2i}
$$
$$
3 \quad\quad 5 \quad\quad\quad -2 \quad\quad\quad 0
$$

By the division algorithm,

$$f(x) = (x - 2 + i)(3x^2 + 5x - 2).$$

Factor $3x^2 + 5x - 2$ as $(3x - 1)(x + 2)$; then a linear factored form of $f(x)$ is

$$f(x) = (x - 2 + i)(3x - 1)(x + 2). \quad ∎$$

7.2 Exercises

Use the factor theorem to decide whether or not the second polynomial is a factor of the first.

1. $4x^2 + 2x + 42$; $x - 3$

2. $-3x^2 - 4x + 2$; $x + 2$

3. $x^3 + 2x^2 - 3$; $x - 1$

4. $2x^3 + x + 2$; $x + 1$

5. $3x^3 - 12x^2 - 11x - 20$; $x - 5$

6. $4x^3 + 6x^2 - 5x - 2$; $x + 2$

7. $2x^4 + 5x^3 - 2x^2 + 5x + 3$; $x + 3$

8. $5x^4 + 16x^3 - 15x^2 + 8x + 16$; $x + 4$

For each of the following, find a polynomial of degree 3 with only real coefficients which satisfies the given conditions.

$a\frac{(x+3)(x+1)}{(x-4)} = 5$

9. Zeros of -3, -1, and 4; $f(2) = 5$ **10.** Zeros of 1, -1, and 0; $f(2) = -3$

11. Zeros of -2, 1, and 0; $f(-1) = -1$ **12.** Zeros of 2, 5, and -3; $f(1) = -4$

13. Zeros of 3, i, and $-i$; $f(2) = 50$ **14.** Zeros of -2, i, and $-i$; $f(-3) = 30$

For each of the following, find a polynomial of lowest degree with only real coefficients having the given zeros.

15. $5 + i$ and $5 - i$ **16.** $3 - 2i$ and $3 + 2i$ **17.** 2, $1 - i$ and $1 + i$

18. -3, $2 - i$, and $2 + i$ **19.** $1 + \sqrt{2}$, $1 - \sqrt{2}$, and 1   **20.** $1 - \sqrt{3}$, $1 + \sqrt{3}$, and -2

21. $2 + i$, $2 - i$, 3, and -1 **22.** $3 + 2i$, $3 - 2i$, -1, and 2

23. 2 and $3 + i$ $(x-2)(x-3+i)(x-3-i)$ **24.** -1 and $4 - 2i$

25. $1 - \sqrt{2}$, $1 + \sqrt{2}$, and $1 - i$ **26.** $2 + \sqrt{3}$, $2 - \sqrt{3}$, and $2 + 3i$

27. $2 - i$, $3 + 2i$ **28.** $5 + i$, $4 - i$

29. 4, $1 - 2i$, $3 + 4i$ **30.** -1, $1 + \sqrt{2}$, $1 - \sqrt{2}$, $1 + 4i$

31. $1 + 2i$, 2 (multiplicity 2)   **32.** $2 + i$, -3 (multiplicity 2)

For each of the following polynomials, one zero is given. Find all others.

33. $f(x) = x^3 - x^2 - 4x - 6$; 3 $3\, | 1 \ -1 \ -4 \ -6$ $find\ other$ $x's$ **34.** $f(x) = x^3 - 5x^2 + 17x - 13$; 1

35. $f(x) = x^3 + x^2 - 4x - 24$; $-2 + 2i$ **36.** $f(x) = x^3 + x^2 - 20x - 50$; $-3 + i$

37. $f(x) = 2x^3 - 2x^2 - x - 6$; 2   **38.** $f(x) = 2x^3 - 5x^2 + 6x - 2$; $1 + i$

39. $f(x) = x^4 + 5x^2 + 4$; $-i$ **40.** $f(x) = x^4 + 10x^3 + 27x^2 + 10x + 26$; i

41. $f(x) = x^4 - 3x^3 + 6x^2 + 2x - 60$; $1 + 3i$ **42.** $f(x) = x^4 - 6x^3 - x^2 + 86x + 170$; $5 + 3i$

43. $f(x) = x^4 - 6x^3 + 15x^2 - 18x + 10$; $2 + i$ **44.** $f(x) = x^4 + 8x^3 + 26x^2 + 72x + 153$; $-3i$

Factor $f(x)$ into linear factors given that k is a zero of $f(x)$.

45. $f(x) = 2x^3 - 3x^2 - 17x + 30$; $k = 2$ **46.** $f(x) = 2x^3 - 3x^2 - 5x + 6$; $k = 1$

47. $f(x) = 6x^3 + 25x^2 + 3x - 4$; $k = -4$ **48.** $f(x) = 8x^3 + 50x^2 + 47x - 15$; $k = -5$

49. $f(x) = x^3 + (7 - 3i)x^2 + (12 - 21i)x - 36i$; $k = 3i$

50. $f(x) = 2x^3 + (3 + 2i)x^2 + (1 + 3i)x + i$; $k = -i$

51. $f(x) = 2x^3 + (3 - 2i)x^2 + (-8 - 5i)x + 3 + 3i$; $k = 1 + i$

52. $f(x) = 6x^3 + (19 - 6i)x^2 + (16 - 7i)x + 4 - 2i$; $k = -2 + i$

53. Show that -2 is a zero of multiplicity 2 of $f(x) = x^4 + 2x^3 - 7x^2 - 20x - 12$ and find all other complex zeros. Then write $f(x)$ in factored form.

54. Show that -1 is a zero of multiplicity 3 of $f(x) = x^5 + 9x^4 + 33x^3 + 55x^2 + 42x + 12$, and find all other complex zeros. Then write $f(x)$ in factored form.

55. The displacement at time t of a particle moving along a straight line is given by

$$s(t) = t^3 - 2t^2 - 5t + 6,$$

where t is in seconds and s is measured in centimeters. The displacement is 0 after 1 second has elapsed. At what other times is the displacement 0?

56. For headphone radios, the cost function (in thousands of dollars) is given by

$$C(x) = 2x^3 - 9x^2 + 17x - 4,$$

and the revenue function (in thousands of dollars) is given by $R(x) = 5x$, where x is the number of items (in hundred thousands) produced. Cost equals revenue if 200,000 items are produced $(x = 2)$. Find all other break-even points.

57. Explain why it is not possible for a polynomial of degree 3 with real coefficients to have zeros of 1, 2, and $1 + i$.

58. Show that the zeros of $f(x) = x^3 + ix^2 - (7 - i)x + (6 - 6i)$ are $1 - i$, 2, and -3. Does the conjugate zeros theorem apply? Why?

59. Show that the equation

$$\frac{1}{x - 3} + \frac{2}{x + 3} - \frac{6}{x^2 - 9} = 0$$

has no solution. Why doesn't this contradict the fundamental theorem of algebra?

60. Show that $f(x) = 0$ and $g(x) = 0$ have exactly the same solutions (with the same multiplicity) if and only if $f(x) = c \cdot g(x)$ for some complex number c.

61. Let $f(x)$ and $q(x)$ be polynomials each of which has c as a zero. Show that $x - c$ is a factor of the remainder when $f(x)$ is divided by $q(x)$.

62. Let $f(x)$ be a polynomial of odd degree having only real coefficients. Show that a real number a exists such that $f(a) = 0$.

If c and d are complex numbers, prove each of the following statements. (*Hint:* Let $c = a + bi$ and $d = m + ni$ and form the conjugates, the sums, and the products.)

63. $\overline{c + d} = \overline{c} + \overline{d}$

64. $\overline{cd} = \overline{c} \cdot \overline{d}$.

65. $\overline{a} = a$ for any real number a.

66. $\overline{c^n} = (\overline{c})^n$

Read

7.3 Rational Zeros of Polynomials

By the fundamental theorem of algebra, every polynomial of degree 1 or more has a zero. However, the fundamental theorem merely says that such a zero exists. It gives no help at all in identifying zeros. Other theorems can be used to find any rational zeros of polynomials with rational coefficients or to find decimal approximations of any irrational zeros.

The *rational zeros theorem* gives a useful method for finding a set of possible zeros of a polynomial with integer coefficients.

Rational Zeros Theorem

Let

$$f(x) = a_n x^n + a_{n-1} x^{n-1} + \ldots + a_1 x + a_0, \qquad a_n \neq 0,$$

a_i are integers

be a polynomial with only integer coefficients. If p/q is a rational number written in lowest terms and if p/q is a zero of $f(x)$, then p is a factor of the constant term a_0 and q is a factor of the leading coefficient a_n.

To prove this theorem, begin with $f(p/q) = 0$ since p/q is a zero of $f(x)$.

$$a_n(p/q)^n + a_{n-1}(p/q)^{n-1} + \ldots + a_1(p/q) + a_0 = 0$$

This can also be written as

$$a_n(p^n/q^n) + a_{n-1}(p^{n-1}/q^{n-1}) + \ldots + a_1(p/q) + a_0 = 0.$$

Multiply both sides of this last result by q^n and add $-a_0 q^n$ to both sides.

$$a_n p^n + a_{n-1}p^{n-1}q + \ldots + a_1 pq^{n-1} = -a_0 q^n$$

Factoring out p gives

$$p(a_n p^{n-1} + a_{n-1}p^{n-2}q + \ldots + a_1 q^{n-1}) = -a_0 q^n.$$

This result shows that $-a_0 q^n$ equals the product of the two factors, p and $(a_n p^{n-1} + \ldots + a_1 q^{n-1})$. For this reason, p must be a factor of $-a_0 q^n$. Since it was given that p/q is written in lowest terms, p and q have no common factor other than 1, so p is not a factor of q^n. Thus p must be a factor of a_0. In a similar way, it can be shown that q is a factor of a_n.

Example I Find all rational zeros of $f(x) = 2x^3 + 3x^2 - 1$.

Since $f(x)$ has only integer coefficients, the rational zeros theorem applies. For this polynomial, $a_0 = -1$ and $a_n = a_3 = 2$. By the rational zeros theorem, p must be a factor of $a_0 = -1$ so that p is either 1 or -1. Also q must be a factor of $a_3 = 2$, so q is ± 1 or ± 2. Any rational zeros are in the form p/q, and so must be either ± 1, or $\pm 1/2$. Use synthetic division to check out these possibilities.

$$
\begin{array}{r|rrrr}
1) & 2 & 3 & 0 & -1 \\
  &   & 2 & 5 & 5 \\
\hline
  & 2 & 5 & 5 & 4
\end{array}
\qquad
\begin{array}{r|rrrr}
-1) & 2 & 3 & 0 & -1 \\
   &   & -2 & -1 & 1 \\
\hline
   & 2 & 1 & -1 & 0
\end{array}
$$

$$
\begin{array}{r|rrrr}
1/2) & 2 & 3 & 0 & -1 \\
    &   & 1 & 2 & 1 \\
\hline
    & 2 & 4 & 2 & 0
\end{array}
\qquad
\begin{array}{r|rrrr}
-1/2) & 2 & 3 & 0 & -1 \\
     &   & -1 & -1 & 1/2 \\
\hline
     & 2 & 2 & -1 & -1/2
\end{array}
$$

The only two rational zeros are -1 and $1/2$. Use synthetic division again, with the quotient from the synthetic division by -1 above. The remainder is zero, which shows that -1 is a zero of multiplicity 2.

$$
\begin{array}{r|rrr}
-1) & 2 & 1 & -1 \\
   &   & -2 & 1 \\
\hline
   & 2 & -1 & 0
\end{array}
\quad \blacksquare
$$

Example 2 Find all rational zeros of $f(x) = 2x^4 - 11x^3 + 14x^2 - 11x + 12$.

All coefficients are integers, so the rational zeros theorem can be used. If p/q is to be a rational zero of $p(x)$, by the rational zeros theorem p must be a factor of $a_0 = 12$ and q must be a factor of $a_4 = 2$. The possible values of p are ± 1, ± 2, ± 3, ± 4, ± 6, or ± 12, while q must be ± 1 or ± 2. The possible rational zeros are found by forming all possible quotients of the form p/q; then any rational zero of

$f(x)$ will come from the list

$$\pm 1, \ \pm 1/2, \ \pm 2, \ \pm 3, \ \pm 3/2, \ \pm 4, \ \pm 6, \text{ or } \pm 12.$$

Though it is not certain whether any of these numbers are zeros, if $f(x)$ has any rational zeros, they will be in the list above. These proposed zeros can be checked by synthetic division. A search of the possibilities leads to the discovery that 4 is a rational zero of $f(x)$.

$$
\begin{array}{r|rrrrr}
4) & 2 & -11 & 14 & -11 & 12 \\
  &   & 8 & -12 & 8 & -12 \\
\hline
  & 2 & -3 & 2 & -3 & 0
\end{array}
$$

As a fringe benefit of this calculation, zeros of the simpler polynomial $q(x) = 2x^3 - 3x^2 + 2x - 3$ can now be sought. Any rational zero of $q(x)$ will have a numerator of ± 3 or ± 1, with a denominator of ± 1 or ± 2. Thus, any rational zeros of $q(x)$ will come from the list

$$\pm 3, \ \pm 3/2, \ \pm 1, \ \pm 1/2.$$

Again use synthetic division and trial and error to find that 3/2 is a zero.

$$
\begin{array}{r|rrrr}
3/2) & 2 & -3 & 2 & -3 \\
     &   & 3 & 0 & 3 \\
\hline
     & 2 & 0 & 2 & 0
\end{array}
$$

The quotient is $2x^2 + 2$, which, by the quadratic formula, has i and $-i$ as zeros. They are, however, complex zeros. The rational zeros of

$$f(x) = 2x^4 - 11x^3 + 14x^2 - 11x + 12$$

are 4 and 3/2. ∎

Example 3 Find all rational zeros of $f(x) = 6x^4 + 7x^3 - 12x^2 - 3x + 2$, and factor the polynomial.

For a rational number p/q to be a zero of $f(x)$, p must be a factor of $a_0 = 2$ and q must be a factor of $a_4 = 6$. Thus, p can be ± 1 or ± 2 and q can be ± 1, ± 2, ± 3, or ± 6. The rational zeros, p/q, must be from the following list.

$$\pm 1, \ \pm 2, \ \pm 1/2, \ \pm 1/3, \ \pm 1/6, \ \pm 2/3$$

Check 1 first because it is easy.

$$
\begin{array}{r|rrrrr}
1) & 6 & 7 & -12 & -3 & 2 \\
   &   & 6 & 13 & 1 & -2 \\
\hline
   & 6 & 13 & 1 & -2 & 0
\end{array}
$$

The 0 remainder shows that 1 is a zero. Now use the quotient polynomial $6x^3 + 13x^2 + x - 2$ and synthetic division to find that -2 is also a zero.

$$
\begin{array}{r|rrrr}
-2) & 6 & 13 & 1 & -2 \\
    &   & -12 & -2 & 2 \\
\hline
    & 6 & 1 & -1 & 0
\end{array}
$$

The new quotient polynomial is $6x^2 + x - 1$. Use the quadratic formula or factor to solve the equation $6x^2 + x - 1 = 0$. The remaining two zeros are $1/3$ and $-1/2$.

Factor the polynomial $f(x)$ in the following way. Since the four zeros of $f(x) = 6x^4 + 7x^3 - 12x^2 - 3x + 2$ are 1, -2, $1/3$, and $-1/2$, the factors are $x - 1$, $x + 2$, $x - 1/3$, and $x + 1/2$, and

$$f(x) = a(x - 1)(x + 2)\left(x - \frac{1}{3}\right)\left(x + \frac{1}{2}\right).$$

Since the leading coefficient of $f(x)$ is 6, let $a = 6$. Then

$$f(x) = 6(x - 1)(x + 2)\left(x - \frac{1}{3}\right)\left(x + \frac{1}{2}\right)$$

$$= (x - 1)(x + 2)(3)\left(x - \frac{1}{3}\right)(2)\left(x + \frac{1}{2}\right)$$

$$= (x - 1)(x + 2)(3x - 1)(2x + 1). \quad \blacksquare$$

To find any rational zeros of a polynomial with fractional coefficients, first multiply the polynomial by a number that will clear it of all fractional coefficients. Then use the rational zeros theorem, which requires only integer coefficients.

Example 4 Find all rational zeros of

$$f(x) = x^4 - \frac{1}{6}x^3 + \frac{2}{3}x^2 - \frac{1}{6}x - \frac{1}{3}.$$

To find the values of x that make $f(x) = 0$, or

$$x^4 - \frac{1}{6}x^3 + \frac{2}{3}x^2 - \frac{1}{6}x - \frac{1}{3} = 0,$$

multiply both sides by 6 to eliminate all fractions. This gives

$$6x^4 - x^3 + 4x^2 - x - 2 = 0.$$

This polynomial will have the same zeros as $f(x)$. The possible rational zeros are of the form p/q where p is ± 1, or ± 2, and q is ± 1, ± 2, ± 3, or ± 6. Then p/q may be

$$\pm 1, \ \pm 2, \ \pm \frac{1}{2}, \ \pm \frac{1}{3}, \ \pm \frac{1}{6}, \ \text{or} \ \pm \frac{2}{3}.$$

Use synthetic division to find that $-1/2$ and $2/3$ are zeros.

$$
-\frac{1}{2}\overline{)\begin{array}{ccccc} 6 & -1 & 4 & -1 & -2 \\ & -3 & 2 & -3 & 2 \\ \hline 6 & -4 & 6 & -4 & 0 \end{array}}
\qquad
\frac{2}{3}\overline{)\begin{array}{cccc} 6 & -4 & 6 & -4 \\ & 4 & 0 & 4 \\ \hline 6 & 0 & 6 & 0 \end{array}}
$$

The final quotient is $q(x) = 6x^2 + 6 = 6(x^2 + 1)$. The zeros of this polynomial can be found by the quadratic formula; the zeros are i and $-i$. Since these zeros are not rational numbers, there are just two rational zeros: $-1/2$ and $2/3$. $\quad \blacksquare$

7.3 Exercises

Give all possible rational zeros for the following polynomials.

1. $f(x) = 6x^3 + 17x^2 - 31x - 1$

2. $f(x) = 15x^3 + 61x^2 + 2x - 1$

3. $f(x) = 12x^3 + 20x^2 - x - 2$

4. $f(x) = 12x^3 + 40x^2 + 41x + 3$

5. $f(x) = 2x^3 + 7x^2 + 12x - 8$

6. $f(x) = 2x^3 + 20x^2 + 68x - 40$

7. $f(x) = x^4 + 4x^3 + 3x^2 - 10x + 50$

8. $f(x) = x^4 - 2x^3 + x^2 + 18$

Find all rational zeros of the following polynomials.

9. $f(x) = x^3 - 2x^2 - 13x - 10$

10. $f(x) = x^3 + 5x^2 + 2x - 8$

11. $f(x) = x^3 + 6x^2 - x - 30$

12. $f(x) = x^3 - x^2 - 10x - 8$

13. $f(x) = x^3 + 9x^2 - 14x - 24$

14. $f(x) = x^3 + 3x^2 - 4x - 12$

15. $f(x) = x^4 + 9x^3 + 21x^2 - x - 30$

16. $f(x) = x^4 + 4x^3 - 7x^2 - 34x - 24$

Find the rational zeros of the following polynomials; then write each polynomial in factored form with each factor having only integer coefficients.

17. $f(x) = 6x^3 + 17x^2 - 31x - 12$

18. $f(x) = 15x^3 + 61x^2 + 2x - 8$

19. $f(x) = 12x^3 + 20x^2 - x - 6$

20. $f(x) = 12x^3 + 40x^2 + 41x + 12$

21. $f(x) = 2x^3 + 7x^2 + 12x - 8$

22. $f(x) = 2x^3 + 20x^2 + 68x - 40$

23. $f(x) = x^4 + 4x^3 + 3x^2 - 10x + 50$

24. $f(x) = x^4 - 2x^3 + x^2 + 18$

25. $f(x) = x^4 + 2x^3 - 13x^2 - 38x - 24$

26. $f(x) = 6x^4 + x^3 - 7x^2 - x + 1$

27. $f(x) = 3x^4 + 4x^3 - x^2 + 4x - 4$

28. $f(x) = x^4 + 8x^3 + 16x^2 - 8x - 17$

29. $f(x) = x^5 + 3x^4 - 5x^3 - 11x^2 + 12$

30. $f(x) = 4x^5 + 4x^4 - 37x^3 - 37x^2 + 9x + 9$

Find all rational zeros of the following polynomials.

31. $f(x) = x^3 - \dfrac{4}{3}x^2 - \dfrac{13}{3}x - 2$

32. $f(x) = x^3 + x^2 - \dfrac{16}{9}x + \dfrac{4}{9}$

33. $f(x) = x^4 + \dfrac{1}{4}x^3 + \dfrac{11}{4}x^2 + x - 5$ $\times \left(4\right)$ *get rid of fraction*

34. $f(x) = \dfrac{10}{7}x^4 - x^3 - 7x^2 + 5x - \dfrac{5}{7}$

35. $f(x) = \dfrac{1}{3}x^5 + x^4 - \dfrac{5}{3}x^3 - \dfrac{11}{3}x^2 + 4$

36. $f(x) = x^5 + x^4 - \dfrac{37}{4}x^2 + \dfrac{9}{4}x + \dfrac{9}{4}$

37. Show that $f(x) = x^2 - 2$ has no rational zeros, so $\sqrt{2}$ must be irrational.

38. Show that $f(x) = x^2 - 5$ has no rational zeros, so $\sqrt{5}$ must be irrational.

39. Show that $f(x) = x^4 + 5x^2 + 4$ has no rational zeros.

40. Show that $f(x) = x^5 - 3x^3 + 5$ has no rational zeros.

41. Show that any integer zeros of a polynomial must be factors of the constant term a_0.

7.4 Approximate Zeros of Polynomials

Every polynomial of degree 1 or more has a zero. However, the polynomial does not necessarily have real zeros. Even if it does have real zeros, often there is no way to find them. In this section methods of approximating any real zeros of a polynomial are discussed. These methods work well with a computer or a calculator.

Much of the work in locating real zeros uses the following result, which is related to the fact that graphs of polynomial functions are unbroken curves, with no gaps or sudden jumps. The proof requires advanced methods, so it is not given here.

Intermediate Value Theorem for Polynomials

If $f(x)$ is a polynomial with only real coefficients, and if for real numbers a and b, the values $f(a)$ and $f(b)$ are opposite in sign, then there exists at least one real zero between a and b.

By the intermediate value theorem, if d is a number between $f(a)$ and $f(b)$, then there *must* be a number c between a and b such that $f(c) = d$. This theorem helps to identify intervals where zeros of polynomials are located. For example, in Figure 1, $f(a)$ and $f(b)$ are opposite in sign, so 0 is between $f(a)$ and $f(b)$. Then, by the intermediate value theorem, there must be a number c between a and b such that $f(c) = 0$.

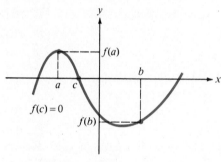

Figure 1

Example 1 Does $f(x) = x^3 - 2x^2 - x + 1$ have any real zeros between 2 and 3?
Use synthetic division to find $f(2)$ and $f(3)$.

$$
\begin{array}{r}
2)\overline{1 \quad -2 \quad -1 \quad 1} \\
2 \quad 0 \quad -2 \\
\hline
1 \quad 0 \quad -1 \quad -1
\end{array}
\qquad
\begin{array}{r}
3)\overline{1 \quad -2 \quad -1 \quad 1} \\
3 \quad 3 \quad 6 \\
\hline
1 \quad 1 \quad 2 \quad 7
\end{array}
$$

Since $f(2)$ is negative but $f(3)$ is positive, there must be a real zero between 2 and 3. ∎

The intermediate value theorem for polynomials is helpful in limiting the search for real zeros to smaller and smaller intervals. In Example 1 the theorem was used to find that there is a real zero between 2 and 3. The theorem could then be used repeatedly to locate the zero more accurately.

As suggested by the graphs of Figure 2, if the values of $|x|$ in a polynomial get larger and larger, then so will the values of $|y|$. This is used in the next theorem, the **boundedness theorem,** which shows how the bottom row of a synthetic division can be used to place upper and lower bounds on the possible real zeros of a polynomial.

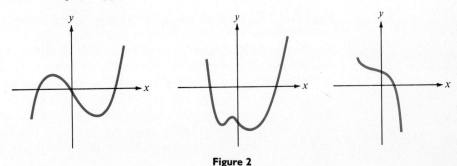

Figure 2

Boundedness Theorem

Let $f(x)$ be a polynomial with real coefficients and with a positive leading coefficient. If $f(x)$ is divided synthetically by $x - c$ and

(a) if $c > 0$ and all numbers in the bottom row of the synthetic division are nonnegative, then $f(x)$ has no zero greater than c;

(b) if $c < 0$ and the numbers in the bottom row of the synthetic division alternate in sign (with 0 considered positive or negative, as needed), then $f(x)$ has no zero less than c.

Example 2   Approximate the real zeros of $f(x) = x^4 - 6x^3 + 8x^2 + 2x - 1$.

First, check for rational zeros. The only possible rational zeros are ± 1.

$$\begin{array}{r|rrrrr} 1)1 & -6 & 8 & 2 & -1 \\ & 1 & -5 & 3 & 5 \\ \hline 1 & -5 & 3 & 5 & 4 \end{array} \qquad \begin{array}{r|rrrrr} -1)1 & -6 & 8 & 2 & -1 \\ & -1 & 7 & -15 & 13 \\ \hline 1 & -7 & 15 & -13 & 12 \end{array}$$

Neither 1 nor -1 is a zero, so the polynomial has no rational zeros. Now use the two theorems of this section to search in some consistent way for the location of irrational real zeros.

The leading coefficient of $f(x)$ is positive and the numbers in the last row of the second synthetic division above alternate in sign. Since $-1 < 0$, by the boundedness theorem -1 is less than any zero of $f(x)$. (It has been shown that -1 is not a zero of $f(x)$.) Also, $f(-1) = 12 > 0$. By substitution, or synthetic division, $f(0) = -1 < 0$. Thus there is at least one real zero between -1 and 0.

Let us try $c = -.5$ in the synthetic division. Divide $f(x)$ by $x + .5$.

$$\begin{array}{r|rrrrr} -.5)1 & -6 & 8 & 2 & -1 \\ & -.5 & 3.25 & -5.625 & 1.8125 \\ \hline 1 & -6.5 & 11.25 & -3.625 & .8125 \end{array}$$

Since $f(-.5) > 0$ and $f(0) < 0$, there is a real zero between $-.5$ and 0.

Try $c = -.4$.

$$
\begin{array}{r|rrrr}
-.4) & 1 & -6 & 8 & 2 & -1 \\
& & -.4 & 2.56 & -4.224 & .8896 \\
\hline
& 1 & -6.4 & 10.56 & -2.224 & -.1104
\end{array}
$$

Since $f(-.5)$ is positive, but $f(-.4)$ is negative, there is a zero between $-.5$ and $-.4$. The value of $f(-.4)$ is closer to zero than $f(-.5)$, so it is probably safe to say that, to one decimal place of accuracy, $-.4$ is an approximation to a real zero of $f(x)$. A more accurate result can be found, if desired, by continuing this process.

To locate the remaining real zeros of $f(x)$, continue in the same way. Use synthetic division to find $f(1)$, $f(2)$, $f(3)$, and so on, until there is a change in sign. It is helpful to use the shortened form of synthetic division shown below. Only the last row of the synthetic division is shown for each line. The first row of the chart is used for each division and the work in the second row of the division is done mentally.

x					$f(x)$	
	1	-6	8	2	-1	
-1	1	-7	15	-13	12	
						←—Zero between -1 and 0
0	1	-6	8	2	-1	
						←—Zero between 0 and 1
1	1	-5	3	5	4	
2	1	-4	0	2	3	
						←—Zero betwen 2 and 3
3	1	-3	-1	-1	-4	
						←—Zero between 3 and 4
4	1	-2	0	2	7	

Since the polynomial is of degree 4, there are no more than 4 zeros. Expand the table to approximate the real zeros to the nearest tenth. For example, for the zero between 0 and 1, the work might go as follows. Start halfway between 0 and 1 with $x = .5$. Since $f(.5) > 0$ and $f(0) < 0$, try $x = .4$ next.

x					$f(x)$	
	1	-6	8	2	-1	
.5	1	-5.5	5.25	4.63	1.31	
.4	1	-5.6	5.76	4.30	.72	
.3	1	-5.7	6.29	3.89	.17	
						←—Zero between .3 and .2
.2	1	-5.8	6.84	3.37	$-.33$	

The value $f(.3) = .17$ is closer to 0 than $f(.2) = -.33$, so to the nearest tenth, the zero is .3. Use synthetic division to verify that the remaining two zeros are approximately 2.4 and 3.7. ■

Descartes' rule of signs, stated below, gives a useful, practical test for finding the number of positive or negative real zeros of a given polynomial.

Descartes' Rule of Signs

Let $f(x)$ be a polynomial with real coefficients and terms in descending order.

(a) The number of positive real zeros of $f(x)$ is either equal to the number of variations in sign occurring in the coefficients of $f(x)$, or else is less than the number of variations by a positive even integer.

(b) The number of negative real zeros of $f(x)$ either equals the number of variations in sign of $f(-x)$, or else is less than the number of variations by a positive even integer.

In the theorem, the number of variations in sign of the coefficients of $f(x)$ or $f(-x)$ refers to changes from positive to negative in successive terms of the polynomial. Missing terms (those with 0 coefficients) are counted as no change in sign and can be ignored.

For the purposes of this theorem, zeros of multiplicity k count as k zeros. For example,

$$f(x) = (x - 1)^4 = + x^4 - 4x^3 + 6x^2 - 4x + 1$$

$$\underbrace{\qquad}_{1} \underbrace{\qquad}_{2} \underbrace{\qquad}_{3} \underbrace{\qquad}_{4}$$

has 4 changes of sign. By Descartes' rule of signs, $f(x)$ has 4, 2, or 0 positive real zeros. In this case, there are 4; each of the 4 positive real zeros is 1.

The polynomial in Example 2,

$$f(x) = x^4 - 6x^3 + 8x^2 + 2x - 1,$$

has 3 variations in sign:

$$+ x^4 - 6x^3 + 8x^2 + 2x - 1$$

$$\underbrace{\qquad}_{1} \underbrace{\qquad}_{2} \underbrace{\qquad}_{3}$$

By Descartes' rule of signs, $f(x)$ has either 3 or $3 - 2 = 1$ positive real zeros. Example 2 showed that $f(x)$ has 3 positive real zeros. Since $f(x)$ is of degree 4 and has 3 positive real zeros, it must have 1 negative real zero, which corresponds to the result in Example 2. This could be verified with part (b) of Descartes' rule of signs. Since

$$f(-x) = (-x)^4 - 6(-x)^3 + 8(-x)^2 + 2(-x) - 1$$
$$= x^4 + 6x^3 + 8x^2 - 2x - 1$$

has only one variation in sign, $f(x)$ has only one negative real zero.

Example 3 Find the number of positive and negative real zeros of

$$f(x) = x^5 + 5x^4 + 3x^2 + 2x + 1.$$

The polynomial $f(x)$ has no variations in sign and so has no positive real zeros.

Here

$$f(-x) = -x^5 + 5x^4 + 3x^2 - 2x + 1,$$

with three variations in sign, so $f(x)$ has either 3 or 1 negative real zeros. The other zeros are complex numbers.   ∎

If a polynomial of degree 3 or more cannot be factored, it can often be graphed by calculating many points to plot. The theorems in this chapter are helpful in deciding which points to plot. (With calculus, other methods are available that make graphing polynomials much easier.)

Example 4   Graph the function defined by $f(x) = 8x^3 - 12x^2 + 2x + 1$.
By Descartes' rule of signs, $f(x)$ has 2 or 0 positive real zeros and 1 negative real zero. To find some ordered pairs belonging to the graph, use synthetic division to evaluate, say $f(3)$, in hopes of finding a number greater than all zeros of $f(x)$.

$$\begin{array}{r|rrrr} 3) & 8 & -12 & 2 & 1 \\ & & 24 & 36 & 114 \\ \hline & 8 & 12 & 38 & 115 \end{array}$$

Since $f(3) = 115$, the point $(3, 115)$ belongs to the graph. Also, since the bottom row is all positive, there are no zeros greater than 3. Now find $f(-1)$.

$$\begin{array}{r|rrrr} -1) & 8 & -12 & 2 & 1 \\ & & -8 & 20 & -22 \\ \hline & 8 & -20 & 22 & -21 \end{array}$$

From this result, the point $(-1, -21)$ belongs to the graph. Since the signs in the last row alternate, -1 is less than any zero of $f(x)$. Using the shortened form of synthetic division shown in the chart below, find several points between -1 and 3, to help in sketching the graph.

x				$f(x)$	Ordered pair	
	8	-12	2	1		
3	8	12	38	115	(3, 115)	
2	8	4	10	21	(2, 21)	
1	8	-4	-2	-1	(1, -1)	←Zero
0	8	-12	2	1	(0, 1)	←Zero
-1	8	-20	22	-21	(-1, -21)	←Zero

All three possible real zeros have been located. Since the numbers in the row for $x = 2$ are all positive and $2 > 0$, 2 is greater than or equal to any zero of $f(x)$ and the top row of the chart is not needed.

By the intermediate value theorem, there is a zero between 0 and 1 and between -1 and 0, as well as between 1 and 2. To get the graph, plot the points from the chart and then draw a continuous curve through them, as in Figure 3.   ∎

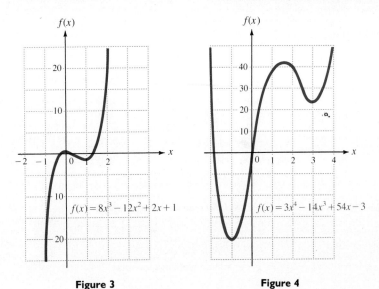

$f(x)$

$f(x) = 8x^3 - 12x^2 + 2x + 1$

Figure 3

$f(x) = 3x^4 - 14x^3 + 54x - 3$

Figure 4

Example 5   Graph the function defined by $f(x) = 3x^4 - 14x^3 + 54x - 3$.

Use Descartes' rule of signs to see that there are 3 positive real zeros or 1 positive real zero and 1 negative real zero. To find points to plot, use synthetic division to make a table like the one shown below. Start with $x = 0$ and work up through the positive integers until a row with all positive numbers is found. Then work down through the negative integers until a row with alternating signs is found.

x					$f(x)$	Ordered pair
	3	-14	0	54	-3	
5	3	1	5	79	392	(5, 392)
4	3	-2	-8	22	85	(4, 85)
3	3	-5	-15	9	24	(3, 24)
2	3	-8	-16	22	41	(2, 41)
1	3	-11	-11	43	40	(1, 40)
0	3	-14	0	54	-3	(0, -3)
-1	3	-17	17	37	-40	(-1, -40)
-2	3	-20	40	-26	49	(-2, 49)

Since the row in the chart for $x = 5$ contains all positive numbers, the polynomial has no zero greater than 5. Also, since the row for $x = -2$ has numbers which alternate in sign, there is no zero less than -2. By the changes in sign of $f(x)$, the polynomial has zeros between 0 and 1 and between -2 and -1. Plotting the points found above and drawing a continuous curve through them gives the graph shown in Figure 4. ∎

7.4 Exercises

Use the intermediate value theorem for polynomials to show that the following polynomials have a real zero between the numbers given.

1. $f(x) = 3x^2 - 2x - 6$; 1 and 2

2. $f(x) = x^3 + x^2 - 5x - 5$; 2 and 3

3. $f(x) = 2x^3 - 8x^2 + x + 16$; 2 and 2.5

4. $f(x) = 3x^3 + 7x^2 - 4$; 1/2 and 1

5. $f(x) = 2x^4 - 4x^2 + 3x - 6$; 2 and 1.5

6. $f(x) = x^4 - 4x^3 - x + 1$; 1 and .3

7. $f(x) = -3x^4 - x^3 + 2x^2 + 4$; 1 and 1.2

8. $f(x) = x^5 + 2x^4 + x^3 + 3$; -1.3 and -2

Show that the real zeros of each of the following polynomials satisfy the given conditions.

9. $f(x) = x^4 - x^3 + 3x^2 - 8x + 8$; no real zero greater than 2

10. $f(x) = 2x^5 - x^4 + 2x^3 - 2x^2 + 4x - 4$; no real zero greater than 1

11. $f(x) = x^4 + x^3 - x^2 + 3$; no real zero less than -2

12. $f(x) = x^5 + 2x^3 - 2x^2 + 5x + 5$; no real zero less than -1

13. $f(x) = 3x^4 + 2x^3 - 4x^2 + x - 1$; no real zero greater than 1

14. $f(x) = 3x^4 + 2x^3 - 4x^2 + x - 1$; no real zero less than -2

15. $f(x) = x^5 - 3x^3 + x + 2$; no real zero greater than 2

16. $f(x) = x^5 - 3x^3 + x + 2$; no real zero less than -3

For each of the following polynomials, **(a)** find the number of positive and negative real zeros; **(b)** approximate each zero as a decimal to the nearest tenth.

17. $f(x) = x^3 + 3x^2 - 2x - 6$

18. $f(x) = x^3 + x^2 - 5x - 5$

19. $f(x) = x^3 - 4x^2 - 5x + 14$

20. $f(x) = x^3 + 9x^2 + 34x + 13$

21. $f(x) = x^3 + 6x - 13$

22. $f(x) = 4x^3 - 3x^2 + 4x - 5$

23. $f(x) = 4x^4 - 8x^3 + 17x^2 - 2x - 14$

24. $f(x) = 3x^4 - 4x^3 - x^2 + 8x - 2$

25. $f(x) = -x^4 + 2x^3 + 3x^2 + 6$

26. $f(x) = -2x^4 - x^2 + x - 5$

The following polynomials have zeros in the given intervals. Approximate these zeros to the nearest hundredth.

27. $f(x) = x^4 + x^3 - 6x^2 - 20x - 16$; [3.2, 3.3] and $[-1.4, -1.1]$

28. $f(x) = x^4 - 2x^3 - 2x^2 - 18x + 5$; [.2, .4] and [3.7, 3.8]

29. $f(x) = x^4 - 4x^3 - 20x^2 + 32x + 12$; $[-4, -3]$, $[-1, 0]$, [1, 2], and [6, 7]

30. $f(x) = x^4 - 4x^3 - 44x^2 + 160x - 80$; $[-7, -6]$, [0, 1], [2, 3], and [7, 8]

Graph each of the functions defined as follows.

31. $f(x) = x^3 - 7x - 6$

32. $f(x) = x^3 + x^2 - 4x - 4$

33. $f(x) = x^4 - 5x^2 + 6$

34. $f(x) = x^3 - 3x^2 - x + 3$

35. $f(x) = 6x^3 + 11x^2 - x - 6$

36. $f(x) = x^4 - 2x^2 - 8$

37. $f(x) = -x^3 + 6x^2 - x - 14$

38. $f(x) = 6x^4 - x^3 - 23x^2 - 4x + 12$

Approximate the solution of each of the following inequalities to the nearest tenth.

39. $2x^3 - 3x^2 - 12x + 1 \leq 0$

40. $3x^3 - 3x^2 + 1 \geq 0$

41. $x^4 - 18x^2 + 5 \geq 0$

42. $x^4 - 8x^2 - 1 \leq 0$

43. Suppose a polynomial $f(x)$ has only rational zeros. Show how to express $f(x)$ with only rational coefficients; with only irrational coefficients.

44. Let $f(x) = x^n + a_{n-1}x^{n-1} + a_{n-2}x^{n-2} + \ldots + a_1 x + a_0$. Let m be the largest of the absolute values of the a's. Show that all real zeros of $f(x)$ are no greater than $m + 1$.

7.5 Partial Fractions

In Section 1.6 sums of rational expressions were found by combining two or more rational expressions into one rational expression. Here the reverse problem is considered: given one rational expression, express it as the sum of two or more rational expressions. The sum of rational expressions is called the **partial fraction decomposition;** each term in the sum is a **partial fraction.** The technique of decomposing a rational expression into partial fractions is useful in calculus and other areas of mathematics.

To form a partial fraction decomposition of a rational expression, use the following steps.

Partial Fraction Decomposition of $\dfrac{f(x)}{g(x)}$

Step 1 If $f(x)/g(x)$ is not a proper fraction (a fraction with the numerator of lower degree than the denominator), divide $f(x)$ by $g(x)$. For example,

$$\frac{x^4 - 3x^3 + x^2 + 5x}{x^2 + 3} = x^2 - 3x - 2 + \frac{14x + 6}{x^2 + 3}.$$

Then apply the following steps to the remainder, which is a proper fraction.

Step 2 Factor $g(x)$ completely into factors of the form $(ax + b)^m$ or $(cx^2 + dx + e)^n$, where $cx^2 + dx + e$ is prime and m and n are integers.

Step 3 For each factor of the form $(ax + b)^m$ the decomposition must include the terms

$$\frac{A_1}{ax + b} + \frac{A_2}{(ax + b)^2} + \ldots + \frac{A_m}{(ax + b)^m}.$$

Step 4 For each factor of the form $(cx^2 + dx + e)^n$, the decomposition must include the terms

$$\frac{B_1 x + C_1}{cx^2 + dx + e} + \frac{B_2 x + C_2}{(cx^2 + dx + e)^2} + \ldots + \frac{B_n x + C_n}{(cx^2 + dx + e)^n}.$$

Step 5 Use algebraic techniques to solve for the constants in the numerators of the decomposition.

The following examples illustrate algebraic techniques used to find the constants in Step 5.

Example 1 Find the partial fraction decomposition of $\dfrac{5x - 2}{x^3 - 4x}$.

The denominator can be factored as $x^3 - 4x = x(x + 2)(x - 2)$. Since the factors are **distinct linear factors,** use Step 3 to write the decomposition as

$$\frac{5x - 2}{x^3 - 4x} = \frac{A}{x} + \frac{B}{x + 2} + \frac{C}{x - 2}, \tag{1}$$

where A, B, and C are constants that need to be found. Multiply both sides of equation (1) by $x(x + 2)(x - 2)$, getting

$$5x - 2 = A(x + 2)(x - 2) + Bx(x - 2) + Cx(x + 2). \tag{2}$$

Equation (1) is an identity, since both sides represent the same rational expression. Thus, equation (2) is also an identity. Equation (1) holds for all values of x except 0, -2, and 2. However, equation (2) holds for all values of x. In particular, substituting 0 for x in equation (2) gives

$$-2 = -4A, \quad \text{so that} \quad A = \frac{1}{2}.$$

Similarly, if $x = -2$, then

$$-12 = 8B, \quad \text{so that} \quad B = -\frac{3}{2}.$$

Finally, if $x = 2$, then

$$8 = 8C, \quad \text{so that} \quad C = 1.$$

The rational expression can be written as the following sum of partial fractions.

$$\frac{5x - 2}{x^3 - 4x} = \frac{1}{2x} + \frac{-3}{2(x + 2)} + \frac{1}{x - 2} \quad \blacksquare$$

Example 2 Find the partial fraction decomposition of $\dfrac{2x}{(x - 1)^3}$.

The denominator is already factored with **repeated linear factors.** Using Step 3 above, write the decomposition as

$$\frac{2x}{(x - 1)^3} = \frac{A}{x - 1} + \frac{B}{(x - 1)^2} + \frac{C}{(x - 1)^3}.$$

Clear the denominators by multiplying both sides of this equation by $(x - 1)^3$.

$$2x = A(x - 1)^2 + B(x - 1) + C$$

Substituting 1 for x leads to $C = 2$, so that

$$2x = A(x - 1)^2 + B(x - 1) + 2. \tag{1}$$

The only root has been substituted and values for A and B still need to be found.

However, *any* number can be substituted for x. For example, if $x = -1$, equation (1) becomes

$$-2 = 4A - 2B + 2$$
$$-4 = 4A - 2B$$
$$-2 = 2A - B. \tag{2}$$

Substituting 0 for x in equation (1) gives

$$0 = A - B + 2$$
$$2 = -A + B. \tag{3}$$

Now solve the system of equations (2) and (3) to get $A = 0$ and $B = 2$. The partial fraction decomposition is

$$\frac{2x}{(x-1)^3} = \frac{2}{(x-1)^2} + \frac{2}{(x-1)^3}.$$

Three substitutions were needed because there were three constants to evaluate, A, B, and C.

To check this result, combine terms on the right. ■

Example 3   Find the partial fraction decomposition of $\dfrac{x^2 + 3x - 1}{(x + 1)(x^2 - 2)}$.

This denominator has **distinct linear and quadratic factors** where neither is repeated. (Since $x^2 - 2$ cannot be factored over the *integers*, it is considered prime here.) The partial fraction decomposition is

$$\frac{x^2 + 3x - 1}{(x + 1)(x^2 - 2)} = \frac{A}{x + 1} + \frac{Bx + C}{x^2 - 2}.$$

Multiply both sides by $(x + 1)(x^2 - 2)$ to get

$$x^2 + 3x - 1 = A(x^2 - 2) + (Bx + C)(x + 1). \tag{1}$$

First substitute -1 for x to get

$$(-1)^2 + 3(-1) - 1 = A[(-1)^2 - 2] + 0$$
$$-3 = -A$$
$$A = 3.$$

Replace A with 3 in equation (1) and let any number substitute for x. If $x = 0$,

$$0^2 + 3(0) - 1 = 3(0^2 - 2) + (B \cdot 0 + C)(0 + 1)$$
$$-1 = -6 + C$$
$$C = 5.$$

Now, letting $A = 3$ and $C = 5$, substitute again in equation (1) using another number for x. For $x = 1$,

$$3 = -3 + (B + 5)(2)$$
$$6 = 2B + 10$$
$$B = -2.$$

Using $A = 3$, $B = -2$, and $C = 5$, the partial fraction decomposition is

$$\frac{x^2 + 3x - 1}{(x + 1)(x^2 - 2)} = \frac{3}{x + 1} + \frac{-2x + 5}{x^2 - 2}.$$

Again, this work can be checked by combining terms on the right. ■

For quadratic factors another method is often more convenient. For instance, in Example 3, after both sides were multiplied by the common denominator, the equation was

$$x^2 + 3x - 1 = A(x^2 - 2) + (Bx + C)(x + 1).$$

Multiplying on the right and collecting like terms gives

$$x^2 + 3x - 1 = Ax^2 - 2A + Bx^2 + Bx + Cx + C$$
$$x^2 + 3x - 1 = (A + B)x^2 + (B + C)x + (C - 2A).$$

Now, equating the coefficients of like powers of x gives the three equations

$$1 = A + B$$
$$3 = B + C$$
$$-1 = C - 2A.$$

Solving this system of equations for A, B, and C would give the partial fraction decomposition. The next example uses a combination of the two methods.

Example 4   Find the partial fraction decomposition of $\dfrac{2x}{(x^2 + 1)^2(x - 1)}$.

This expression has both a linear factor and a **repeated quadratic factor.** By Steps 3 and 4 from the box at the beginning of this section,

$$\frac{2x}{(x^2 + 1)^2(x - 1)} = \frac{Ax + B}{x^2 + 1} + \frac{Cx + D}{(x^2 + 1)^2} + \frac{E}{x - 1}.$$

Multiplication of both sides by $(x^2 + 1)^2(x - 1)$ leads to

$$2x = (Ax + B)(x^2 + 1)(x - 1) + (Cx + D)(x - 1) + E(x^2 + 1)^2. \qquad \textbf{(1)}$$

If $x = 1$, equation (1) reduces to

$$2 = 4E, \quad \text{or} \quad E = \frac{1}{2}.$$

Substituting 1/2 for E in equation (1) and combining terms on the right gives

$$2x = (A + 1/2)x^4 + (-A + B)x^3 + (A - B + C + 1)x^2 +$$
$$(-A + B + D - C)x + (-B - D + 1/2).$$

Setting corresponding coefficients of x^4 equal,

$$0 = A + \frac{1}{2} \quad \text{or} \quad A = -\frac{1}{2}.$$

From the corresponding coefficients of x^3,

$$0 = -A + B.$$

Since $A = -1/2$, then $B = -1/2$.

Using the coefficients of x^2,

$$0 = A - B + C + 1.$$

Since $A = -1/2$ and $B = -1/2$,

$$C = -1.$$

Finally, from the coefficients of x,

$$2 = -A + B + D - C.$$

Substituting for A, B, and C gives

$$D = 1.$$

With $A = -1/2$, $B = -1/2$, $C = -1$, $D = 1$, and $E = 1/2$, the given fraction has the partial fraction decomposition

$$\frac{2x}{(x^2 + 1)^2(x - 1)} = \frac{-\dfrac{1}{2}x - \dfrac{1}{2}}{x^2 + 1} + \frac{-x + 1}{(x^2 + 1)^2} + \frac{\dfrac{1}{2}}{x - 1}$$

or

$$\frac{2x}{(x^2 + 1)^2(x - 1)} = \frac{-(x + 1)}{2(x^2 + 1)} + \frac{-x + 1}{(x^2 + 1)^2} + \frac{1}{2(x - 1)}. \quad \blacksquare$$

The two methods discussed in this section are summarized here.

To form a partial fraction decomposition, use either of the following methods or a combination of the two.

Method 1   For linear factors

1. Multiply both sides of the rational expression by the common denominator.
2. Substitute the root of each factor in the resulting equation. For repeated linear factors, substitute as many other numbers as necessary to find all the constants in the numerators.

Method 2   For quadratic factors

1. Multiply both sides of the rational expression by the common denominator.
2. Collect terms on the right side of the resulting equation.
3. Equate the coefficients of like terms to get a system of equations.
4. Solve the system to find the constants in the numerators.

7.5 Exercises

Find the partial fraction decomposition for the following rational expressions.

1. $\dfrac{3x - 1}{x(x + 1)}$

2. $\dfrac{5}{3x(2x + 1)}$

3. $\dfrac{2}{x(x + 3)}$

4. $\dfrac{1}{3x(x - 1)}$

5. $\dfrac{x + 2}{(x + 1)(x - 1)}$

6. $\dfrac{4x + 2}{(x + 2)(2x - 1)}$

7. $\dfrac{-7}{x(3x - 1)(x + 1)}$

8. $\dfrac{-2}{x^2(x - 2)}$

9. $\dfrac{2}{x^2(x + 3)}$

10. $\dfrac{2x}{(x + 1)(x + 2)^2}$

11. $\dfrac{x - 1}{(x + 2)(x - 3)^2}$

12. $\dfrac{x + 2}{(x - 1)^2(x + 5)}$

13. $\dfrac{2x + 1}{(x + 2)^3}$

14. $\dfrac{4x^2 - x - 15}{x(x + 1)(x - 1)}$

15. $\dfrac{3}{x^2 + 4x + 3}$

16. $\dfrac{1}{x^2 + 5x + 4}$

17. $\dfrac{x}{2x^2 + 5x + 2}$

18. $\dfrac{2x}{3x^2 + 2x - 1}$

19. $\dfrac{x^3 + 2}{x^3 - 3x^2 + 2x}$

20. $\dfrac{x^3 + 4}{9x^3 - 4x}$

21. $\dfrac{5}{x(x^2 + 3)}$

22. $\dfrac{-3}{x^2(x^2 + 5)}$

23. $\dfrac{2x + 1}{(x + 1)(x^2 + 2)}$

24. $\dfrac{3x - 2}{(x + 4)(3x^2 + 1)}$

25. $\dfrac{3}{x(x + 1)(x^2 + 1)}$

26. $\dfrac{1}{x(2x + 1)(3x^2 + 4)}$

27. $\dfrac{x^4 + 1}{x(x^2 + 1)^2}$

28. $\dfrac{3x - 1}{x(2x^2 + 1)^2}$

29. $\dfrac{3x^4 + x^3 + 5x^2 - x + 4}{(x - 1)(x^2 + 1)^2}$

30. $\dfrac{-x^4 - 8x^2 + 3x - 10}{(x + 2)(x^2 + 4)^2}$

31. $\dfrac{11x - 24}{5(x^2 + x - 2)}$

32. $\dfrac{25x - 25}{3(x^2 - 2x - 15)}$

33. $\dfrac{x^2}{x^4 - 1}$

34. $\dfrac{3x}{x^4 + x^2}$

35. $\dfrac{4x^2 + 13x - 9}{x^3 + 2x^2 - 3x}$

36. $\dfrac{2x^2 - 12x + 4}{x^3 - 4x^2}$

Find constants a, b, and c so that the following statements are true.

37. $\dfrac{x - 1}{x^3 - x^2 - 4x} = \dfrac{a}{x} + \dfrac{bx + c}{x^2 - x - 4}$

38. $\dfrac{x + 2}{x^3 + 2x^2 + x} = \dfrac{a}{x} + \dfrac{bx + c}{x^2 + 2x + 1}$

39. $\dfrac{x + 3}{x^3 + 64} = \dfrac{a}{x + 4} + \dfrac{bx + c}{x^2 - 4x + 16}$

40. $\dfrac{2x + 1}{(x + 1)(x^2 - 3x + 5)} = \dfrac{a}{x + 1} + \dfrac{bx + c}{x^2 - 3x + 5}$

Chapter 7 Summary

Key Words

synthetic division
remainder theorem
factor theorem

zero
fundamental theorem of
algebra

multiplicity
Descartes' rule of signs
partial fractions

Review Exercises

For each of the following divisions, find $q(x)$ and $r(x)$.

1. $\dfrac{4x^3 - 2x^2 + 3x + 5}{2x - 1}$

2. $\dfrac{3x^3 + 5x^2 - 5x + 2}{3x + 2}$

3. $\dfrac{x^4 - 3x^2 + 5x - 1}{x^2 + 2}$

4. $\dfrac{2x^4 - x^3 + x^2 + 5x}{2x^2 - x}$

Use synthetic division to find $q(x)$ and r for each of the following.

5. $\dfrac{2x^3 + 3x^2 - 4x + 1}{x - 1}$ *use synthetic division*

6. $\dfrac{4x^3 + 3x^2 + 3x + 5}{x - 5}$

7. $\dfrac{2x^3 - 4x + 6}{x - 2}$

8. $\dfrac{3x^4 - x^2 + x - 1}{x + 1}$

Use synthetic division to find $f(2)$ for each of the following.   *find remainder*

9. $f(x) = x^3 + 3x^2 - 5x + 1$

10. $f(x) = 2x^3 - 4x^2 + 3x - 10$

11. $f(x) = 5x^4 - 12x^2 + 2x - 8$

12. $f(x) = x^5 - 3x^2 + 2x - 4$

Find a polynomial of lowest degree having the following zeros.

13. $-1, 4, 7$

14. $8, 2, 3$

15. $-\sqrt{7}, \sqrt{7}, 2, -1$

16. $1 + \sqrt{5}, 1 - \sqrt{5}, -4, 1$

17. Is -1 a zero of $f(x) = 2x^4 + x^3 - 4x^2 + 3x + 1$?

18. Is -2 a zero of $f(x) = 2x^4 + x^3 - 4x^2 + 3x + 1$?

19. Is $x + 1$ a factor of $f(x) = x^3 + 2x^2 + 3x - 1$?

20. Is $x + 1$ a factor of $f(x) = 2x^3 - x^2 + x + 4$?

21. Find a polynomial of degree 3 with -2, 1, and 4 as zeros, and $f(2) = 16$.

22. Find a polynomial of degree 4 with 1, -1, and $3i$ as zeros, and $f(2) = 39$.

23. Find a lowest-degree polynomial with real coefficients having zeros 2, -2, and $-i$.

24. Find a lowest-degree polynomial with real coefficients having zeros of 2, -3, and $5i$.

25. Find the polynomial of lowest degree with real coefficients having -3 and $1 - i$ as zeros.

26. Find all zeros of $f(x) = x^4 - 3x^3 - 8x^2 + 22x - 24$, given that $1 - i$ is a zero.

27. Find all zeros of $f(x) = x^4 - 6x^3 + 14x^2 - 24x + 40$, given that $3 + i$ is a zero.

28. Find all zeros of $f(x) = x^4 + x^3 - x^2 + x - 2$, given that 1 is a zero.

Find all rational zeros of the following.

29. $f(x) = 2x^3 - 9x^2 - 6x + 5$

30. $f(x) = 3x^3 - 10x^2 - 27x + 10$ *FIND*

31. $f(x) = x^3 - \dfrac{17}{6}x^2 - \dfrac{13}{3}x - \dfrac{4}{3}$

32. $f(x) = 8x^4 - 14x^3 - 29x^2 - 4x + 3$

33. $f(x) = -x^4 + 2x^3 - 3x^2 + 11x - 7$

34. $f(x) = x^5 - 3x^4 - 5x^3 + 15x^2 + 4x - 12$

(handwritten:)

$f(2) = a(x+2)(x-1)(x-4) = 16$

$a = -2$

$(-2)(x+2)(x-1)(x-4)$

List poss. rat. zeros

$p = \pm 1, 5$

$q = \pm 1, 2$

$p/q = \frac{1}{2}, 1, \frac{5}{2}, 5$

35. Use a polynomial to show that $\sqrt{11}$ is irrational.

36. Show that $f(x) = x^3 - 9x^2 + 2x - 5$ has no rational zeros.

Show that the following polynomials have real zeros satisfying the given conditions.

37. $f(x) = 3x^3 - 8x^2 + x + 2$, zero in $[-1, 0]$ and $[2, 3]$

38. $f(x) = 4x^3 - 37x^2 + 50x + 60$, zero in $[2, 3]$ and $[7, 8]$

39. $f(x) = x^3 + 2x^2 - 22x - 8$, zero in $[-1, 0]$ and $[-6, -5]$

40. $f(x) = 2x^4 - x^3 - 21x^2 + 51x - 36$ has no real zero greater than 4

41. $f(x) = 6x^4 + 13x^3 - 11x^2 - 3x + 5$ has no real zero greater than 1 or less than -3

Approximate the real zeros of each of the following as a decimal to the nearest tenth. Then graph the function defined by the given expression.

42. $f(x) = 2x^3 - 11x^2 - 2x + 2$

43. $f(x) = x^4 - 4x^3 - 5x^2 + 14x - 15$

APPROX.

Graph each function defined below.

44. $f(x) = 3x^4 + 4x^2 - x$

45. $f(x) = 2x^5 - 3x^4 + x^2 - 2$

46. **(a)** Find the number of positive and negative zeros of $f(x) = x^3 + 3x^2 - 4x - 2$.

   (b) Show that $f(x)$ has a zero between -4 and -3. Approximate this zero to the nearest tenth.

   (c) Graph $y = f(x)$.

Find the partial fraction decomposition for each rational expression.

47. $\dfrac{5x - 2}{x^2 - 4}$

48. $\dfrac{1}{x^3 + 3x^2}$

49. $\dfrac{8}{x^2(x^2 + 2)}$

50. $\dfrac{3x - 1}{(x + 1)^2(x^2 + 5)}$

51. $\dfrac{3x^3 - 18x^2 + 29x - 4}{(x + 1)(x - 2)^3}$

52. $\dfrac{x^2 - x - 21}{(x^2 + 4)(2x - 1)^2}$

53. $\dfrac{5x^3 - x^2 + 7x + 4}{(x^2 + 1)^2}$

54. $\dfrac{14x^2 + 12x + 3}{2x^3 + 3x^2 + x}$

8

Further Topics in Algebra

Mathematical Induction

Many results in mathematics are claimed to be true for every positive integer. Any of these results could be checked for $n = 1$, $n = 2$, $n = 3$, and so on, but since the set of positive integers is infinite it would be impossible to check every possible case. For example, let S_n represent the statement that the sum of the first n positive integers is $n(n + 1)/2$.

$$S_n: 1 + 2 + 3 + \ldots + n = \frac{n(n + 1)}{2}.$$

The truth of this statement is easily verified for the first few values of n:

If $n = 1$, then S_1 is $\qquad\qquad\qquad 1 = \dfrac{1(1 + 1)}{2}$ which is true.

If $n = 2$, then S_2 is $\qquad\quad 1 + 2 = \dfrac{2(2 + 1)}{2}$ which is true.

If $n = 3$, then S_3 is $\qquad 1 + 2 + 3 = \dfrac{3(3 + 1)}{2}$ which is true.

If $n = 4$, then S_4 is $\quad 1 + 2 + 3 + 4 = \dfrac{4(4 + 1)}{2}$ which is true.

Continuing in this way for any amount of time would still not prove that S_n is true for *every* positive integer value of n. To prove that such statements are true for every positive integer value of n, the following principle is often used.

Principle of Mathematical Induction

Let S_n be a statement concerning the positive integer n. Suppose that

(a) S_1 is true;

(b) for any arbitrary positive integer k, the truth of S_k implies the truth of S_{k+1}.

Then S_n is true for every positive integer value of n.

A proof by mathematical induction can be explained as follows. By (a) above, the statement is true when $n = 1$. By (b) above, the fact that the statement is true for $n = 1$ implies that it is true for $n = 1 + 1 = 2$. Using (b) again, it is thus true for $2 + 1 = 3$, for $3 + 1 = 4$, for $4 + 1 = 5$, and so on. By continuing in this way, the statement must be true for *every* positive integer, no matter how large.

The situation is similar to that of a number of dominoes lined up as shown in Figure 1. If the first domino is pushed over, it pushes the next, which pushes the next, and so on until all are down.

Figure I

Another example of the principle of mathematical induction is an infinite ladder. Suppose the rungs are spaced so that, whenever you are on a rung, you know you can move to the next rung. Then *if* you can get to the first rung, you can go as high up the ladder as you wish.

Two separate steps are required for a proof by mathematical induction:

Proof by Mathematical Induction

1. Prove that the statement S_n is true for $n = 1$.
2. Show that, for any positive integer k, S_k implies the truth of S_{k+1}.

In the next example mathematical induction is used to prove the statement S_n mentioned at the beginning of this section.

Example I   Let S_n represent the statement

$$1 + 2 + 3 + \ldots + n = \frac{n(n + 1)}{2}.$$

Prove that S_n is true for every positive integer n.

The proof by mathematical induction is as follows.

Step 1   Show that the statement is true when $n = 1$. If $n = 1$, S_1 becomes

$$1 = \frac{1(1 + 1)}{2}$$

which is true.

Step 2   Assume that S_n is true for the statement $n = k$. That is, assume that

$$1 + 2 + 3 + \ldots + k = \frac{k(k + 1)}{2}$$

is true. Show that this implies the truth of S_{k+1}, where S_{k+1} is the statement

$$1 + 2 + 3 + \ldots + k + (k + 1) = \frac{(k + 1)[(k + 1) + 1]}{2}.$$

Given

$$1 + 2 + 3 + \ldots + k = \frac{k(k + 1)}{2},$$

adding $k + 1$ to both sides of the expression gives

$$1 + 2 + 3 + \ldots + k + (k + 1) = \frac{k(k + 1)}{2} + (k + 1).$$

Then, factor on the right to get

$$= (k + 1)\left(\frac{k}{2} + 1\right)$$

$$= (k + 1)\left(\frac{k + 2}{2}\right)$$

$$1 + 2 + 3 + \ldots + k + (k + 1) = \frac{(k + 1)[(k + 1) + 1]}{2}.$$

This final result is S_{k+1} the statement to be established for $n = k + 1$; therefore the truth of S_k implies the truth of S_{k+1}. The two steps required for a proof by mathematical induction have now been completed, so that the statement S_n is true for every positive integer value of n.   ∎

Example 2   Prove that if x is a real number between 0 and 1, then for every positive integer n, $0 < x^n < 1$.

Here S_1 is: if $0 < x < 1$, then $0 < x^1 < 1$, which is true. Assume now that S_k is true, that is

$$\text{if } 0 < x < 1, \text{ then } 0 < x^k < 1.$$

It must now be shown that this implies the truth of S_{k+1}. Multiply all members of $0 < x^k < 1$ by x to get

$$x \cdot 0 < x \cdot x^k < x \cdot 1.$$

(Here the fact that $0 < x$ is used.) Simplify to get

$$0 < x^{k+1} < x.$$

Since

$$x < 1,$$
$$x^{k+1} < x < 1,$$

and

$$0 < x^{k+1} < 1.$$

This work shows that the truth of S_k implies the truth of S_{k+1}, so that the given statement is true for every positive integer n.   ■

Some statements S_n are not true for the first few values of n, but are true for all values of n that are at least equal to some fixed integer j. The following slightly generalized form of the principle of mathematical induction takes care of these cases.

Generalized Principle of Mathematical Induction

> Let S_n be a statement concerning the positive integer n. Let j be a fixed positive integer. Suppose that
>
> **(a)** S_j is true;
> **(b)** for any positive integer k ($k \geq j$), the truth of S_k implies the truth of S_{k+1}.
>
> Then S_n is true for all positive integers n, where $n \geq j$.

Example 3   Let S_n represent the statement $2^n > 2n + 1$. Show that S_n is true for all values of n such that $n \geq 3$.

(Check that S_n is false for $n = 1$ and $n = 2$.) As before, the proof requires two steps.

Step 1   Show that S_n is true for $n = 3$. If $n = 3$, S_n is

$$2^3 > 2 \cdot 3 + 1$$

or

$$8 > 7$$

which is true.

Step 2   Assume that S_n is true for some positive integer k, where $k \geq 3$. That is, assume the truth of

$$S_k: 2^k > 2k + 1.$$

Now show that the truth of S_k implies the truth of S_{k+1}, where S_{k+1} is

$$2^{k+1} > 2(k + 1) + 1.$$

Use the assumption that $2^k > 2k + 1$; multiply both sides by 2, obtaining

$$2 \cdot 2^k > 2(2k + 1)$$

or

$$2^{k+1} > 4k + 2.$$

Rewrite $4k + 2$ as $2(k + 1) + 2k$, giving

$$2^{k+1} > 2(k + 1) + 2k. \tag{1}$$

Since k is a positive integer,

$$2k > 1. \tag{2}$$

Adding $2(k + 1)$ to both sides of inequality (2) gives

$$2(k + 1) + 2k > 2(k + 1) + 1. \tag{3}$$

From inequalities (1) and (3),

$$2^{k+1} > 2(k + 1) + 2k > 2(k + 1) + 1$$

or

$$2^{k+1} > 2(k + 1) + 1.$$

Thus, the truth of S_k implies the truth of S_{k+1}, and this, together with the fact that S_3 is true, shows that S_n is true for every positive integer value of n greater than or equal to 3. ■

Example 4   There is a flaw in the following "proof." See if you can spot it.

Prove: In every set of n dogs, if at least one is male, then they are all male.

S_1 is obviously true: in a set of one dog, if one is male then all are male. Assume that S_k is true. Let us show that S_{k+1} is true. Suppose

$$\{D_1, D_2, \ldots, D_k, D_{k+1}\}$$

is any set of $k + 1$ dogs, and suppose that one of them is male; say D_1 is male (if it happens that the male was D_2 or D_6, or whatever, just rearrange them so that D_1 is a male). Thus $\{D_1, D_2, \ldots, D_k\}$ is a set of k dogs and one of them is a male, so *all* are males, because S_k is assumed true. But this means that $\{D_2, D_3, \ldots, D_k, D_{k+1}\}$ is now a set of k dogs (we left D_1 out) and one is male (in fact, D_2, $\ldots, D_k$ are males) so they *all* are, including D_{k+1}. Thus, all $k + 1$ dogs are male. This completes the induction argument. (The error appears as a footnote.*) ■

8.1 Exercises

Write out in full and verify each of the statements $S_1, S_2, S_3, S_4,$ and S_5 for each of the following. Then use mathematical induction to prove that each of the given statements is true for every positive integer n.

1. $2 + 4 + 6 + \ldots + 2n = n(n + 1)$ **2.** $1 + 3 + 5 + \ldots + (2n - 1) = n^2$

Use the method of mathematical induction to prove that each of the following statements is true for every positive integer n.

3. $2 + 4 + 8 + \ldots + 2^n = 2^{n+1} - 2$

4. $1^2 + 2^2 + 3^2 + \ldots + n^2 = \dfrac{n(n + 1)(2n + 1)}{6}$

---

*Answer to the question posed in Example 4: The deduction that S_k implies S_{k+1} is correct whenever $k \geq 2$. However, *it does not hold for $k = 1$*; that is, the argument fails to prove that if S_1 is true, then S_2 is also true.

Adapted from *Algebra and Trigonometry with Analytic Geometry* by Arthur B. Simon. Copyright © 1979 by W. H. Freeman and Company. Reprinted by permission.

5. $1^3 + 2^3 + 3^3 + \ldots + n^3 = \dfrac{n^2(n+1)^2}{4}$

6. $3 + 3^2 + 3^3 + \ldots + 3^n = \dfrac{3(3^n - 1)}{2}$

7. $5 \cdot 6 + 5 \cdot 6^2 + 5 \cdot 6^3 + \ldots + 5 \cdot 6^n = 6(6^n - 1)$

8. $\dfrac{1}{1 \cdot 2} + \dfrac{1}{2 \cdot 3} + \dfrac{1}{3 \cdot 4} + \ldots + \dfrac{1}{n(n+1)} = \dfrac{n}{n+1}$

9. $\dfrac{1}{1 \cdot 4} + \dfrac{1 \cdot}{4 \cdot 7} + \dfrac{1}{7 \cdot 10} + \ldots + \dfrac{1}{(3n-2)(3n+1)} = \dfrac{n}{3n+1}$

10. $\dfrac{1}{2} + \dfrac{1}{2^2} + \dfrac{1}{2^3} + \ldots + \dfrac{1}{2^n} = 1 - \dfrac{1}{2^n}$

11. $1 \cdot 2 + 2 \cdot 3 + 3 \cdot 4 + \ldots + n(n+1) = \dfrac{n(n+1)(n+2)}{3}$

12. $1 \cdot 4 + 2 \cdot 9 + 3 \cdot 16 + \ldots + n(n+1)^2 = \dfrac{n(n+1)(n+2)(3n+5)}{12}$

13. $x^{2n} + x^{2n-1}y + \ldots + xy^{2n-1} + y^{2n} = \dfrac{x^{2n+1} - y^{2n+1}}{x - y}$

14. $x^{2n-1} + x^{2n-2}y + \ldots + xy^{2n-2} + y^{2n-1} = \dfrac{x^{2n} - y^{2n}}{x - y}$

15. $(a^m)^n = a^{mn}$ (Assume a and m are constant.)   **16.** $(ab)^n = a^n b^n$ (Assume a and b are constant.)

17. $2^n > 2n$ if $n \geq 3$ **18.** $3^n > 2n + 1$, if $n \geq 2$ **19.** $3n^3 + 6n$ is divisible by 9

20. $n^2 + n$ is divisible by 2   **21.** $3^{2n} - 1$ is divisible by 8   **22.** If $a > 1$, then $a^n > 1$

23. If $a > 1$, then $a^n > a^{n-1}$ **24.** If $0 < a < 1$, then $a^n < a^{n-1}$ **25.** $2^n > n^2$, for $n > 4$

26. If $n \geq 4$, $n! > 2^n$, where $n! = n(n-1)(n-2) \ldots (3)(2)(1)$

27. $4^n > n^4$, for $n \geq 5$

28. $(1 + x)^n \geq 1 + nx$ for every $n > 1$ and fixed $x \geq -1$

29. What is wrong with the following "proof" by mathematical induction?

Prove: Any natural number equals the next natural number.
    To begin, we assume the statement true for some natural number k:

$$k = k + 1.$$

We must now show that the statement is true for $n = k + 1$. If we add 1 to both sides, we have

$$k + 1 = k + 1 + 1$$
$$k + 1 = k + 2.$$

Hence, if the statement is true for $n = k$, it is also true for $n = k + 1$. Thus, the theorem is proved.

30. In the country of Pango, the government prints only three-glok banknotes and five-glok notes. Prove that any purchase of 8 gloks or more can be paid for with only three-glok notes and five-glok notes. (*Hint:* Consider two cases: k gloks being paid with only three-glok notes, and k gloks requiring at least one five-glok note.)

31. Suppose that n straight lines (with $n \geq 2$) are drawn in a plane, where no two lines are parallel and no three lines pass through the same point. Show that the number of points of intersection of the lines is $(n^2 - n)/2$.

32. The series of sketches below starts with an equilateral triangle having sides of length 1. In the following steps, equilateral triangles are constructed on each side of the preceding figure. The lengths of the sides of these new triangles is 1/3 the length of the sides of the preceding triangles. Develop a formula for the number of sides of the *n*th figure. Use mathematical induction to prove your answer.

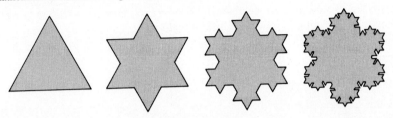

33. Find the perimeter of the *n*th figure in Exercise 32.

34. Show that the area of the *n*th figure in Exercise 32 is

$$\sqrt{3}\left[\frac{2}{5} - \frac{3}{20}\left(\frac{4}{9}\right)^{n-1}\right].$$

35. A pile of *n* rings, each smaller than the one below it, is on a peg. Two other pegs are attached to a board with this peg. In the game called the *Tower of Hanoi* puzzle, all the rings must be moved to a different peg, with only one ring moved at a time, and with no ring ever placed on top of a smaller ring. Find the least number of moves that would be required. Prove your result with mathematical induction.

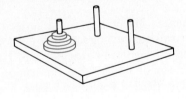

READ
8.1

8.2 The Binomial Theorem

This section introduces a method of expanding the binomial expression $(x + y)^n$. Expansions of $(x + y)^n$ for various positive integer values of n are listed below. For example,

$$(x + y)^1 = x + y$$
$$(x + y)^2 = x^2 + 2xy + y^2$$
$$(x + y)^3 = x^3 + 3x^2y + 3xy^2 + y^3$$
$$(x + y)^4 = x^4 + 4x^3y + 6x^2y^2 + 4xy^3 + y^4$$
$$(x + y)^5 = x^5 + 5x^4y + 10x^3y^2 + 10x^2y^3 + 5xy^4 + y^5.$$

These results indicate a pattern for writing a general expression for $(x + y)^n$. First, notice that each expression begins with x raised to the same power as the binomial itself. That is, the expansion of $(x + y)^1$ has a first term of x^1, $(x + y)^2$ has a first term of x^2, $(x + y)^3$ has a first term of x^3, and so on. Also, the last term in each expansion is y to the same power as the binomial. This means that the expansion of $(x + y)^n$ should begin with the term x^n and end with the term y^n.

Also, the exponents on x decrease by one in each term after the first, while the exponents on y, beginning with y in the second term, increase by one in each succeeding term. Thus, the *variables* in the expansion of $(x + y)^n$ should have the following pattern.

$$x^n, \ x^{n-1}y, \ x^{n-2}y^2, \ x^{n-3}y^3, \ \ldots, \ xy^{n-1}, \ y^n$$

In this pattern, the sum of the exponents on x and y in each term is n. For example, in the third term in the list above, the variable is $x^{n-2}y^2$, and the sum of the exponents is $n - 2 + 2 = n$.

Now examine the *coefficients* in the terms of the expansions shown above. Writing the coefficients alone gives the following pattern.

Pascal's Triangle

$$\begin{array}{ccccccccccc}
 & & & & & 1 & & 1 & & & \\
 & & & & 1 & & 2 & & 1 & & \\
 & & & 1 & & 3 & & 3 & & 1 & \\
 & & 1 & & 4 & & 6 & & 4 & & 1 \\
 & 1 & & 5 & & 10 & & 10 & & 5 & & 1 \\
\end{array}$$

With the coefficients arranged in this way it can be seen that each number in the triangle is the sum of the two numbers directly above it (one to the right and one to the left.) For example, in the fourth row from the top, 1 is the sum of 1, the only number above it, 4 is the sum of 1 and 3, 6 is the sum of 3 and 3, and so on. This triangular array of numbers is called **Pascal's triangle,** in honor of the seventeenth-century mathematician Blaise Pascal, one of the first to use it extensively.

To get the coefficients for $(x + y)^6$, include a sixth row in the array of numbers given above. By adding adjacent numbers the sixth row is found to be

$$1 \quad 6 \quad 15 \quad 20 \quad 15 \quad 6 \quad 1.$$

Using these coefficients the expansion of $(x + y)^6$ is

$$(x + y)^6 = x^6 + 6x^5y + 15x^4y^2 + 20x^3y^3 + 15x^2y^4 + 6xy^5 + y^6.$$

Although it is possible to use Pascal's triangle to find the coefficients of $(x + y)^n$ for any positive integer value of n, this becomes impractical for large values of n due to the need to write out all the preceding rows. A more efficient way of finding these coefficients uses factorial notation. The number $n!$ (read "n-factorial"), is defined as follows.

n-factorial

For any positive integer n,

$$n! = n(n - 1)(n - 2) \ldots (3)(2)(1).$$

For example, $5! = 5 \cdot 4 \cdot 3 \cdot 2 \cdot 1 = 120$, $7! = 7 \cdot 6 \cdot 5 \cdot 4 \cdot 3 \cdot 2 \cdot 1 = 5040$, $2! = 2 \cdot 1 = 2$, and so on. To simplify certain formulas later on, $0!$ is defined as

$$0! = 1.$$

Now look at the coefficients of the expression

$$(x + y)^5 = x^5 + 5x^4y + 10x^3y^2 + 10x^2y^3 + 5xy^4 + y^5.$$

The coefficient on the second term, $5x^4y$, is 5, and the exponents on the variables are 4 and 1. Note that

$$5 = \frac{5!}{4!1!}.$$

The coefficient on the third term is 10, with exponents of 3 and 2, and

$$10 = \frac{5!}{3!2!}.$$

The last term (the sixth term) can be written as $y^5 = 1x^0y^5$, with coefficient 1, and exponents of 0 and 5. Since $0! = 1$, check that

$$1 = \frac{5!}{0!5!}.$$

Generalizing from these examples, the coefficient for the term of the expansion of $(x + y)^n$ in which the variable part is x^ry^{n-r} (where $r \leq n$) will be

$$\frac{n!}{r!(n - r)!}.$$

This number, called a **binomial coefficient,** is often symbolized $\binom{n}{r}$ (read "*n* above *r*").

Binomial Coefficient

For nonnegative integers n and r, with $r \leq n$, the symbol $\binom{n}{r}$ is defined as

$$\binom{n}{r} = \frac{n!}{r!(n - r)!}.$$

These binomial coefficients are just numbers from Pascal's triangle. For example, $\binom{3}{0}$ is the first number in the third row, and $\binom{7}{4}$ is the fifth number in the seventh row.

Example 1 (a) $\dbinom{6}{2} = \dfrac{6!}{2!(6-2)!} = \dfrac{6!}{2!4!} = \dfrac{6 \cdot 5 \cdot 4 \cdot 3 \cdot 2 \cdot 1}{2 \cdot 1 \cdot 4 \cdot 3 \cdot 2 \cdot 1} = 15$

(b) $\dbinom{8}{0} = \dfrac{8!}{0!(8-0)!} = \dfrac{8!}{0!8!} = \dfrac{8!}{1 \cdot 8!} = 1$

(c) $\dbinom{10}{10} = \dfrac{10!}{10!(10-10)!} = \dfrac{10!}{10!0!} = 1$ ∎

Our conjectures about the expansion of $(x + y)^n$ are summarized below.

Observations on the Expansion of $(x + y)^n$

1. There are $(n + 1)$ terms in the expansion.
2. The first term is x^n and the last term is y^n.
3. The exponent on x decreases by 1 and the exponent on y increases by 1 in each succeeding term.
4. The sum of the exponents on x and y in any term is n.
5. The coefficient of the second term and of the next to last term is n.

These observations about the expansion of $(x + y)^n$ for any positive integer value of n suggest the **binomial theorem.**

Binomial Theorem

For any positive integer n and any complex numbers x and y,

$$(x + y)^n = x^n + \binom{n}{n-1}x^{n-1}y + \binom{n}{n-2}x^{n-2}y^2 +$$
$$\binom{n}{n-3}x^{n-3}y^3 + \ldots + \binom{n}{n-r}x^{n-r}y^r + \ldots$$
$$+ \binom{n}{1}xy^{n-1} + y^n. \tag{1}$$

The binomial theorem can be proved by mathematical induction. Let S_n be statement (1) above. Begin by verifying S_n for $n = 1$.

$$S_1: \quad (x + y)^1 = x^1 + y^1,$$

which is true.

Now assume that S_n is true for the positive integer k. Statement S_k becomes (using the definition of the binomial coefficient)

$$S_k: \quad (x + y)^k = x^k + \frac{k!}{1!(k-1)!}x^{k-1}y + \frac{k!}{2!(k-2)!}x^{k-2}y^2$$
$$+ \ldots + \frac{k!}{(k-1)!1!}xy^{k-1} + y^k. \tag{2}$$

Multiply both sides of equation (2) by $x + y$.

$$(x + y)^k \cdot (x + y)$$
$$= x(x + y)^k + y(x + y)^k$$
$$= \left[x \cdot x^k + \frac{k!}{1!(k - 1)!} x^k y + \frac{k!}{2!(k - 2)!} x^{k-1} y^2 + \ldots + \frac{k!}{(k - 1)!1!} x^2 y^{k-1} + xy^k \right]$$
$$+ \left[x^k \cdot y + \frac{k!}{1!(k - 1)!} x^{k-1} y^2 + \ldots + \frac{k!}{(k - 1)!1!} xy^k + y \cdot y^k \right].$$

Rearrange terms to get

$$(x + y)^{k+1}$$
$$= x^{k+1} + \left[\frac{k!}{1!(k - 1)!} + 1 \right] x^k y + \left[\frac{k!}{2!(k - 2)!} + \frac{k!}{1!(k - 1)!} \right] x^{k-1} y^2 + \ldots$$
$$+ \left[1 + \frac{k!}{(k - 1)!1!} \right] xy^k + y^{k+1}. \tag{3}$$

The first expression in brackets in equation (3) simplifies to $\binom{k + 1}{1}$. To see this, note that

$$\binom{k + 1}{1} = \frac{(k + 1)(k)(k - 1)(k - 2) \ldots 1}{1 \cdot (k)(k - 1)(k - 2) \ldots 1} = k + 1.$$

Also $$\frac{k!}{1!(k - 1)!} + 1 = \frac{k!}{(k - 1)!} + 1$$

$$= \frac{k! + (k - 1)!}{(k - 1)!} = \frac{k(k - 1)! + (k - 1)!}{(k - 1)!}$$

$$= \frac{(k + 1)(k - 1)!}{(k - 1)!} = k + 1.$$

The second expression becomes $\binom{k + 1}{2}$, the last $\binom{k + 1}{k}$, and so on. The result

of equation (3) is just equation (2) with every k replaced by $k + 1$. Thus, the truth of S_n when $n = k$ implies the truth of S_n for $n = k + 1$, which completes the proof of the theorem by mathematical induction.

Example 2   Write out the binomial expansion of $(x + y)^9$.
        Using the binomial theorem,

$$(x + y)^9 = x^9 + \binom{9}{8} x^8 y + \binom{9}{7} x^7 y^2 + \binom{9}{6} x^6 y^3 + \binom{9}{5} x^5 y^4$$
$$+ \binom{9}{4} x^4 y^5 + \binom{9}{3} x^3 y^6 + \binom{9}{2} x^2 y^7 + \binom{9}{1} xy^8 + y^9.$$

Now evaluate each of the binomial coefficients.

$$(x + y)^9 = x^9 + \frac{9!}{1!8!} x^8 y + \frac{9!}{2!7!} x^7 y^2 + \frac{9!}{3!6!} x^6 y^3 + \frac{9!}{4!5!} x^5 y^4$$

$$+ \frac{9!}{5!4!} x^4 y^5 + \frac{9!}{6!3!} x^3 y^6 + \frac{9!}{7!2!} x^2 y^7 + \frac{9!}{8!1!} xy^8 + y^9$$

$$= x^9 + 9x^8 y + 36x^7 y^2 + 84x^6 y^3 + 126x^5 y^4 + 126x^4 y^5$$

$$+ 84x^3 y^6 + 36x^2 y^7 + 9xy^8 + y^9 \quad \blacksquare$$

Example 3 Expand $\left(a - \dfrac{b}{2} \right)^5$.

Again use the binomial theorem. Let $x = a$, $y = -\dfrac{b}{2}$, and $n = 5$, to get

$$\left(a - \frac{b}{2} \right)^5 = a^5 + \binom{5}{4} a^4 \left(-\frac{b}{2} \right) + \binom{5}{3} a^3 \left(-\frac{b}{2} \right)^2 + \binom{5}{2} a^2 \left(-\frac{b}{2} \right)^3$$

$$+ \binom{5}{1} a \left(-\frac{b}{2} \right)^4 + \left(-\frac{b}{2} \right)^5$$

$$= a^5 + 5a^4 \left(-\frac{b}{2} \right) + 10a^3 \left(-\frac{b}{2} \right)^2 + 10a^2 \left(-\frac{b}{2} \right)^3$$

$$+ 5a \left(-\frac{b}{2} \right)^4 + \left(-\frac{b}{2} \right)^5$$

$$= a^5 - \frac{5}{2} a^4 b + \frac{5}{2} a^3 b^2 - \frac{5}{4} a^2 b^3 + \frac{5}{16} ab^4 - \frac{1}{32} b^5. \quad \blacksquare$$

As Example 3 illustrates, any expansion of the *difference* of two terms has alternating signs.

Example 4 Expand $\left(\dfrac{3}{m^2} - 2\sqrt{m} \right)^4$. (Assume $m > 0$.)

By the binomial theorem,

$$\left(\frac{3}{m^2} - 2\sqrt{m} \right)^4 = \left(\frac{3}{m^2} \right)^4 + \binom{4}{1} \left(\frac{3}{m^2} \right)^3 (-2\sqrt{m})^1 + \binom{4}{2} \left(\frac{3}{m^2} \right)^2 (-2\sqrt{m})^2$$

$$+ \binom{4}{3} \left(\frac{3}{m^2} \right)^1 (-2\sqrt{m})^3 + (-2\sqrt{m})^4$$

$$= \frac{81}{m^8} + 4 \left(\frac{27}{m^6} \right) (-2m^{1/2}) + 6 \left(\frac{9}{m^4} \right) (4m)$$

$$+ 4 \left(\frac{3}{m^2} \right) (-8m^{3/2}) + 16m^2.$$

Here the fact that $\sqrt{m} = m^{1/2}$ was used. Finally,

$$\left(\frac{3}{m^2} - 2\sqrt{m} \right)^4 = \frac{81}{m^8} - \frac{216}{m^{11/2}} + \frac{216}{m^3} - \frac{96}{m^{1/2}} + 16m^2. \quad \blacksquare$$

Our observations on the binomial theorem make it possible to write any single term of a binomial expansion. For example, to find the tenth term of $(x + y)^n$, where $n \geq 9$, first notice that in the tenth term y is raised to the ninth power (since y has the power 1 in the second term, the power 2 in the third term, and so on). Because the exponents on x and y in any term must have a sum of n, the exponent on x in the tenth term is $n - 9$. Thus, the tenth term of the expansion is

$$\binom{n}{n-9} x^{n-9} y^9 = \frac{n!}{(n-9)!9!} x^{n-9} y^9.$$

This same idea can be used to obtain the result given in the following theorem.

rth Term of the Binomial Expansion

The rth term of the binomial expansion of $(x + y)^n$, where $n \geq r - 1$, is

$$\binom{n}{n-(r-1)} x^{n-(r-1)} y^{r-1}.$$

Example 5 Find the seventh term of $(a + 2b)^{10}$.

In the seventh term $2b$ has an exponent of 6, while a has an exponent of $10 - 6$, or 4. The seventh term is

$$\binom{10}{4} a^4 (2b)^6 = 210 a^4 (64 b^6)$$

$$= 13{,}440 a^4 b^6.$$

8.2 Exercises

Write out the binomial expansion for each of the following.

1. $(x + y)^6$

2. $(m + n)^4$

3. $(p - q)^5$

4. $(a - b)^7$

5. $(r^2 + s)^5$

6. $(m + n^2)^4$

7. $(p + 2q)^4$

8. $(3r - s)^6$

9. $\left(\dfrac{m}{2} - 1\right)^6$

10. $\left(3 + \dfrac{y}{3}\right)^5$

11. $\left(2p + \dfrac{q}{3}\right)^4$

12. $\left(\dfrac{r}{6} - \dfrac{m}{2}\right)^3$

13. $(m^{-2} + m^2)^4$

14. $\left(\sqrt{x} + \dfrac{4}{\sqrt{x}}\right)^5$

15. $(p^{2/3} - 5p^{4/3})^5$

16. $(3y^{3/2} + 5y^{1/2})^4$

Write only the first four terms in each of the following expansions.

17. $(x + 6)^{21}$

18. $(y - 8)^{17}$

19. $(3m^{-2} + 5m^{-4})^9$

20. $\left(\dfrac{8}{k^2} + 6k\right)^{17}$

For each of the following write the indicated term of the binomial expansion.

21. 5th term of $(m - 2p)^{12}$

22. 4th term of $(3x + y)^6$

23. 9th term of $(2m + n)^{10}$

24. 7th term of $(3r - 5s)^{12}$

25. 17th term of $(p^2 + q)^{20}$

26. 10th term of $(2x^2 + y)^{14}$

27. 8th term of $(x^3 + 2y)^{14}$ $7 - 1 = 12 - 6$

28. 13th term of $(a + 2b^3)^{12}$

29. Find the middle term of $(3x^7 + 2y^3)^8$.

$12 \, (3r)^6 (s_5)^6$

30. Find the two middle terms of $(-2m^{-1} + 3n^{-2})^{11}$.

$\binom{12}{6} 729 r^6$

Use the binomial expansion to evaluate each of the following to three decimal places.

31. $(1.01)^{10}$ (*Hint:* $1.01 = 1 + .01$)   **32.** $(0.99)^{15}$ (*Hint:* $0.99 = 1 - 0.01$)

33. $(1.99)^8$ **34.** $(2.99)^3$ **35.** $(3.02)^6$ **36.** $(1.01)^9$

In later courses, it is shown that

$$(1 + x)^n = 1 + nx + \frac{n(n-1)}{2!}x^2 + \frac{n(n-1)(n-2)}{3!}x^3 + \cdots$$

for any real number n (not just positive integer values) and any real number x where $|x| < 1$. This result, a generalized binomial theorem, may be used to find approximate values of powers and roots. For example,

$$\sqrt[4]{630} = (625 + 5)^{1/4} = \left[625\left(1 + \frac{5}{625}\right)\right]^{1/4} = 625^{1/4}\left(1 + \frac{5}{625}\right)^{1/4}.$$

37. Use the result above to approximate $(1 + 5/625)^{1/4}$ to the nearest thousandth. Then approximate $\sqrt[4]{630}$.

38. Approximate $\sqrt[3]{9.42}$, using the method described above.

39. Approximate $(1.02)^{-3}$.

40. Approximate $1/(1.04)^5$.

41. In the formula above, let $n = -1$ and find $(1 + x)^{-1}$.

42. Find the first four terms when $1 + x$ is divided into 1. Compare with the results of Exercise 41.

43. Work out the first four terms in the expansion of $(1 + 3)^{1/2}$ using $x = 3$ and $n = 1/2$ in the formula above. Is the result close to $(1 + 3)^{1/2} = 4^{1/2} = 2$? Why not?

44. Use the result above to show that for small values of x, $\sqrt{1 + x} \approx 1 + \frac{1}{2}x$.

Prove that each of the following properties of factorial notation are true for any positive integer n.

45. $n(n - 1)! = n!$ **46.** $\dfrac{n!}{n(n-1)} = (n - 2)!$

Find all values of n for which the following are true.

47. $(n + 2)! = 56 \cdot n!$ **48.** $(n - 4)! = n - 4$

8.3 Arithmetic Sequences

A **sequence** is a function having as domain a set of positive integers. Instead of using $f(x)$ notation to indicate a sequence, it is customary to use a_n, where n is used to represent a positive integer. The elements in the range of a sequence, called the **terms** of the sequence, are $a_1, a_2, a_3, \ldots$. The **general term**, or **nth term**, of the sequence is a_n.

Example 1   Write the first five terms for each of the following sequences.

(a) $a_n = \dfrac{n + 1}{n + 2}$

Replacing n, in turn, with 1, 2, 3, 4, and 5 gives

$$\frac{2}{3}, \frac{3}{4}, \frac{4}{5}, \frac{5}{6}, \frac{6}{7}.$$

(b) $a_n = (-1)^n \cdot n$

Replace n, in turn, with 1, 2, 3, 4, and 5, to get

$$a_1 = (-1)^1 \cdot 1 = -1$$
$$a_2 = (-1)^2 \cdot 2 = 2$$
$$a_3 = (-1)^3 \cdot 3 = -3$$
$$a_4 = (-1)^4 \cdot 4 = 4$$
$$a_5 = (-1)^5 \cdot 5 = -5.$$

(c) $b_n = \dfrac{(-1)^n}{2^n}$

Here $b_1 = -1/2$, $b_2 = 1/4$, $b_3 = -1/8$, $b_4 = 1/16$, and $b_5 = -1/32$.   ∎

A sequence is a **finite sequence** if the domain is the set $\{1, 2, 3, 4, \ldots, n\}$, where n is a positive integer. An **infinite sequence** has the set of all positive integers as its domain.

Example 2   The sequence of positive even integers,

$$2, 4, 6, 8, 10, 12, 14, \ldots,$$

is infinite, but the sequence of days in June,

$$1, 2, 3, 4, \ldots, 29, 30,$$

is finite.   ∎

Example 3   Find the first four terms for the sequence defined as follows: $a_1 = 4$, and for $n > 1$, $a_n = 2 \cdot a_{n-1} + 1$.

We know $a_1 = 4$. Since $a_n = 2 \cdot a_{n-1} + 1$,

$$a_2 = 2 \cdot a_1 + 1 = 2 \cdot 4 + 1 = 9$$
$$a_3 = 2 \cdot a_2 + 1 = 2 \cdot 9 + 1 = 19$$
$$a_4 = 2 \cdot a_3 + 1 = 2 \cdot 19 + 1 = 39. \quad ∎$$

The definition of this sequence is an example of a **recursive definition,** one in which each term is defined as an expression involving the previous term.

A sequence in which each term after the first is obtained by adding a fixed number to the previous term is called an **arithmetic sequence** (or **arithmetic progression**).

The fixed number that is added is called the **common difference.** The sequence

$$5, \ 9, \ 13, \ 17, \ 21, \ . \ . \ .$$

is an arithmetic sequence since each term after the first is obtained by adding 4 to the previous term. That is,

$$9 = 5 + 4$$
$$13 = 9 + 4$$
$$17 = 13 + 4$$
$$21 = 17 + 4,$$

and so on. The common difference is 4.

Example 4   Find the common difference, d, for the arithmetic sequence

$$-9, \ -7, \ -5, \ -3, \ -1, \ . \ . \ . \ .$$

Since this sequence is arithmetic, d can be found by choosing any two adjacent terms and subtracting the first from the second. If we choose -7 and -5,

$$d = -5 - (-7) = 2.$$

Choosing -9 and -7 would give $d = -7 - (-9) = 2$, the same result.   ■

Example 5   Write the first five terms for each arithmetic sequence.

(a) $a_1 = 7, d = -3$

The first term is 7, and each succeeding term is found by adding -3 to the preceding term.

$$a_1 = 7$$
$$a_2 = 7 + (-3) = 4$$
$$a_3 = 4 + (-3) = 1$$
$$a_4 = 1 + (-3) = -2$$
$$a_5 = -2 + (-3) = -5.$$

(b) $a_1 = -12, d = 5$

Use the definition of an arithmetic sequence.

$$a_1 = -12$$
$$a_2 = -12 + d = -12 + 5 = -7$$
$$a_3 = -7 + d = -7 + 5 = -2$$
$$a_4 = -2 + d = -2 + 5 = 3$$
$$a_5 = 3 + d = 3 + 5 = 8.$$ ■

If a_1 is the first term of an arithmetic sequence and d is the common difference, then the terms of the sequence are given by

$$a_1 = a_1$$
$$a_2 = a_1 + d$$
$$a_3 = a_2 + d = a_1 + d + d = a_1 + 2d$$
$$a_4 = a_3 + d = a_1 + 2d + d = a_1 + 3d$$
$$a_5 = a_1 + 4d$$
$$a_6 = a_1 + 5d,$$

and, by this pattern,

$$a_n = a_1 + (n - 1)d.$$

This result could be proven by mathematical induction (see Exercise 95); a summary is given below.

nth Term of an Arithmetic Sequence	In an arithmetic sequence with first term a_1 and common difference d, the nth term, a_n, is given by 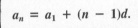 $$a_n = a_1 + (n - 1)d.$$

Example 6   Find a_{13} and a_n for the arithmetic sequence

$$-3, \ 1, \ 5, \ 9, \ \ldots \ .$$

Here $a_1 = -3$ and $d = 1 - (-3) = 4$. To find a_{13}, substitute 13 for n in the formula above.

$$a_{13} = a_1 + (13 - 1)d$$
$$a_{13} = -3 + (12)4$$
$$a_{13} = -3 + 48$$
$$a_{13} = 45$$

To find a_n, substitute values for a_1 and d in the formula for a_n.

$$a_n = -3 + (n - 1) \cdot 4$$
$$a_n = -3 + 4n - 4$$
$$a_n = 4n - 7 \quad \blacksquare$$

Example 7   Find a_{18} and a_n for the arithmetic sequence having $a_2 = 9$ and $a_3 = 15$.
Find d: $d = a_3 - a_2 = 15 - 9 = 6$.

Since
$$a_2 = a_1 + d,$$
$$9 = a_1 + 6 \quad \text{and} \quad a_1 = 3.$$

Then
$$a_{18} = 3 + (18 - 1) \cdot 6$$
$$a_{18} = 105$$

and
$$a_n = 3 + (n - 1) \cdot 6$$
$$a_n = 3 + 6n - 6$$
$$a_n = 6n - 3. \quad \blacksquare$$

Example 8 A child building a tower with blocks uses 15 for the first row. Each row has 2 blocks less than the previous row. If there are 8 rows in the tower, how many blocks are used for the top row?

The number of blocks in each row forms an arithmetic sequence with $a_1 = 15$ and $d = -2$. We want to find a_n for $n = 8$. Using the formula

$$a_n = a_1 + (n - 1)d$$

gives
$$a_8 = 15 + (8 - 1)(-2) = 1.$$

There is just one block in the top row. $\blacksquare$

Example 9 Suppose that an arithmetic sequence has $a_8 = -16$ and $a_{16} = -40$. Find a_1.

Since $a_8 = a_1 + (8 - 1)d$, replacing a_8 with -16 gives $-16 = a_1 + 7d$ or $a_1 = -16 - 7d$. Similarly, $-40 = a_1 + 15d$ or $a_1 = -40 - 15d$. From these two equations, using the substitution method given earlier, $-16 - 7d = -40 - 15d$, so $d = -3$. To find a_1, substitute -3 for d in $-16 = a_1 + 7d$:

$$-16 = a_1 + 7d$$
$$-16 = a_1 + 7(-3)$$
$$a_1 = 5. \quad \blacksquare$$

[handwritten: $a_8 = a_1 + 7d = -16$; $a_{16} = a_1 + 15d = -40$]

It is often necessary to add the terms of an arithmetic sequence. For example, suppose that a man borrows $3000 and agrees to pay $100 per month plus interest of 1% per month on the unpaid balance until the loan is paid off. The first month he pays $100 to reduce the loan plus interest of $(.01)3000 = 30$ dollars. The second month he pays another $100 toward the loan and interest of $(.01)2900 = 29$ dollars. Since the loan is reduced by $100 each month, his interest payments decrease by $(.01)100 = 1$ dollar each month, forming the arithmetic sequence

$$30, 29, 28, \ldots, 3, 2, 1.$$

The total amount of interest paid is given by the sum of the terms of this sequence. A formula will be developed here to find this sum without adding all thirty numbers directly.

To find this formula, suppose a sequence has terms $a_1, a_2, a_3, a_4 \ldots$. Then S_n is defined as the sum of the first n terms of the sequence. That is,

$$S_n = a_1 + a_2 + a_3 + \ldots + a_n.$$

To find a formula for S_n, assume the sequence is arithmetic and write the sum of the first n terms as follows.

$$S_n = a_1 + [a_1 + d] + [a_1 + 2d] + \ldots + [a_1 + (n - 1)d]$$

The formula for the general term was used in the last expression. Now write the same sum in reverse order, beginning with a_n and *subtracting d*.

$$S_n = a_n + [a_n - d] + [a_n - 2d] + \ldots + [a_n - (n - 1)d]$$

Adding respective sides of these two equations term by term gives

$$S_n + S_n = (a_1 + a_n) + (a_1 + a_n) + \ldots + (a_1 + a_n)$$

or $$2S_n = n(a_1 + a_n),$$

since there are n terms of $a_1 + a_n$ on the right. Now solve for S_n to get

$$S_n = \frac{n}{2}(a_1 + a_n).$$

Using the formula $a_n = a_1 + (n - 1)d$, this result for S_n can also be written as

$$S_n = \frac{n}{2}[a_1 + a_1 + (n - 1)d]$$

or $$S_n = \frac{n}{2}[2a_1 + (n - 1)d],$$

an alternate formula for the sum of the first n terms of an arithmetic sequence.

Sum of the First n Terms of an Arithmetic Sequence

Sum of the First n Terms of an Arithmetic Sequence

If an arithmetic sequence has first term a_1 and common difference d, then the sum of the first n terms, S_n, is given by

$$S_n = \frac{n}{2}(a_1 + a_n)$$

or

$$S_n = \frac{n}{2}[2a_1 + (n - 1)d].$$

The first formula of the theorem is used when the first and last terms are known; otherwise the second formula is used.

Either one of these formulas can be used to find the amount of interest the man above will pay on the $3000 loan. In the sequence of interest payments $a_1 = 30$, $d = -1$, $n = 30$, and $a_n = 1$. Choosing the first formula

$$S_n = \frac{n}{2}(a_1 + a_n)$$

gives $$S_n = \frac{30}{2}(30 + 1) = 15(31) = 465,$$

so a total of $465 interest will be paid over 30 months.

Example 10 (a) Find S_{12} for the arithmetic sequence $-9, -5, -1, 3, 7, \ldots$.
Here $a_1 = -9$, $d = 4$, and $n = 12$. Use the second formula above.

Using $a_1 = -9$ and $d = 4$ in the second formula,

$$S_n = \frac{n}{2}[2a_1 + (n-1)d],$$

$$S_{12} = \frac{12}{2}[2(-9) + 11(4)] = 6(-18 + 44) = 156.$$

(b) Find the sum of the first 60 positive integers.

In this example, $n = 60$, $a_1 = 1$, and $a_{60} = 60$, so it is more convenient to use the first of the two formulas:

$$S_n = \frac{n}{2}(a_1 + a_n)$$

$$S_{60} = \frac{60}{2}(1 + 60) = 30 \cdot 61 = 1830. \quad \blacksquare$$

Example 11 The sum of the first 17 terms of an arithmetic sequence is 187. If $a_{17} = -13$, find a_1 and d.

Use the formula for S_n, with $n = 17$, to find a_1.

$$187 = \frac{17}{2}(a_1 - 13)$$

$$374 = 17(a_1 - 13)$$

$$22 = a_1 - 13$$

$$a_1 = 35.$$

From the formula for a_n,

$$a_{17} = a_1 + (17 - 1)d,$$

and

$$-13 = 35 + 16d$$

$$-48 = 16d$$

$$d = -3. \quad \blacksquare$$

Sigma Notation Sometimes a special shorthand notation is used for the sum of n terms of a sequence. The symbol Σ, the Greek capital letter *sigma*, is used to indicate a sum. For example,

$$\sum_{i=1}^{n} a_i$$

represents the sum

$$\sum_{i=1}^{n} a_i = a_1 + a_2 + a_3 + \ldots + a_n.$$

The letter i is called the **index of summation.** Do not confuse this use of i with the use of i to represent an imaginary number.

Example 12 Find each of the following sums.

(a) $\displaystyle\sum_{i=1}^{10} (4i + 8)$

This sum can be written as

$$\sum_{i=1}^{10} (4i + 8) = [4(1) + 8] + [4(2) + 8] + [4(3) + 8] + \ldots + [4(10) + 8]$$
$$= 12 + 16 + 20 + \ldots + 48,$$

the sum of the first ten terms of the arithmetic sequence having

$$a_1 = 4 \cdot 1 + 8 = 12,$$
$$n = 10,$$

and $$a_n = a_{10} = 4 \cdot 10 + 8 = 48.$$

By a formula for S_n,

$$\sum_{i=1}^{10} (4i + 8) = S_{10} = \frac{10}{2}(12 + 48) = 5(60) = 300.$$

(b) $\displaystyle\sum_{i=1}^{15} (9 - i) = S_{15} = \frac{15}{2}[8 + (-6)]$

$$= \frac{15}{2}[2]$$

$$= 15. \quad \blacksquare$$

8.3 Exercises

Write the first five terms of each of the following sequences.

1. $a_n = 6n + 4$
2. $a_n = 3n - 2$
3. $a_n = 2^{-n+1}$

4. $a_n = (-1)^n(n + 2)$
5. $a_n = \dfrac{2}{n + 3}$
6. $a_n = \dfrac{8n - 4}{2n + 1}$

7. $a_n = (-2)^n(n)$
8. $a_n = (-\frac{1}{2})^n(n^{-1})$
9. $a_n = \dfrac{n^2 + 1}{n^2 + 2}$

10. $a_n = \dfrac{(-1)^{n-1}(n + 1)}{n + 2}$
11. a_n is the nth prime number

12. b_n is the number of positive integers that divide into n (with remainder 0)

Find the first ten terms for the sequences defined as follows.

13. $a_1 = 4$, $a_n = a_{n-1} + 5$, for $n > 1$
14. $a_1 = -3$, $a_n = a_{n-1} + 2$, for $n > 1$

15. $a_1 = 2$, $a_n = 2 \cdot a_{n-1}$, for $n > 1$
16. $a_1 = 3$, $a_n = -2 \cdot a_{n-1}$, for $n > 1$

17. $a_1 = 1$, $a_2 = 1$, $a_n = a_{n-1} + a_{n-2}$, for $n \geq 3$ (the Fibonacci sequence)

18. $a_1 = 1$, $a_n = n \cdot \cos(\pi \cdot a_{n-1})$, for $n > 1$ (for students who have studied trigonometry)

Write the terms of the arithmetic sequences satisfying each of the following conditions.

19. $a_1 = 4$, $d = 2$, $n = 5$

20. $a_1 = 6$, $d = 8$, $n = 4$

21. $a_2 = 9$, $d = -2$, $n = 4$

22. $a_3 = 7$, $d = -4$, $n = 4$

23. $a_1 = 4 - \sqrt{5}$, $a_2 = 4$, $n = 5$

24. $a_1 = -8$, $a_2 = -8 + \sqrt{7}$, $n = 5$

For each of the following arithmetic sequences, find d and a_n.

25. 12, 17, 22, 27, 32, 37, . . .

26. 8, 17, 26, 35, 44, 53, . . .

27. 18, 15, 12, 9, 6, . . .

28. 30, 24, 18, 12, . . .

29. $-6 + \sqrt{2}$, $-6 + 2\sqrt{2}$, $-6 + 3\sqrt{2}$, . . .

30. $4 - \sqrt{11}$, $5 - \sqrt{11}$, $6 - \sqrt{11}$, . . .

31. x, $x + m$, $x + 2m$, $x + 3m$, $x + 4m$, . . .

32. $k + p$, $k + 2p$, $k + 3p$, $k + 4p$, . . .

33. $2z + m$, $2z$, $2z - m$, $2z - 2m$, $2z - 3m$, . . .

34. $3r - 4z$, $3r - 3z$, $3r - 2z$, $3r - z$, $3r$, $3r + z$, . . .

Find a_8 and a_n for each of the following arithmetic sequences.

35. $a_1 = 5$, $d = 2$

36. $a_1 = -3$, $d = -4$

37. $a_3 = 2$, $d = 1$

38. $a_4 = 5$, $d = -2$

39. $a_1 = 8$, $a_2 = 6$

40. $a_1 = 6$, $a_2 = 3$

41. $a_1 = 12$, $a_3 = 6$

42. $a_2 = 5$, $a_4 = 1$

43. $a_5 = 4.2$, $a_6 = 4.5$

44. $a_3 = -9.5$, $a_4 = -10.6$

45. $a_1 = x$, $a_2 = x + 3$

46. $a_2 = y + 1$, $d = -3$

47. $a_6 = 2m$, $a_7 = 3m$

48. $a_5 = 4p + 1$, $a_7 = 6p + 7$

Find the sum of the first ten terms for each of the following arithmetic sequences.

49. $a_1 = 8$, $d = 3$

50. $a_1 = 2$, $d = 6$

51. $a_3 = 5$, $a_4 = 8$

52. $a_2 = 9$, $a_4 = 13$

53. 5, 9, 13, . . .

54. 8, 6, 4, . . .

55. $a_1 = 9.428$, $d = -1.723$

56. $a_1 = -3.119$, $d = 2.422$

57. $a_4 = 2.556$, $a_5 = 3.004$

58. $a_7 = 11.192$, $a_9 = 4.812$

Evaluate each of the following sums.

59. $\displaystyle\sum_{i=1}^{3} (i + 4)$

60. $\displaystyle\sum_{i=1}^{5} (i - 8)$

61. $\displaystyle\sum_{i=1}^{10} (2i + 3)$

62. $\displaystyle\sum_{i=1}^{15} (5i - 9)$

63. $\displaystyle\sum_{i=1}^{12} (-5 - 8i)$

64. $\displaystyle\sum_{i=1}^{19} (-3 - 4i)$

65. $\displaystyle\sum_{i=1}^{1000} i$

66. $\displaystyle\sum_{i=1}^{2000} i$

67. $\displaystyle\sum_{i=6}^{15} (4i - 2)$

68. $\displaystyle\sum_{i=9}^{20} (8i + 3)$

69. $\displaystyle\sum_{i=7}^{12} (6 - 2i)$

70. $\displaystyle\sum_{i=4}^{17} (-3 - 5i)$

Find a_1 for each of the following arithmetic sequences.

71. $a_9 = 47$, $a_{15} = 77$

72. $a_{10} = 50$, $a_{20} = 110$

73. $a_{15} = 168$, $a_{16} = 180$

74. $a_{10} = -54$, $a_{17} = -89$

75. $S_{20} = 1090$, $a_{20} = 102$

76. $S_{31} = 5580$, $a_{31} = 360$

77. Find the sum of all the integers from 51 through 71.

78. Find the sum of all the integers from -8 through 30.

79. Find the sum of the first n positive integers.

80. Find the sum of the first n odd positive integers.

81. If a clock strikes the proper number of bongs each hour on the hour, how many bongs will it bong in a month of 30 days?

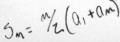

$S_n = \frac{n}{2}(a_1 + a_n)$

$5(5+23)$
$5(28)$ 140

82. A stack of telephone poles has 30 in the bottom row, 29 in the next, and so on, with one pole in the top row. How many poles are in the stack?

83. A sky diver falls 5 meters during the first second, 15 meters during the second, 25 meters during the third, and so on. How many meters will the diver fall during the tenth second? During the first ten seconds?

84. Deepwell Drilling Company charges a flat $500 set-up charge, plus $5 for the first foot of well drilled, $6 for the second, $7 for the third, and so on. Find the total charge for a 70-foot well.

85. An object falling under the force of gravity falls about 16 feet the first second, 48 feet during the second, 80 feet during the third, and so on. How far would the object fall during the eighth second? What is the total distance the object would fall in eight seconds?

86. The population of a city was 49,000 five years ago. Each year, the zoning commission permits an increase of 580 in the population of the city. What will the population be five years from now?

87. The sum of four terms in an arithmetic sequence is 66. The sum of the squares of the terms is 1214. Find the terms.

88. The sum of five terms of an arithmetic sequence is 5. If the product of the first and second is added to the product of the fourth and fifth, the result is 326. Find the terms.

89. A super slide of uniform slope is to be built on a level piece of land. There are to be twenty equally spaced supports, with the longest support 15 m in length, and the shortest 2 m in length. Find the total length of all the supports.

90. How much material would be needed for the rungs of a ladder of 31 rungs, if the rungs taper from 18 inches to 28 inches? Assume that the lengths of the rungs form the terms of an arithmetic sequence.

91. Find all arithmetic sequences $a_1, a_2, a_3, \ldots$, such that $a_1{}^2, a_2{}^2, a_3{}^2, \ldots$, is also an arithmetic sequence.

92. Suppose that $a_1, a_2, a_3, \ldots$ and $b_1, b_2, b_3, \ldots$ are each arithmetic sequences. Let $d_n = a_n + c \cdot b_n$, for any real number c and every positive integer n. Show that $d_1, d_2, d_3, \ldots$ is an arithmetic sequence.

Explain why each of the following sequences is arithmetic.

93. $\log 2, \log 4, \log 8, \log 16, \log 32, \ldots$

94. $\log 12, \log 36, \log 108, \log 324, \ldots$

Prove each of the following results by mathematical induction. Assume $a_1, a_2, \ldots$ are the terms of an arithmetic sequence having first term a_1 and common difference d.

95. $a_n = a_1 + (n - 1)d$

96. $S_n = \dfrac{n}{2}(a_1 + a_n)$

Decide which of the following statements are true for every real number x_i, real constant k, and natural number n.

97. $\displaystyle\sum_{i=1}^{n} k \cdot x_i = k^n \cdot \sum_{i=1}^{n} x_i$

98. $\displaystyle\sum_{i=1}^{n} k \cdot x_i = k \cdot \sum_{i=1}^{n} x_i$

99. $\displaystyle\sum_{i=1}^{n} k = nk$

100. $\displaystyle\sum_{i=1}^{n} k = k^n$

101. $\left[\displaystyle\sum_{i=1}^{n} x_i\right]^2 = \sum_{i=1}^{n} x_i{}^2$

102. $\displaystyle\sum_{i=1}^{n} (x_i - k) = \sum_{i=1}^{n} x_i - nk$

8.4 Geometric Sequences

A **geometric sequence** (or **geometric progression**) is a sequence in which each term after the first is obtained by multiplying the preceding term by a constant nonzero real number, called the **common ratio.** An example of a geometric sequence is 2, 8, 32, 128, . . . in which the first term is 2 and the common ratio is 4.

If the common ratio of a geometric sequence is r, then by the definition of a geometric sequence

$$r = \frac{a_{n+1}}{a_n}$$

for every positive integer n. Therefore, the common ratio can be found by choosing any term except the first and dividing it by the preceding term.

In the geometric sequence

$$2, \ 8, \ 32, \ 128, \ . \ . \ .$$

$r = 4$. Notice that

$$8 = 2 \cdot 4$$
$$32 = 8 \cdot 4 = (2 \cdot 4) \cdot 4 = 2 \cdot 4^2$$
$$128 = 32 \cdot 4 = (2 \cdot 4^2) \cdot 4 = 2 \cdot 4^3.$$

To generalize this, assume that a geometric sequence has first term a_1 and common ratio r. The second term can be written as $a_2 = a_1 r$, the third as $a_3 = a_2 r = (a_1 r)r = a_1 r^2$, and so on. Following this pattern, the nth term is $a_n = a_1 r^{n-1}$. Again, this result is proven by mathematical induction. (See Exercise 77.)

***n*th Term of a Geometric Sequence**	In the geometric sequence with first term a_1 and common ratio r, the nth term is $$a_n = a_1 r^{n-1}.$$

Example I Find a_5 and a_n for each of the following geometric sequences.

(a) 4, 12, 36, 108, . . .

The first term, a_1, is 4. To find r, choose any term except the first and divide it by the preceding term. For example,

$$r = 36/12 = 3.$$

To find the fifth term, a_5, start with $a_n = a_1 r^{n-1}$ and replace n with 5, r with 3, and a_1 with 4.

$$a_5 = 4 \cdot (3)^{5-1} = 4 \cdot 3^4 = 324$$

Also, $\qquad a_n = 4 \cdot 3^{n-1}.$

(b) 64, 32, 16, 8, . . .

Here $r = 8/16 = 1/2$, and $a_1 = 64$, so

$$a_5 = 64\left(\frac{1}{2}\right)^{5-1} = 64\left(\frac{1}{16}\right) = 4.$$

Also,

$$a_n = 64\left(\frac{1}{2}\right)^{n-1}. \quad \blacksquare$$

Example 2   Find a_1 and r for the geometric sequence with third term 20 and sixth term 160.
Use the formula for the nth term of a geometric sequence.

$$\text{For } n = 3, \quad a_3 = a_1 r^2 = 20;$$

$$\text{for } n = 6, \quad a_6 = a_1 r^5 = 160.$$

Since $a_1 r^2 = 20$, then $a_1 = 20/r^2$. Substituting this in the second equation gives

$$a_1 r^5 = 160$$

$$\left(\frac{20}{r^2}\right) r^5 = 160$$

$$20 r^3 = 160$$

$$r^3 = 8$$

$$r = 2.$$

Since $a_1 r^2 = 20$ and $r = 2$, then $a_1 = 5$. $\blacksquare$

Example 3   An insect population is growing in such a way that each generation is 1.5 times as large as the last generation. Suppose there were 100 insects in the first generation. How many would there be in the fourth generation?

The number of insects in each generation can be described as a geometric sequence with a_1 as the first-generation population, a_2 the second-generation population, and so on. Then the fourth-generation population will be a_4. Using the formula for a_n, where $n = 4$, $r = 1.5$, and $a_1 = 100$, gives

$$a_4 = a_1 r^3 = 100(1.5)^3 = 100(3.375) = 337.5.$$

In the fourth generation, the population will number about 338 insects. $\blacksquare$

In applications of geometric sequences, it is often necessary to know the sum of the first n terms for the sequence. For example, a scientist might want to know the total number of insects in four generations of the population discussed in Example 3.

To find a formula for the sum of the first n terms of a geometric sequence, S_n, first write the sum as

$$S_n = a_1 + a_2 + a_3 + \ldots + a_n$$

or

$$S_n = a_1 + a_1 r + a_1 r^2 + \ldots + a_1 r^{n-1}. \tag{1}$$

If $r = 1$, $S_n = na_1$, which is a correct formula for this case. If $r \neq 1$, multiply both sides of equation (1) by r, obtaining

$$rS_n = a_1 r + a_1 r^2 + a_1 r^3 + \ldots + a_1 r^n. \tag{2}$$

If (2) is subtracted from (1),

$$S_n - rS_n = a_1 - a_1r^n$$

or

$$S_n(1 - r) = a_1(1 - r^n),$$

which finally gives

$$S_n = \frac{a_1(1 - r^n)}{1 - r}, \quad (r \neq 1).$$

This discussion is summarized below.

Sum of n Terms of a Geometric Sequence

If a geometric sequence has first term a_1 and common ratio r, then the sum of the first n terms, S_n, is

$$S_n = \frac{a_1(1 - r^n)}{1 - r} \quad (r \neq 1).$$

This formula can be used to find the total insect population in Example 3 over the four-generation period. With $n = 4$, $a_1 = 100$, and $r = 1.5$,

$$S_4 = \frac{100(1 - 1.5^4)}{1 - 1.5} = \frac{100(1 - 5.0625)}{-.5} = 812.5,$$

so the total population for the four generations will amount to about 813 insects.

Example 4   Find S_5 for the geometric sequence 3, 6, 12, 24, 48.
Here $a_1 = 3$ and $r = 2$. From the formula above,

$$S_5 = \frac{3(1 - 2^5)}{1 - 2} = \frac{3(1 - 32)}{-1} = \frac{3(-31)}{-1} = 93. \quad \blacksquare$$

Example 5   Find $\sum_{i=1}^{6} 2 \cdot 3^i$.

This sum is the sum of the first six terms of a geometric sequence having $a_1 = 2 \cdot 3^1 = 6$ and $r = 3$. From the formula for S_n,

$$\sum_{i=1}^{6} 2 \cdot 3^i = S_6 = \frac{6(1 - 3^6)}{1 - 3} = \frac{6(1 - 729)}{-2} = \frac{6(-728)}{-2} = 2184. \quad \blacksquare$$

Now the discussion of sums of sequences will be extended to include infinite geometric sequences such as the infinite sequence

$$2, \quad 1, \quad \frac{1}{2}, \quad \frac{1}{4}, \quad \frac{1}{8}, \quad \frac{1}{16}, \quad \cdots$$

with first term 2 and common ratio 1/2. Using the formula above gives the following sequence.

$$S_1 = 2, \quad S_2 = 3, \quad S_3 = \frac{7}{2}, \quad S_4 = \frac{15}{4}, \quad S_5 = \frac{31}{8}, \quad S_6 = \frac{63}{16}$$

These sums seem to be getting closer and closer to the number 4. In fact, by selecting a value of n large enough, we can make S_n as close as desired to 4. This is expressed as

$$\lim_{n \to \infty} S_n = 4.$$

(Read: "the limit of S_n as n increases without bound is 4.") For no value of n is $S_n = 4$. However, if n is large enough, then S_n is as close to 4 as desired.*

Since

$$\lim_{n \to \infty} S_n = 4,$$

the number 4 is called the **sum** of the infinite geometric sequence

$$2, \quad 1, \quad \frac{1}{2}, \quad \frac{1}{4}, \quad \ldots$$

and

$$2 + 1 + \frac{1}{2} + \frac{1}{4} + \frac{1}{8} + \ldots = 4.$$

Example 6   Find $1 + \dfrac{1}{3} + \dfrac{1}{9} + \dfrac{1}{27} + \ldots$.

Use the formula for the first n terms of a geometric sequence to get

$$S_1 = 1, \quad S_2 = \frac{4}{3}, \quad S_3 = \frac{13}{9}, \quad S_4 = \frac{40}{27},$$

and in general

$$S_n = \frac{1\left[1 - \left(\dfrac{1}{3}\right)^n\right]}{1 - \dfrac{1}{3}}.$$

The following chart shows the value of $(1/3)^n$ for larger and larger values of n.

n	1	10	100	200
$\left(\dfrac{1}{3}\right)^n$	$\dfrac{1}{3}$	.0000169	1.9×10^{-48}	3.76×10^{-96}

As n gets larger and larger, $(1/3)^n$ gets closer and closer to 0. That is,

$$\lim_{n \to \infty} \left(\frac{1}{3}\right)^n = 0,$$

*These phrases "large enough" and "as close as desired" are not nearly precise enough for mathematicians; much of a standard calculus course is devoted to making them more precise.

making it reasonable that

$$\lim_{n \to \infty} S_n = \lim_{n \to \infty} \frac{1\left[1 - \left(\frac{1}{3}\right)^n\right]}{1 - \frac{1}{3}} = \frac{1(1 - 0)}{1 - \frac{1}{3}} = \frac{1}{\frac{2}{3}} = \frac{3}{2}.$$

Hence,

$$1 + \frac{1}{3} + \frac{1}{9} + \frac{1}{27} + \ldots = \frac{3}{2}. \quad \blacksquare$$

If a geometric sequence has a first term a_1 and a common ratio r, then

$$S_n = \frac{a_1(1 - r^n)}{1 - r}$$

for every positive integer n. If $-1 < r < 1$, then $\lim_{n \to \infty} r^n = 0$, and

$$\lim_{n \to \infty} S_n = \frac{a_1(1 - 0)}{1 - r} = \frac{a_1}{1 - r}.$$

This quotient, $a_1/(1 - r)$ is called the **sum of the infinite geometric sequence.** The limit, $\lim_{n \to \infty} S_n$, is often expressed as S_∞ or $\sum_{i=1}^{\infty} a_i$.

These results lead to the following definition.

Sum of an Infinite Geometric Sequence

The sum of the infinite geometric sequence with first term a_1 and common ratio r, where $-1 < r < 1$, is given by

$$S_\infty = \sum_{i=1}^{\infty} a_i = \lim_{n \to \infty} S_n = \frac{a_1}{1 - r}.$$

Example 7 (a) Find the sum $-\frac{3}{4} + \frac{3}{8} - \frac{3}{16} + \frac{3}{32} - \frac{3}{64} + \ldots$

The first term is $a_1 = -3/4$. To find r, divide any term by the preceding term. For example,

$$r = \frac{-\frac{3}{16}}{\frac{3}{8}} = -\frac{1}{2}.$$

Since $-1 < r < 1$, the formula above applies, and

$$S_\infty = \frac{a_1}{1 - r} = \frac{-\frac{3}{4}}{1 - \left(-\frac{1}{2}\right)} = -\frac{1}{2}.$$

(b) $\displaystyle\sum_{i=1}^{\infty} \left(\frac{3}{5}\right)^i = \frac{\dfrac{3}{5}}{1 - \dfrac{3}{5}} = \frac{3}{2}$ ■

Example 8   A population of 100 fruit flies is placed in a large glass jar. Each month the population increases by 10%. What is the population after 5 months?

After 1 month the population will increase to

$$100 + 100(.10) = 100(1.10) = 110$$

fruit flies. Each month thereafter the population will again increase by a factor of 1.10, so the population at the end of each month is a geometric sequence with $a_1 = 110$ and $r = 1.10$. To find the population after 5 months, find a_5:

$$a_5 = 110(1.10)^4 \approx 161.$$

After 5 months the population will grow to about 161 fruit flies.   ■

8.4 Exercises

Write the terms of the geometric sequences that satisfy each of the following conditions.

1. $a_1 = 2, r = 3, n = 4$ **2.** $a_1 = 4, r = 2, n = 5$ **3.** $a_1 = 1/2, r = 4, n = 4$

4. $a_1 = 2/3, r = 6, n = 3$ **5.** $a_3 = 6, a_4 = 12, n = 5$ **6.** $a_2 = 9, a_3 = 3, n = 4$

Find a_5 and a_n for each of the following geometric sequences.

7. $a_1 = 4, r = 3$ **8.** $a_1 = 8, r = 4$ **9.** $a_1 = -3, r = -5$

10. $a_1 = -4, r = -2$ **11.** $a_2 = 3, r = 2$ **12.** $a_3 = 6, r = 3$

13. $a_4 = 64, r = -4$ **14.** $a_4 = 81, r = -3$

For each of the following sequences that are geometric, find r and a_n.

15. $6, 12, 24, 48, \ldots$ **16.** $4, 16, 64, 256, \ldots$

17. $3/4, 3/2, 3, 6, 12, \ldots$ **18.** $-7, -5, -3, -1, 1, 3, \ldots$

19. $a_3 = 9, r = 2$ **20.** $a_5 = 6, r = 1/2$

21. $a_3 = -2, r = 3$ **22.** $a_2 = 5, r = 1/2$

Use the formula for S_n to find the sum of the first five terms for each of the following geometric sequences.

23. $3, 6, 12, 24, \ldots$ **24.** $5, 20, 80, 320, \ldots$ **25.** $12, -6, 3, -3/2, \ldots$

26. $18, -3, 1/2, -1/12, \ldots$ **27.** $a_1 = 8.423, r = 2.859$ **28.** $a_1 = -3.772, r = -1.553$

Find each of the following sums.

29. $\displaystyle\sum_{i=1}^{4} 2^i$ **30.** $\displaystyle\sum_{i=1}^{6} 3^i$ **31.** $\displaystyle\sum_{i=1}^{8} 64(1/2)^i$

32. $\displaystyle\sum_{i=1}^{6} 81(2/3)^i$ **33.** $\displaystyle\sum_{i=3}^{6} 2^i$ **34.** $\displaystyle\sum_{i=4}^{7} 3^i$

Find r for each of the following infinite geometric sequences. Identify any whose sums would exist.

35. 9, 18, 36, 72, 144, . . .

36. 3, 9, 27, 81, . . .

37. 10, 100, 1000, 10,000, . . .

38. $-8, -16, -32, -64, . . .$

39. 12, 6, 3, 3/2, . . .

40. $1, -0.9, 0.81, -0.729, . . .$

Find each of the following sums which exist by using the formula of this section where it applies.

41. $16 + 4 + 1 + . . .$

42. $81 + 27 + 9 + 3 + 1 + . . .$

43. $100 + 10 + 1 + . . .$

44. $128 + 64 + 32 + . . .$

45. $\dfrac{3}{4} + \dfrac{3}{8} + \dfrac{3}{16} + . . .$

46. $\dfrac{4}{5} + \dfrac{2}{5} + \dfrac{1}{5} + . . .$

47. $\dfrac{1}{3} - \dfrac{2}{9} + \dfrac{4}{27} - \dfrac{8}{81} + . . .$

48. $1 + \dfrac{1}{1.01} + \dfrac{1}{(1.01)^2} + . . .$

49. $\displaystyle\sum_{i=1}^{\infty} (1/4)^i$

50. $\displaystyle\sum_{i=1}^{\infty} (-1/4)^i$

51. $\displaystyle\sum_{i=1}^{\infty} (.3)^i$

52. $\displaystyle\sum_{i=1}^{\infty} 10^{-i}$

53. Mitzi drops a ball from a height of 10 meters and notices that on each bounce the ball returns to about 3/4 of its previous height. About how far will the ball travel before it comes to rest? (*Hint:* Consider the sum of two sequences).

54. A sugar factory receives an order for 1000 units of sugar. The production manager thus orders production of 1000 units of sugar. He forgets, however, that the production of sugar requires some sugar (to prime the machines, for example), and so he ends up with only 900 units of sugar. He then orders an additional 100 units, and receives only 90 units. A further order for 10 units produces 9 units. Finally seeing he is wrong, the manager decides to try mathematics. He views the production process as an infinite geometric progression with $a_1 = 1000$ and $r = .1$. Using this, find the number of units of sugar that he should have ordered originally.

55. After a person pedaling a bicycle removes his or her feet from the pedals, the wheel rotates 400 times the first minute. As it continues to slow down, it rotates in each minute only 3/4 as many times as in the previous minute. How many times will the wheel rotate before coming to a complete stop?

56. A pendulum bob swings through an arc 40 cm long on its first swing. Each swing thereafter, it swings only 80% as far as on the previous swing. How far will it swing altogether before coming to a complete stop?

57. Suppose you could save $1 on January 1, $2 on January 2, $4 on January 3, and so on. What amount would you save on January 31? What would be the total amount of your savings during January? (*Hint:* $2^{31} = 2,147,483,648$.)

58. Richland Oil has a well which produced $4,000,000 of income its first year. Each year thereafter, the well produced half as much income as the previous year. What is the total amount of income produced by the well in six years?

59. The final step in processing a black and white photographic print is to immerse the print in a chemical called "fixer." The print is then washed in running water. Under certain conditions, 98% of the fixer in a print will be removed with 15 minutes of washing. How much of the original fixer would be left after one hour of washing?

60. A sequence of equilateral triangles is constructed. The first triangle has sides 2 m in length. To get the second triangle, midpoints of the sides of the original triangle are connected. What is the length of the side of the eighth such triangle? See the figure below.

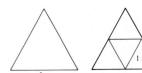

61. A sequence of equilateral triangles is constructed. The first triangle has sides 2 m in length. To get the second triangle, midpoints of the sides of the original triangle are connected. (See Exercise 60.) If this process could be continued indefinitely, what would be the total perimeter of all the triangles?

62. What would be the total area of all the triangles of Exercise 61, disregarding the overlapping?

63. A scientist has a vat containing 100 liters of a pure chemical. Twenty liters are drained and replaced with water. After complete mixing, twenty liters of the mixture is drained and replaced with water. What will be the strength of the mixture after nine such drainings?

64. The half-life of a radioactive substance is the time it takes for half the substance to decay. Suppose the half-life of a substance is 3 years, and that 10^{15} molecules of the substance are present initially. How many molecules will be present after 15 years?

65. Each year a machine loses 20% of the value it had at the beginning of the year. Find the value of the machine at the end of 6 years if it cost $100,000 new.

66. A bicycle wheel rotates 400 times in one minute. If the rider removes his or her feet from the pedals, the wheel will start to slow down. Each minute, it will rotate only 3/4 as many times as in the preceding minute. How many times will the wheel rotate in the fifth minute after the rider's feet were removed from the pedals? (Compare your answer to that of Exercise 55.)

67. A piece of paper is .008 inch thick. Suppose the paper is folded in half, so that its thickness doubles, for 12 times in a row. How thick would the final stack of paper be?

68. Fruit-and-vegetable dealer Greg Odjakjian paid 10¢ per pound for 10,000 pounds of onions. Each week the price he charges increases by .1¢ per pound, while the onions lose 5% of their weight. If he sells the onions after six weeks, did he make or lose money? How much?

69. Find three numbers x, y, and z that are consecutive terms of both an arithmetic sequence and a geometric sequence.

70. Let $a_1, a_2, a_3, \ldots$ and $b_1, b_2, b_3, \ldots$ be geometric sequences. Let $d_n = c \cdot a_n \cdot b_n$ for any real number c and every positive integer n. Show that $d_1, d_2, d_3, \ldots$ is a geometric sequence.

Find a_1 and r for each of the following geometric sequences.

71. $a_2 = 6$, $a_6 = 486$

72. $a_3 = -12$, $a_6 = 96$

73. $a_2 = 64$, $a_8 = 1$

74. $a_2 = 100$, $a_5 = .1$

Explain why the following sequences are geometric.

75. log 6, log 36, log 1296, log 1,679,616, . . .   **76.** log 2, log 4, log 16, log 256, . . .

Prove each of the following results by mathematical induction. Assume that $a_1, a_2, a_3, \ldots$ is a geometric sequence with first term a_1.

77. $a_n = a_1 \cdot r^{n-1}$

78. $S_n = \dfrac{a_1(1 - r^n)}{1 - r}$

8.5 Series

In the previous two sections partial sums of sequences were discussed. These sums are also called *series*.

Series

> A **finite series** is an expression of the form
>
> $$a_1 + a_2 + a_3 + \ldots + a_n = \sum_{i=1}^{n} a_i,$$
>
> and an **infinite series** is an expression of the form
>
> $$a_1 + a_2 + a_3 + \ldots + a_n + \ldots = \sum_{i=1}^{\infty} a_i.$$

Recall, in the summation notation, $\displaystyle\sum_{i=1}^{n} a_i$, i is the index of summation.

Given an expression in summation notation, it is usually not difficult to write out the first few terms of the series. For example,

$$\sum_{i=1}^{4} (2i - 5) = [2(1) - 5] + [2(2) - 5] + [2(3) - 5] + [2(4) - 5]$$

$$= -3 + (-1) + 1 + 3 = 0.$$

Example 1 Write the terms for each of the following sums. Evaluate each sum if possible.

(a) $\displaystyle\sum_{i=1}^{4} a_i = a_1 + a_2 + a_3 + a_4$

(b) $\displaystyle\sum_{i=1}^{3} (6x_i - 2)$ if $x_1 = 2$, $x_2 = 4$, $x_3 = 6$

Let $i = 1, 2,$ and 3 respectively to get

$$\sum_{i=1}^{3} (6x_i - 2) = (6x_1 - 2) + (6x_2 - 2) + (6x_3 - 2).$$

Now substitute the given values for x_1, x_2, and x_3.

$$\sum_{i=1}^{3} (6x_i - 2) = (6 \cdot 2 - 2) + (6 \cdot 4 - 2) + (6 \cdot 6 - 2)$$
$$= 10 + 22 + 34 = 66$$

(c) $\sum_{i=1}^{4} f(x_i)\Delta x$ if $f(x) = x^2$, $x_1 = 0$, $x_2 = 2$, $x_3 = 4$, $x_4 = 6$, and $\Delta x = 2$.

$$\sum_{i=1}^{4} f(x_i)\Delta x = f(x_1)\Delta x + f(x_2)\Delta x + f(x_3)\Delta x + f(x_4)\Delta x$$
$$= x_1^2\Delta x + x_2^2\Delta x + x_3^2\Delta x + x_4^2\Delta x$$
$$= 0^2(2) + 2^2(2) + 4^2(2) + 6^2(2)$$
$$= 0 + 8 + 32 + 72 = 112 \quad \blacksquare$$

Sometimes it is necessary to work in the opposite direction. That is, when given the terms of a series, write the series in summation notation.

Example 2   Write each series in summation notation.

(a) $1 + 4 + 7 + \ldots + 28$

To write the series in summation notation, the general term a_n must be determined. There are no rules that can be used to find a_n. However, the terms of this series appear to increase by three, so it may be the sum of an arithmetic sequence with $a_1 = 1$ and $d = 3$. If so, the formula $a_n = a_1 + (n - 1)d$ can be used to find the number n of the last term. Let $a_n = 28$, $a_1 = 1$, and $d = 3$, to get

$$a_n = a_1 + (n - 1)d$$
$$28 = 1 + (n - 1)3$$
$$27 = 3n - 3$$
$$30 = 3n$$
$$n = 10.$$

Since n is a counting number, the sequence is arithmetic with general term

$$a_n = 1 + (n - 1)3 = 1 + 3n - 3 = 3n - 2.$$

The series has 10 terms and can be written in summation notation as

$$1 + 4 + 7 + \ldots + 28 = \sum_{i=1}^{10} (3i - 2).$$

(b) $5 - 15 + 45 - \ldots + 3645$

From the first three terms, successive terms appear to be those of a geometric sequence with a common ratio of -3. Use the formula for the nth term of a geometric sequence with $a_n = 3645$ to find n.

$$a_n = a_1(r)^{n-1}$$
$$3645 = 5(-3)^{n-1}$$
$$729 = (-3)^{n-1}$$

Since $729 = (3)^6 = (-3)^6$,

$$(-3)^6 = (-3)^{n-1}$$
$$6 = n - 1$$
$$n = 7,$$

and the general term is

$$a_n = 5(-3)^{n-1}.$$

In summation notation,

$$5 - 15 + 45 - \ldots + 3645 = \sum_{i=1}^{7} 5(-3)^{i-1}. \quad \blacksquare$$

A given series can be represented by summation notation in more than one way as shown in the next example.

Example 3   Use summation notation to rewrite each series with the index of summation starting at the indicated number.

(a) $\displaystyle\sum_{i=1}^{8} (3i - 4); \quad 0$

Write out the first few terms.

$$\sum_{i=1}^{8} (3i - 4) = -1 + 2 + 5 + \ldots$$

Although the new index will start at 0 instead of 1, the coefficient of i must stay the same, so, instead of $3i - 4$, the new expression will be $3i + k$, where

$$3(0) + k = 3(1) - 4 = -1,$$

or $$k = -1.$$

Check the next term using $3i + k = 3i - 1$; with $i = 1$, the next term is

$$3(1) - 1 = 2,$$

which is correct for the second term. If i goes from 1 to 8, there are 8 terms, so if i starts at 0 and there are 8 terms, the upper index is 7, and the alternative expression is

$$\sum_{i=0}^{7} (3i - 1).$$

(b) $\displaystyle\sum_{i=2}^{10} i^2; \quad -1$

Again, begin by writing out a few terms.

$$\sum_{i=2}^{10} i^2 = 4 + 9 + 16 + \ldots$$

Since the terms are all squares, let the new expression be $(i + k)^2$. For $i = -1$, which is to be the first value of i, this becomes $(-1 + k)^2$ which must equal 4.

Solve the equation

$$(-1 + k)^2 = 4.$$

Take square roots on both sides to get

$$-1 + k = \pm 2$$
$$k = 3 \text{ or } -1.$$

Check the next term; i will equal 0.

$$(0 + k)^2 = 9$$
$$k = \pm 3.$$

The solution that works in both cases is 3, so k is 3. When i goes from 2 to 10, there are 9 terms, so in the new notation, i must go from -1 to 7. The new summation notation is

$$\sum_{i=-1}^{7} (i + 3)^2. \quad \blacksquare$$

Polynomial functions, defined by expressions of the form

$$f(x) = a_n x^n + a_{n-1} x^{n-1} + \ldots + a_1 x + a_0,$$

can be written in compact form, using summation notation, as

$$f(x) = \sum_{i=0}^{n} a_i x^i.$$

The binomial theorem also looks much more manageable written in summation notation. The theorem can be summarized as:

$$(x + y)^n = \sum_{r=0}^{n} \binom{n}{r} x^{n-r} y^r.$$

Several properties of summation are given below. These provide useful shortcuts for evaluating summations.

Properties of Summation

If $a_1, a_2, a_3, \ldots, a_n$ and $b_1, b_2, b_3, \ldots, b_n$ are two sequences, and c is a constant, then for every positive integer n,

(a) $\displaystyle\sum_{i=1}^{n} c = nc$

(b) $\displaystyle\sum_{i=1}^{n} ca_i = c\sum_{i=1}^{n} a_i$

(c) $\displaystyle\sum_{i=1}^{n} (a_i + b_i) = \sum_{i=1}^{n} a_i + \sum_{i=1}^{n} b_i$

(d) $\displaystyle\sum_{i=1}^{n} (a_i - b_i) = \sum_{i=1}^{n} a_i - \sum_{i=1}^{n} b_i.$

To prove property (a), expand the series to get

$$c + c + c + c + \ldots + c,$$

where there are n terms of c, so the sum is nc.

Property (c) can also be proved by first expanding the series:

$$\sum_{i=1}^{n} (a_i + b_i) = (a_1 + b_1) + (a_2 + b_2) + \ldots + (a_n + b_n).$$

Now use the commutative and associative properties to rearrange the terms.

$$\sum_{i=1}^{n} (a_i + b_i) = (a_1 + a_2 + \ldots + a_n) + (b_1 + b_2 + \ldots + b_n)$$

$$= \sum_{i=1}^{n} a_i + \sum_{i=1}^{n} b_i$$

Proofs of the other two properties are left for the exercises.

Example 4 Use the properties of summation to evaluate $\sum_{i=1}^{6} (i^2 + 3i + 5)$.

$$\sum_{i=1}^{6} (i^2 + 3i + 5) = \sum_{i=1}^{6} i^2 + \sum_{i=1}^{6} 3i + \sum_{i=1}^{6} 5 \qquad \text{Property (c)}$$

$$= \sum_{i=1}^{6} i^2 + 3\sum_{i=1}^{6} i + \sum_{i=1}^{6} 5 \qquad \text{Property (b)}$$

$$= \sum_{i=1}^{6} i^2 + 3\sum_{i=1}^{6} i + 6(5) \qquad \text{Property (a)}$$

The following summations were found in the text and exercises of the first section of this chapter.

$$\sum_{i=1}^{n} i^2 = 1^2 + 2^2 + \ldots + n^2 = \frac{n(n + 1)(2n + 1)}{6}$$

and

$$\sum_{i=1}^{n} i = 1 + 2 + \ldots + n = \frac{n(n + 1)}{2}$$

For example, with $n = 6$,

$$\sum_{i=1}^{6} i^2 = \frac{6(6 + 1)(2 \cdot 6 + 1)}{6} = 7 \cdot 13 = 91$$

and

$$\sum_{i=1}^{6} i = \frac{6(6 + 1)}{2} = 3 \cdot 7 = 21.$$

Substituting these results into the work above gives

$$\sum_{i=1}^{6} (i^2 + 3i + 5) = \sum_{i=1}^{6} i^2 + 3\sum_{i=1}^{6} i + 6(5)$$

$$= 91 + 3(21) + 6(5) = 184. \qquad \blacksquare$$

8.5 Exercises

Write out the terms for each of the following sums where $x_1 = -1$, $x_2 = 0$, $x_3 = 1$, $x_4 = 2$, $x_5 = 3$.

1. $\displaystyle\sum_{i=1}^{4} (3x_i - 2)$

2. $\displaystyle\sum_{i=1}^{5} x_i^2$

3. $\displaystyle\sum_{i=1}^{3} (2x_i - x_i^2)$

4. $\displaystyle\sum_{i=1}^{4} (x_i^2 + x_i)$

5. $\displaystyle\sum_{i=1}^{5} \frac{x_i - 1}{x_i + 3}$

6. $\displaystyle\sum_{i=1}^{4} \frac{x_i}{x_i + 2}$

Write out the terms of $\displaystyle\sum_{i=1}^{4} f(x_i)\Delta x$ with $x_1 = 0$, $x_2 = 2$, $x_3 = 4$, $x_4 = 6$, and $\Delta x = .5$ for the functions defined as follows.

7. $f(x) = 2x - 5$

8. $f(x) = 4x + 3$

9. $f(x) = x^2 - 1$

10. $f(x) = 3 - x^2$

11. $f(x) = \dfrac{4}{x + 1}$

12. $f(x) = \dfrac{3}{2x + 1}$

Write each of the following sums using summation notation.

13. $8 + 6 + 4 + \ldots - 20$

14. $-2 + 5 + 12 + \ldots + 61$

15. $10 + 15 + 20 + \ldots + 100$

16. $5 + 2 - 1 - 4 - \ldots - 31$

17. $7 + 14 + 28 + \ldots + 1792$

18. $-4 - 12 - 36 - \ldots - 972$

19. $9 - 3 + 1 - \ldots + \dfrac{1}{81}$

20. $16 + 8 + 4 + \ldots + \dfrac{1}{16}$

21. $4 + 9 + 16 + \ldots + 169$

22. $5 + 10 + 17 + \ldots + 170$

23. $1 + \dfrac{1}{2} + \dfrac{1}{3} + \ldots + \dfrac{1}{10}$

24. $\dfrac{1}{2} + \dfrac{2}{3} + \dfrac{3}{4} + \ldots + \dfrac{14}{15}$

25. $2 + 4 + 8 + \ldots + 64$

26. $\dfrac{2}{1} + \dfrac{3}{2} + \dfrac{4}{3} + \ldots + \dfrac{12}{11}$

Use summation notation to rewrite each series with the index of summation starting at the indicated number.

27. $\displaystyle\sum_{i=1}^{5} (6 - 3i); \quad 3$

28. $\displaystyle\sum_{i=1}^{7} (5i + 2); \quad -2$

29. $\displaystyle\sum_{i=1}^{10} 2(3)^i; \quad 0$

30. $\displaystyle\sum_{i=-1}^{6} 5(2)^i; \quad 3$

31. $\displaystyle\sum_{i=-1}^{9} (i^2 - 2i); \quad 0$

32. $\displaystyle\sum_{i=3}^{11} (2i^2 + 1); \quad 0$

Use the properties of summation to evaluate each summation. The following sums from the first section of this chapter may be needed.

$$\sum_{i=1}^{n} i = \frac{n(n + 1)}{2} \qquad \sum_{i=1}^{n} i^2 = \frac{n(n + 1)(2n + 1)}{6} \qquad \sum_{i=1}^{n} i^3 = \frac{n^2(n + 1)^2}{4}$$

33. $\displaystyle\sum_{i=1}^{5} (5i + 3)$

34. $\displaystyle\sum_{i=1}^{5} (8i - 1)$

35. $\displaystyle\sum_{i=1}^{5} (4i^2 - 2i + 6)$

36. $\displaystyle\sum_{i=1}^{6} (2 + i - i^2)$

37. $\displaystyle\sum_{i=1}^{4} (3i^3 + 2i - 4)$

38. $\displaystyle\sum_{i=1}^{6} (i^2 + 2i^3)$

Use the summation properties to rewrite each of the following summations. (*Hint:* Think of n as a constant.)

39. $\displaystyle\sum_{i=1}^{n}\left[4+\left(\frac{2}{n}\right)i\right]\frac{2}{n}$

40. $\displaystyle\sum_{i=1}^{n}\left[\left(\frac{2}{n}\right)i+1\right]\frac{2}{n}$

41. $\displaystyle\sum_{i=1}^{n}\left[3+\left(\frac{1}{n}\right)i\right]^2\frac{1}{n}$

42. $\displaystyle\sum_{i=1}^{n}\left[5+\left(\frac{3}{n}\right)i\right]^2\frac{3}{n}$

43. Prove that $\displaystyle\sum_{i=1}^{n}ca_i=c\sum_{i=1}^{n}a_i.$

44. Prove that $\displaystyle\sum_{i=1}^{n}(a_i-b_i)=\sum_{i=1}^{n}a_i-\sum_{i=1}^{n}b_i.$

8.6 Counting Problems

If there are 3 roads from Albany to Baker and 2 roads from Baker to Creswich, in how many ways can one travel from Albany to Creswich by way of Baker? For each of the 3 roads from Albany to Baker, there are 2 different roads from Baker to Creswich. Hence, there are $3 \cdot 2 = 6$ different ways to make the trip, as shown in the **tree diagram** of Figure 2.

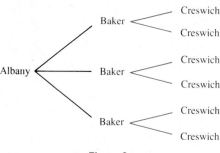

Figure 2

This example illustrates the following fundamental principle of counting.

Fundamental Principle of Counting	If one event can occur in m ways and a second event can occur in n ways, then both events can occur in mn ways, provided the outcome of the first event does not influence the outcome of the second.

The fundamental principal of counting can be extended to any finite number of events, provided the outcome of no one event influences the outcome of another. Such events are called **independent events.**

Example 1 A restaurant offers a choice of 3 salads, 5 main dishes, and 2 desserts. Use the fundamental principle of counting to find the number of different 3-course meals that can be selected.

Three independent events are involved: selecting a salad, selecting a main dish, and selecting a dessert. The first event can occur in 3 ways, the second event can occur in 5 ways, and the third event can occur in 2 ways; thus there are

$$3 \cdot 5 \cdot 2 = 30 \text{ possible meals.} \quad \blacksquare$$

Example 2 Janet Branson has 5 different books that she wishes to arrange on her desk. How many different arrangements are possible?

Five events are involved: selecting a book for the first spot, selecting a book for the second spot, and so on. Here the outcome of the first event *does* influence the outcome of the other events (since one book has already been chosen). For the first spot Branson has 5 choices, for the second spot 4 choices, for the third spot 3 choices, and so on. Now use the fundamental principle of counting to find that there are

$$5 \cdot 4 \cdot 3 \cdot 2 \cdot 1 \text{ or } 120 \text{ different arrangements.} \quad \blacksquare$$

In using the fundamental principle of counting, products such as $5 \cdot 4 \cdot 3 \cdot 2 \cdot 1$ from Example 2 occur often. For convenience in writing these products, use the symbol $n!$ (read "n factorial"), which was defined earlier for any counting number n, as follows.

$$n! = n(n - 1)(n - 2)(n - 3) \ldots (2)(1)$$

Thus $5 \cdot 4 \cdot 3 \cdot 2 \cdot 1$ is written as $5!$. Also, $3! = 3 \cdot 2 \cdot 1 = 6$. By the definition of $n!$, $n[(n - 1)!] = n!$ for all natural numbers $n \geq 2$. It is convenient to have this relation hold also for $n = 1$, so, by definition,

$$0! = 1.$$

Example 3 Suppose Branson wishes to place only 3 of the 5 books on her desk. How many arrangements of 3 books are possible?

She again has 5 ways to fill the first spot, 4 ways to fill the second spot, and 3 ways to fill the third. Since she wants to use only 3 books, there are only 3 spots to be filled (3 events) instead of 5, with

$$5 \cdot 4 \cdot 3 = 60 \text{ arrangements.} \quad \blacksquare$$

The number 60 in the example above is called the number of permutations of 5 things taken 3 at a time, written $P(5, 3) = 60$. The number of ways of arranging 5 elements from a set of 5 elements, written $P(5, 5) = 120$, was found in Example 2.

Permutations of n Elements Taken n at a Time

If $P(n, n)$ denotes the number of permutations of n elements taken n at a time, then

$$P(n, n) = n!$$

A **permutation** of n elements taken r at a time is the number of ways of *arranging* r elements from a set of n elements. Generalizing from the examples above,

$$P(n, r) = n(n - 1)(n - 2) \ldots (n - r + 1)$$

$$= \frac{n(n - 1)(n - 2) \ldots (n - r + 1)(n - r) \ldots (2)(1)}{(n - r)(n - r - 1) \ldots (2)(1)}$$

$$= \frac{n!}{(n - r)!}.$$

Permutations of n Elements Taken r at a Time

> If $P(n, r)$ denotes the number of permutations of n elements taken r at a time, then
>
> $$P(n, r) = \frac{n!}{(n - r)!}.$$

Example 4   Suppose 8 people enter an event in a swim meet. In how many ways could the gold, silver, and bronze prizes be awarded?

Using the fundamental principle of counting, there are 3 choices to be made giving $8 \cdot 7 \cdot 6 = 336$. However, the formula for $P(n, r)$ can also be used to get the same result.

$$P(8, 3) = \frac{8!}{5!} = \frac{8 \cdot 7 \cdot 6 \cdot 5 \cdot 4 \cdot 3 \cdot 2 \cdot 1}{5 \cdot 4 \cdot 3 \cdot 2 \cdot 1} = 8 \cdot 7 \cdot 6 = 336 \quad \blacksquare$$

Example 5   In how many ways can 6 students be seated in a row of 6 desks?

Use $P(n, n)$ with $n = 6$ to get

$$P(6, 6) = 6! = 6 \cdot 5 \cdot 4 \cdot 3 \cdot 2 \cdot 1 = 720. \quad \blacksquare$$

The formulas for permutations given above assumed that the elements were *distinct*. However, if some of the elements are indistinguishable, the formula must be adjusted. For example, if the 7 letters in the word ELEMENT were all different, there would be $7! = 5040$ arrangements of the letters, but the three E's are indistinguishable. There are $3! = 6$ ways to arrange the three E's. Since these permutations are not distinguishable, there are only

$$\frac{7!}{3!} = \frac{5040}{6} = 840$$

distinguishable ways to arrange the 7 letters.

In a set with more than one group of identical elements, the number of permutations of each such group must be considered, with the following result.

Distinguishable Permutations

> The number of distinguishable permutations of n elements with n_1 alike of one kind, n_2 alike of another kind, . . . , and n_r alike of another kind is
>
> $$\frac{n!}{n_1! n_2! \ldots n_r!}.$$

In the work above, a method for finding the number of ways to arrange r elements from a set of n elements was discussed. Sometimes the arrangement (or order) of the elements is not important. For example, suppose three people, Ms. Opelka, Mr. Adams, and Ms. Jacobs, apply for 2 identical jobs. Ignoring all other factors, in how many ways can the personnel officer select 2 people from the 3 applicants? Here the arrangement (or order) of the people is unimportant. Selecting Ms. Opelka and Mr. Adams is the same as selecting Mr. Adams and Ms. Opelka. Therefore, there are only 3 ways to select 2 of the three applicants:

> Ms. Opelka, Mr. Adams
>
> Ms. Opelka, Ms. Jacobs
>
> Mr. Adams, Ms. Jacobs.

These three choices are called the combinations of 3 elements taken 2 at a time. **Combinations** refer to the number of r-element subsets that can be taken from a set of n elements.

Each combination or subset of r elements forms $r!$ permutations. Therefore, the number of combinations of n elements taken r at a time can be found by dividing the number of permutations, $P(n, r)$, by $r!$ to get

$$\frac{P(n, r)}{r!}$$

combinations. This expression can be rewritten as

$$\frac{P(n, r)}{r!} = \frac{\dfrac{n!}{(n - r)!}}{r!} = \frac{n!}{(n - r)!r!}.$$

The symbol $\binom{n}{r}$ is used to represent the number of combinations of n things taken r at a time. With this symbol the results above are stated as follows.

Combinations of n Elements Taken r at a Time	If $\binom{n}{r}$ represents the number of combinations of n things taken r at a time, with $r \le n$, then $$\binom{n}{r} = \frac{n!}{(n - r)!r!}.$$

This same result was obtained in work with the binomial theorem in Section 2 of this chapter. There the values of $\binom{n}{r}$ were the coefficients in the expansion of a binomial.*

In the discussion above, the number of ways to select two of three applicants was 3. The same result can now be found using the formula.

$$\binom{3}{2} = \frac{3!}{(3 - 2)!2!} = \frac{3 \cdot 2 \cdot 1}{1 \cdot 2 \cdot 1} = 3.$$

*Alternate notations for the number of combinations of n elements taken r at a time are $C(n, r)$, C_r^n, and $_nC_r$.

Example 6   How many different committees of 3 people can be chosen from a group of 8 people?

Since the order in which the committee is chosen is not of interest, use combinations to get

$$\binom{8}{3} = \frac{8!}{5!3!} = \frac{8 \cdot 7 \cdot 6 \cdot 5 \cdot 4 \cdot 3 \cdot 2 \cdot 1}{5 \cdot 4 \cdot 3 \cdot 2 \cdot 1 \cdot 3 \cdot 2 \cdot 1} = 56. \quad \blacksquare$$

Example 7   From a group of 30 employees, 3 are to be selected to work on a special project.

(a) In how many different ways can the employees be selected?

Use combinations, not permutations, here because the order within the group of three does not matter. The number of 3-element combinations from a set of 30 elements, using the formula, is

$$\binom{30}{3} = \frac{30!}{27!3!} = 4060.$$

There are 4060 ways to select the project group.

(b) In how many different ways can the group of 3 be selected if it has already been decided that a certain employee must work on the project?

Since one employee has already been selected to work on the project, the problem is reduced to selecting 2 more from the remaining 29 employees. Thus,

$$\binom{29}{2} = \frac{29!}{27!2!} = 406.$$

In this case, the project group can be selected in 406 different ways. $\blacksquare$

Both permutations and combinations give the number of r-element subsets that can be selected from a set of n elements. The differences between permutations and combinations are outlined below.

Permutations	Combinations
The subsets are ordered; subsets with the same elements arranged in different orders are counted separately.	The subsets are not ordered; subsets with the same elements are the same subset.
$$P(n, r) = \frac{n!}{(n - r)!}$$	$$\binom{n}{r} = \frac{n!}{(n - r)!r!}$$

8.6 Exercises

Evaluate each of the following.

1. $P(7, 7)$ **2.** $P(5, 3)$ **3.** $P(6, 5)$ **4.** $P(4, 2)$

5. $P(10, 2)$ **6.** $P(8, 2)$ **7.** $P(8, 3)$ **8.** $P(11, 4)$

9. $P(7, 1)$ **10.** $P(18, 0)$ **11.** $P(9, 0)$ **12.** $P(14, 1)$

13. $\binom{6}{5}$ 14. $\binom{4}{2}$ 15. $\binom{8}{5}$ 16. $\binom{10}{2}$

17. $\binom{15}{4}$ 18. $\binom{9}{3}$ 19. $\binom{10}{7}$ 20. $\binom{10}{3}$

21. $\binom{14}{1}$ 22. $\binom{20}{2}$ 23. $\binom{18}{0}$ 24. $\binom{13}{0}$

25. In how many ways can 6 people be seated in a row of 6 seats?

26. In how many ways can 7 out of 10 people be assigned to 7 seats?

27. In how many ways can 5 bank tellers be assigned to 5 different windows? In how many ways can 10 tellers be assigned to the 5 windows?

28. A couple has narrowed down their choice of names for a new baby to 3 first names and 5 middle names. How many different first- and middle-name arrangements are possible?

29. How many different homes are available if a builder offers a choice of 5 basic plans, 3 roof styles, and 2 types of siding?

30. An automaker produces 7 models, each available in 6 colors, with 4 upholstery fabrics and 5 interior colors. How many varieties of the auto are available?

31. A concert is to consist of 5 works: two modern, two romantic, and one classical. In how many ways can the program be arranged?

32. If the program in Exercise 31 must be shortened to 3 works chosen from the 5, how many arrangements are possible?

33. In Exercise 31, how many different programs are possible if the two modern works are to be played first, then the two romantic, and then the classical?

34. How many 4-letter radio-station call letters can be made if the first letter must be K or W and no letter may be repeated? How many if repeats are allowed?

35. How many of the 4-letter call letters in Exercise 34 with no repeats end in K?

36. A business school gives courses in typing, shorthand, transcription, business English, technical writing, and accounting. How many ways can a student arrange his program if he takes 3 courses?

37. In how many ways can the letters in the word MISSISSIPPI be arranged?

38. Find the number of permutations of the letters in each of the following words: (a) initial   (b) decreed.

39. Find the number of permutations of the letters in each of the following words: (a) little   (b) statistics.

40. A printer has 5 A's, 4 B's, 2 C's, and 2 D's. How many different "words" are possible which use all these letters? (A "word" does not have to have any meaning here.)

41. Mike has 4 blue, 3 green, and 2 red books to arrange on a shelf.
    (a) In how many ways can this be done if they can be arranged in any order?
    (b) In how many distinguishable ways if books of the same color are identical and must be grouped together?
    (c) In how many distinguishable ways if books of the same color are identical but need not be grouped together?

42. A child has a set of differently shaped plastic objects. There are 3 pyramids, 4 cubes, and 7 spheres.
    (a) In how many ways can she arrange them in a row if they are all different colors?
    (b) In how many ways if the same shapes must be grouped?
    (c) In how many distinguishable ways can they be arranged in a row if blocks of the same shape are also the same color, but need not be grouped?

43. A club has 30 members. If a committee of 4 is to be selected at random, how many different committees are possible?

44. How many different samples of 3 apples can be drawn from a crate of 25 apples?

45. A group of 3 students is to be selected randomly from a group of 12 to participate in an experimental class. In how many ways can this be done? In how many ways can the group which will not participate be selected?

46. Hal's Hamburger Heaven sells hamburgers with cheese, relish, lettuce, tomato, mustard, or ketchup. How many different hamburgers can be made using any 3 of the extras?

47. How many different 2-card hands can be dealt from a deck of 52 cards?

48. How many different 13-card bridge hands can be dealt from a deck of 52 cards?

49. Five cards are marked with the numbers 1, 2, 3, 4, and 5, shuffled, and 2 cards are then drawn. How many different 2-card combinations are possible?

50. If a bag contains 15 marbles, how many samples of 2 marbles can be drawn from it? How many samples of 4 marbles?

51. In Exercise 50, if the bag contains 3 yellow, 4 white, and 8 blue marbles, how many samples of 2 can be drawn in which both marbles are blue?

52. In Exercise 44, if it is known that there are 5 rotten apples in the crate,
    (a) how many samples of 3 could be drawn in which all 3 are rotten?
    (b) how many samples of 3 could be drawn in which there are 1 rotten apple and 2 good apples?

53. Glendale Heights City Council is composed of 5 liberals and 4 conservatives. Three members are to be selected randomly as delegates to a convention.
    (a) How many delegations are possible?
    (b) How many delegations could have all liberals?
    (c) How many delegations could have 2 liberals and 1 conservative?
    (d) If 1 member of the council serves as mayor, how many delegations are possible which include the mayor?

54. Seven factory workers decide to send a delegation of 2 to their supervisor to discuss their grievances.
    (a) How many different delegations are possible?
    (b) If it is decided that a certain employee must be in the delegation, how many different delegations are possible?
    (c) If there are 2 women and 5 men in the group, how many delegations would include a woman?

A poker hand is made up of 5 cards drawn at random from a deck of 52 cards. Any 5 cards in one suit are called a flush. The 5 highest cards, that is, the A, K, Q, J, and 10 of any one suit are called a royal flush. Use combinations to set up each of the following. Do not evaluate.

55. Find the total number of all possible poker hands.

56. How many royal flushes in hearts are possible?

57. How many royal flushes in any of the four suits are possible?

58. How many flushes in hearts are possible?

59. How many flushes in any of the four suits are possible?

60. How many combinations of 3 aces and 2 eights are possible?

Solve the following problems by using either combinations or permutations.

61. In how many ways can the letters of the word TOUGH be arranged?

62. If Matthew has 8 courses to choose from, how many ways can he arrange his schedule if he must pick 4 of them?

63. How many samples of 3 pineapples can be drawn from a crate of 12?

64. Velma specializes in making different vegetable soups with carrots, celery, beans, peas, mushrooms, and potatoes. How many different soups can she make using any 4 ingredients?

Solve each of the following equations for n.

65. $P(n, 3) = 8 \cdot P(n - 1, 2)$

66. $30 \cdot P(n, 2) = P(n + 2, 4)$

67. $\binom{n + 2}{4} = 15 \cdot \binom{n}{4}$

68. $\binom{n}{n - 2} = 66$

Prove each of the following statements for positive integers n and r, with $r \le n$.

69. $P(n, n - 1) = P(n, n)$

70. $P(n, 1) = n$

71. $P(n, 0) = 1$

72. $\binom{n}{n} = 1$

73. $\binom{n}{0} = 1$

74. $\binom{n}{n - 1} = n$

75. $\binom{n}{n - r} = \binom{n}{r}$

76. $\binom{2n}{r} \le \binom{2n}{n}$ for all r, $0 \le r \le 2n$

77. $\binom{n}{r} = \binom{n - 2}{r - 2} + 2 \cdot \binom{n - 2}{r - 1} + \binom{n - 2}{r}$ for $2 \le r \le n - 2$.

78. If $n > 1$, then $r \cdot \binom{n}{r} = n \cdot \binom{n - 1}{r - 1}$.

79. $\binom{n}{1} + \binom{n}{2} + \ldots + \binom{n}{n} = 2^n - 1$

80. Suppose m and n are any positive integers greater than 1. Decide which is larger, $(mn)!$ or $(m!)(n!)$.

8.7 Basics of Probability

The study of probability theory has become increasingly popular because of the wide range of practical applications. In this section, the basic ideas of probability are introduced.

Consider an experiment which has one or more possible **outcomes,** each of which is equally likely to occur. For example, the experiment of tossing a coin has two equally likely possible outcomes: landing heads up (*H*) or landing tails up (*T*). Also, the experiment of rolling a die has six equally likely outcomes: landing so the face which is up shows 1, 2, 3, 4, 5, or 6 points.

The set S of all possible outcomes of a given experiment is called the **sample space** for the experiment. (In this text all sample spaces are finite.) A sample space for the experiment of tossing a coin consists of the outcomes H and T. This sample space can be written in set notation as

$$S = \{H, T\}.$$

Similarly, a sample space for the experiment of rolling a single die is

$$S = \{1, 2, 3, 4, 5, 6\}.$$

Any subset of the sample space is called an **event.** In the experiment with the die, for example, "the number showing is a three" is an event, say E_1, such that $E_1 = \{3\}$. "The number showing is greater than three" is also an event, say E_2, such that $E_2 = \{4, 5, 6\}$. To represent the number of outcomes which belong to event E, the notation $n(E)$ is used. In the experiment with the die, $n(E_1) = 1$ and $n(E_2) = 3$. The number of outcomes in an event is used to determine the probability of the event.

Probability of Event E	The **probability** of an event E, written $P(E)$, is the ratio of *the number of outcomes in sample space S which belong to event E* compared to *the total number of outcomes in sample space S.* $$P(E) = \frac{n(E)}{n(S)}$$

To use this definition to find the probability of the event E_1 given above, start with the sample space for the experiment, $S = \{1, 2, 3, 4, 5, 6\}$, and the desired event, $E_1 = \{3\}$. Since $n(E_1) = 1$ and since there are 6 outcomes in the sample space,

$$P(E_1) = \frac{n(E_1)}{n(S)} = \frac{1}{6}.$$

Example 1 A single die is rolled. Write the following events in set notation and give the probability for each event.

(a) E_3: the number showing is even

Use the definition above. Since $E_3 = \{2, 4, 6\}$, $n(E_3) = 3$. As shown above, $n(S) = 6$, so

$$P(E_3) = \frac{3}{6} = \frac{1}{2}.$$

(b) E_4: the number showing is greater than 4

Again $n(S) = 6$. Event $E_4 = \{5, 6\}$, with $n(E_4) = 2$. By the definition,

$$P(E_4) = \frac{2}{6} = \frac{1}{3}.$$

(c) $E_5 = \{1, 2, 3, 4, 5, 6\}$ and $P(E_5) = \dfrac{6}{6} = 1$

(d) $E_6 = \emptyset$ and $P(E_6) = \dfrac{0}{6} = 0.$ ∎

In part (c), $E_5 = S$. Therefore the event E_5 is a **certain event,** sure to occur every time the experiment is performed. An event which is certain to occur, such as E_5, always has a probability of 1. On the other hand, $E_6 = \emptyset$ and $P(E_6)$ is 0. The probability of an **impossible event,** such as E_6, is always 0, since none of the outcomes in the sample space satisfy the event. For any event E, $P(E)$ is between 0 and 1 inclusive.

The set of all outcomes in the sample space which do *not* belong to event E is called the **complement** of E, written E'. For example, in the experiment of drawing a single card from a standard deck of 52 cards, let E be the event "the card is an ace." Then E' is the event "the card is not an ace." From the definition of E', for any event E,

$$E \cup E' = S \qquad \text{and} \qquad E \cap E' = \emptyset.^*$$

Probability concepts can be illustrated using **Venn diagrams,** as in Figure 3. The rectangle in Figure 3 represents the sample space in an experiment. The area inside the circle represents event E, while the area inside the rectangle, but outside the circle, represents event E'.

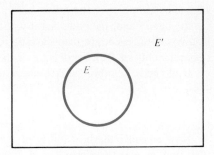

Figure 3

Example 2   In the experiment of drawing a card from a well-shuffled deck, find the probability of events E, the card is an ace, and E'.

Since there are four aces in the deck of 52 cards, $n(E) = 4$, and $n(S) = 52$. Therefore, $P(E) = \dfrac{n(E)}{n(S)} = \dfrac{4}{52} = \dfrac{1}{13}$. Of the 52 cards, 48 are not aces, so

$$P(E') = \dfrac{n(E')}{n(S)} = \dfrac{48}{52} = \dfrac{12}{13}.$$ ∎

---

*The **union** of two sets A and B is the set $A \cup B$ made up of all the elements from either A or B, or both. The **intersection** of sets A and B, written $A \cap B$, is made up of all the elements that belong to both sets at the same time.

In Example 2, $P(E) + P(E') = (1/13) + (12/13) = 1$. This is always true for any event E and its complement E'. That is,

$$P(E) + P(E') = 1.$$

This can be restated as

$$P(E) = 1 - P(E') \qquad \text{or} \qquad P(E') = 1 - P(E).$$

These two equations suggest an alternate way to compute the probability of an event. For example, if it is known that $P(E) = 1/10$, then

$$P(E') = 1 - \frac{1}{10} = \frac{9}{10}.$$

Sometimes probability statements are expressed in terms of odds, a comparison of $P(E)$ with $P(E')$. The **odds** in favor of an event E are expressed as the ratio of $P(E)$ to $P(E')$ or as the fraction $P(E)/P(E')$. For example, if the probability of rain can be established as 1/3, the odds that it will rain are

$$P(\text{rain}) \text{ to } P(\text{no rain}) = \frac{1}{3} \text{ to } \frac{2}{3}$$

$$= \frac{1/3}{2/3}$$

$$= \frac{1}{2} \quad \text{or} \quad 1 \text{ to } 2.$$

On the other hand, the odds that it will not rain are 2 to 1 (or 2/3 to 1/3). If the odds in favor of an event are, say, 3 to 5, then the probability of the event is 3/8, while the probability of the complement of the event is 5/8.

If the odds favoring event E are m to n, then

$$P(E) = \frac{m}{m + n} \qquad \text{and} \qquad P(E') = \frac{n}{m + n}.$$

Now consider the probability of a **compound event** which involves an *alternative*, such as E or F, where E and F are simple events. For example, in the experiment of rolling a die, suppose H is the event "the result is a 3," and K is the event "the result is an even number." What is the probability of "the result is a 3 or an even number"? We have

$$H = \{3\} \qquad\qquad P(H) = \frac{1}{6}$$

$$K = \{2, 4, 6\} \qquad\qquad P(K) = \frac{3}{6} = \frac{1}{2};$$

therefore $\qquad H \text{ or } K = \{2, 3, 4, 6\} \qquad P(H \text{ or } K) = \frac{4}{6} = \frac{2}{3}.$

Notice that $P(H) + P(K) = P(H \text{ or } K)$.

Before assuming that this relationship is true in general, consider another event for this experiment, "the result is a 2," event G. Now

$$G = \{2\} \qquad\qquad P(G) = \frac{1}{6}$$

$$K = \{2, 4, 6\} \qquad P(K) = \frac{3}{6} = \frac{1}{2};$$

therefore $\qquad K \text{ or } G = \{2, 4, 6\} \quad P(K \text{ or } G) = \frac{3}{6} = \frac{1}{2}.$

In this case $P(K) + P(G) \neq P(K \text{ or } G)$.

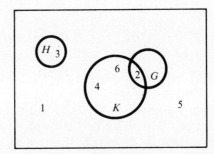

Figure 4

As Figure 4 shows, the difference in the two examples above comes from the fact that events H and K cannot occur simultaneously. Such events are called **mutually exclusive events.** For these events, $H \cap K = \emptyset$, which is true in general for any two mutually exclusive events. Events K and G, however, can occur simultaneously. Both are satisfied if the result of the roll is a 2, the element in their intersection ($K \cap G = \{2\}$). This example suggests the following property.

Probability of Alternate Events	For any events E and F, $$P(E \text{ or } F) = P(E \cup F) = P(E) + P(F) - P(E \cap F).$$

Example 3 One card is drawn from a well-shuffled deck of 52 cards. What is the probability of the following outcomes?

(a) The card is an ace or a spade.

The events "drawing an ace" and "drawing a spade" are not mutually exclusive since it is possible to draw the ace of spades, an outcome satisfying both events. The probability is

$$P(\text{ace or spade}) = P(\text{ace}) + P(\text{spade}) - P(\text{ace and spade})$$

$$= \frac{4}{52} + \frac{13}{52} - \frac{1}{52} = \frac{16}{52} = \frac{4}{13}.$$

(b) The card is a three or a king.

"Drawing a 3" and "drawing a king" are mutually exclusive events because it is impossible to draw one card that is both a 3 and a king.

$$P(3 \text{ or } K) = P(3) + P(K) - P(3 \text{ and } K)$$

$$= \frac{4}{52} + \frac{4}{52} - 0 = \frac{8}{52} = \frac{2}{13} \quad \blacksquare$$

Example 4 For the experiment consisting of one roll of a pair of dice, find the probability that the sum of the points showing is at most 4.

"At most 4" can be written as "2 or 3 or 4." (A sum of 1 is meaningless here.) Then

$$P(\text{at most } 4) = P(2 \text{ or } 3 \text{ or } 4)$$

$$= P(2) + P(3) + P(4), \tag{1}$$

since the events represented by "2," "3," and "4" are mutually exclusive.

The sample space for this experiment includes the 36 possible pairs of numbers from 1 to 6: (1, 1), (1, 2), (1, 3), (1, 4), (1, 5), (1, 6), (2, 1), (2, 2), and so on. The pair (1, 1) is the only one with a sum of 2, so $P(2) = 1/36$. Also $P(3) = 2/36$ since both (1, 2) and (2, 1) give a sum of 3. The pairs, (1, 3), (2, 2), and (3, 1) have a sum of 4, so $P(4) = 3/36$. Substituting into equation (1) above gives

$$P(\text{at most } 4) = \frac{1}{36} + \frac{2}{36} + \frac{3}{36} = \frac{6}{36} = \frac{1}{6}. \quad \blacksquare$$

8.7 Exercises

Write a sample space with equally likely outcomes for each of the following experiments.

1. A two-headed coin is tossed once.

2. Two ordinary coins are tossed.

3. Three ordinary coins are tossed.

4. Slips of paper marked with the numbers 1, 2, 3, 4, and 5 are placed in a box. After mixing well, two slips are drawn.

5. An unprepared student takes a three-question true/false quiz in which he guesses the answer to all three questions.

6. A die is rolled and then a coin is tossed.

Write the following events in set notation and give the probability of each event.

7. In the experiment of Exercise 2:
   (a) both coins show the same face;
   (b) at least one coin turns up heads.

8. In Exercise 1:
   (a) the result of the toss is heads;
   (b) the result of the toss is tails.

9. In Exercise 4:
   (a) both slips are marked with even numbers;
   (b) both slips are marked with odd numbers;
   (c) both slips are marked with the same number;
   (d) one slip is marked with an odd number, the other with an even number.

10. In Exercise 5:
    (a) the student gets all three answers correct;
    (b) he gets all three answers wrong;
    (c) he gets exactly two answers correct;
    (d) he gets at least one answer correct.

11. A marble is drawn at random from a box containing 3 yellow, 4 white, and 8 blue marbles.
    (a) a yellow marble is drawn
    (b) a blue marble is drawn
    (c) a black marble is drawn.
    (d) What are the odds in favor of drawing a yellow marble?
    (e) What are the odds against drawing a blue marble?

12. A baseball player with a batting average of .300 comes to bat. What are the odds in favor of his getting a hit?

13. In Exercise 4, what are the odds that the sum of the numbers on the two slips of paper is 5?

14. If the odds that it will rain are 4 to 5, what is the probability of rain?

15. If the odds that a candidate will win an election are 3 to 2, what is the probability that the candidate will lose?

16. A card is drawn from a well-shuffled deck of 52 cards. Find the probability that the card is (a) a 9 (b) black (c) a black 9 (d) a heart (e) the 9 of hearts (f) a face card (K, Q, J of any suit).

17. Mrs. Elliott invites 10 relatives to a party: her mother, two uncles, three brothers, and four cousins. If the chances of any one guest arriving first are equally likely, find the following probabilities.
    (a) The first guest is an uncle or a brother.
    (b) The first guest is a brother or cousin.
    (c) The first guest is a brother or her mother.

18. One card is drawn from a standard deck of 52 cards. What is the probability that the card is (a) a 9 or a 10? (b) red or a 3? (c) a heart or black? (d) less than a 4? (Consider aces as 1's.)

19. Two dice are rolled. Find the probability that
    (a) the sum of the points is at least 10;
    (b) the sum of the points is either 7 or at least 10;
    (c) the sum of the points is 2 or the dice both show the same number.

20. A student estimates that his probability of getting an A in a certain course is .4; a B, .3; a C, .2; and a D, .1.
    (a) Assuming that only the grades A, B, C, D, and F are possible, what is the probability that he will fail the course?
    (b) What is the probability that he will receive a grade of C or better?
    (c) What is the probability that he will receive at least a B in the course?
    (d) What is the probability that he will get at most a C in the course?

21. If a marble is drawn from a bag containing 3 yellow, 4 white, and 8 blue marbles, what is the probability that
   (a) the marble is either yellow or white?
   (b) it is either yellow or blue?
   (c) it is either red or white?

The table below gives a certain golfer's probabilities of scoring in various ranges on a par-70 course.

Range	Probability
below 60	.01
60–64	.08
65–69	.15
70–74	.28
75–79	.22
80–84	.08
85–89	.06
90–94	.04
95–99	.02
100 or more	.06

In a given round, find the probability that the golfer's score will be as in Exercises 22–27.

22. 90 or higher

23. Below par of 70

24. In the 70's

25. In the 90's

26. Not in the 60's

27. Not in the 60's or 70's

28. Find the odds in favor of the golfer shooting below par.

29. Find the odds against the golfer shooting in the 70s.

Francisco has set up the following table of probabilities for the number of hours it will take him to finish his homework.

Hours	1	2	3	4	5	6
Probability	.05	.10	.20	.40	.10	.15

Find the probability that the number of hours needed to finish his homework will be as follows.

30. Fewer than 3 hours

31. 3 hours or less

32. More than 2 hours

33. At least 2 hours

34. More than 1 hour and less than 5 hours

35. 8 hours.

The table below shows the probability that a customer of a department store will make a purchase in the indicated range.

Cost	Probability
Below $2	.07
$2–$4.99	.18
$5–$9.99	.21
$10–$19.99	.16
$20–$39.99	.11
$40–$69.99	.09
$70–$99.99	.07
$100–$149.99	.08
$150 or over	.03

Find the probability that a customer makes a purchase which is in the following ranges.

36. $10 to $69.99   **37.** $20 or more   **38.** More than $4.99

39. Less than $100   **40.** $100 or more

In most animals and plants, it is very unusual for the number of main parts of the organism (arms, legs, toes, flower petals, etc.) to vary from generation to generation. Some species, however, have *meristic variability,* in which the number of certain body parts varies from generation to generation. One researcher studied the front feet of certain guinea pigs and produced the following probabilities.*

$$P(\text{only four toes, all perfect}) = .77$$
$$P(\text{one imperfect toe and four good ones}) = .13$$
$$P(\text{exactly five good toes}) = .10$$

Find the probability of each of the following events.

41. No more than four good toes   **42.** Five toes, whether perfect or not

The probabilities for the outcomes of an experiment having sample space $S = \{s_1, s_2, s_3, s_4, s_5, s_6\}$ are shown here.

Outcomes	s_1	s_2	s_3	s_4	s_5	s_6
Probability	.17	.03	.09	.46	.21	.04

Let $E = \{s_1, s_2, s_5\}$, and let $F = \{s_4, s_5\}$. Find each of the following probabilities.

43. $P(E)$ **44.** $P(F)$ **45.** $P(E \cap F)$

46. $P(E \cup F)$ **47.** $P(E' \cup F')$ **48.** $P(E' \cap F)$

49. Let E, F, and G be events from a sample space S. Show that

$$P(E \cup F \cup G) = P(E) + P(F) + P(G) - P(E \cap F) - P(E \cap G) - P(F \cap G) + P(E \cap F \cap G).$$

(*Hint:* Let $H = E \cup F$ and use the formula for the probability of alternate events.)

Chapter 8 Summary

Key Words

mathematical induction	index of summation	permutations
factorial notation	geometric sequence	combinations
binomial theorem	common ratio	sample space
sequence	partial sum	event
nth term of a sequence	infinite geometric sequence	probability
arithmetic sequence	fundamental principle of	
common difference	counting	

*From ''An Analysis of Variability in Guinea Pigs'' by J. R. Wright, in *Genetics,* v. 19 (1934), 506–36.

Review Exercises

Use mathematical induction to prove that each of the following is true for every positive integer n.

1. $1 + 3 + 5 + 7 + \ldots + (2n - 1) = n^2$

2. $2 + 6 + 10 + 14 + \ldots + (4n - 2) = 2n^2$

3. $2^2 + 4^2 + 6^2 + \ldots + (2n)^2 = \dfrac{2n(n + 1)(2n + 1)}{3}$

4. $2 + 2^2 + 2^3 + \ldots + 2^n = 2(2^n - 1)$

5. $1 \cdot 4 + 2 \cdot 9 + 3 \cdot 16 + \ldots + n(n + 1)^2 = \dfrac{n(n + 1)(n + 2)(3n + 5)}{12}$

6. $1^3 + 3^3 + 5^3 + \ldots + (2n - 1)^3 = n^2(2n^2 - 1)$

Use the binomial theorem to expand each of the following.

7. $(x + 2y)^4$

8. $(3z - 5w)^3$

9. $\left(3\sqrt{x} - \dfrac{1}{\sqrt{x}}\right)^5$

10. $(m^3 - m^{-2})^4$

Find the indicated term or terms for each of the following expansions.

11. Fifth term of $(3x - 2y)^6$

$5\text{-}1 \quad \dbinom{6}{2}(3x)^2(-2y)^4$

12. Eighth term of $(2m + n^2)^{12}$

13. First four terms of $(3 + x)^{16}$

14. Last three terms of $(2m - 3n)^{15}$

Write the first five terms for each of the following sequences.

15. $a_n = 2(n + 3)$

16. $a_n = n(n + 1)$

17. $a_1 = 5, a_2 = 3, a_n = a_{n-1} - a_{n-2}$ for $n \geq 3$

18. $b_1 = -2, b_2 = 2, b_3 = -4, b_n = -2 \cdot b_{n-2}$ if n is even, and $b_n = 2 \cdot b_{n-2}$ if n is odd.

19. Arithmetic, $a_2 = 6, d = -4$

20. Arithmetic, $a_3 = 9, a_4 = 7$

21. Arithmetic, $a_1 = 3 - \sqrt{5}, a_2 = 4$

22. Arithmetic, $a_3 = \pi, a_4 = 0$

23. Geometric, $a_1 = 4, r = 2$

24. Geometric, $a_4 = 8, r = 1/2$

25. Geometric, $a_1 = -3, a_2 = 4$

26. Geometric, $a_3 = 8, a_5 = 72$

27. A certain arithmetic sequence has $a_6 = -4$ and $a_{17} = 51$. Find a_1 and a_{20}.

28. For a given geometric sequence, $a_1 = 4$ and $a_5 = 324$. Find a_6.

Find a_8 for each of the following arithmetic sequences. $\quad a_n = a_1 + (n-1)d$

29. $a_1 = 6, d = 2$

30. $a_1 = -4, d = 3$

31. $a_1 = 6x - 9, a_2 = 5x + 1$

32. $a_3 = 11m, a_5 = 7m - 4$

Find S_{12} for each of the following arithmetic sequences. $\quad S_n = \dfrac{n}{2}(a_1 + a_n)$

33. $a_1 = 2, d = 3$

34. $a_2 = 6, d = 10$

35. $a_1 = -4k, d = 2k$

Find a_5 for each of the following geometric sequences. $\quad a_n = a_1 r^{n-1}$

36. $a_1 = 3, r = 2$

37. $a_2 = 3125, r = 1/5$

38. $a_1 = 6, a_3 = 24$

39. $a_1 = 5x, a_2 = x^2$

40. $a_2 = \sqrt{6}, a_4 = 6\sqrt{6}$

Find S_4 for each of the following geometric sequences. $\quad S_n = \dfrac{a_1(1 - r^n)}{1 - r}$

41. $a_1 = 1, r = 2$

42. $a_1 = 3, r = 3$

43. $a_1 = 2k, a_2 = -4k$

Evaluate each of the following sums which exist.

44. $18 + 9 + 9/2 + 9/4 + \ldots$

45. $20 + 15 + 45/4 + 135/16 + \ldots$

46. $-5/6 + 5/9 - 10/27 + \ldots$

47. $1/16 + 1/8 + 1/4 + 1/2 + \ldots$

48. $.9 + .09 + .009 + .0009 + \ldots$

Evaluate each of the following sums which exist.

49. $\displaystyle\sum_{i=1}^{4} \frac{2}{i}$

50. $\displaystyle\sum_{i=1}^{7} (-1)^{i+1} \cdot 6$

51. $\displaystyle\sum_{i=4}^{8} 3i(2i - 5)$

52. $\displaystyle\sum_{i=1}^{6} i(i + 2)$

53. $\displaystyle\sum_{i=1}^{4} \frac{i + 1}{i}$

54. $\displaystyle\sum_{i=1}^{12} (8i + 2)$

55. $\displaystyle\sum_{i=1}^{10,000} i$

56. $\displaystyle\sum_{i=1}^{6} 4 \cdot 3^{i}$

57. $\displaystyle\sum_{i=1}^{4} 8 \cdot 2^{i}$

58. $\displaystyle\sum_{i=1}^{\infty} \left(\frac{5}{8}\right)^{i}$

59. $\displaystyle\sum_{i=1}^{\infty} -10\left(\frac{5}{2}\right)^{i}$

60. $\displaystyle\sum_{i=1}^{\infty} 6\left(\frac{2}{3}\right)^{i}$

Evaluate each of the following sums where $x_1 = 0$, $x_2 = 2$, $x_3 = 4$, $x_4 = 6$, $x_5 = 8$, $x_6 = 10$.

61. $\displaystyle\sum_{i=1}^{5} (x_i^2 - 4)$

62. $\displaystyle\sum_{i=1}^{3} (3x_i^2 + 2x_i)$

63. $\displaystyle\sum_{i=1}^{4} f(x_i)\, \Delta x$; $f(x) = x^2 - x$; $\Delta x = .2$

64. $\displaystyle\sum_{i=1}^{6} f(x_i)\Delta x$; $f(x) = (x - 2)^3$; $\Delta x = .1$

Write each of the following sums using summation notation.

65. $4 - 1 - 6 - \ldots - 66$

66. $10 + 14 + 18 + \ldots + 86$

67. $4 + 12 + 36 + \ldots + 972$

68. $\dfrac{5}{6} + \dfrac{6}{7} + \dfrac{7}{8} + \ldots + \dfrac{12}{13}$

Use summation notation to rewrite each sum with the index of summation starting at the indicated number.

69. $\displaystyle\sum_{i=1}^{8} (3 + 2i)$; -2

70. $\displaystyle\sum_{i=2}^{9} (4 - 6i)$; 0

71. $\displaystyle\sum_{i=1}^{12} (i^2 + 1)$; 2

72. $\displaystyle\sum_{i=-2}^{10} i^3$; 1

Use the properties of summation, along with sums given earlier, to evaluate each summation.

73. $\displaystyle\sum_{i=1}^{7} (6i + 2)$

74. $\displaystyle\sum_{i=1}^{5} (3i - 8)$

75. $\displaystyle\sum_{i=1}^{4} (i^2 + 2i)$

76. $\displaystyle\sum_{i=1}^{6} (8 + i^3)$

Find each of the following.

77. $P(9, 2)$

78. $P(8, 3)$

79. $P(6, 0)$

80. $\dbinom{8}{3}$

81. $\dbinom{10}{5}$

82. Four students are to be assigned to 4 different summer jobs. Each student is qualified for all 4 jobs. In how many ways can the jobs be assigned?

83. Nine football teams are competing for 1st, 2nd, and 3rd place titles in a statewide tournament. In how many ways can the winners be determined?

84. How many different license-plate numbers can be formed using 3 letters followed by 3 digits if no repeats are allowed? How many if there are no repeats and either letters or numbers come first?

Write sample spaces for the following.

85. A die is rolled.

86. A card is drawn from a deck containing only the thirteen spades.

87. The weight of a person is measured to the nearest half pound; the scale will not measure more than 300 pounds.

88. A coin is tossed four times.

An urn contains five balls labeled 3, 5, 7, 9, and 11, respectively, while a second urn contains four red and two green balls. An experiment consists of pulling one ball from each urn, in turn. Write an expression for the events in Exercises 89–91.

89. The sample space

90. Event E, the first ball is greater than 5

91. Event F, the second ball is green

92. Are the outcomes in the sample space equally likely?

A company sells typewriters and copiers. Let E be the event "a customer buys a typewriter," and let F be the event "a customer buys a copier." Write each of the following using $\cap$, $\cup$, or $'$ as necessary.

93. A customer buys neither

94. A customer buys at least one

Find the odds in favor of a card drawn from an ordinary deck being

95. A club

96. A black jack

97. A red face card or a queen.

A sample shipment of five swimming pool filters is chosen at random. The probability of exactly 0, 1, 2, 3, 4, or 5 filters being defective is given in the following table.

Number defective	0	1	2	3	4	5
Probability	.31	.25	.18	.12	.08	.06

Find the probability that the following number of filters is defective.

98. No more than 3

99. At least 3

A card is drawn from a standard deck of 52 cards. Find the probability that the card is as described in Exercises 100–105.

100. A black king

101. A face card or an ace

102. An ace or a diamond

103. Not a diamond

104. Not a diamond or not black

105. A diamond and black

106. Prove each of the following statements for real numbers a and b.
   (a) $|a| < |b|$ if and only if $a^2 < b^2$ (b) $|a - b| \geq ||a| - |b||$
   (c) $|a_1 + a_2 + \ldots + a_n| \leq |a_1| + |a_2| + \ldots + |a_n|$
   (d) $|a_1 + a_2 + \ldots + a_n| \geq |a_1| - |a_2| - \ldots - |a_n|$

Exercise 106 from Thomas and Finney, *Calculus and Analytic Geometry,* copyright © 1979, Addison-Wesley, Reading, Massachusetts. P. 58. Reprinted with permission.

Table IA   Natural Logarithms and Powers of e

x	e^x	e^{-x}	ln x	x	e^x	e^{-x}	ln x
0.00	1.00000	1.00000		1.60	4.95302	0.20189	0.4700
0.01	1.01005	0.99004	−4.6052	1.70	5.47394	0.18268	0.5306
0.02	1.02020	0.98019	−3.9120	1.80	6.04964	0.16529	0.5878
0.03	1.03045	0.97044	−3.5066	1.90	6.68589	0.14956	0.6419
0.04	1.04081	0.96078	−3.2189	2.00	7.38905	0.13533	0.6931
0.05	1.05127	0.95122	−2.9957				
0.06	1.06183	0.94176	−2.8134	2.10	8.16616	0.12245	0.7419
0.07	1.07250	0.93239	−2.6593	2.20	9.02500	0.11080	0.7885
0.08	1.08328	0.92311	−2.5257	2.30	9.97417	0.10025	0.8329
0.09	1.09417	0.91393	−2.4079	2.40	11.02316	0.09071	0.8755
0.10	1.10517	0.90483	−2.3026	2.50	12.18248	0.08208	0.9163
				2.60	13.46372	0.07427	0.9555
0.11	1.11628	0.89583	−2.2073	2.70	14.87971	0.06720	0.9933
0.12	1.12750	0.88692	−2.1203	2.80	16.44463	0.06081	1.0296
0.13	1.13883	0.87810	−2.0402	2.90	18.17412	0.05502	1.0647
0.14	1.15027	0.86936	−1.9661	3.00	20.08551	0.04978	1.0986
0.15	1.16183	0.86071	−1.8971				
0.16	1.17351	0.85214	−1.8326	3.50	33.11545	0.03020	1.2528
0.17	1.18530	0.84366	−1.7720	4.00	54.59815	0.01832	1.3863
0.18	1.19722	0.83527	−1.7148	4.50	90.01713	0.01111	1.5041
0.19	1.20925	0.82696	−1.6607				
				5.00	148.41316	0.00674	1.6094
0.20	1.22140	0.81873	−1.6094	5.50	224.69193	0.00409	1.7047
0.30	1.34985	0.74081	−1.2040				
0.40	1.49182	0.67032	−0.9163	6.00	403.42879	0.00248	1.7918
0.50	1.64872	0.60653	−0.6931	6.50	665.14163	0.00150	1.8718
0.60	1.82211	0.54881	−0.5108				
0.70	2.01375	0.49658	−0.3567	7.00	1096.63316	0.00091	1.9459
0.80	2.22554	0.44932	−0.2231	7.50	1808.04241	0.00055	2.0149
0.90	2.45960	0.40656	−0.1054	8.00	2980.95799	0.00034	2.0794
				8.50	4914.76884	0.00020	2.1401
1.00	2.71828	0.36787	0.0000				
1.10	3.00416	0.33287	0.0953	9.00	8130.08392	0.00012	2.1972
1.20	3.32011	0.30119	0.1823	9.50	13359.72683	0.00007	2.2513
1.30	3.66929	0.27253	0.2624				
1.40	4.05519	0.24659	0.3365	10.00	22026.46579	0.00005	2.3026
1.50	4.48168	0.22313	0.4055				

Table IB   Additional Natural Logarithms

x	ln x	x	ln x	x	ln x	x	ln x	x	ln x
		4.5	1.5041	6.0	1.7918	7.5	2.0149	9.0	2.1972
3.1	1.1314	4.6	1.5261	6.1	1.8083	7.6	2.0281	9.1	2.2083
3.2	1.1632	4.7	1.5476	6.2	1.8245	7.7	2.0412	9.2	2.2192
3.3	1.1939	4.8	1.5686	6.3	1.8405	7.8	2.0541	9.3	2.2300
3.4	1.2238	4.9	1.5892	6.4	1.8563	7.9	2.0669	9.4	2.2407
3.5	1.2528	5.0	1.6094	6.5	1.8718	8.0	2.0794	9.5	2.2513
3.6	1.2809	5.1	1.6292	6.6	1.8871	8.1	2.0919	9.6	2.2618
3.7	1.3083	5.2	1.6487	6.7	1.9021	8.2	2.1041	9.7	2.2721
3.8	1.3350	5.3	1.6677	6.8	1.9169	8.3	2.1163	9.8	2.2824
3.9	1.3610	5.4	1.6864	6.9	1.9315	8.4	2.1282	9.9	2.2925
4.0	1.3863	5.5	1.7047	7.0	1.9459	8.5	2.1401	10.0	2.3026
4.1	1.4110	5.6	1.7228	7.1	1.9601	8.6	2.1518	11.0	2.3979
4.2	1.4351	5.7	1.7405	7.2	1.9741	8.7	2.1633	12.0	2.4849
4.3	1.4586	5.8	1.7579	7.3	1.9879	8.8	2.1748	13.0	2.5649
4.4	1.4816	5.9	1.7750	7.4	2.0015	8.9	2.1861	14.0	2.6391

Table 2 Common Logarithms

n	0	1	2	3	4	5	6	7	8	9
1.0	.0000	.0043	.0086	.0128	.0170	.0212	.0253	.0294	.0334	.0374
1.1	.0414	.0453	.0492	.0531	.0569	.0607	.0645	.0682	.0719	.0755
1.2	.0792	.0828	.0864	.0899	.0934	.0969	.1004	.1038	.1072	.1106
1.3	.1139	.1173	.1206	.1239	.1271	.1303	.1335	.1367	.1399	.1430
1.4	.1461	.1492	.1523	.1553	.1584	.1614	.1644	.1673	.1703	.1732
1.5	.1761	.1790	.1818	.1847	.1875	.1903	.1931	.1959	.1987	.2014
1.6	.2041	.2068	.2095	.2122	.2148	.2175	.2201	.2227	.2253	.2279
1.7	.2304	.2330	.2355	.2380	.2405	.2430	.2455	.2480	.2504	.2529
1.8	.2553	.2577	.2601	.2625	.2648	.2672	.2695	.2718	.2742	.2765
1.9	.2788	.2810	.2833	.2856	.2878	.2900	.2923	.2945	.2967	.2989
2.0	.3010	.3032	.3054	.3075	.3096	.3118	.3139	.3160	.3181	.3201
2.1	.3222	.3243	.3263	.3284	.3304	.3324	.3345	.3365	.3385	.3404
2.2	.3424	.3444	.3464	.3483	.3502	.3522	.3541	.3560	.3579	.3598
2.3	.3617	.3636	.3655	.3674	.3692	.3711	.3729	.3747	.3766	.3784
2.4	.3802	.3820	.3838	.3856	.3874	.3892	.3909	.3927	.3945	.3962
2.5	.3979	.3997	.4014	.4031	.4048	.4065	.4082	.4099	.4116	.4133
2.6	.4150	.4166	.4183	.4200	.4216	.4232	.4249	.4265	.4281	.4298
2.7	.4314	.4330	.4346	.4362	.4378	.4393	.4409	.4425	.4440	.4456
2.8	.4472	.4487	.4502	.4518	.4533	.4548	.4564	.4579	.4594	.4609
2.9	.4624	.4639	.4654	.4669	.4683	.4698	.4713	.4728	.4742	.4757
3.0	.4771	.4786	.4800	.4814	.4829	.4843	.4857	.4871	.4886	.4900
3.1	.4914	.4928	.4942	.4955	.4969	.4983	.4997	.5011	.5024	.5038
3.2	.5051	.5065	.5079	.5092	.5105	.5119	.5132	.5145	.5159	.5172
3.3	.5185	.5198	.5211	.5224	.5237	.5250	.5263	.5276	.5289	.5302
3.4	.5315	.5328	.5340	.5353	.5366	.5378	.5391	.5403	.5416	.5428
3.5	.5441	.5453	.5465	.5478	.5490	.5502	.5514	.5527	.5539	.5551
3.6	.5563	.5575	.5587	.5599	.5611	.5623	.5635	.5647	.5658	.5670
3.7	.5682	.5694	.5705	.5717	.5729	.5740	.5752	.5763	.5775	.5786
3.8	.5798	.5809	.5821	.5832	.5843	.5855	.5866	.5877	.5888	.5899
3.9	.5911	.5922	.5933	.5944	.5955	.5966	.5977	.5988	.5999	.6010
4.0	.6021	.6031	.6042	.6053	.6064	.6075	.6085	.6096	.6107	.6117
4.1	.6128	.6138	.6149	.6160	.6170	.6180	.6191	.6201	.6212	.6222
4.2	.6232	.6243	.6253	.6263	.6274	.6284	.6294	.6304	.6314	.6325
4.3	.6335	.6345	.6355	.6365	.6375	.6385	.6395	.6405	.6415	.6425
4.4	.6435	.6444	.6454	.6464	.6474	.6484	.6493	.6503	.6513	.6522
4.5	.6532	.6542	.6551	.6561	.6571	.6580	.6590	.6599	.6609	.6618
4.6	.6628	.6637	.6646	.6656	.6665	.6675	.6684	.6693	.6702	.6712
4.7	.6721	.6730	.6739	.6749	.6758	.6767	.6776	.6785	.6794	.6803
4.8	.6812	.6821	.6830	.6839	.6848	.6857	.6866	.6875	.6884	.6893
4.9	.6902	.6911	.6920	.6928	.6937	.6946	.6955	.6964	.6972	.6981
5.0	.6990	.6998	.7007	.7016	.7024	.7033	.7042	.7050	.7059	.7067
5.1	.7076	.7084	.7093	.7101	.7110	.7118	.7126	.7135	.7143	.7152
5.2	.7160	.7168	.7177	.7185	.7193	.7202	.7210	.7218	.7226	.7235
5.3	.7243	.7251	.7259	.7267	.7275	.7284	.7292	.7300	.7308	.7316
5.4	.7324	.7332	.7340	.7348	.7356	.7364	.7372	.7380	.7388	.7396
n	0	1	2	3	4	5	6	7	8	9

Table 2   453

Table 2   Common Logarithms (continued)

n	0	1	2	3	4	5	6	7	8	9
5.5	.7404	.7412	.7419	.7427	.7435	.7443	.7451	.7459	.7466	.7474
5.6	.7482	.7490	.7497	.7505	.7513	.7520	.7528	.7536	.7543	.7551
5.7	.7559	.7566	.7574	.7582	.7589	.7597	.7604	.7612	.7619	.7627
5.8	.7634	.7642	.7649	.7657	.7664	.7672	.7679	.7686	.7694	.7701
5.9	.7709	.7716	.7723	.7731	.7738	.7745	.7752	.7760	.7767	.7774
6.0	.7782	.7789	.7796	.7803	.7810	.7818	.7825	.7832	.7839	.7846
6.1	.7853	.7860	.7868	.7875	.7882	.7889	.7896	.7903	.7910	.7917
6.2	.7924	.7931	.7938	.7945	.7952	.7959	.7966	.7973	.7980	.7987
6.3	.7993	.8000	.8007	.8014	.8021	.8028	.8035	.8041	.8048	.8055
6.4	.8062	.8069	.8075	.8082	.8089	.8096	.8102	.8109	.8116	.8122
6.5	.8129	.8136	.8142	.8149	.8156	.8162	.8169	.8176	.8182	.8189
6.6	.8195	.8202	.8209	.8215	.8222	.8228	.8235	.8241	.8248	.8254
6.7	.8261	.8267	.8274	.8280	.8287	.8293	.8299	.8306	.8312	.8319
6.8	.8325	.8331	.8338	.8344	.8351	.8357	.8363	.8370	.8376	.8382
6.9	.8388	.8395	.8401	.8407	.8414	.8420	.8426	.8432	.8439	.8445
7.0	.8451	.8457	.8463	.8470	.8476	.8482	.8488	.8494	.8500	.8506
7.1	.8513	.8519	.8525	.8531	.8537	.8543	.8549	.8555	.8561	.8567
7.2	.8573	.8579	.8585	.8591	.8597	.8603	.8609	.8615	.8621	.8627
7.3	.8633	.8639	.8645	.8651	.8657	.8663	.8669	.8675	.8681	.8686
7.4	.8692	.8698	.8704	.8710	.8716	.8722	.8727	.8733	.8739	.8745
7.5	.8751	.8756	.8762	.8768	.8774	.8779	.8785	.8791	.8797	.8802
7.6	.8808	.8814	.8820	.8825	.8831	.8837	.8842	.8848	.8854	.8859
7.7	.8865	.8871	.8876	.8882	.8887	.8893	.8899	.8904	.8910	.8915
7.8	.8921	.8927	.8932	.8938	.8943	.8949	.8954	.8960	.8965	.8971
7.9	.8976	.8982	.8987	.8993	.8998	.9004	.9009	.9015	.9020	.9025
8.0	.9031	.9036	.9042	.9047	.9053	.9058	.9063	.9069	.9074	.9079
8.1	.9085	.9090	.9096	.9101	.9106	.9112	.9117	.9122	.9128	.9133
8.2	.9138	.9143	.9149	.9154	.9159	.9165	.9170	.9175	.9180	.9186
8.3	.9191	.9196	.9201	.9206	.9212	.9217	.9222	.9227	.9232	.9238
8.4	.9243	.9248	.9253	.9258	.9263	.9269	.9274	.9279	.9284	.9289
8.5	.9294	.9299	.9304	.9309	.9315	.9320	.9325	.9330	.9335	.9340
8.6	.9345	.9350	.9355	.9360	.9365	.9370	.9375	.9380	.9385	.9390
8.7	.9395	.9400	.9405	.9410	.9415	.9420	.9425	.9430	.9435	.9440
8.8	.9445	.9450	.9455	.9460	.9465	.9469	.9474	.9479	.9484	.9489
8.9	.9494	.9499	.9504	.9509	.9513	.9518	.9523	.9528	.9533	.9538
9.0	.9542	.9547	.9552	.9557	.9562	.9566	.9571	.9576	.9581	.9586
9.1	.9590	.9595	.9600	.9605	.9609	.9614	.9619	.9624	.9628	.9633
9.2	.9638	.9643	.9647	.9652	.9657	.9661	.9666	.9671	.9675	.9680
9.3	.9685	.9689	.9694	.9699	.9703	.9708	.9713	.9717	.9722	.9727
9.4	.9731	.9736	.9741	.9745	.9750	.9754	.9759	.9763	.9768	.9773
9.5	.9777	.9782	.9786	.9791	.9795	.9800	.9805	.9809	.9814	.9818
9.6	.9823	.9827	.9832	.9836	.9841	.9845	.9850	.9854	.9859	.9863
9.7	.9868	.9872	.9877	.9881	.9886	.9890	.9894	.9899	.9903	.9908
9.8	.9912	.9917	.9921	.9926	.9930	.9934	.9939	.9943	.9948	.9952
9.9	.9956	.9961	.9965	.9969	.9974	.9978	.9983	.9987	.9991	.9996
n	0	1	2	3	4	5	6	7	8	9

Answers To Selected Exercises

Chapter 1

Section 1.1 (page 8)
1. commutative   **3.** commutative   **5.** identity   **7.** associative   **9.** closure   **11.** commutative property of addition and multiplication   **13.** false (the identity property says $8 \cdot 1 = 8$)   **15.** true   **17.** false ($9/0$ is not a rational number) **19.** false ($0 - 1$ is not in the set)   **21.** true   **23.** yes   **25.** no [$\sqrt{2} + (-\sqrt{2}) = 0$ is not irrational]   **27.** no [$\sqrt{2}(1/\sqrt{2}) = 1$ is not irrational]   **29.** yes   **31.** no ($1 + 1 = 2$ is not in the set)   **33.** 24   **35.** 17   **37.** -21 **39.** $-6/7$ **41.** meaningless   **43.** 2.22   **45.** natural, whole, integer, rational, real   **47.** integer, rational, real **49.** rational, real   **51.** irrational, real   **53.** ($-\sqrt{36} = -6$) integer, rational, real   **55.** meaningless   **57.** -20 **59.** -54 **61.** $-1/2$ **63.** meaningless   **65.** 2   **67.** 14/15   **69.** no; for example, $6 + (5 \cdot 4) \neq (6 + 5)(6 + 4)$ **71.** commutative; distributive; substitution; commutative; substitution

Section 1.2 (page 15)
1. $-9, -4, -2, 3, 8$ **3.** $-|9|, -|-6|, |-8|$ **5.** $-5, -4, -2, -\sqrt{3}, \sqrt{6}, \sqrt{8}, 3$ **7.** 3/4, 7/5, $\sqrt{2}$, 22/15, 8/5 **9.** $-|-8| - |-6|, -|-2| + (-3), -|3|, -|-2|, |-8 + 2|$ **11.** 8   **13.** 6   **15.** 4   **17.** 16   **19.** 8   **21.** -16 **23.** 6   **25.** -6 **27.** $8 - \sqrt{50}$ **29.** $5 - \sqrt{7}$ **31.** $\pi - 3$ **33.** $x - 4$ **35.** $8 - 2k$ **37.** $56 - 7m$ **39.** $8 + 4m$ or $4m + 8$ **41.** $y - x$ **43.** $3 + x^2$ **45.** $1 + p^2$ **47.** $6 - \pi$ **49.** $\sqrt{7} - 1$ **51.** 1 **53. (a)** 1   **(b)** 1   **(c)** 13   **(d)** 14   **(e)** 2   **55. (a)** 2   **(b)** 7   **(c)** 2   **(d)** 0   **(e)** 9   **57.** first part of multiplication property **59.** second part of multiplication property   **61.** first part of multiplication property   **63.** triangle inequality, $|a + b| \leq |a| + |b|$ **65.** property of absolute value, $|a| \cdot |b| = |ab|$ **67.** property of absolute value, $|a/b| = |a|/|b|$ **69.** trichotomy property **71.** if $x = y$ or $x = -y$ **73.** if $y = 0$ or if $|x| \geq |y|$ and x and y have opposite signs   **75.** if $y = 0$ or $|x| \geq |y|$ and x and y are either both positive or both negative   **77.** -1 if $x < 0$ and 1 if $x > 0$ **79.** 1   **91.** x must satisfy $-9 \leq x \leq 9$ **93.** $-1 < x < 0$ or $x > 1$

Section 1.3 (page 22)
1. 3375   **3.** 1/8   **5.** 31/108   **7.** $-1/64$ **9.** 8   **11.** 125,000   **13.** 1200   **15.** .001042   **17.** $-72m^5$ **19.** $1/6^4$ **21.** $1/5^4$ **23.** 3^7 **25.** $1/d^8$ **27.** $x^2/(2y)$ **29.** $(4 + s)^6$ **31.** $1/(m^8n^4)$ **33.** a^3b^6 **35.** $x^6/(4y^4)$ **37.** $9m^5/n^3$ **39.** $1/(r^{10}s^8t^{12})$ **41.** $k^2/(p + q)^7$ **43.** 5   **45.** 6   **47.** 0   **49.** 73/8   **51.** 9/8   **53.** 1   **55.** 1/6   **57.** $-13/66$ **59.** 6.93×10^4 **61.** 6×10^9 **63.** 7.92×10^{-3} **65.** 4×10^7 **67.** 9×10^{-8} **69.** 3.8×10^{-6} **71.** 820,000 **73.** 170,000,000   **75.** .00615   **77.** 11,400   **79.** .0809   **81.** .00096   **83.** 4.232   **85.** 1.8×10^{-9} **87.** 2×10^{-3} **89.** 7×10^4 **91.** 4.86×10^7 **93.** 1.29×10^{-4} **95.** -6.04×10^{11} **97.** -2.51×10^5 **99.** 3.2×10^4 hr, or about 3 yr and 8 mo   **101.** about 8.33 min   **103.** 80×10^3 **105.** 47×10^{-6} **107.** 690×10^9 **109.** 90.5×10^{-9} **111.** $2k$ **113.** $30p^{3-r}$ **115.** $15x^{p+2}$ **117.** k^{-7} or $1/k^7$ **119.** $1/(8b^y)$ **121.** $-2m^{2-n}$

Section 1.4 (page 27)
1. $x^2 - x + 3$ **3.** $3x^3 - 3x^2 + 6x + 2$ **5.** $9y^2 - 4y + 4$ **7.** $p^3 - 7p^2 - p - 7$ **9.** $6a^2 + a - 1$ **11.** $3b^2 - 8b - 1$ **13.** $-14q^2 + 11q - 14$ **15.** $28r^2 + r - 2$ **17.** $18p^2 - 27pq - 35q^2$ **19.** $15x^2 - \dfrac{7}{3}x - \dfrac{2}{9}$ **21.** $\dfrac{6}{25}y^2 + \dfrac{11}{40}yz + \dfrac{1}{16}z^2$ **23.** $25r^2 - 4$ **25.** $16x^2 - 9y^2$ **27.** $36k^2 - 36k + 9$ **29.** $16m^2 + 16mn + 4n^2$ **31.** $8z^3 - 12z^2 + 6z - 1$ **33.** $12x^5 + 8x^4 - 20x^3 + 4x^2$ **35.** $15m^4 - 10m^3 + 5m^2 - 5m$ **37.** $-2z^3 + 7z^2 - 11z + 4$ **39.** $27p^3 - 1$ **41.** $8m^3 + 1$

43. $m^2 + mn - 2n^2 - 2km + 5kn - 3k^2$ **45.** $a^2 - 2ab + b^2 + 4ac - 4bc + 4c^2$ **47.** $x^{-2} - 4x^{-1} + 4$
49. $12m^{-2} - 5m^{-1}n^{-1} - 2n^{-2}$ **51.** $8 + 2\sqrt{30}$ **53.** $-3 - 3\sqrt[3]{20} + 3\sqrt[3]{50}$ **55.** $2m^2 - 4m + 8$
57. $5x^3 + 10x^2 + 4x - 3/x$ **59.** $5y^2 - 3xy + 8x^2$ **61.** -8 **63.** 1 **65.** -24 **67.** $k^{2m} - 4$
69. $b^{2r} + b^r - 6$ **71.** $3p^{2x} - 5p^x - 2$ **73.** $m^{2x} - 4m^x + 4$ **75.** $27k^{3a} - 54k^{2a} + 36k^a - 8$
77. (a) 0, 1, 2, or 3 (b) 0, 1, 2, or 3 (c) 6 **79.** (a) m (b) m (c) $m + n$ **81.** $x^2 + y^2 + z^2 + 2xy + 2xz + 2yz$

Section 1.5 (page 33)

1. $4m(3n - 2)$ **3.** $3r(4r^2 + 2r - 1)$ **5.** $2px(3x - 4x^2 - 6)$ **7.** $2(a + b)(1 + 2m)$ **9.** $(x - 3)(x - 8)$
11. $(4p - 1)(p + 1)$ **13.** $(2z - 5)(z + 6)$ **15.** $3(2r - 1)(2r + 5)$ **17.** $(6r - 5s)(3r + 2s)$
19. $(3x - 4y)(5x + 2y)$ **21.** $x^2(3 - x)^2$ **23.** $z(2r - 3)(r + 1)$ **25.** $(2m + 5)(2m - 5)$ **27.** $9(4r + 3s)(4r - 3s)$
29. $(11p^2 + 3q^2)(11p^2 - 3q^2)$ **31.** $(4x^2 + y^2)(2x + y)(2x - y)$ **33.** $(p^4 + 1)(p^2 + 1)(p + 1)(p - 1)$
35. $(2m - 3n)(4m^2 + 6mn + 9n^2)$ **37.** $(2 - x)(2 + x)(4 + 2x + x^2)(4 - 2x + x^2)$ **39.** $(4 + m)(x + y)$
41. $(q + 3 + p)(q + 3 - p)$ **43.** $(a + b + x + y)(a + b - x - y)$ **45.** $(2x - y)y$ **47.** $(11m - 7)(3m - 7)$
49. $(x + y + 5z)(x + y - 3z)$ **51.** $(3a + 3b + 8c)(2a + 2b - 5c)$ **53.** $4pq$ **55.** $r(r^2 + 18r + 108)$
57. $(3 - m - 2n)(9 + 3m + 6n + m^2 + 4mn + 4n^2)$ **59.** $(5m^2 - 2z - 2x^2)(25m^4 + 10m^2z + 10m^2x^2 + 4z^2 + 8zx^2 + 4x^4)$
61. $(m^n + 4)(m^n - 4)$ **63.** $(x^n - y^{2n})(x^{2n} + x^ny^{2n} + y^{4n})$ **65.** $(2x^n + 3y^n)(x^n - 13y^n)$ **67.** $(2y^a - 3)^2$
69. $(5q^r - 3t^p)^2$ **71.** $(5r^{2k} + 6s^{4k})(25r^{4k} - 30r^{2k}s^{4k} + 36s^{8k})$ **73.** $[3(m + p)^k + 5][2(m + p)^k - 3]$
75. $9.44(x^3 + 5.1x^2 - 3.2x + 4)$ **77.** $7.88(8.2x^4 - 3z^2 + z + 14.5)$ **79.** $p^{-4}(1 - p^2)$ **81.** $4k^{-3}(3 + k - 2k^2)$
83. $25p^{-6}(4 - 2p^4 + 3p^8)$ **85.** $2(3 - z)^2(z^2 + 3z + 5)$ **87.** $-2(xy + xz + yz)$ **89.** $(x^2 + 8 + 4x)(x^2 + 8 - 4x)$
91. $(p^2 + 9 + 3p)(p^2 + 9 - 3p)$ **93.** $(z^2 - 5 + z)(z^2 - 5 - z)$ **95.** $2(x^3 + 5)(5x - 11)^3(25x^3 - 33x^2 + 50)$
97. $4(x^2 + 3)^{-3}(4x + 7)^{-4}(-7x^2 - 7x - 9)$

Section 1.6 (page 39)

1. $\{x | x \neq -6\}$ **3.** $\{x | x \neq 3/5\}$ **5.** set of all real numbers **7.** $5p/2$ **9.** $8/9$ **11.** $3/(t - 3)$ **13.** $(2x + 4)/x$
15. $(m - 2)/(m + 3)$ **17.** $(2m + 3)/(4m + 3)$ **19.** $(y + 2)/2$ **21.** $25p^2/9$ **23.** $2/9$ **25.** $3/10$ **27.** $5x/y$
29. $2(a + 4)/(a - 3)$ **31.** 1 **33.** $(k + 2)/(k + 3)$ **35.** $(m + 6)/(m + 3)$ **37.** $(m - 3)/(2m - 3)$ **39.** $(x^2 - 1)/x^2$
41. $(x + y)/(x - y)$ **43.** $(x^2 - xy + y^2)/(x^2 + xy + y^2)$ **45.** $14/r$ **47.** $19/(6k)$ **49.** $5/(12y)$ **51.** 1
53. $(6 + p)/(2p)$ **55.** $(8 - y)/(4y)$ **57.** $137/(30m)$ **59.** $(2y + 1)/[y(y + 1)]$ **61.** $3/[2(a + b)]$
63. $-2/[(a + 1)(a - 1)]$ **65.** $(2m^2 + 2)/[(m - 1)(m + 1)]$ **67.** $4/(a - 2)$ **69.** $(3x + y)/(2x - y)$
71. $5/[(a - 2)(a - 3)(a + 2)]$ **73.** $(x - 11)/[(x + 4)(x - 4)(x - 3)]$ **75.** $a(a + 5)/[(a + 6)(a + 1)(a - 1)]$
77. $(6x^2 - 6x - 5)/[(x + 5)(x - 2)(x - 1)]$ **79.** $(p + 5)/[p(p + 1)]$ **81.** $-1/[x(x + h)]$, **83.** $(x + 1)/(x - 1)$
85. $-1/(x + 1)$ **87.** $(2 - b)(1 + b)/[b(1 - b)]$ **89.** $(m^3 - 4m - 1)/(m - 2)$
91. $(3z^2 - 75z + 8)(z + 1)/[5(z - 25)]$ **93.** 6 **95.** $1/3$ **97.** $a + b$ **99.** -1 **101.** $1/(ab)$
103. $y(xy - 9)/(x^2y^2 - 9)$ **105.** $4(x^3 + 2)^3(2x^3 - 15x^2 - 2)/(x - 5)^5$

Section 1.7 (page 47)

1. $5\sqrt{2}$ **3.** $5\sqrt[3]{2}$ **5.** $-3\sqrt{5}/5$ **7.** $4\sqrt{5}$ **9.** $9\sqrt{3}$ **11.** $\dfrac{-5\sqrt{7}}{7}$ **13.** $-\sqrt[3]{12}/2$ **15.** $\sqrt[4]{24}/2$ **17.** $2\sqrt[3]{2}$
19. $7\sqrt[3]{3}$ **21.** $2\sqrt{3}$ **23.** $13\sqrt[3]{4}/6$ **25.** $xyz^2\sqrt{2x}$ **27.** $2zx^2y\sqrt[3]{2z^2x^2y}$ **29.** $ab\sqrt{ab}(b - 2a^2 + b^3)$ **31.** -7
33. 10 **35.** $11 + 4\sqrt{6}$ **37.** $5\sqrt{6}$ **39.** $2\sqrt[3]{9} - 7\sqrt[3]{3} - 4$ **41.** $\sqrt{6x}/(3x)$ **43.** $x^2y\sqrt{xy}/z$ **45.** $2\sqrt[3]{x}/x$
47. $-km\sqrt[3]{k^2}/r^2$ **49.** $h\sqrt[4]{9g^3hr^2}/(3r^2)$ **51.** $m\sqrt[3]{n^2}/n$ **53.** $2\sqrt[4]{x^3y^3}$ **55.** $\sqrt[6]{4}$ **57.** $\sqrt[18]{x}$ **59.** 1
61. $-3(1 + \sqrt{2})$ **63.** $(4\sqrt{3} - 3)/13$ **65.** $4(2 + \sqrt{y})/(4 - y)$ **67.** $(\sqrt{m} + \sqrt{p})/(m - p)$
69. $-2(\sqrt{5} + 3\sqrt{3} - 2\sqrt{15} - 7)/11$ **71.** $6\sqrt[3]{2}$ **73.** $11\sqrt[3]{49}/7$
75. $y(\sqrt{x} - y - z)/[x - (y + z)^2]$ or $y(\sqrt{x} - y - z)/(x - y^2 - 2yz - z^2)$ **77.** $2(3 - \sqrt{1 + k})/(8 - k)$
79. $m\sqrt{p/p} + p\sqrt{m/m}$, or $(m^2\sqrt{p} + p^2\sqrt{m})/(pm)$ **81.** $-2x - 2\sqrt{x(x + 1)} - 1$
83. $5(\sqrt[3]{a^2} - \sqrt[3]{ab} + \sqrt[3]{b^2})/(a + b)$ **85.** $-5(2 + \sqrt{3})(1 - \sqrt{2})$ **87.** $(6 - \sqrt{3})(5 + \sqrt{2})(3 - \sqrt{5})/92$
89. $(\sqrt{3} + \sqrt{2})/2$ **91.** $(\sqrt{7} + \sqrt{2})/7$ **93.** $3/(2\sqrt{3})$ **95.** $-1/[2(1 - \sqrt{2})]$ or $1/(2\sqrt{2} - 2)$ **97.** $x/(\sqrt{x} + x)$
99. $-1/(2x - 2\sqrt{x(x + 1)} + 1)$ **101.** 5.36×10^3 **103.** 1.40×10^{-1} **105.** 2.64×10^{-1} **107.** 9.93 or 9.93×10^0
109. 4 **111.** $\sqrt{5z} - \sqrt{26}$ **115.** approximately 3.146

Section 1.8 (page 51)

1. 2 **3.** 4 **5.** $1/9$ **7.** $27/8$ **9.** $1/27$ **11.** $1,000,000$ **13.** $4p^2$ **15.** $9x^4$ **17.** $x^{2/3}$ **19.** $z^{5/3}$ **21.** $y^{9/2}$
23. $m^{5/6}$ **25.** $m^{7/3}$ **27.** $(1 + n)^{5/4}$ **29.** $6yz^{2/3}$ **31.** $r^{3/14}s^{7/20}$ **33.** $a^{2/3}b^2$ **35.** $h^{1/3}t^{1/5}/k^{2/5}$ **37.** $x^7/625$
39. $16,384a^2/b^2$ **41.** $m/(6p + 5)$ **43.** $4z^2/(x^5y)$ **45.** r^{6+p} **47.** $m^{3/2}$ **49.** $p^{(m+n+m^2)/(mn)}$ **51.** $y - 10y^2$
53. $-4k^{10/3} + 24k^{4/3}$ **55.** $x^2 - x$ **57.** $r - 2 + r^{-1}$ or $r - 2 + 1/r$ **59.** $x^{1/2} + x^{-1/2}$ or $x^{1/2} + 1/x^{1/2}$ or $(x + 1)/x^{1/2}$
61. $r\sqrt[6]{r^5}$ **63.** $p\sqrt[10]{p^3}$ **65.** $p^2q^2\sqrt[6]{p}$ **67.** $m^2n\sqrt[6]{m^2n}$ **69.** $mn^2\sqrt[15]{m^{14}n^{13}}$ **71.** $2(p - 1)/(p - 3)^{1/2}$

73. $2(3a - 8)/(2a - 5)^{3/2}$ **75.** $1/(m - 1)^{3/2}$ **77.** $k^{3/4}(4k + 1)$ **79.** $z^{-1/2}(9 + 2z)$ **81.** $p^{-1/4}(p^{-1/2} - 2)$
83. $(p + 4)^{-3/2}(p^2 + 9p + 21)$ **85.** $(p + 2)/[2(p + 1)^{3/2}]$ **87.** $(2x^2 + 15)/(2x^2 + 5)^{4/3}$
89. $x^{-3}(x^{-1} - 5)^2(2 - x^{-2})(7x^{-1} - 20 - 6x)$ **91.** Let $a = 2$ and $b = 1/2$; then $a^b = 2^{1/2} = \sqrt{2}$ is irrational.
93. 64 hundred dollars   **95.** \$64,000,000   **97.** about \$10,000,000   **99.** about 86.3 mi   **101.** about 211 mi
103. 29   **105.** 177

Section 1.9 (page 59)

1. imaginary and complex   **3.** real and complex   **5.** imaginary and complex   **7.** complex   **9.** $0 + 10i$ **11.** $0 - 20i$
13. $0 - i\sqrt{39}$ **15.** $5 + 2i$ **17.** $-6 - 14i$ **19.** $9 - 5i\sqrt{2}$ **21.** -5 **23.** $-4 + 0i$ **25.** $2 + 0i$
27. $-2 + 0i$ **29.** $7 - i$ **31.** $2 + 0i$ **33.** $1 - 10i$ **35.** $-10 + 5i$ **37.** $8 - i$ **39.** $-14 + 2i$ **41.** $5 - 12i$
43. $-8 - 6i$ **45.** $13 + 0i$ **47.** $7 + 0i$ **49.** $0 + 25i$ **51.** $12 + 9i$ **53.** $0 + i$ **55.** $7/25 - (24/25)i$
57. $26/29 + (7/29)i$ **59.** $-2 + i$ **61.** $-2i$ **63.** $[(3 - 2\sqrt{5}) + (-2 - 3\sqrt{5})i]/13$ **65.** i **67.** i **69.** 1
71. $-i$ **73.** 1   **75.** i **77.** $5/2 + i$ **79.** $-16/65 - (37/65)i$ **81.** $27/10 + (11/10)i$ **83.** $17/10 - (11/10)i$
85. $37/34 + (165/34)i$ **87.** $a = 23; b = 5$ **89.** $a = 18; b = -3$ **91.** $a = -5; b = 1$
93. $a = -3/4; b = 3$ **95.** $-18 + 19i$ **109.** 25   **111.** $a = 0$ or $b = 0$

Chapter 1 Review Exercises (page 61)

1. commutative   **3.** inverse   **5.** distributive   **7.** no   **9.** 20/3   **11.** 9/4   **13.** 6   **15.** $-12, -6, 0, 6$
17. $-\sqrt{7}, \pi/4, \sqrt{11}$ **19.** natural, whole, integer, rational, real   **21.** ($\sqrt{36} = 6$) natural, whole, integer, rational, real
23. integer, rational, real   **25.** irrational, real   **27.** irrational, real   **29.** $-|3 - (-2)|, -|-2|, |6 - 4|, |8 + 1|$
31. $3 - \sqrt{7}$ **33.** $m - 3$ **35.** (a) 1   (b) 12   **37.** if $x \geq 0$ **39.** if $x = 0$ and $y = 0$ **41.** if A and B are the same
point   **43.** $-8x^2 - 15x + 16$ **45.** $24k^2 - 5k - 14$ **47.** $x^2 + 4xy + 4y^2 - 2xz - 4yz + z^2$ **49.** $x^6/2$
51. $z(7z - 9z^2 + 1)$ **53.** $(3m + 1)(2m - 5)$ **55.** $5m^3(3m - 5n)(2m + n)$ **57.** $(13y^2 + 1)(13y^2 - 1)$
59. $(x + 1)(x - 3)$ **61.** $(-r^3 + 2)(7r^6 - 10r^3 + 4)$ **63.** $(m - n)(3 + 4k)$ **65.** $(2x - 1)(b + 3)$
67. $(y^k + 3)(y^k - 3)$ **69.** $(m - 3)(2m + 3)/(5m)$ **71.** $(x + 1)/(x + 4)$ **73.** $(p + 6q)^2(p + q)/(5p)$ **75.** $37/(20y)$
77. $1 - (9/p^2)$ or $(p^2 - 9)/p^2$ **79.** $(q + p)/(pq - 1)$ **81.** $(3m^2 + 2m - 12)/[5(m + 2)]$ **83.** p^8 **85.** $s/(36r)$
87. $2/(27x^{12}y^3)$ **89.** $10\sqrt{2}$ **91.** $\sqrt{21}r/(3r)$ **93.** $-r^2m\sqrt{m^2z}/z$ **95.** 66   **97.** $m(-9\sqrt{2m} + 5\sqrt{m})$
99. $23\sqrt{5}/30$ **101.** $(z\sqrt{z} - 1)/(z - 1)$ **103.** $\sqrt{5} - 2$ **105.** $1/216$ **107.** $16r^{9/4}s^{7/3}$ **109.** $1/p^{3/2}$ **111.** $1/m^{2p}$
113. $10z^{7/3} - 4z^{1/3}$ **115.** $3p^2 + 3p^{3/2} - 5p - 5p^{1/2}$ **117.** $-2 - 3i$ **119.** $5 + 4i$ **121.** $29 + 37i$
123. $-32 + 24i$ **125.** $-2 - 2i$ **127.** $0 + i$ **129.** $8/5 + (6/5)i$ **131.** $-5/26 + (1/26)i$ **133.** $7/2 - (11/4)i$
135. $0 + 2i\sqrt{3}$ **137.** 0   **141.** 0   **143.** $<$ **145.** $1/b < 1/a < 1 < \sqrt{a} < \sqrt{b} < a < b < a^2 < b^2$

Chapter 2

Section 2.1 (page 69)

1. identity   **3.** conditional   **5.** identity   **7.** identity   **9.** not equivalent   **11.** equivalent   **13.** not equivalent   **15.** not
equivalent   **17.** $\{4\}$ **19.** $\{12\}$ **21.** $\{-2/7\}$ **23.** $\{-7/8\}$ **25.** $\{-1\}$ **27.** $\{3\}$ **29.** $\{4\}$ **31.** $\{12\}$ **33.** $\{3/4\}$
35. $\{-12/5\}$ **37.** $\varnothing$ **39.** $\{27/7\}$ **41.** $\{-59/6\}$ **43.** $\{-9/4\}$ **45.** $\varnothing$ **47.** $\{-4\}$ **49.** $\{-19/75\}$
51. $x = -3a + b$ **53.** $x = (3a + b)/(3 - a)$ **55.** $x = (3 - 3a)/(a^2 - a - 1)$ **57.** $x = 2a^2/(a^2 + 3)$
59. $x = (m + 4)/(5 + 2m)$ **61.** \$205.41   **63.** \$66.50   **65.** Dividing out $(x - 3)$ is improper.   **67.** $k = 6$
69. $k = -2$ **71.** $\{.72\}$ **73.** $\{6.53\}$ **75.** $\{-13.26\}$

Section 2.2 (page 76)

1. $V = k/P$ **3.** $l = V/(wh)$ **5.** $g = (V - V_0)/t$ **7.** $g = 2s/t^2$ **9.** $B = 2A/h - b$ **11.** $r_1 = S/(2\pi h) - r_2$
13. $l = gt^2/(4\pi^2)$ **15.** $h = (S - 2\pi r^2)/(2\pi r)$ **17.** $f = AB(p + 1)/24$ **19.** $R = r_1r_2/(r_2 + r_1)$ **21.** 6 cm
23. 7 quarters   **25.** 2 liters   **27.** 38/3 kg   **29.** 400/3 liters   **31.** about 840 mi   **33.** 15 min   **35.** 35 kph
37. 18/5 hr   **39.** 78 hr   **41.** 40/3 hr   **43.** \$57.65   **45.** \$12,000 at 13% and \$8000 at 16%   **47.** \$54,000 at 8 1/2% and
\$6000 at 16%   **49.** \$70,000 for land that makes a profit and \$50,000 for land that produces a loss   **51.** \$267.57   **53.** no
solution; given numbers are inconsistent   **55.** \$20,000 at 12% and \$40,000 at 10%   **57.** $(10n - v)/5$ fives; $(v - 5n)/5$ tens
59. $100b/c$

Section 2.3 (page 87)

1. $\{\pm 4\}$ **3.** $\{\pm 3\sqrt{3}\}$ **5.** $\{3 \pm \sqrt{5}\}$ **7.** $\{(1 \pm 2\sqrt{3})/3\}$ **9.** $\{2, 3\}$ **11.** $\{-5/2, 10/3\}$ **13.** $\{-3/2, -1/4\}$
15. $\{3, 5\}$ **17.** $\{1 \pm \sqrt{5}\}$ **19.** $\{(-1 \pm i)/2\}$ **21.** $\{(1 \pm \sqrt{5})/2\}$ **23.** $\{(-1 \pm \sqrt{7})/2\}$ **25.** $\{3 \pm \sqrt{2}\}$
27. $\{(3 \pm i\sqrt{7})/2\}$ **29.** $\{(1 \pm i\sqrt{7})/4\}$ **31.** $\{1 \pm 2i\}$ **33.** $\{3, -1/4\}$ **35.** $\{3/2, 1\}$ **37.** $\{1.243, -.643\}$

39. $\{1.281, -.781\}$ **41.** $\{3.562, -.562\}$ **43.** $\{(\sqrt{2} \pm \sqrt{6})/2\}$ **45.** $\{\sqrt{2}, \sqrt{2}/2\}$ **47.** $\{(-i \pm i\sqrt{5})/2\}$
49. $\{(1 \pm \sqrt{2})/i\}$ or $\{-i \pm i\sqrt{2}\}$ **51.** $\{[(1 - i) \pm (\sqrt{7} + i\sqrt{7})]/4\}$ **53.** $\{1, (-1 \pm i\sqrt{3})/2\}$
55. $\{-3, (3 \pm 3i\sqrt{3})/2\}$ **57.** $\{4, -2 \pm 2i\sqrt{3}\}$ **59.** $\{-5/2, (5 \pm 5i\sqrt{3})/4\}$ **61.** 0; one rational solution **63.** 1; two
rational solutions **65.** 84; two irrational solutions **67.** -23; two complex solutions **69.** $t = \sqrt{2sg}/g$
71. $h = \sqrt{d^4kL}/L$ or $d^2\sqrt{kL}/L$ **73.** $t = \sqrt{(s - s_0 - k)g}/g$ **75.** $r = (-\pi h \pm \sqrt{\pi^2h^2 + 2\pi S})/(2\pi)$
77. 16, 18 or $-18, -16$ **79.** $12, -2$ **81.** 50 m by 100 m **83.** 9 ft by 12 ft **85.** 50 mph **87.** 16.95 cm,
20.95 cm, 26.95 cm **89.** ± 12 **91.** $121/4$ **93.** $1, 9$ **99.** (a) $x = (y \pm \sqrt{8 - 11y^2})/4$ (b) $y = (x \pm \sqrt{6 - 11x^2})/3$

Section 2.4 (page 93)
1. $\{\pm\sqrt{3}, \pm\sqrt{5}\}$ **3.** $\{\pm 1, \pm\sqrt{10}/2\}$ **5.** $\{4, 6\}$ **7.** $\{(-5 \pm \sqrt{21})/2\}$ **9.** $\{(-6 \pm 2\sqrt{3})/3, (-4 \pm \sqrt{2})/2\}$
11. $\{7/2, -1/3\}$ **13.** $\{-63, 28\}$ **15.** $\{\pm\sqrt{5}/5\}$ **17.** $\{\pm 1, \pm 2\sqrt{6}/3\}$ **19.** $\{4, 5^{2/3}\}$ **21.** $\{\pm\sqrt{5}\}$ **23.** $\{25/16\}$
25. $\{(-9 \pm \sqrt{129})/6, (-3 \pm \sqrt{33})/2\}$ **27.** $\{(7 \pm \sqrt{177})/8, (-1 \pm \sqrt{19})/3\}$ **29.** $\{1/2\}$ **31.** $\{-1\}$ **33.** $\{5\}$
35. $\{9\}$ **37.** $\{9\}$ **39.** $\{5/4\}$ **41.** $\{2, -2\}$ **43.** $\{2, -2/9\}$ **45.** $\{0, 9\}$ **47.** $\{-2\}$ **49.** $\{2\}$ **51.** $\{3/2\}$
53. $\{31\}$ **55.** $\{-3, 1\}$ **57.** $\{-27, 3\}$ **59.** $\{1/4, 1\}$ **61.** $\{0, 8\}$ **63.** $\{-1\}$ **65.** $\{1/5\}$ **67.** $\{\pm 2.121, \pm 1.528\}$
69. $\{\pm.917, \pm.630i\}$ **71.** $h = (d/k)^2$ **73.** $L = P^2g/4$ **75.** $y = (a^{2/3} - x^{2/3})^{3/2}$

Section 2.5 (page 100)
1. $(-1, 4)$

$-1 \quad 0 \qquad\qquad 4$

3. $(-\infty, 0)$

$-3 \qquad 0$

5. $[1, 2)$

$0 \quad 1 \quad 2 \quad 3$

7. $(-\infty, -9)$

$-9 \; -6 \; -3 \quad 0 \quad 3$

9. $-4 < x < 3$ **11.** $x \le -1$ **13.** $-2 \le x < 6$ **15.** $x \le -4$
17. $(-\infty, 4]$ **19.** $[-1, +\infty)$ **21.** $(-\infty, 6]$ **23.** $(-\infty, 4)$ **25.** $[-11/5, +\infty)$ **27.** $[1, 4]$ **29.** $(-6, -4)$
31. $(-16, 19]$ **33.** $(-1/3, 5/3)$ **35.** $[4, 40/3)$ **37.** $(0, 10)$ **39.** $[-2, 3]$ **41.** $(-\infty, 1/2) \cup (4, +\infty)$
43. $((-5 - \sqrt{33})/2, (-5 + \sqrt{33})/2)$ **45.** $(-\infty, -2] \cup [0, 2]$ **47.** $(-2, 0) \cup (1/4, +\infty)$ **49.** $(-5, 3]$
51. $(-\infty, -2)$ **53.** $(-\infty, 6) \cup [15/2, +\infty)$ **55.** $(-\infty, 1) \cup (9/5, +\infty)$ **57.** $(-\infty, -3/2) \cup [-1/2, +\infty)$
59. $(-2, +\infty)$ **61.** $(0, 4/11) \cup (1/2, +\infty)$ **63.** $(-\infty, -2] \cup (1, 2)$ **65.** $(-\infty, 5)$ **67.** $[3/2, +\infty)$ **69.** $(5/2, +\infty)$
71. $[-8/3, 3/2] \cup (6, +\infty)$ **73.** if k is in $(-\infty, -4\sqrt{2}] \cup [4\sqrt{2}, +\infty)$ **75.** if k is in $(-\infty, 0] \cup [8, +\infty)$
77. $[500, +\infty)$ **79.** $R < C$ for all positive x; this product will never break even **81.** $(0, 5/4), (6, +\infty)$
83. for any x in $[0, 5/3) \cup (10, +\infty)$ **85.** 32°F to 86°F **87.** $(9/2, 6)$ **89.** $1/10 < x < 1/(15b^2)$ if $b < \sqrt{6}/3$
95. when $b > 1$ **99.** $[n - 1, +\infty)$

Section 2.6 (page 108)
1. $\{1, 3\}$ **3.** $\{-1/3, 1\}$ **5.** $\{2/3, 8/3\}$ **7.** $\{-6, 14\}$ **9.** $\{5/2, 7/2\}$ **11.** $\{-4/3, 2/9\}$ **13.** $\{-7/3, -1/7\}$
15. $\{-3/5, 11\}$ **17.** $\{-1/2\}$ **19.** $[-3, 3]$ **21.** $(-\infty, -1) \cup (1, +\infty)$ **23.** $\emptyset$ **25.** $[-10, 10]$ **27.** $(-4, -1)$
29. $(-\infty, -2/3) \cup (2, +\infty)$ **31.** $(-\infty, -8/3] \cup [2, +\infty)$ **33.** $(-3/2, 13/10)$ **35.** $(-3, -1)$ **37.** $(1, 5) \cup (5, +\infty)$
39. $(-\infty, 3/2] \cup [5/2, +\infty)$ **41.** $[7/4, 2) \cup (2, 5/2]$ **43.** $(-1, 0) \cup (0, 5/11)$ **45.** $(-\infty, 3/8) \cup (7/16, 1/2) \cup (1/2, +\infty)$
47. $\{1\}$ **49.** $\{3, -1/3\}$ **51.** $\{0, \pm 1\}$ **53.** $\{5/3, -4/3\}$ **55.** $\{-7/6, -3/2\}$ **57.** $(-2, -\sqrt{2}) \cup (\sqrt{2}, 2)$
59. $(-\infty, -\sqrt{15}] \cup [-1, 1] \cup [\sqrt{15}, +\infty)$ **61.** $(-\infty, -1/5] \cup [1, +\infty)$ **63.** $|x - 2| \le 4$ **65.** $|z - 12| \ge 2$
67. $|k - 1| = 6$ **69.** if $|x - 2| \le .0004$, then $|y - 7| \le .00001$ **71.** $m = 2, n = 20$ **75.** $[-4, 4]$

Chapter 2 Review Exercises (page 109)
1. $\{6\}$ **3.** $\{-11/3\}$ **5.** $\{-96/7\}$ **7.** $\{-2\}$ **9.** $\{13\}$ **11.** $x = -6b - a - 6$ **13.** $x = (6 - 3m)/(1 + 2k - km)$
15. $y = 6a/(4 - a)$ **17.** $C = (5/9)(F - 32)$ **19.** $j = n - nA/l$ **21.** $r_1 = kr_2/(r_2 - k)$ **23.** $L = V/(\pi r^2)$
25. $x = -4/(y^2 - 5y - 6p)$ **27.** $500 **29.** 12 hours **31.** $\{-7 \pm \sqrt{5}\}$ **33.** $\{5/2, -3\}$ **35.** $\{7, -3/2\}$
37. $\{3, -1/2\}$ **39.** $\{-2i \pm i\sqrt{5}\}$ **41.** -188; two complex solutions **43.** 484; two rational solutions **45.** 0; one rational
solution **47.** 50 m by 225 m, or 112.5 m by 100 m **49.** $\{\pm i, \pm 1/2\}$ **51.** $\{-2, 3\}$ **53.** $\{5/2, -15\}$ **55.** $\{63/2\}$
57. $\{3\}$ **59.** $\{-2, -1\}$ **61.** $\{1, -4\}$ **63.** $\{-1\}$ **65.** $\{6\}$ **67.** $\{-1\}$ **69.** $(-7/13, +\infty)$ **71.** $(-\infty, 1]$
73. $(1, +\infty)$ **75.** $[4, 5]$ **77.** $[-5/2, 8)$ **79.** $[-4, 1]$ **81.** $(-2/3, 5/2)$ **83.** $(-\infty, -4] \cup [0, 4]$

85. $(-2, 0)$ **87.** $(-3, 1) \cup [7, +\infty)$ **89.** 0 to 30 (remember, x must start at 0)   **91.** $\{3, -11\}$ **93.** $\{-1, 5\}$
95. $\{11/27, 25/27\}$ **97.** $\{4/3, -2/7\}$ **99.** $[-7, 7]$ **101.** $\varnothing$ **103.** $(-2/7, 8/7)$ **105.** $(-\infty, -4) \cup (-2/3, +\infty)$
107. $[1/2, 1) \cup (1, +\infty)$ **109.** (a) $(-\infty, +\infty)$ (b) none   **111.** $(x + 2)(36/x + 3)$ or $42 + 3x + 72/x$ **113.** for $x \le 2$,
$x = y - 2$ and for $x > 2$, $x = (y + 2)/3$ **117.** $(-\infty, -7) \cup (-7, -3/2)$ **119.** $M = 1/2$
121. One answer is $27 + 5 = 32$.

Cumulative Review Exercises (page 113)
1. rational, real   **3.** $(\sqrt{49} = 7)$ whole, integer, rational, real   **5.** closed   **7.** not closed   **9.** $-|-7|, -|2|, |-3|$
11. $-|-8| - |-6|, -|-2| + (-3), -|3|, -|-2|, |-8 + 2|$ **13.** 0   **15.** $7 - \sqrt{5}$ **17.** $m - 3y$ **19.** $-1/125$
21. 1   **23.** 3^4 or 81   **25.** p^{11}/q^4 **27.** $-5m^2 + 3m + 5$ **29.** $3k^2 - 29k + 56$ **31.** $27k^3 - 135k^2 + 225k - 125$
33. $3y^2 + yz + 11y - 2z^2 + z + 10$ **35.** $5x^3 + 10x^2 + 4x - 3/x$ **37.** $3m(m^2 + 3 + 5m^4)$ **39.** $(2q - 3)(3q + 4)$
41. $(2a + 5)(4a^2 - 10a + 25)$ **43.** $(r - p)(s + t)$ **45.** $-16z$ **47.** $4/(x - 1)$ **49.** $(3z - 2)/(3z + 2)$
51. $(3m + 11)/[2(m - 7)]$ **53.** $n - m$ **55.** $10\sqrt{10}$ **57.** $-2\sqrt[4]{2}$ **59.** $15\sqrt[3]{75}$ **61.** $5hyq^2\sqrt[4]{3hy^2q}$
63. $-34\sqrt{2}$ **65.** $-(2 + \sqrt{7})/3$ **67.** 1/64   **69.** $1/(a - b)$ **71.** $24a^{4/3}/b^{5/3}$ **73.** $rt^2/s^{1/3}$ **75.** $6 + 2i$
77. $17 + i$ **79.** $-12 - 5i$ **81.** $7/2 - (1/2)i$ **83.** $2/13 + (3/13)i$ **85.** $20i$ **87.** $-\sqrt{21}$ **89.** $-i$ **91.** $\{-6/11\}$
93. $\{-10/13\}$ **95.** $\{-19/75\}$ **97.** $t = (v - v_0)/g$ **99.** $B = 2A/h - b$ or $B = (2A - bh)/h$ **101.** 20 cm, 20 cm, 14 cm
103. \$60,000   **105.** 10 pounds   **107.** 7 km per hour   **109.** $\{(3 \pm \sqrt{7})/5\}$ **111.** $\{-3/2, -1/4\}$ **113.** $\{(1 \pm \sqrt{6})/2\}$
115. $\{1 \pm i\}$ **117.** $\{(1 \pm 2i\sqrt{2})/6\}$ **119.** $\{i \pm i\sqrt{2}\}$ **121.** $a = \pm\sqrt{10zyb}/(10yb)$ **123.** $-1, 1$ **125.** 7-1/2 hours
127. $\{\pm\sqrt{3}, \pm\sqrt{2}/2\}$ **129.** $\{-124, 9\}$ **131.** $\{12\}$ **133.** $\{0, 18\}$ **135.** $\{0, \pm\sqrt{7}\}$ **137.** $[6, +\infty)$ **139.** $(2, 9)$
141. $(-\infty, -6] \cup [0, +\infty)$ **143.** $(-\infty, -2] \cup [0, 3]$ **145.** $(-6, 2) \cup [6, +\infty)$ **147.** $(-3, 3)$
149. $(-\infty, -4) \cup (-1, +\infty)$ **151.** $[-3, 1/5]$ **153.** $\{1, 3\}$ **155.** $\{6, -4/5\}$

Chapter 3

Section 3.1 (page 123)

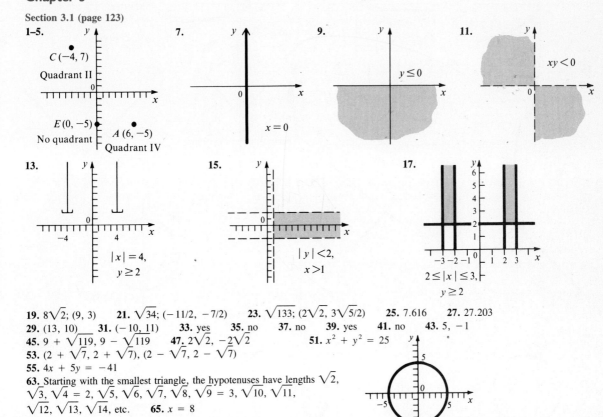

1–5. $C(-4, 7)$ Quadrant II; $E(0, -5)$ No quadrant; $A(6, -5)$ Quadrant IV

7. $x = 0$

9. $y \le 0$

11. $xy < 0$

13. $|x| = 4$, $y \ge 2$

15. $|y| < 2$, $x > 1$

17. $2 \le |x| \le 3$, $y \ge 2$

19. $8\sqrt{2}$; $(9, 3)$ **21.** $\sqrt{34}$; $(-11/2, -7/2)$ **23.** $\sqrt{133}$; $(2\sqrt{2}, 3\sqrt{5}/2)$ **25.** 7.616   **27.** 27.203
29. $(13, 10)$ **31.** $(-10, 11)$ **33.** yes   **35.** no   **37.** no   **39.** yes   **41.** no   **43.** $5, -1$
45. $9 + \sqrt{119}, 9 - \sqrt{119}$ **47.** $2\sqrt{2}, -2\sqrt{2}$ **51.** $x^2 + y^2 = 25$
53. $(2 + \sqrt{7}, 2 + \sqrt{7}), (2 - \sqrt{7}, 2 - \sqrt{7})$
55. $4x + 5y = -41$
63. Starting with the smallest triangle, the hypotenuses have lengths $\sqrt{2}$, $\sqrt{3}, \sqrt{4} = 2, \sqrt{5}, \sqrt{6}, \sqrt{7}, \sqrt{8}, \sqrt{9} = 3, \sqrt{10}, \sqrt{11}$, $\sqrt{12}, \sqrt{13}, \sqrt{14}$, etc.   **65.** $x = 8$

Section 3.2 (page 131)

1.
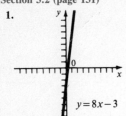
$y = 8x - 3$

3.
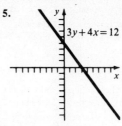
$y = 3x$

5.
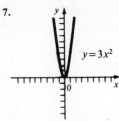
$3y + 4x = 12$

7.
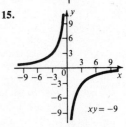
$y = 3x^2$

9.
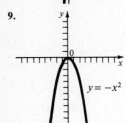
$y = -x^2$

11.
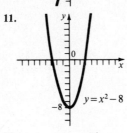
$y = x^2 - 8$
-8

13.
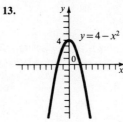
$y = 4 - x^2$
4

15.
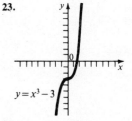
$xy = -9$

17.
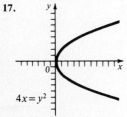
$4x = y^2$

19.
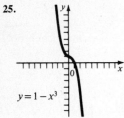
$16y^2 = -x$
-16 -8

21.
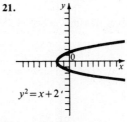
$y^2 = x + 2$

23.
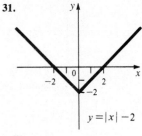
$y = x^3 - 3$

25.
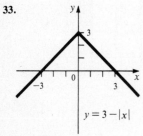
$y = 1 - x^3$

27.
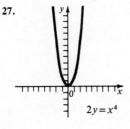
$2y = x^4$

29.
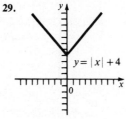
$y = |x| + 4$

31.
$y = |x| - 2$
-2 2
-2

33.
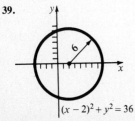
$y = 3 - |x|$
3
-3 3

35.
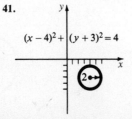
$y = |x + 3|$
3
-3

37.
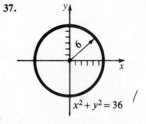
$x^2 + y^2 = 36$
6

39.
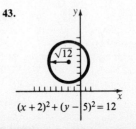
$(x - 2)^2 + y^2 = 36$
6

41.
$(x - 4)^2 + (y + 3)^2 = 4$
2

43.
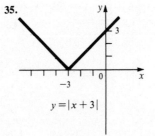
$(x + 2)^2 + (y - 5)^2 = 12$
$\sqrt{12}$

45. $(x - 1)^2 + (y - 4)^2 = 9$ **47.** $(x + 8)^2 + (y - 6)^2 = 25$ **49.** $(x + 1)^2 + (y - 2)^2 = 25$
51. $(x + 3)^2 + (y + 2)^2 = 4$ **53.** $(-3, -4); r = 4$ **55.** $(6, -5); r = 6$ **57.** $(-4, 7); r = 0$ (a point)
59. $(0, 1); r = 7$ **61.** $(1.42, -.7); r = .8$ **63.** $(x + 2)^2 + (y + 3/2)^2 = 149/4$ **65.** yes   **67.** $C = 10\pi, A = 25\pi$
69. $C = 4\pi, A = 4\pi$ **77.** (a) inside (b) outside (c) on (d) outside

Section 3.3 (page 138)

1.
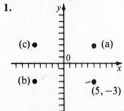

3.
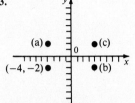

5.
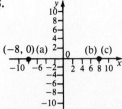

7.
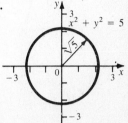

9.
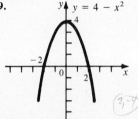

11.
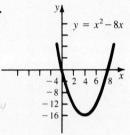

13.
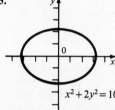

15.
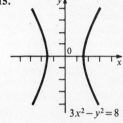

17.
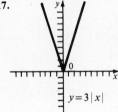

19.
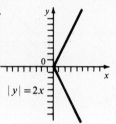

21.
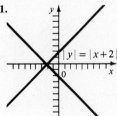

23.
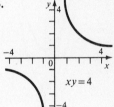

25. The graph is a straight line with a point missing at the origin.

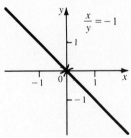

27.
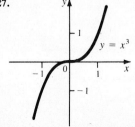

29.
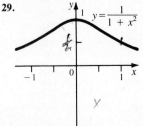

31.
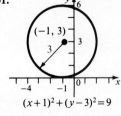

33.
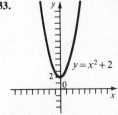

35.

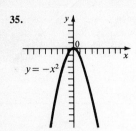

$y = -x^2$

37.

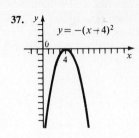

$y = -(x+4)^2$

39.

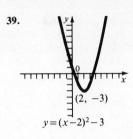

$y = (x-2)^2 - 3$

(2, −3)

41.

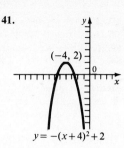

(−4, 2)

$y = -(x+4)^2 + 2$

43. reflected about the *x*-axis

Section 3.4 (page 148)

1. (a) 0 (b) 4 (c) 2 (d) 4   **3.** (a) −3 (b) −2 (c) 0 (d) 2   **5.** −1   **7.** −8   **9.** 4   **11.** $3a^2 - 1$ **13.** 9
15. 289   **17.** $-6m - 1$ **19.** $15a - 7$ **21.** $|25p^2 - 20p - 4|$ **23.** $3p - 1 + |4p^2 - 8|$ **25.** $|m^2 - 8| \cdot (3m - 1)$
27. $(3 - m - p)/(m + p)$ **29.** −66.8   **31.** −42.464   **33.** $(-\infty, +\infty); (-\infty, +\infty)$ **35.** $(-\infty, +\infty); [0, +\infty)$
37. $[-8, +\infty); [0, +\infty)$ **39.** $[-4, 4]; [0, 4]$ **41.** $[4, +\infty); [0, +\infty)$ **43.** $(-\infty, +\infty); (0, 1/5]$
45. $(-\infty, 1) \cup (1, 2) \cup (2, +\infty)$ **47.** $(-\infty, -1] \cup [5, +\infty); [0, +\infty)$ **49.** $(-\infty, +\infty); [0, +\infty)$
51. $[-5/2, +\infty); (-\infty, 0]$ **53.** $[-6, 6]; [0, 6]$ **55.** $(-\infty, -3] \cup [3, +\infty); (-\infty, 0]$ **57.** $[-5, 4]; [-2, 6]$
59. $(-\infty, +\infty); (-\infty, 12]$ **61.** $[-3, 4]; [-6, 8]$ **63.** (a) $x^2 + 2xh + h^2 - 4$ (b) $2xh + h^2$ (c) $2x + h$
65. (a) $6x + 6h + 2$ (b) $6h$ (c) 6   **67.** (a) $2x^3 + 6x^2h + 6xh^2 + 2h^3 + x^2 + 2xh + h^2$
(b) $6x^2h + 6xh^2 + 2h^3 + 2xh + h^2$ (c) $6x^2 + 6xh + 2h^2 + 2x + h$ **69.** increasing on $(-\infty, -3]$ and $[3, +\infty)$; decreasing
on $[-3, 3]$ **71.** decreasing on $(-\infty, 3/2]$; increasing on $[3/2, +\infty)$ **73.** increasing on $(-\infty, +\infty)$; never decreasing
75. never increasing; decreasing on $(-\infty, +\infty)$ **77.** increasing on $[0, +\infty)$; decreasing on
$(-\infty, 0]$ **79.** increasing on $(-\infty, -2]$; decreasing on $[-2, +\infty)$ **81.** increasing on $[0, +\infty)$;
never decreasing   **83.** even   **85.** even   **87.** neither   **89.** neither

91.

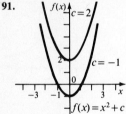

$c = 2$
$c = -1$
$f(x) = x^2 + c$

93.

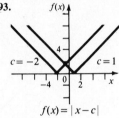

$c = -2$ $c = 1$
$f(x) = |x - c|$

95.

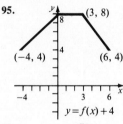

(3, 8)
(−4, 4)   (6, 4)
$y = f(x) + 4$

97.

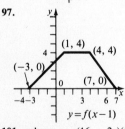

(1, 4)   (4, 4)
(−3, 0)
(7, 0)
$y = f(x - 1)$

99.

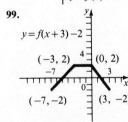

$y = f(x + 3) - 2$
(−3, 2)   (0, 2)
(−7, −2)   (3, −2)

101. volume $= x(16 - 2x)(12 - 2x)$ **103.** $t = 1, t = -2$ **105.** does not hold   **107.** does not hold   **111.** no
113. If *r* is the radius of the circle, then $A = \pi r^2$ and $C = 2\pi r$.   **115.** If *x* is the length of one side of the rectangle, then the
area of the rectangle is $A = 2x\sqrt{r^2 - x^2/4}$.

Section 3.5 (page 155)

1. $10x + 2$; $-2x - 4$; $24x^2 + 6x - 3$; $(4x - 1)/(6x + 3)$; all domains are $(-\infty, +\infty)$, except for f/g, which is $(-\infty, -1/2) \cup$
$(-1/2, +\infty)$ **3.** $4x^2 - 4x + 1$; $2x^2 - 1$; $(3x^2 - 2x)(x^2 - 2x + 1)$; $(3x^2 - 2x)/(x^2 - 2x + 1)$; all domains are $(-\infty, +\infty)$,
except for f/g, which is $(-\infty, -1) \cup (-1, +\infty)$
5. $\sqrt{2x + 5} + \sqrt{4x - 9}$; $\sqrt{2x + 5} - \sqrt{4x - 9}$; $\sqrt{(2x + 5)(4x - 9)}$; $\sqrt{(2x + 5)/(4x - 9)}$; all domains are $[9/4, +\infty)$,
except f/g, which is $(9/4, +\infty)$ **7.** $5x^2 - 11x + 7$; $3x^2 - 11x - 3$; $(4x^2 - 11x + 2)(x^2 + 5)$; $(4x^2 - 11x + 2)/(x^2 + 5)$; all
domains (including f/g) are $(-\infty, +\infty)$ **9.** 55   **11.** 1848   **13.** −6/7   **15.** $4m^2 + 6m + 1$ **17.** 1122   **19.** 97

21. $256k^2 + 48k + 2$ **23.** $24x + 4$; $24x + 35$ **25.** $-5x^2 + 20x + 18$; $-25x^2 - 10x + 6$ **27.** $-64x^3 + 2$; $-4x^3 + 8$
29. $1/x^2$; $1/x^2$ **31.** $\sqrt{8x^2 - 4}$; $8x + 10$ **33.** $x/(2 - 5x)$; $2(x - 5)$ **35.** $\sqrt{(x - 1)/x}$; $-1\sqrt{x + 1}$ **37.** 4   **39.** 0   **41.** 1
43. 2   **57.** $18a^2 + 24a + 9$ **59.** $16\pi t^2$ **65.** Choose any real values for a and b. Then c and d are any values that satisfy $d(a - 1) = b(c - 1)$.

Section 3.6 (page 163)
1. one-to-one   **3.** one-to-one   **5.** not one-to-one   **7.** one-to-one   **9.** one-to-one   **11.** not one-to-one   **13.** not one-to-one   **15.** not one-to-one   **17.** one-to-one   **19.** one-to-one   **21.** one-to-one   **23.** not one-to-one   **25.** inverses
27. not inverses   **29.** not inverses   **31.** inverses   **33.** inverses   **35.** not inverses   **37.** not inverses   **39.** inverses
41. not inverses
43.

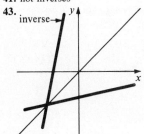

45.

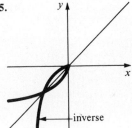

47.

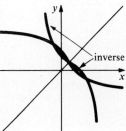

49. $f^{-1}(x) = (x + 5)/4$

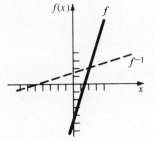

51. $f^{-1}(x) = -\dfrac{5}{2}x$

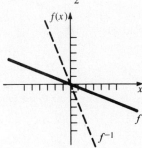

53. $f^{-1}(x) = \sqrt[3]{-x - 2}$

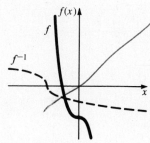

55. $f^{-1}(x) = (9 - x)/3$

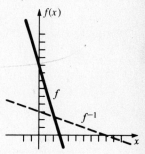

57. not one-to-one   **59.** $f^{-1}(x) = 4/x$

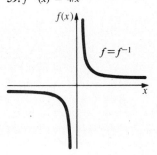

61. $f^{-1}(x) = (x + 5)/(3x + 6)$
domain f:
$(-\infty, 1/3) \cup (1/3, +\infty)$
domain f^{-1}:
$(-\infty, -2) \cup (-2, +\infty)$

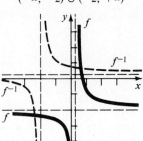

63. $f^{-1}(x) = x^2 - 6$
domain: $[0, +\infty)$

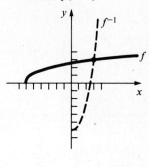

65. 4   **67.** 2   **69.** -2 **71.** 1.14   **73.** $(f \circ g)^{-1}(x) = (\sqrt[3]{19 - x})/2$ **75.** $(g^{-1} \circ f^{-1})(x) = (\sqrt[3]{19 - x})/2$

Section 3.7 (page 169)
1. $a = kb$ **3.** $x = k/y$ **5.** $r = kst$ **7.** $w = kx^2/y$ **9.** 220/7   **11.** 32/15   **13.** 18/125   **15.** 1075
17. m increases by $2^2 \cdot 3^4 = 324$ times   **19.** 2304 ft   **21.** .0444 ohms   **23.** 140 kg per cm^2 **25.** \$1375
27. 8/9 metric tons   **29.** 7500 lbs   **31.** 150 kg per m^2 **33.** 4,000,000   **35.** 1600   **37.** about 4.94   **39.** about 7.4 km
41. 263 pounds

Chapter 3 Review Exercises (page 172)
1.

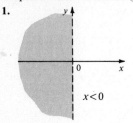

$x < 0$

3.

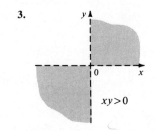

$xy > 0$

5. $d(P, Q) = \sqrt{85}$; $(-1/2, 2)$
7. $(7, -13)$
9. $-7, -1, 8, 23$
11. no such points exist

15. domain: $(-\infty, +\infty)$ **17.** domain: $(-\infty, +\infty)$ **19.** domain: $(-\infty, 0) \cup (0, +\infty)$ **21.** domain: $[7, +\infty)$

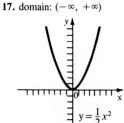

$x + y = 4$

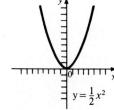

$y = \frac{1}{2}x^2$

$y = -\frac{8}{x}$

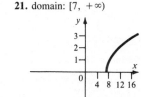

$y = \sqrt{x - 7}$

23. $(x + 2)^2 + (y - 3)^2 = 25$ **25.** $(x + 8)^2 + (y - 1)^2 = 289$ **27.** $(2, -3)$; $r = 1$ **29.** $(-7/2, -3/2)$; $r = 3\sqrt{6}/2$
31. yes; yes; yes   **33.** no; no; no   **35.** yes; no; no   **37.** yes; yes; yes
39.

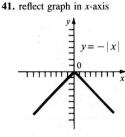

$y = |x|$

41. reflect graph in x-axis

$y = -|x|$

43. translate 1 unit to left; reflect in x-axis; translate 3 units up

$(-1, 3)$

$y = -|x + 1| + 3$

45. $(-\infty, +\infty)$ **47.** $(-\infty, +\infty)$ **49.** $(-\infty, 8) \cup (8, +\infty)$ **51.** $[-7, 7]$
53.

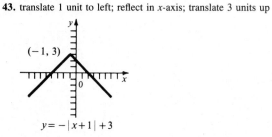

$(-4, 3)$ $(8, 3)$
$(4, 1)$
$y = f(x) + 3$

55.

$y = f(x - 2)$
$(-2, 0)$ $(2, 0)$ $(10, 0)$
$(6, -2)$

57.

$y = f(x + 3) - 2$
$(-5, 0)$
$(-7, -2)$ $(5, -2)$
$(-3, -2)$ $(1, -4)$

59.

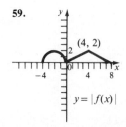

$(4, 2)$
$y = |f(x)|$

61. $-3x^2 - 3xh - h^2 + 4x + 2h$ **63.** $1/(\sqrt{x + h} + \sqrt{x})$ **65.** odd   **67.** odd   **69.** $4x^2 - 3x - 8$ **71.** 44
73. $16k^2 - 6k - 8$ **75.** $-23/4$ **77.** $(-\infty, +\infty)$ **79.** $\sqrt{x^2 - 2}$ **81.** $\sqrt{34}$ **83.** 1   **85.** increasing on $(-\infty, -1]$,
decreasing on $[-1, +\infty)$ **87.** never increasing or decreasing   **89.** not one-to-one   **91.** not one-to-one   **93.** not one-to-one
95. not one-to-one

97. $f^{-1}(x) = (x - 3)/12$ **99.** $f^{-1}(x) = \sqrt[3]{x + 3}$ **101.** $f^{-1}(x) = \sqrt{25 - x^2}$; domain: [0, 5]

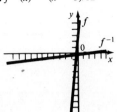

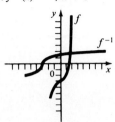

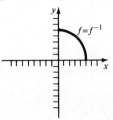

103. $m = kz^2$ **105.** $Y = (kMN^2)/X^3$ **107.** 27/2   **109.** 1372/729   **111.** 36 in   **113.** The conversion of u U.S. dollars into c Canadian dollars is $c = V(u) = (1 + .12)u = 1.12u$, and the conversion of d Canadian dollars into t U.S. dollars is $t = H(d) = (1 - .12)d = .88d$; therefore $H[V(u)] = .88(1.12)u = .9856u \neq u$.

117.
$$(f \circ g)(x) = \begin{cases} 2 \text{ if } x < 0 \\ x \text{ if } 0 \le x \le 1 \\ 2 \text{ if } x > 1 \end{cases}$$

119.
$$(f \circ g)(x) = \begin{cases} 1 \text{ if } x < 0 \\ 4x^2 \text{ if } 0 \le x \le 1/2 \\ 0 \text{ if } 1/2 < x \le 1 \\ 1 \text{ if } x > 1 \end{cases}$$

121.
$$f^{-1}(x) = \begin{cases} x \text{ if } x < 1 \\ \sqrt{x} \text{ if } 1 \le x \le 81 \\ x^2/729 \text{ if } x > 81 \end{cases}$$

Chapter 4

Section 4.1 (page 182)

1.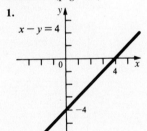

3.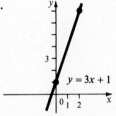

5.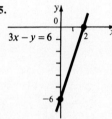

7.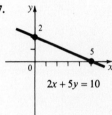

9.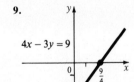

11.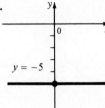

13.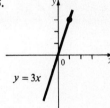

15.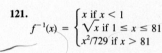

17.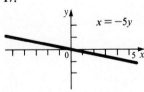

19.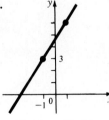

21.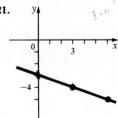

23.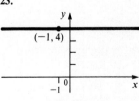

25.

27. 1/5   **29.** 7/9   **31.** 3/5   **33.** undefined slope   **35.** 2

37. 4   **39.** −3/4   **41.** 0   **43.** 0   **45.** .5785   **47.** yes   **49.** no

51.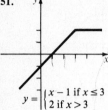

$$y = \begin{cases} x - 1 \text{ if } x \le 3 \\ 2 \text{ if } x > 3 \end{cases}$$

53.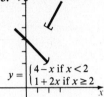

$$y = \begin{cases} 4 - x \text{ if } x < 2 \\ 1 + 2x \text{ if } x \ge 2 \end{cases}$$

55.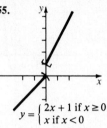

$$y = \begin{cases} 2x + 1 \text{ if } x \ge 0 \\ x \text{ if } x < 0 \end{cases}$$

57.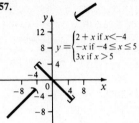

$$y = \begin{cases} 2 + x \text{ if } x < -4 \\ -x \text{ if } -4 \le x \le 5 \\ 3x \text{ if } x > 5 \end{cases}$$

59.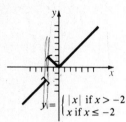

$$y = \begin{cases} |x| \text{ if } x > -2 \\ x \text{ if } x \le -2 \end{cases}$$

61. (a) 140   (b) 220   (c) 220   (d) 220   (e) 220   (f) 60   (g) 60   (h)

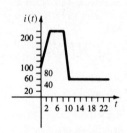

63.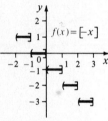

$f(x) = [-x]$

65.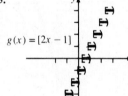

$g(x) = [2x - 1]$

67.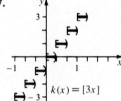

$k(x) = [3x]$

69.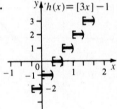

$h(x) = [3x] - 1$

71. (a) $11 (b) $18 (c) $32
(e) domain: $(0, +\infty)$ (at least in theory);
range: $\{11, 18, 25, 32, 39, \ldots\}$

(d)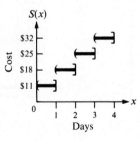

73. $-[-x]$ ounces

75.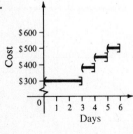

77. (a) 16   (b) 11   (c) 6   (d) 8   (e) 4   (f) 0   (g) and (k) are graphed below
(h) 0   (i) 40/3   (j) 80/3   (l) 8   (m) 6

$p = \frac{3}{4}x$

$p = 16 - \frac{5}{4}x$

Section 4.2 (page 189)
1. $2x + y = 5$ **3.** $3x + 2y = -7$ **5.** $3x + 4y = 6$ **7.** $y = 2$ **9.** $x = -8$ **11.** $x + 3y = 10$
13. $-x + 4y = 13$ **15.** $3x + 4y = 12$ **17.** $2x - 3y = 6$ **19.** $x = -5$ **21.** $x = -6$
23. $5.081x + y = -4.69256$ **25.** $x + 3y = 11$ **27.** $2x - y = 9$ **29.** $x - y = 7$ **31.** $5x - 3y = -13$
33. $-2x + y = 4$ **35.** $x = -5$ **37.** no   **39.** (a) $-1/2$ (b) $-7/2$ **51.** $2x + h$ **53.** $-1/[x(x + h)]$
55. $x = (b_2 - b_1)/(m_1 - m_2)$ **57.** $y = 640x + 1100; m = 640$ **59.** $y = -1000x + 40,000; m = -1000$
61. $y = 2.5x - 70; m = 2.5$

Section 4.3 (page 197)

1.
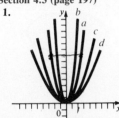

3.
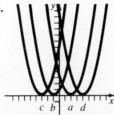

5. vertex: $(2, 0)$;
axis: $x = 2$
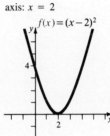
$f(x) = (x-2)^2$

7. vertex: $(-3, -4)$;
axis: $x = -3$
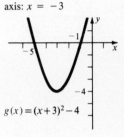
$g(x) = (x+3)^2 - 4$

9. vertex: $(-3, 2)$;
axis: $x = -3$
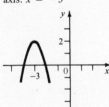
$k(x) = -2(x+3)^2 + 2$

11. vertex: $(-1, -3)$;
axis: $x = -1$
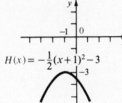
$H(x) = -\frac{1}{2}(x+1)^2 - 3$

13. vertex: $(1, 2)$;
axis: $x = 1$
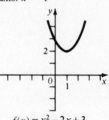
$f(x) = x^2 - 2x + 3$

15. vertex: $(-2, 6)$;
axis: $x = -2$

$g(x) = -x^2 - 4x + 2$

17. vertex: $(1, 3)$;
axis: $x = 1$
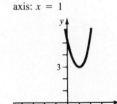
$f(x) = 2x^2 - 4x + 5$

19.
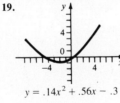
$y = .14x^2 + .56x - .3$

21. $y = -.09x^2 - 1.8x + .5$
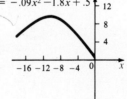

23. $[2, +\infty)$ **25.** $(-\infty, 0]$ **27.** $(-\infty, -5/4]$ **29.** $-4; x = -4$ **31.** $1/3; x = 1/3$ **33.** $(-4 - \sqrt{3}, -4 + \sqrt{3})$
35. $(-\infty, (1 - \sqrt{19})/3] \cup [(1 + \sqrt{19})/3, +\infty)$ **37.** $5\sqrt{2}$ **39.** $\sqrt{13}$ **41.** $\sqrt{181}/4$ **43.** 80 by 160 ft   **45.** 5 in
47. 10, 10   **49.** (a) $x(500 - x) = 500x - x^2$ (b)   **(c)** 250 **(d)** \$62,500

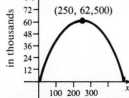
R(x)
(250, 62,500)
in thousands
72, 60, 48, 36, 24, 12
100 200 300
x

51. $10\sqrt{3}$ m ≈ 17.32 m   **53.** 25   **55.** 1/2   **57.** Any x-intercepts are given by $(-b + \sqrt{b^2 - 4ac})/(2a)$ or
$(-b - \sqrt{b^2 - 4ac})/(2a)$.   **59.** if $b^2 - 4ac > 0$ **61. (a)** $|y + p|$ **(b)** The distances from the origin to the line and to the
point are the same.   **(c)** $x^2 = 4py$ **63.** $x^2 = -20y$ **65.** $y^2 = 32x$ **67.** $(x + 5)^2 = -12(y - 5)$
69. $(y + 4)^2 = -12(x - 6)$ **73. (a)** 3   **(b)** 1/9

Section 4.4 (page 210)

1. (3, 0), (−3, 0); (0, 0)   **3.** (3, 0), (−3, 0); (0, 0)   **5.** (0, 3), (0, −3); (0, 0)   **7.** (4, 0), (−4, 0); (0, 0)

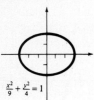
$\frac{x^2}{9} + \frac{y^2}{4} = 1$

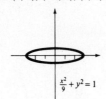
$\frac{x^2}{9} + y^2 = 1$

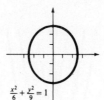
$\frac{x^2}{6} + \frac{y^2}{9} = 1$

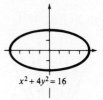
$x^2 + 4y^2 = 16$

9. (0, 0); $y = \pm x$ **11.** (0, $2\sqrt{2}$), (0, $-2\sqrt{2}$); (0, 0)   **13.** (0, 0); $y = \pm(5/2)x$ **15.** (1/3, 0), (−1/3, 0); (0, 0)

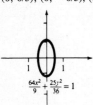
$x^2 = 9 + y^2$

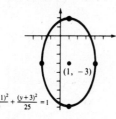
$2x^2 + y^2 = 8$

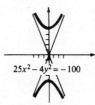
$25x^2 - 4y^2 = -100$

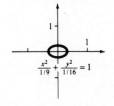
$\frac{x^2}{1/9} + \frac{y^2}{1/16} = 1$

17. (0, 6/5), (0, −6/5); (0, 0)   **19.** (1, 2), (1, −8); (1, −3)   **21.** (3, −2); $y + 2 = \pm(7/4)(x - 3)$ **23.** (3, −1); $y + 1 = \pm(5/6)(x - 3)$

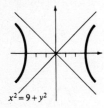
$\frac{64x^2}{9} + \frac{25y^2}{36} = 1$

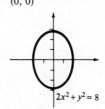
(1, −3)
$\frac{(x-1)^2}{9} + \frac{(y+3)^2}{25} = 1$

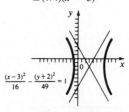
$\frac{(x-3)^2}{16} - \frac{(y+2)^2}{49} = 1$

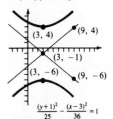
(3, 4)   (9, 4)
(3, −1)
(3, −6)   (9, −6)
$\frac{(y+1)^2}{25} - \frac{(x-3)^2}{36} = 1$

25.   **27.** function   **29.**   **31.** function

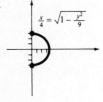
$\frac{x}{4} = \sqrt{1 - \frac{y^2}{9}}$

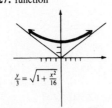
$\frac{y}{3} = \sqrt{1 + \frac{x^2}{16}}$

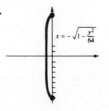
$x = -\sqrt{1 - \frac{y^2}{64}}$

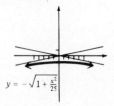
$y = -\sqrt{1 + \frac{x^2}{25}}$

33. $x^2/16 + y^2/12 = 1$ **35.** $x^2/36 + y^2/20 = 1$ **37.** $\dfrac{(x - 3)^2}{16} + \dfrac{(y + 2)^2}{25} = 1$ **39.** $x^2/9 - y^2/7 = 1$
41. $y^2/9 - x^2/25 = 1$ **43.** about 141.6 million miles

Section 4.5 (page 218)

1.
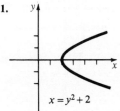
$x = y^2 + 2$

3.
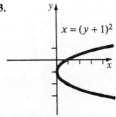
$x = (y + 1)^2$

5.
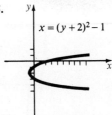
$x = (y + 2)^2 - 1$

7.
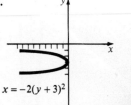
$x = -2(y + 3)^2$

9.
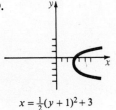
$x = \frac{1}{2}(y + 1)^2 + 3$

11.
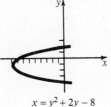
$x = y^2 + 2y - 8$

13.
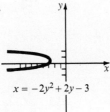
$x = -2y^2 + 2y - 3$

15. ellipse
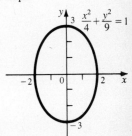
$\frac{x^2}{4} + \frac{y^2}{9} = 1$

17. hyperbola
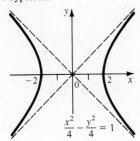
$\frac{x^2}{4} - \frac{y^2}{4} = 1$

19. line
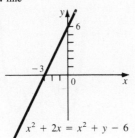
$x^2 + 2x = x^2 + y - 6$

21. hyperbola
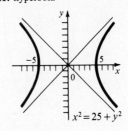
$x^2 = 25 + y^2$

23. ellipse
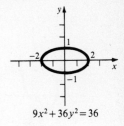
$9x^2 + 36y^2 = 36$

25. circle
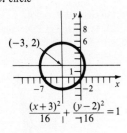
$(-3, 2)$
$\frac{(x+3)^2}{16} + \frac{(y-2)^2}{16} = 1$

27. parabola
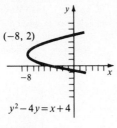
$(-8, 2)$
$y^2 - 4y = x + 4$

29. empty set (no graph)

31. circle
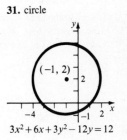
$(-1, 2)$
$3x^2 + 6x + 3y^2 - 12y = 12$

33. parabola
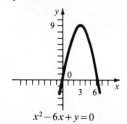
$x^2 - 6x + y = 0$

35. hyperbola
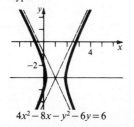
$4x^2 - 8x - y^2 - 6y = 6$

37. ellipse
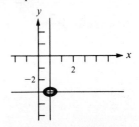
$4x^2 - 8x + 9y^2 + 54y = -84$

39. empty set (no graph)

Section 4.6 (page 224)

1. (a) (b) y-axis   **3.** (a) (b) origin   **5.** (a) (b) (0, 1)   **7.** (a) (b) (−1, 0)

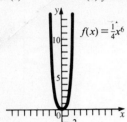

$f(x) = \frac{1}{4}x^6$

$f(x) = \frac{-5}{4}x^5$

$f(x) = \frac{1}{2}x^3 + 1$

$f(x) = -(x+1)^3$

9. (a) (b) $x = 1$ **11.** $f(x) = 2x(x-3)(x+2)$ **13.** $f(x) = x^2(x-2)(x+3)^2$ **15.**

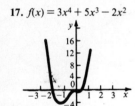

$f(x) = (x-1)^4 + 2$

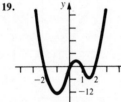
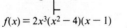

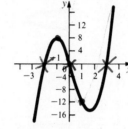

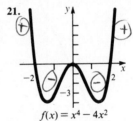

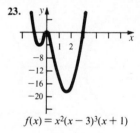

$f(x) = x^3 - x^2 - 2x$

17. $f(x) = 3x^4 + 5x^3 - 2x^2$ **19.**   **21.**   **23.**

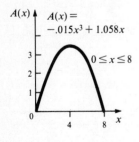

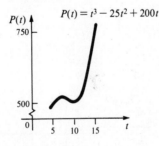

$f(x) = 2x^3(x^2 - 4)(x-1)$

$f(x) = x^4 - 4x^2$

$f(x) = x^2(x-3)^3(x+1)$

25. (a) (b) between 4 and 5 hr, closer to 5 hr (c) from about 1 hr to about 8 hr

27. (a) (b) increasing from $t = 0$ to $t = 6\,2/3$ and from $t = 10$ on; decreasing from $t = 6\,2/3$ to $t = 10$

29. odd   **31.** even   **33.** odd   **35.** neither   **37.** even   **39.** neither   **41.** Maximum is 26.136 when $x = -3.4$; minimum is 25 when $x = -3$.   **43.** Maximum is 1.048 when $x = -.1$; minimum is -5 when $x = -1$.   **45.** Maximum is 84 when $x = -2$; minimum is -13 when $x = -1$.

$A(x) = -.015x^3 + 1.058x$

$0 \le x \le 8$

$P(t) = t^3 - 25t^2 + 200t$

47.

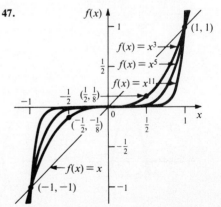

$f(x) = x^3$

$f(x) = x^5$

$f(x) = x^{11}$

$(\frac{1}{2}, \frac{1}{8})$

$(-\frac{1}{2}, -\frac{1}{8})$

$f(x) = x$

$(1, 1)$

$(-1, -1)$

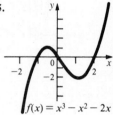

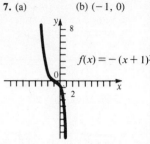

Section 4.7 (page 233)
1. vertical asymptote: $x = 5$; horizontal asymptote: $y = 0$ **3.** vertical asymptote: $x = 7/3$; horizontal asymptote: $y = 0$
5. vertical asymptote: $x = -2$; horizontal asymptote: $y = -1$ **7.** vertical asymptote: $x = -9/2$; horizontal asymptote: $y = 3/2$
9. vertical asymptotes: $x = 3$, $x = 1$; horizontal asymptote: $y = 0$ **11.** oblique asymptote: $y = x - 3$; vertical asymptote
$x = -3$ **13.** vertical asymptotes: $x = -2$, $x = 5/2$; horizontal asymptote: $y = 1/2$ **15.** no vertical asymptotes; horizontal
asymptote: $y = 2$

17.
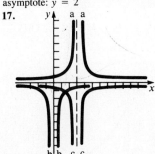

19.
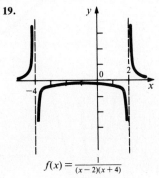
$$f(x) = \frac{1}{(x-2)(x+4)}$$

21.
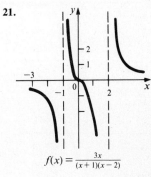
$$f(x) = \frac{3x}{(x+1)(x-2)}$$

23.
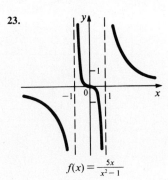
$$f(x) = \frac{5x}{x^2-1}$$

25.
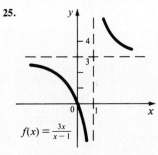
$$f(x) = \frac{3x}{x-1}$$

27.
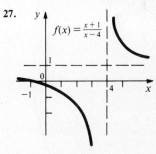
$$f(x) = \frac{x+1}{x-4}$$

29.
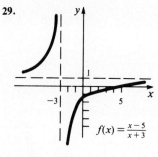
$$f(x) = \frac{x-5}{x+3}$$

31.
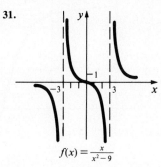
$$f(x) = \frac{x}{x^2-9}$$

33.
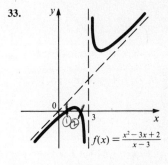
$$f(x) = \frac{x^2-3x+2}{x-3}$$

35.
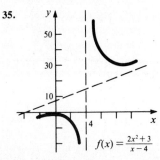
$$f(x) = \frac{2x^2+3}{x-4}$$

37.
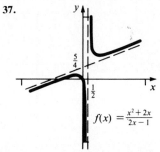
$$f(x) = \frac{x^2+2x}{2x-1}$$

39.
$$f(x) = \frac{(x-3)(x+1)}{(x-1)^2}$$
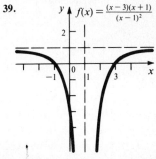

41.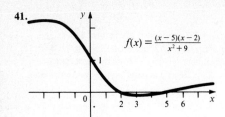
$$f(x) = \frac{(x-5)(x-2)}{x^2+9}$$

43.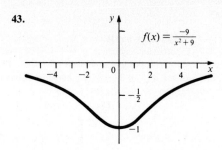
$$f(x) = \frac{-9}{x^2+9}$$

45.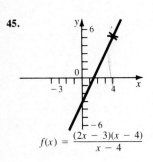
$$f(x) = \frac{(2x-3)(x-4)}{x-4}$$

47.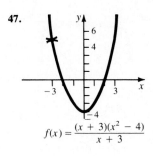
$$f(x) = \frac{(x+3)(x^2-4)}{x+3}$$

49. (a) 440, 400, 338, 259, 210, 176

(b)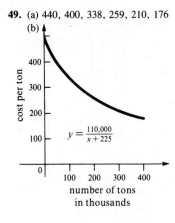
$$y = \frac{110,000}{x+225}$$

cost per ton

number of tons in thousands

51. (a) 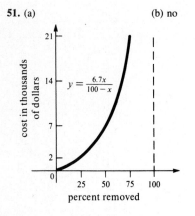 (b) no
$$y = \frac{6.7x}{100-x}$$

cost in thousands of dollars

percent removed

53. (a) \$65.5 tens of millions, or \$655,000,000 (b) \$64 tens of millions, or \$640,000,000 (c) \$60 tens of millions or \$600,000,000 (d) \$40 tens of millions or \$400,000,000 (e) \$0
(f)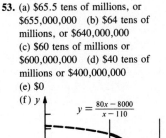
$$y = \frac{80x-8000}{x-110}$$

Chapter 4 Review Exercises (page 236)
1. 6/5 **3.** not defined **5.** 9/4 **7.** 1/5 **9.** 0

11.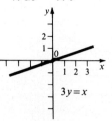
$3x+7y=14$

13.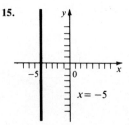
$3y=x$

15.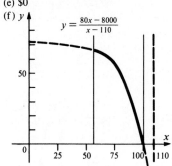
$x=-5$

17. $x + 3y = 10$
19. $2x + y = 1$
21. $15x + 30y = 13$
23. $y = 3/4$
25. $5x - 8y = -40$
27. $y = -5$

29.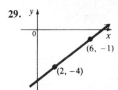

31.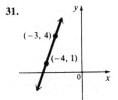

33.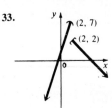

35.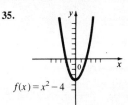

$f(x) = x^2 - 4$

37.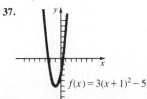

$f(x) = 3(x+1)^2 - 5$

39.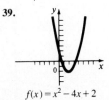

$f(x) = x^2 - 4x + 2$

41. $(-3, -9)$; $x = -3$
43. $(7/2, -41/4)$; $x = 7/2$
45. $(-1, 4)$; $x = -1$
47. 5.5 and 5.5
49. a square, 45 m on a side

51.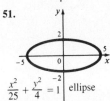

$\dfrac{x^2}{25} + \dfrac{y^2}{4} = 1$ ellipse

53.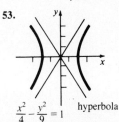

$\dfrac{x^2}{4} - \dfrac{y^2}{9} = 1$ hyperbola

55.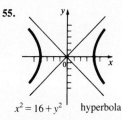

$x^2 = 16 + y^2$ hyperbola

57.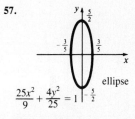

$\dfrac{25x^2}{9} + \dfrac{4y^2}{25} = 1$ ellipse

59. ellipse

$\dfrac{(x-2)^2}{9} + \dfrac{(y+3)^2}{4} = 1$

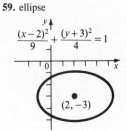

61. hyperbola

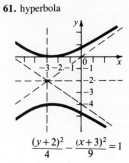

$\dfrac{(y+2)^2}{4} - \dfrac{(x+3)^2}{9} = 1$

63. semi-ellipse

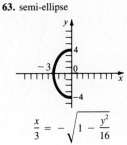

$\dfrac{x}{3} = -\sqrt{1 - \dfrac{y^2}{16}}$

65. semi-ellipse

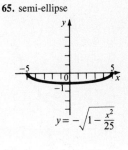

$y = -\sqrt{1 - \dfrac{x^2}{25}}$

67.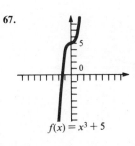

$f(x) = x^3 + 5$

69. $f(x) = x^2(2x+1)(x-2)$

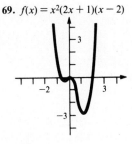

71. $f(x) = 2x^3 + 13x^2 + 15x$

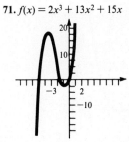

73.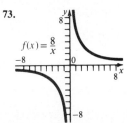

$f(x) = \dfrac{8}{x}$

75.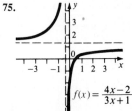

$f(x) = \dfrac{4x-2}{3x+1}$

77.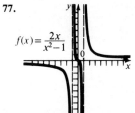

$f(x) = \dfrac{2x}{x^2-1}$

79.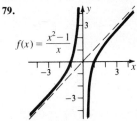

$f(x) = \dfrac{x^2-1}{x}$

81. $2x - 3y = 14$ **83.** -7 **85.** One square has vertices at $(-5, 19)$, $(0, 7)$, $(12, 12)$, $(7, 24)$; another has vertices at $(5, -5)$, $(0, 7)$, $(12, 12)$, $(17, 0)$; a third has vertices at $(0, 7)$, $(17/2, 7/2)$, $(12, 12)$, $(7/2, 31/2)$.

Chapter 5

Section 5.1 (page 245)

1. yes   **3.** no   **5.** 4/9   **7.** $\sqrt{6}/3$

9. (a)

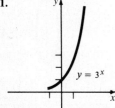

$y = 2^x + 1$

(b)

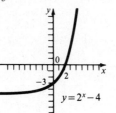

$y = 2^x - 4$

(c)

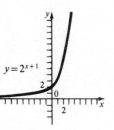

$y = 2^{x+1}$

(d)

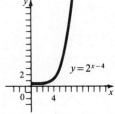

$y = 2^{x-4}$

11.

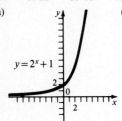

$y = 3^x$

13.

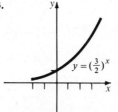

$y = \left(\frac{3}{2}\right)^x$

15.

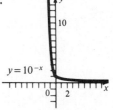

$y = 10^{-x}$

17.

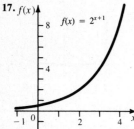

$f(x) = 2^{x+1}$

19.

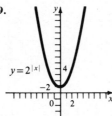

$y = 2^{|x|}$

21.

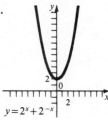

$y = 2^x + 2^{-x}$

23.

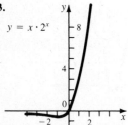

$y = x \cdot 2^x$

25.

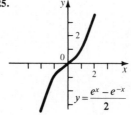

$y = \dfrac{e^x - e^{-x}}{2}$

27.

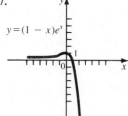

$y = (1 - x)e^x$

29. $\{1/2\}$ **31.** $\{-2\}$ **33.** $\{0\}$ **35.** $\{3\}$ **37.** $\{8\}$ **39.** $\{1/5\}$.   **41.** $\{0\}$ **43.** $\{3/5\}$ **45.** $\{-2/3\}$ **47.** $\{-3, 3\}$
49. $\{-1, 5\}$ **51.** $\{0\}$ **55.** \$3,516.56   **57.** \$6,799.21   **59.** \$8,158.66   **61. (a)** \$12,075.32   **(b)** \$12,348.76
(c) \$12,650.78   **63. (a)** 500 g   **(b)** 409 g   **(c)** 335 g   **(d)** 184 g   **(e)** $Q(t)$

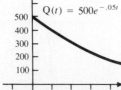

$Q(t) = 500e^{-.05t}$

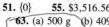

65. (a) about 2,690,000 (b) about 3,020,000 (c) about 9,600,000 (d) about 38,400,000   **67.** 11.6 pounds per square inch
71. 0   **73.** 2   **75.** neither   **79.** 2.718   **83.** $4/(e^{-x} - e^x)^2$

Section 5.2 (page 254)
1. $\log_3 81 = 4$ **3.** $\log_{10} 10,000 = 4$ **5.** $\log_{1/2} 16 = -4$ **7.** $\log_{10} .0001 = -4$ **9.** $6^2 = 36$ **11.** $(\sqrt{3})^8 = 81$
13. $10^{-4} = .0001$ **15.** $m^n = k$ **17.** 2   **19.** 1   **21.** -3 **23.** 1/2   **25.** $-1/6$ **27.** 9   **29.** 8   **31.** 4
33. 1/2   **35.** 2   **37.** x **39.** 9   **41.** {1/5}   **43.** {3/2}   **45.** {16}   **47.** {1}   **49.** $\log_3 2 - \log_3 5$
51. $\log_2 6 + \log_2 x - \log_2 y$ **53.** $1 + (1/2) \log_5 7 - \log_5 3$ **55.** cannot be simplified using the properties of logarithms
57. $\log_k p + 2 \cdot \log_k q - \log_k m$ **59.** $(1/2)(\log_m 5 + 3 \cdot \log_m r - 5 \cdot \log_m z)$ **61.** $\log_a (xy)/m$ **63.** $\log_m (a^2/b^6)$
65. $\log_x (a^{3/2}b^{-4})$ or $\log_x (a^{3/2}/b^4)$ **67.** $\log_b [7x(x + 2)/8]$ **69.** $\log_a [(z + 1)^2(3z + 2)]$
71. $\log_5 (5^{1/3} m^{-1/3})$ or $\log_5(5^{1/3}/m^{1/3})$

73.
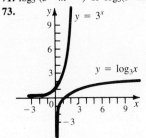

75. (a)
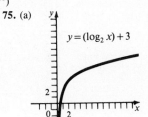

(b)
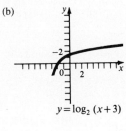

(c)
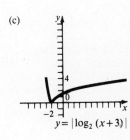

77. $f(x)$
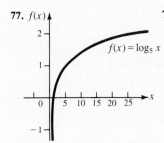

79.
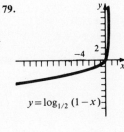

81.
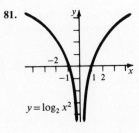

83.
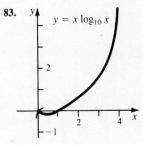

85. .7781   **87.** .9542   **89.** 1.4771   **91.** (a) about 239 (b) about 477 (c) about 759 (d)
93. (a) 21 (b) 70 (c) 91 (d) 120 (e) 140   **95.** 398,000,000 I_0
99.
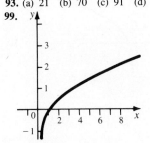

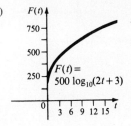

Section 5.3 (page 260)
1. 1   **3.** 4   **5.** -3 **7.** 1.2   **9.** 1.3863
11. 2.8332   **13.** 5.8579   **15.** 3.8918   **17.** 3.7377
19. 11.3022   **21.** 1.43   **23.** .59   **25.** -1.58
27. .96   **29.** 1.89   **31.** -5.05

33.
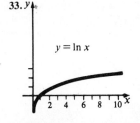

35.
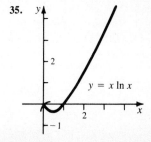

37. 1 **39.** (a) 2.03 (b) 2.28 (c) 2.17 (d) 1.21 **41.** (a) about 495 (b) about 6 **43.** 378 feet
45. (a) 7,298,000; 77,338,000; 168,192,000; 175,139,000 (b) According to the given equation, the population can never increase to 197,273,000. **47.** .86

Section 5.4 (page 267)
1. {1.631} **3.** {1.069} **5.** {−.535} **7.** {−.080} **9.** {3.797} **11.** {2.386} **13.** {−.123} **15.** ∅ **17.** {1.104}
19. {17.475} **21.** {−2.141, 2.141} **23.** {1.386} **25.** {11} **27.** {5} **29.** {10} **31.** {5} **33.** {4} **35.** ∅
37. {11} **39.** {3} **41.** ∅ **43.** {−2, 2} **45.** {1, 10} **47.** {2.423} **49.** {−.204} **51.** 15 seconds **53.** 46 days
55. {0} **57.** {0, ln 5} **59.** $x = \ln[y/(y + 1)]$ **61.** $x = \ln(y + \sqrt{y^2 + 1})$ **63.** $t = (−2/R)\ln(1 − RI/E)$
65. $n = \ln[A/(A − Pi)]/\ln(1 + i)$ **67.** $x = (1 − y)/2$ **69.** (1/3, +∞) **71.** (0, 1) ∪ (1, 10) **73.** (0, 1)
75. $[−\sqrt[6]{3}, −1) ∪ (−1, 1) ∪ (1, \sqrt[6]{3}]$ **77.** 89 decibels is about twice as loud as 86 decibels, which is a 100% increase.
79. $n = [\ln(A/P)]/[m\ln(1 + i/m)]$ **81.** $t = −[\ln(M/434)]/.08$

Section 5.5 (page 274)
1. about 14 months **3.** about 283 grams **5.** about 11.3° **7.** 5600 years **11.** about 15,000 years **13.** $2,166.57
15. $17,143.21 **17.** $12,652.54 **19.** about 13.9 years **21.** $i = [\ln(A/P)]/n$ **23.** 81.25°C **25.** 117.5°C
27. about 1611 **29.** in about 1.8 decades or 18 years

Section 5.6 (page 280)
1. 2 **3.** 6 **5.** −4 **7.** −5 **9.** −3.0491 or .9509 − 4 **11.** 4.8338 **13.** .8825 **15.** 4.6331 **17.** −2.0918 or
.9082 − 3 **19.** 34.4 **21.** .0747 **23.** 3120 **25.** .00637 **27.** 4.83 **29.** −.02 **31.** 5.62 **33.** 3.32
35. 4.73 **37.** 1.79 **39.** {.40} **41.** {−.19} **43.** {−1.47} **45.** {.38} **47.** 2.01 **49.** 10.6 **51.** 21.5
53. (a) about 350 years (b) about 4000 years (c) about 2300 years **55.** 3.2 **57.** 1.8 **59.** $2.0 \times 10^{−3}$
61. $1.6 \times 10^{−5}$ **63.** 8706 **65.** (a) about 78 (b) about 38 (c) about 2800

Chapter 5 Appendix (page 283)
1. 3.3701 **3.** 1.6836 **5.** .7975 − 2 **7.** 1.4322 **9.** 62.26 **11.** .004534 **13.** 493,200 **15.** .0001667
17. 1.396 **19.** 1.550 **21.** .9308 **23.** 15.57 **25.** .1405

Chapter 5 Review Exercises (page 284)
1. **3.** **5.** **7.**

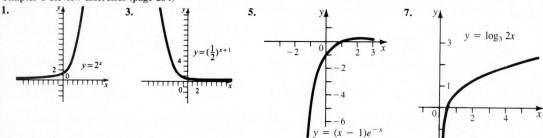

9. {5/3} **11.** {3/2} **13.** 800 g **15.** $1790.19 **17.** about 12 years **19.** $\log_2 32 = 5$ **21.** $\log_{1/16}(1/2) = 1/4$
23. $\log 3 = .4771$ **25.** $\ln 1.1052 = .1$ **27.** $10^{−3} = .001$ **29.** $10^{.537819} = 3.45$ **31.** $\log_3 m + \log_3 n − \log_3 p$
33. $2\log_5 x + 4\log_5 y + (3/5)\log_5 m + (1/5)\log_5 p$ **35.** .9279 **37.** 3.0212 **39.** 3150 **41.** .0446 **43.** 5.97
45. 1.44 **47.** .8 m **49.** 1.5 m **51.** (a) 207 (b) 235 (c) 249 (d)
53. {1.490} **55.** {1.303} **57.** {.747} **59.** {4} **61.** {2} **63.** {2}
65. $x = [\ln(y + \sqrt{y^2 + 1})]/\ln 5$ **67.** $x = (4 + a \pm \sqrt{a^2 + 12a})/2$
69. $c = de^{(N − a)/b}$ **71.** −5.3 **73.** 4.4886 **75.** 2.0794
77. −3.2403 **79.** 1.2785 **81.** −2.0082 or .9918 − 3
83. $7,182.87 **85.** $1,241.99 **87.** about 28 quarters or 7 years
89. about 5.3 years **91.** $16,095.74 **93.** $580,792.63
95. (a) about 286 grams (b) about 28.9 days **97.** (a) −.223; .182
(b) .405 + .288 = .693 (c) .405 + .693 = 1.098; 3(.693) = 2.079
99. 100! ≈ 9.32×10^{157}; 200! ≈ 7.88×10^{374}

Chapter 6

Section 6.1 (page 296)

1. $\{(-32, -8)\}$ **3.** $\{(1, 3)\}$ **5.** $\{(-1, 3)\}$ **7.** $\{(2, 18)\}$ **9.** $\{(4, 1, -2)\}$ **11.** $\{(1, 0, 2)\}$ **13.** $\{(-2, 0)\}$
15. $\{(1, 3)\}$ **17.** $\{(4, -2)\}$ **19.** $\{(2, -2)\}$ **21.** $\{(12, 6)\}$ **23.** $\{(4, 3)\}$ **25.** $\{(-4, -1)\}$ **27.** $\{(.38, -.47)\}$
29. $\emptyset$ **31.** $\{(5, 2)\}$ **33.** $\{(1, 2, -1)\}$ **35.** $\{(2, 0, 3)\}$ **37.** $\emptyset$ **39.** $\{(2, 2, 0)\}$ **41.** $\emptyset$ **43.** $\{(2, 2)\}$
45. $\{(1/5, 1)\}$ **47.** $\{(1/5, 14/3)\}$ **49.** $\{(4, 6, 1)\}$ **51.** $\{(x, 2x - 5, (3x - 4)/3)\}$ **53.** $\{(x, -x, -9x)\}$
55. $\{(x, 2x, 6 - 6x)\}$ **57.** goats cost \$30 and sheep cost \$25   **59.** 32 days at \$14   **61.** 18 liters of 70% and
12 liters of 20%   **63.** \$10,000 at 8% and \$15,000 at 10%   **65.** \$5000 at 8% and \$15,000 at 12%   **67.** 200 gallons
69. 72 kph and 92 kph   **71.** 15 cm, 12 cm, 6 cm   **73.** 20, 18, 12   **75.** $\{(a + 1, 2)\}$ **77.** 37

Section 6.2 (page 305)

1. $\begin{bmatrix} 2 & 3 & | & 11 \\ 1 & 2 & | & 8 \end{bmatrix}$ **3.** $\begin{bmatrix} 1 & 5 & | & 6 \\ 0 & 1 & | & 1 \end{bmatrix}$ **5.** $\begin{bmatrix} 2 & 1 & 1 & | & 3 \\ 3 & -4 & 2 & | & -7 \\ 1 & 1 & 1 & | & 2 \end{bmatrix}$ **7.** $\begin{bmatrix} 1 & 1 & 0 & | & 2 \\ 0 & 2 & 1 & | & -4 \\ 0 & 0 & 1 & | & 2 \end{bmatrix}$ **9.** $2x + y = 1, 3x - 2y = -9$

11. $x = 2, y = 3, z = -2$ **13.** $3x + 2y + z = 1, 2y + 4z = 22, -x - 2y + 3z = 15$ **15.** $\{(2, 3)\}$ **17.** $\{(-3, 0)\}$
19. $\{(7/2, -1)\}$ **21.** $\{(5, 0)\}$ **23.** $\{(-2, 1, 3)\}$ **25.** $\{(-1, 23, 16)\}$ **27.** $\{(3, 2, -4)\}$ **29.** $\{(2, 1, -1)\}$ **31.** $\emptyset$
33. $\{(3, 3 - z, z)\}$ **35.** $\left\{\left(\dfrac{-12}{23} - \dfrac{15}{23}z, \dfrac{1}{23}z - \dfrac{13}{23}, z\right)\right\}$ **37.** $\left\{\left(-6 - \dfrac{1}{2}z, \dfrac{1}{2}z + 2, z\right)\right\}$ **39.** $\{(-1, 2, 5, 1)\}$
41. $\{(1 - 4w, 1 - w, 2 + w, w)\}$ **43.** $\{(1.7, 2.4)\}$ **45.** $\{(.5, -.7, 6)\}$ **47.** wife 40 days, husband 32 days

49. 5 model 201, 8 model 301

51. \$10,000 at 8%, \$7000 at 10%, \$8000 at 9%   **53.** $\begin{bmatrix} 1 & 0 & 0 & 1 & | & 1000 \\ 1 & 1 & 0 & 0 & | & 1100 \\ 0 & 1 & 1 & 0 & | & 700 \\ 0 & 0 & 1 & 1 & | & 600 \end{bmatrix}; \begin{bmatrix} 1 & 0 & 0 & 1 & | & 1000 \\ 0 & 1 & 0 & -1 & | & 100 \\ 0 & 0 & 1 & 1 & | & 600 \\ 0 & 0 & 0 & 0 & | & 0 \end{bmatrix}$

55. 1000, 1000   **57.** 600, 600   **59.** inconsistent system, no solution   **61.** $\{(-15 + z, 5 - 2z, z)\}$ **63.** $A = 1/2, B = 1/2$
65. $A = 3, B = -1, C = 2$ **67.** $A = 2, B = 1, C = 3$

Section 6.3 (page 314)

1. $x = 2, y = 4, z = 8$ **3.** $x = -15, y = 5, k = 3$ **5.** $z = 18, r = 3, s = 3, p = 3, a = 3/4$
7. $\begin{bmatrix} 14 & -11 & -3 \\ -2 & 4 & -1 \end{bmatrix}$ **9.** $\begin{bmatrix} -6 & 8 \\ 4 & 2 \end{bmatrix}$ **11.** can't be done   **13.** $\begin{bmatrix} -12x + 8y & -x + y \\ x & 8x - y \end{bmatrix}$ **15.** $\begin{bmatrix} 7 & 2 \\ 9 & 0 \\ 8 & 6 \end{bmatrix}; \begin{bmatrix} 7 & 9 & 8 \\ 2 & 0 & 6 \end{bmatrix}$

17. $\begin{bmatrix} -4 & 8 \\ 0 & 6 \end{bmatrix}$ **19.** $\begin{bmatrix} 2 & 6 \\ -4 & 6 \end{bmatrix}$ **21.** $\begin{bmatrix} -1 & -3 \\ 2 & -3 \end{bmatrix}$ **23.** $\begin{bmatrix} 13 \\ 25 \end{bmatrix}$ **25.** $\begin{bmatrix} -2 & 5 & 0 \\ 6 & 6 & 1 \\ 12 & 2 & -3 \end{bmatrix}$ **27.** $\begin{bmatrix} 20 & 0 & 8 \\ 8 & 0 & 4 \end{bmatrix}$ **29.** can't be

multiplied   **31.** $[-8]$ **33.** $\begin{bmatrix} -6 & 12 & 18 \\ 4 & -8 & -12 \\ -2 & 4 & 6 \end{bmatrix}$ **35.** (a) $\begin{bmatrix} 17\frac{3}{4} & 23\frac{3}{4} \\ 9 & 12\frac{3}{4} \\ 33 & 42 \\ 6 & 8 \end{bmatrix}$; (b) [220 890 105 125 70]; (c) [4165 5605]

37. $\begin{bmatrix} .018 & -.984 \\ 1.002 & -.056 \end{bmatrix}$ **39.** $\begin{bmatrix} -7.88 & .7 & -.068 \\ -4.87 & -.324 & .259 \\ -1.54 & -.803 & -.192 \end{bmatrix}$ **41.** always true   **43.** always true   **45.** always true

47. always true   **49.** always true   **51.** always true   **53.** Not true. $(A + B)(A - B) = A^2 + BA - AB - B^2$. Since
multiplication is not commutative, $BA - AB$ is not always 0.

Section 6.4 (page 323)

1. yes   **3.** no   **5.** no   **7.** yes   **9.** $\begin{bmatrix} 0 & 1/2 \\ -1 & 1/2 \end{bmatrix}$ **11.** $\begin{bmatrix} 2 & 1 \\ 5 & 3 \end{bmatrix}$ **13.** none   **15.** $\begin{bmatrix} 1 & 0 & 0 \\ 0 & -1 & 0 \\ -1 & 0 & 1 \end{bmatrix}$

17. $\begin{bmatrix} 15 & 4 & -5 \\ -12 & -3 & 4 \\ -4 & -1 & 1 \end{bmatrix}$ **19.** none **21.** $\begin{bmatrix} 7/4 & 5/2 & 3 \\ -1/4 & -1/2 & 0 \\ -1/4 & -1/2 & -1 \end{bmatrix}$ **23.** $\begin{bmatrix} 1/2 & 1/2 & -1/4 & 1/2 \\ -1 & 4 & -1/2 & -2 \\ -1/2 & 5/2 & -1/4 & -3/2 \\ 1/2 & -1/2 & 1/4 & 1/2 \end{bmatrix}$ **25.** $\{(-1, 4)\}$

27. $\{(2, 1)\}$ **29.** $\{(2, 3)\}$ **31.** no inverse, $\{(-8y - 12, y)\}$ **33.** $\{(-8, 6, 1)\}$ **35.** $\{(15, -5, -1)\}$

37. $\{(-31, 24, -4)\}$ **39.** no inverse, $\emptyset$ **41.** $\{(-7, -34, -19, 7)\}$ **47.** (a) $\begin{bmatrix} 72 \\ 48 \\ 60 \end{bmatrix}$ (b) $\begin{bmatrix} 2 & 4 & 2 \\ 2 & 1 & 2 \\ 2 & 1 & 3 \end{bmatrix}\begin{bmatrix} x_1 \\ x_2 \\ x_3 \end{bmatrix} = \begin{bmatrix} 72 \\ 48 \\ 60 \end{bmatrix}$ (c) 8, 8, 12

51. $ab \neq 0$ **53.** $\begin{bmatrix} 1/a & 0 & 0 \\ 0 & 1/b & 0 \\ 0 & 0 & 1/c \end{bmatrix}$

Section 6.5 (page 332)

1. -26 **3.** -16 **5.** $y^2 - 16$ **7.** $x^2 - y^2$ **9.** $0, -5, 0$ **11.** $-6, 0, -6$ **13.** -1 **15.** 0 **17.** 0 **19.** $4x$
21. $10i + 17j - 6k$ **23.** -88 **25.** -5.5 **27.** two rows identical **29.** one column all zeros **31.** multiply each
element of second row by $-1/4$; then two rows are identical **33.** multiply each element of third column by $1/2$; then two columns
are identical **35.** rows and columns exchanged **37.** two rows exchanged **39.** multiply each element of second column by 3
41. multiply elements of first row by 1; add products to elements of second row **43.** multiply elements of second row by 2
45. 0 **47.** 0 **49.** 5 **51.** 32 **53.** -32 **65.** 5/2 **67.** 7 **69.** 8 **75.** a^3

Section 6.6 (page 339)

1. $\{(2, 2)\}$ **3.** $\{(2, -5)\}$ **5.** $\{(2, 3)\}$ **7.** $\{(9/7, 2/7)\}$ **9.** $D = 0, \emptyset$ **11.** $\{(-1, 2, 1)\}$ **13.** $\{(-3, 4, 2)\}$
15. $\{(0, 0, -1)\}$ **17.** $D = 0, \emptyset$ **19.** $\{(197/91, -118/91, -23/91)\}$ **21.** $\{(2, 3, 1)\}$ **23.** $\{(0, 4, 2)\}$
25. $\{(31/5, 19/10, -29/10)\}$ **27.** $\{(1, 0, -1, 2)\}$ **29.** $\{(-.2, -.8)\}$ **31.** $\{(.2, .5, -.2)\}$
33. shirts are \$12, pants are \$18 **35.** 17.5 g of 12-carat, 7.5 g of 22-carat **37.** 104/19 lb **39.** 164 fives, 86 tens, 40 twenties
43. $\{[(a^2 + ab + b^2)/(a + b), -ab/(a + b)]\}$ **45.** $\{([(2b)(b + 4)]/(b^2 + 4), [(4b)(b - 1)]/(b^2 + 4))\}$

Section 6.7 (page 343)

1. $\{(-1, -7), (4, 3)\}$ **3.** $\{(2, 4)\}$ **5.** $\{(5, 26), (-2, 5)\}$ **7.** $\{(1, 1), (-2, 4)\}$ **9.** $\{(2, 1), (1/3, 4/9)\}$
11. $\{(-3/5, 7/5), (-1, 1)\}$ **13.** $\{(2, 2), (2, -2), (-2, 2), (-2, -2)\}$ **15.** $\{(0, 0)\}$ **17.** $\{(1, 1), (1, -1), (-1, 1),$
$(-1, -1)\}$ **19.** same circles, or $\{(x, y)|x^2 + y^2 = 10\}$ **21.** $\{(2, 3), (3, 2)\}$ **23.** $\{(-3, 5), (15/4, -4)\}$
25. $\{(4, -1/8), (-2, 1/4)\}$ **27.** $\{(3, 2), (-3, -2), (4, 3/2), (-4, -3/2)\}$ **29.** $\{(3, 5), (-3, -5)\}$
31. $\{(\sqrt{5}, 0), (-\sqrt{5}, 0), (\sqrt{5}, \sqrt{5}), (-\sqrt{5}, -\sqrt{5})\}$ **33.** $\{(1, 3), (31/14, -9/14)\}$ **35.** $\{(0, 1)\}$ **37.** $\{(12, 1)\}$
39. 6 and 6 **41.** 9 and 15, or -15 and -9 **43.** 5 m and 12 m **45.** yes **47.** $a \neq 5$
49. $\{(-1, 0), (6, -\sqrt{7})\}$ **51.** $\{(\sqrt{17}/17, 1), (-\sqrt{17}/17, 1), (\sqrt{17}/17, -1), (-\sqrt{17}/17, -1)\}$
53. $\{((18 + \sqrt{2a^3 - 108})/6, (-18 + \sqrt{2a^3 - 108})/6), ((18 - \sqrt{2a^3 - 108})/6, (-18 - \sqrt{2a^3 - 108})/6)\}$

Section 6.8 (page 349)

1.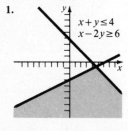
$x + y \leq 4$
$x - 2y \geq 6$

3.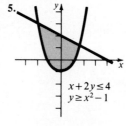
$4x + 3y < 12$
$y + 4x > -4$

5.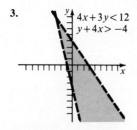
$x + 2y \leq 4$
$y \geq x^2 - 1$

7.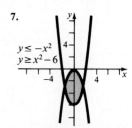
$y \leq -x^2$
$y \geq x^2 - 6$

9.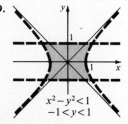
$x^2 - y^2 < 1$
$-1 < y < 1$

11.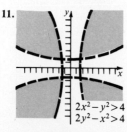
$2x^2 - y^2 > 4$
$2y^2 - x^2 > 4$

13.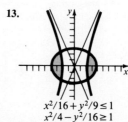
$x^2/16 + y^2/9 \leq 1$
$x^2/4 - y^2/16 \geq 1$

15.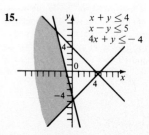
$x + y \leq 4$
$x - y \leq 5$
$4x + y \leq -4$

17.
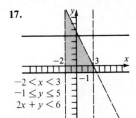
$-2 < x < 3$
$-1 \le y \le 5$
$2x + y < 6$

19.
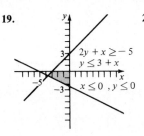
$2y + x \ge -5$
$y \le 3 + x$
$x \le 0 , y \le 0$

21.
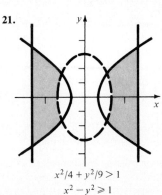
$x^2/4 + y^2/9 > 1$
$x^2 - y^2 \ge 1$
$-4 \le x \le 4$

23.
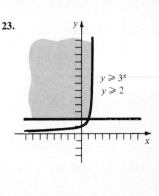
$y \ge 3^x$
$y \ge 2$

25.
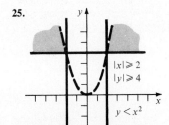
$|x| \ge 2$
$|y| \ge 4$
$y < x^2$

27.
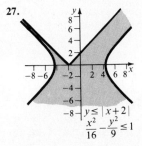
$y \le |x + 2|$
$\dfrac{x^2}{16} - \dfrac{y^2}{9} \le 1$

29.
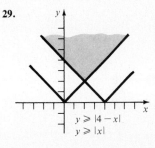
$y \ge |4 - x|$
$y \ge |x|$

31. Let x = number of units of basic, y = number of units of plain; $x \ge 3$, $y \ge 2$, $5x + 4y \le 50$, $2x + y \le 16$.

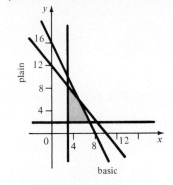

33. Let x = number of red pills, y = number of green pills; $8x + 2y \ge 16$, $x + y \ge 5$, $2x + 7y \ge 20$, $x \ge 0$, $y \ge 0$.

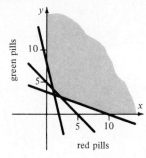

35. Let x = number shipped to warehouse A and y = number shipped to B; $0 \le x \le 75$, $0 \le y \le 80$, $x + y \ge 100$, $12x + 10y \le 1200$.

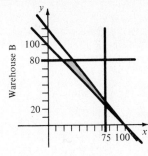

Warehouse A

Section 6.9 (page 354)
1. maximum of 65 at (5, 10); minimum of 8 at (1, 1)   **3.** maximum of 900 at (0, 12); minimum of 0 at (0, 0)   **5.** (6/5, 6/5)
7. (17/3, 5)   **9.** (105/8, 25/8)   **11.** (a) maximum of 204 at (18, 2); (b) 117 3/5 at (12/5, 39/5); (c) 102 at (0, 17/2)
13. $112, with 4 pigs, 12 geese   **15.** 8 of #1, 3 of #2   **17.** 6.4 million gallons of gasoline and 3.2 million gallons of fuel oil, for maximum revenue of $16,960,000   **19.** 250,000 hectares of each for a maximum profit of $132,500,000   **21.** 6000 Type 1 and 10,000 Type 2 for maximum revenue of $1,800

Chapter 6 Review Exercises (page 357)
1. $\{(-1, 3)\}$ **3.** $\{(6, -12)\}$ **5.** $\{(2, 5)\}$ **7.** $\{(-1, 2, 3)\}$ **9.** 10 at 25¢, 12 at 50¢   **11.** $10,000 at 6%, $20,000 at 7%, $20,000 at 9%   **13.** $\{[(15z + 6)/11, (14z - 1)/11, z]\}$ **15.** $\{(3\sqrt{3}, \sqrt{2}), (-3\sqrt{3}, \sqrt{2}), (3\sqrt{3}, -\sqrt{2}), (-3\sqrt{3}, -\sqrt{2})\}$
17. $\{(-2, 0), (1, 1)\}$ **19.** yes, $\{((8 - 8\sqrt{41})/5, (16 + 4\sqrt{41})/5), ((8 + 8\sqrt{41})/5, (16 - 4\sqrt{41})/5)\}$

21. $a = 5, x = 3/2, y = 0, z = 9$ **23.** $\begin{bmatrix} 0 & -2 & 7 \\ 6 & -1 & 10 \end{bmatrix}$ **25.** cannot be done   **27.** $\begin{bmatrix} -9 & 3 \\ 10 & 6 \end{bmatrix}$ **29.** $\{(-4, 6)\}$

31. $\{(0, 1, 0)\}$ **33.** $\begin{bmatrix} 3 & -1 \\ -5 & 2 \end{bmatrix}$ **35.** $\begin{bmatrix} 2/3 & 0 & -1/3 \\ 1/3 & 0 & -2/3 \\ -2/3 & 1 & 1/3 \end{bmatrix}$ **37.** $\{(2, 1)\}$ **39.** $\{(-3, 5, 1)\}$ **41.** -25

43. -44 **45.** 138   **47.** rows and columns interchanged   **49.** two identical rows   **51.** columns 2 and 3 exchanged
53. $\{(-4, 2)\}$ **55.** $\{(16/9, (8 - 9z)/18, z)\}$, dependent system   **57.**
59. Let x = number of batches of cakes,
let y = number of batches of cookies;
$2x + \dfrac{3}{2}y \le 16; 3x + \dfrac{2}{3}y \le 12; x \ge 0;$
$y \ge 0.$

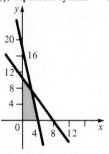

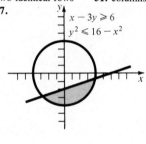

61. maximum of 210 at $(0, 30)$ **63.** minimum of 92/11 at $(12/11, 40/11)$ **65.** The maximum profit is $200, which occurs when 2 batches of cakes and 7 batches of cookies are made.   **67.** $x = (-mb + \sqrt{r^2 + r^2m^2 - b^2})/(1 + m^2)$ or $(-mb - \sqrt{r^2 + r^2m^2 - b^2})/(1 + m^2)$

Chapter 7

Section 7.1 (page 367)
1. $x^2 - 3x - 2$ **3.** $m^3 - m^2 - 6m$ **5.** $3x^2 + 4x + 3/(x - 5)$
7. $4m^2 - 4m + 1 + [-3/(m + 1)]$ **9.** $x^4 + x^3 + 2x - 1 + 3/(x + 2)$
11. $(1/3)x^2 - (1/9)x + 1/(x - 1/3)$ **13.** $9z^2 + 10z + 27 + 52/(z - 2)$ **15.** $y^2 + y + 1$
17. $x^3 + x^2 + x + 1$ **19.** $f(x) = (x - 1)(x^2 + 2x + 3) - 5$
21. $f(x) = (x + 2)(-x^2 + 4x - 8) + 20$ **23.** $f(x) = (x - 3)(x^3 + 2x + 5) + 15$
25. $f(x) = (x + 1)(3x^3 + x^2 - 11x + 11) + 4$ **27.** -8 **29.** 2   **31.** -6 **33.** $-6 - i$ **35.** $-2 - 3i$ **37.** yes
39. yes   **41.** yes   **43.** yes   **45.** yes   **47.** no   **49.** yes   **51.** no   **53.** yes   **55.** no   **63.** 1.51

Section 7.2 (page 373)
1. no   **3.** yes   **5.** yes   **7.** no   **9.** $f(x) = -(1/6)x^3 + (13/6)x + 2$
11. $f(x) = (-1/2)x^3 - (1/2)x^2 + x$ **13.** $f(x) = -10x^3 + 30x^2 - 10x + 30$
15. $f(x) = x^2 - 10x + 26$ **17.** $f(x) = x^3 - 4x^2 + 6x - 4$ **19.** $f(x) = x^3 - 3x^2 + x + 1$
21. $f(x) = x^4 - 6x^3 + 10x^2 + 2x - 15$ **23.** $f(x) = x^3 - 8x^2 + 22x - 20$
25. $f(x) = x^4 - 4x^3 + 5x^2 - 2x - 2$ **27.** $f(x) = x^4 - 10x^3 + 42x^2 - 82x + 65$
29. $f(x) = x^5 - 12x^4 + 74x^3 - 248x^2 + 445x - 500$ **31.** $f(x) = x^4 - 6x^3 + 17x^2 - 28x + 20$
33. $-1 + i, -1 - i$ **35.** $3, -2 - 2i$ **37.** $(-1 + i\sqrt{5})/2, (-1 - i\sqrt{5})/2$ **39.** $i, 2i, -2i$
41. $3, -2, 1 - 3i$ **43.** $2 - i, 1 + i, 1 - i$ **45.** $f(x) = (x - 2)(2x - 5)(x + 3)$
47. $f(x) = (x + 4)(3x - 1)(2x + 1)$ **49.** $f(x) = (x - 3i)(x + 4)(x + 3)$
51. $f(x) = (x - 1 - i)(2x - 1)(x + 3)$ **53.** Zeros are $-2, -1, 3; f(x) = (x + 2)^2(x + 1)(x - 3)$.
55. at $t = 3$ and $t = -2$

Section 7.3 (page 379)
1. $\pm 1, \pm 1/2, \pm 1/3, \pm 1/6$ **3.** $\pm 1, \pm 1/2, \pm 1/3, \pm 1/4, \pm 1/6, \pm 1/12, \pm 2, \pm 2/3$
5. $\pm 1, \pm 1/2, \pm 2, \pm 4, \pm 8$ **7.** $\pm 1, \pm 2, \pm 5, \pm 10, \pm 25, \pm 50$ **9.** $-1, -2, 5$
11. $2, -3, -5$ **13.** no rational zeros   **15.** $1, -2, -3, -5$

17. $-4, 3/2, -1/3; f(x) = (x + 4)(2x - 3)(3x + 1)$
19. $-3/2, -2/3, 1/2; f(x) = 2(x - 1/2)(3x + 2)(2x + 3) = (2x - 1)(3x + 2)(2x + 3)$
21. $1/2; f(x) = 2(x - 1/2)(x^2 + 4x + 8) = (2x - 1)(x^2 + 4x + 8)$
23. no rational zeros; prime   **25.** $-1, -2, -3, 4; f(x) = (x + 1)(x + 2)(x + 3)(x - 4)$
27. $-2, 2/3; f(x) = (x + 2)(3x - 2)(x^2 + 1)$ **29.** $1; f(x) = (x - 1)(x^4 + 4x^3 - x^2 - 12x - 12)$
31. $-2/3, -1, 3$ **33.** $1, -5/4$ **35.** 1

Section 7.4 (page 386)
17. (a) positive: 1; negative: 2 or 0   (b) $-3, -1.4, 1.4$ **19.** (a) positive: 2 or 0; negative: 1   (b) $-2, 1.6, 4.4$
21. (a) positive: 1; negative: 0   (b) 1.5 **23.** (a) positive: 3 or 1; negative: 1   (b) $1.1, -.7$ **25.** (a) positive: 1; negative: 1
(b) $-1.5, 3.1$ **27.** $3.24, -1.24$ **29.** $-3.65, -.32, 1.65, 6.32$
31. 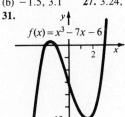   **33.** 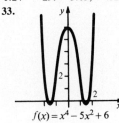   **35.** 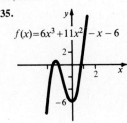   **37.**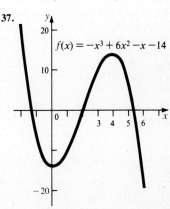

39. $(-\infty, -1.9) \cup (.1, 3.3)$ **41.** $(-\infty, -4.2) \cup (-.5, .5) \cup (4.2, +\infty)$

Section 7.5 (page 392)
1. $-1/x + 4/(x + 1)$ **3.** $2/(3x) - 2/[3(x + 3)]$ **5.** $-1/[2(x + 1)] + 3/[2(x - 1)]$
7. $7/x - 63/[4(3x - 1)] - 7/[4(x + 1)]$ **9.** $-2/(9x) + 2/(3x^2) + 2/[9(x + 3)]$
11. $-3/[25(x + 2)] + 3/[25(x - 3)] + 2/[5(x - 3)^2]$ **13.** $2/(x + 2)^2 - 3/(x + 2)^3$
15. $3/[2(x + 1)] - 3/[2(x + 3)]$ **17.** $-1/[3(2x + 1)] + 2/[3(x + 2)]$ **19.** $1 + 1/x + 5/(x - 2) - 3/(x - 1)$
21. $5/(3x) - 5x/[3(x^2 + 3)]$ **23.** $-1/[3(x + 1)] + (x + 5)/[3(x^2 + 2)]$
25. $3/x - 3/[2(x + 1)] - 3(x + 1)/[2(x^2 + 1)]$ **27.** $1/x - 2x/(x^2 + 1)^2$ **29.** $3/(x - 1) + 1/(x^2 + 1) - 2/(x^2 + 1)^2$
31. $-13/[15(x - 1)] + 46/[15(x + 2)]$ **33.** $1/(x^2 + 1) - 1/[2(x + 1)] + 1/[2(x - 1)]$
35. $3/x - 1/(x + 3) + 2/(x - 1)$ **37.** $a = 1/4; b = -1/4; c = 5/4$ **39.** $a = -1/48; b = 1/48; c = 5/6$

Chapter 7 Review Exercises (page 393)
1. $q(x) = 2x^2 + 3/2; r(x) = 13/2$ **3.** $q(x) = x^2 - 5; r(x) = 5x + 9$ **5.** $q(x) = 2x^2 + 5x + 1; r = 2$
7. $q(x) = 2x^2 + 4x + 4; r = 14$ **9.** 11 **11.** 28 **13.** $f(x) = x^3 - 10x^2 + 17x + 28$
15. $f(x) = x^4 - x^3 - 9x^2 + 7x + 14$ **17.** no   **19.** no   **21.** $f(x) = -2x^3 + 6x^2 + 12x - 16$
23. $f(x) = x^4 - 3x^2 - 4$ **25.** $f(x) = x^3 + x^2 - 4x + 6$ **27.** $3 + i, 3 - i, 2i, -2i$ **29.** $1/2, -1, 5$
31. $4, -1/2, -2/3$ **33.** none
43. $-2.3, 4.6$ **45.**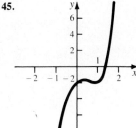

47. $3/(x + 2) + 2/(x - 2)$
49. $4/x^2 - 4/(x^2 + 2)$
51. $2/(x + 1) + 1/(x - 2) + (-3)/(x - 2)^2 + 2/(x - 2)^3$
53. $(5x - 1)/(x^2 + 1) + (2x + 5)/(x^2 + 1)^2$

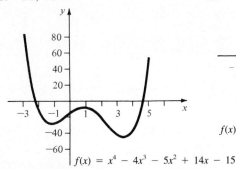

Chapter 8

Section 8.1 (page 399)

1. S_1: $2 = 1(1 + 1)$ S_2: $2 + 4 = 2(2 + 1)$ S_3: $2 + 4 + 6 = 3(3 + 1)$ S_4: $2 + 4 + 6 + 8 = 4(4 + 1)$ S_5: $2 + 4 + 6 + 8 + 10 = 5(5 + 1)$

Section 8.2 (page 407)

1. $x^6 + 6x^5y + 15x^4y^2 + 20x^3y^3 + 15x^2y^4 + 6xy^5 + y^6$ **3.** $p^5 - 5p^4q + 10p^3q^2 - 10p^2q^3 + 5pq^4 - q^5$
5. $r^{10} + 5r^8s + 10r^6s^2 + 10r^4s^3 + 5r^2s^4 + s^5$ **7.** $p^4 + 8p^3q + 24p^2q^2 + 32pq^3 + 16q^4$
9. $m^6/64 - 3m^5/16 + 15m^4/16 - 5m^3/2 + 15m^2/4 - 3m + 1$ **11.** $16p^4 + 32p^3q/3 + 8p^2q^2/3 + 8pq^3/27 + q^4/81$
13. $m^{-8} + 4m^{-4} + 6 + 4m^4 + m^8$ **15.** $p^{10/3} - 25p^4 + 250p^{14/3} - 1250p^{16/3} + 3125p^6 - 3125p^{20/3}$
17. $x^{21} + 126x^{20} + 7560x^{19} + 287,280x^{18}$ **19.** $3^9m^{-18} + 45 \cdot 3^8m^{-20} + 900 \cdot 3^7m^{-22} + 10,500 \cdot 3^6m^{-24}$
21. $7920m^8p^4$ **23.** $180m^2n^8$ **25.** $4845p^8q^{16}$ **27.** $439,296x^{21}y^7$ **29.** $90,720x^{28}y^{12}$ **31.** 1.105
33. 245.937 **35.** 758.650 **37.** $1.002; 5.010$ **39.** $.942$ **41.** $1 - x + x^2 - x^3 + \ldots$ **47.** 6

Section 8.3 (page 415)

1. $10, 16, 22, 28, 34$ **3.** $1, 1/2, 1/4, 1/8, 1/16$ **5.** $1/2, 2/5, 1/3, 2/7, 1/4$ **7.** $-2, 8, -24, 64, -160$
9. $2/3, 5/6, 10/11, 17/18, 26/27$ **11.** $2, 3, 5, 7, 11$ **13.** $4, 9, 14, 19, 24, 29, 34, 39, 44, 49$
15. $2, 4, 8, 16, 32, 64, 128, 256, 512, 1024$ **17.** $1, 1, 2, 3, 5, 8, 13, 21, 34, 55$ **19.** $4, 6, 8, 10, 12$
21. $11, 9, 7, 5$ **23.** $4 - \sqrt{5}, 4, 4 + \sqrt{5}, 4 + 2\sqrt{5}, 4 + 3\sqrt{5}$ **25.** $d = 5, a_n = 7 + 5n$
27. $d = -3, a_n = 21 - 3n$ **29.** $d = \sqrt{2}, a_n = -6 + n\sqrt{2}$ **31.** $d = m, a_n = x + nm - m$
33. $d = -m, a_n = 2z + 2m - mn$ **35.** $a_8 = 19, a_n = 3 + 2n$ **37.** $a_8 = 7, a_n = n - 1$
39. $a_8 = -6, a_n = 10 - 2n$ **41.** $a_8 = -9, a_n = 15 - 3n$ **43.** $a_8 = 5.1, a_n = 2.7 + .3n$
45. $a_8 = x + 21, a_n = x + 3n - 3$ **47.** $a_8 = 4m, a_n = mn - 4m$ **49.** 215 **51.** 125 **53.** 230
55. 16.745 **57.** 32.28 **59.** 18 **61.** 140 **63.** -684 **65.** $500,500$ **67.** 400 **69.** -78 **71.** $a_1 = 7$
73. $a_1 = 0$ **75.** $a_1 = 7$ **77.** 1281 **79.** $n(n + 1)/2$ **81.** 4680 **83.** 95 m, 500 m   **85.** 240 ft, 1024 ft
87. $9, 14, 19, 24$ **89.** 170 m   **91.** All terms are 0; or all terms are 1.   **97.** false   **99.** true   **101.** false

Section 8.4 (page 423)

1. $2, 6, 18, 54$ **3.** $1/2, 2, 8, 32$ **5.** $3/2, 3, 6, 12, 24$ **7.** $a_5 = 324; a_n = 4 \cdot 3^{n-1}$ **9.** $a_5 = -1875; a_n = -3(-5)^{n-1}$
11. $a_5 = 24; a_n = (3/2) \cdot 2^{n-1}$ or $3 \cdot 2^{n-2}$ **13.** $a_5 = -256; a_n = -1 \cdot (-4)^{n-1}$ or $-(-4)^{n-1}$ **15.** $r = 2; a_n = 6 \cdot 2^{n-1}$
17. $r = 2; a_n = (3/4) \cdot 2^{n-1}$ or $3 \cdot 2^{n-3}$ **19.** $a_n = (9/4) \cdot 2^{n-1}$ or $9 \cdot 2^{n-3}$ **21.** $a_n = (-2/9) \cdot 3^{n-1}$ or $-2 \cdot 3^{n-3}$ **23.** 93
25. $33/4$ **27.** 860.95 **29.** 30 **31.** $255/4$ **33.** 120 **35.** $r = 2$ **37.** $r = 10$ **39.** $r = 1/2$; sum would exist
41. $64/3$ **43.** $1000/9$ **45.** $3/2$ **47.** $1/5$ **49.** $1/3$ **51.** $3/7$ **53.** 70 m   **55.** 1600 rotations
57. $\$2^{30}$, or $\$1,073,741,824$; $\$2^{31} - \$1$ **59.** $.000016\%$ **61.** 12 m   **63.** $\approx 13.4\%$ **65.** $\$26,214.40$
67. $.016 \times 2^{11}$ inches, 32.768 inches   **69.** any sequence of the form a, a, a **71.** $a_1 = 2, r = 3$, or $a_1 = -2, r = -3$
73. $a_1 = 128, r = 1/2$, or $a_1 = -128, r = -1/2$

Section 8.5 (page 431)

1. $-5 - 2 + 1 + 4$ **3.** $-3 + 0 + 1$ **5.** $-1 - 1/3 + 0 + 1/5 + 1/3$
7. $-5(.5) + (-1)(.5) + 3(.5) + 7(.5) = -2.5 - .5 + 1.5 + 3.5$ **9.** $-1(.5) + 3(.5) + 15(.5) + 35(.5) =$
$-.5 + 1.5 + 7.5 + 17.5$ **11.** $4(.5) + (4/3)(.5) + (4/5)(.5) + (4/7)(.5) = 2 + 2/3 + 2/5 + 2/7$ **13.** $\sum_{i=1}^{15} (10 - 2i)$
15. $\sum_{i=1}^{19} (5 + 5i)$ **17.** $\sum_{i=1}^{9} 7(2)^{i-1}$ **19.** $\sum_{i=1}^{7} 9\left(-\frac{1}{3}\right)^{i-1}$ **21.** $\sum_{i=2}^{13} i^2$ or $\sum_{i=1}^{12} (i + 1)^2$ **23.** $\sum_{i=1}^{10} (1/i)$ **25.** $\sum_{i=1}^{6} 2^i$
27. $\sum_{i=3}^{7} (12 - 3i)$ **29.** $\sum_{i=0}^{9} 2(3)^{i+1}$ **31.** $\sum_{i=0}^{10} [(i - 1)^2 - 2(i - 1)]$ or $\sum_{i=0}^{10} (i^2 - 4i + 3)$ **33.** 90 **35.** 220 **37.** 304
39. $8 + [2(n + 1)]/n$ **41.** $9 + [3(n + 1)]/n + [(n + 1)(2n + 1)]/(6n^2)$

Section 8.6 (page 436)

1. 5040 **3.** 720 **5.** 90 **7.** 336 **9.** 7 **11.** 1 **13.** 6 **15.** 56 **17.** 1365 **19.** 120 **21.** 14 **23.** 1
25. 720 **27.** $120; 30,240$ **29.** 30 **31.** 120 **33.** 4 **35.** 552 **37.** $34,650$ **39.** (a) 180 (b) $50,400$
41. (a) $362,880$ (b) 6 (c) 1260 **43.** $27,405$ **45.** $220; 220$ **47.** 1326 **49.** 10 **51.** 28 **53.** (a) 84;   (b) 10;
(c) 40;   (d) 28 **55.** $\binom{52}{5} = 2,598,960$ **57.** 4 **59.** $4 \cdot \binom{13}{5} = 5148$ **61.** $P(5, 5) = 120$
63. $\binom{12}{3} = 220$ **65.** 8 **67.** 4

Section 8.7 (page 444)

1. $S = \{H\}$ **3.** $S = \{HHH, HHT, HTH, THH, HTT, THT, TTH, TTT\}$ **5.** Let c = correct, w = wrong. $S = \{ccc, ccw,$ cwc, wcc, wwc, wcw, cww, www$\}$ **7.** (a) $\{HH, TT\}$, 1/2 (b) $\{HH, HT, TH\}$, 3/4   **9.** (a) $\{2$ and $4\}$, 1/10 (b) $\{1$ and 3, 1 and 5, 3 and 5$\}$, 3/10 (c) $\emptyset$, 0 (d) $\{1$ and 2, 1 and 4, 2 and 3, 2 and 5, 3 and 4, 4 and 5$\}$, 3/5   **11.** (a) 1/5 (b) 8/15 (c) 0 (d) 1 to 4 (e) 7 to 8   **13.** 1 to 4   **15.** 2/5   **17.** (a) 1/2 (b) 7/10 (c) 2/5   **19.** (a) 1/6 (b) 1/3 (c) 1/6 **21.** (a) 7/15 (b) 11/15 (c) 4/15   **23.** .24   **25.** .06   **27.** .27   **29.** 1 to 1   **31.** .35   **33.** .95   **35.** 0 **37.** .38   **39.** .89   **41.** .90   **43.** .41   **45.** .21   **47.** .79

Chapter 8 Review Exercises (page 448)

7. $x^4 + 8x^3y + 24x^2y^2 + 32xy^3 + 16y^4$ **9.** $243x^{5/2} - 405x^{3/2} + 270x^{1/2} - 90x^{-1/2} + 15x^{-3/2} - x^{-5/2}$ **11.** $2160x^2y^4$ **13.** $3^{16} + 16 \cdot 3^{15}x + 120 \cdot 3^{14}x^2 + 560 \cdot 3^{13}x^3$ **15.** 8, 10, 12, 14, 16   **17.** 5, 3, -2, -5, -3 **19.** 10, 6, 2, -2, -6,   **21.** $3 - \sqrt{5}$, 4, $5 + \sqrt{5}$, $6 + 2\sqrt{5}$, $7 + 3\sqrt{5}$ **23.** 4, 8, 16, 32, 64 **25.** -3, 4, $-16/3$, 64/9, $-256/27$ **27.** $a_1 = -29$, $a_{20} = 66$ **29.** 20   **31.** $-x + 61$ **33.** 222   **35.** $84k$ **37.** 25   **39.** $x^5/125$ **41.** 15   **43.** $-10k$ **45.** 80   **47.** does not exist   **49.** 25/6   **51.** 690   **53.** 73/12 **55.** 50,005,000   **57.** 240   **59.** $r > 1$ so sum does not exist   **61.** $-4 + 0 + 12 + 32 + 60 = 100$ **63.** $0 + .4 + 2.4 + 6 = 8.8$ **65.** $\sum\limits_{i=1}^{15} (9 - 5i)$ **67.** $\sum\limits_{i=1}^{6} 4(3)^{i-1}$ **69.** $\sum\limits_{i=-2}^{5} (9 + 2i)$ **71.** $\sum\limits_{i=2}^{13} [(i - 1)^2 + 1]$ **73.** 182   **75.** 50   **77.** 72   **79.** 1   **81.** 252   **83.** 504   **85.** $S = \{1, 2, 3, 4, 5, 6\}$ **87.** $S = \{0, .5, 1, 1.5, 2, \ldots, 299.5, 300\}$ **89.** $S = \{(3, R), (3, G), (5, R), (5, G), (7, R), (7, G), (9, R), (9, G), (11, R), (11, G)\}$ **91.** $F = \{(3, G), (5, G), (7, G), (9, G), (11, G)\}$ **93.** $E' \cap F'$ **95.** 1 to 3   **97.** 2 to 11   **99.** .26   **101.** 4/13 **103.** 3/4   **105.** 0

Index

Interval notation

Set	Interval	Graph
$\{x \mid a < x\}$	$(a, +\infty)$	
$\{x \mid a < x < b\}$	(a, b)	
$\{x \mid x < b\}$	$(-\infty, b)$	
$\{x \mid a \leq x\}$	$[a, +\infty)$	
$\{x \mid a \leq x \leq b\}$	$[a, b]$	

Conic sections

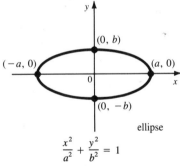

ellipse

$$\frac{x^2}{a^2} + \frac{y^2}{b^2} = 1$$

x intercepts are a and $-a$
y intercepts are b and $-b$

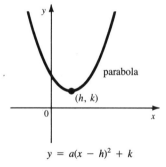

parabola

$$y = a(x - h)^2 + k$$

opens upward if $a > 0$, downward if
$a < 0$, vertex is at (h, k)

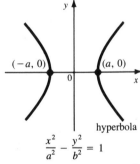

hyperbola

$$\frac{x^2}{a^2} - \frac{y^2}{b^2} = 1$$

x intercepts are a and $-a$

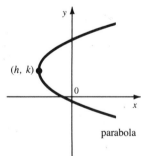

parabola

$$x = a(y - k)^2 + h$$

opens to right if $a > 0$, to left if
$a < 0$, vertex is at (h, k)

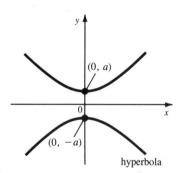

hyperbola

$$\frac{y^2}{a^2} - \frac{x^2}{b^2} = 1$$

y intercepts are a and $-a$

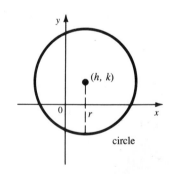

circle

$$(x - h)^2 + (y - k)^2 = r^2$$

center at (h, k), radius r